Synthese Library

Studies in Epistemology, Logic, Methodology, and Philosophy of Science

Volume 527

The aim of *Synthese Library* is to provide a forum for the best current work in the methodology and philosophy of science and in epistemology, all broadly understood. A wide variety of different approaches have traditionally been represented in the Library, and every effort is made to maintain this variety, not for its own sake, but because we believe that there are many fruitful and illuminating approaches to the philosophy of science and related disciplines.

Special attention is paid to methodological studies which illustrate the interplay of empirical and philosophical viewpoints and to contributions to the formal (logical, set-theoretical, mathematical, information-theoretical, decision-theoretical, etc.) methodology of empirical sciences. Likewise, the applications of logical methods to epistemology as well as philosophically and methodologically relevant studies in logic are strongly encouraged. The emphasis on logic will be tempered by interest in the psychological, historical, and sociological aspects of science. In addition to monographs *Synthese Library* publishes thematically unified anthologies and edited volumes with a well-defined topical focus inside the aim and scope of the book series. The contributions in the volumes are expected to be focused and structurally organized in accordance with the central theme(s), and should be tied together by an extensive editorial introduction or set of introductions if the volume is divided into parts. An extensive bibliography and index are mandatory.

Juan M. Durán • Giorgia Pozzi
Editors

Philosophy of Science for Machine Learning

Core Issues and New Perspectives

Springer

Editors
Juan M. Durán
Technology, Policy and Management
Delft University of Technology
Delft, The Netherlands

Giorgia Pozzi
Technology, Policy and Management
Delft University of Technology
Delft, The Netherlands

ISSN 0166-6991 ISSN 2542-8292 (electronic)
Synthese Library
ISBN 978-3-032-03082-5 ISBN 978-3-032-03083-2 (eBook)
https://doi.org/10.1007/978-3-032-03083-2

This work was supported by Delft University of Technology, Delft, The Netherlands

This Springer imprint is published by the registered company Springer Nature Switzerland AG
The registered company address is: Gewerbestrasse 11, 6330 Cham, Switzerland

Contents

To the loving memory of Paul Humphreys, in thankful recognition of his outstanding contributions and for paving the way that many of us are taking.

JMD and GP

A Diego y Kass. Gracias por todo el amor que me dan.

JMD

An Stefan - für deine bedingungslose Liebe und Unterstützung.

GP

Introduction

Juan M. Durán and Giorgia Pozzi

In 2004, Paul Humphreys observed that the influence of computational methods in many areas of the sciences was transforming their theory development and practice. The core argument, according to Humphreys, was that human abilities could no longer be considered the ultimate criteria of epistemic success. An exclusively anthropocentric epistemology, as offered by the diverse forms of empiricism in the philosophy of science, is no longer an attainable position. In 2009, he called this the *anthropocentric predicament* and advocated for an epistemology that addresses the displacement of humans from the center of production of knowledge. Here lies, to our mind, the basis for many of the debates that will follow in the intersection between the philosophy of science and algorithms. In this sense, this book largely builds on Humphreys' work. But it is also in great debt to the many philosophers who have extensively worked in the philosophy of computer science and received less recognition over these recent years—Amnon Eden, Raymond Turner, Aaron Sloman, Paul Thagard, Zenon Pylyshyn, Margaret Boden, Nils Nilsson, and so many others.

We have divided this volume into seven parts, each including a handful of chapters from scholars in the field. With this division, we intend to articulate key debates in the intersection between the philosophy of science and machine learning (ML). This is done, of course, to the extent possible and without any pretense of being exhaustive. Now, as the reader will quickly note, several chapters exceed the exclusivity of the philosophy of science and ML, encroaching into areas such as epistemology, ethics, and the sociology of science, as well as scientific modeling and mathematics. Beyond this minor deviation from the title, the book offers a variety of viewpoints on core issues pertaining to the philosophy of science—and related topics—that are at the basis of current debates on ML. There are not, we must recognize, unifying themes across chapters. Rather, the reader will find a wealth of common ideas and concepts of mutual dependence. The value of this book lies, therefore, not so much in its organization but in the merits of each contributing chapter. We take the opportunity to praise the authors for their magnificent work, as well as their commitment to reviewing their chapters and those of others to a great extent. It goes without much dissension that much more can and must be discussed in the future. Our hope is that this volume will pave the way for and motivate a more systematic discussion of central topics at the intersection of the philosophy of science and ML.

As mentioned, the volume is divided into seven different parts. The first five deal with theoretical issues in connection with ML. Issues about epistemic opacity, knowledge and justification, explanation and understanding, and scientific modeling and representation are at the center of attention. The questions addressed in these parts cover a wide range of old and new philosophical concerns: what is epistemic opacity, and how can explanation and understanding be intelligibly articulated within ML? Can ML outputs be justified and if so, what is a suitable epistemology? What forms of representation—if any at all—can we claim from ML? The last two parts in the volume—parts six and seven—complement the previous whole by looking into scientific practice with and by ML. Part 6 focuses on the role of ML in scientific practice and on which values do and should influence it. Questions here revolve around whether scientists are transitioning into a novel form of scientific inquiry through ML and, if so, how, under what conditions, and with which foreseeable implications. Moreover, questions pertaining to scientific values and inductive risk take a prominent place. For example, questions along the following lines receive full attention: what role do epistemic and non-epistemic values play in constructing and evaluating ML? How can inductive risk in ML be effectively evaluated and managed? Part 7, on the other hand, works toward visualizing the philosophical nuances required for analyzing ML for its applicability in scientific fields. Of particular interest are the applicability of ML in astrophysics and particle physics, neuroscience, climate science, and public health. Two goals are anticipated within these parts. On the one hand, as it was mentioned before, they showcase ML in different fields of scientific inquiry. On the other, they exemplify how previously discussed theoretical issues, such as explanation and understanding, relate to scientific practice with ML.

Before entering into a brief description of each part and chapter individually, we would like to entertain some ideas of what comes next. We predict that the debates addressed here are likely to continue in the foreseeable future due to a number of outstanding issues that were not able to be addressed here. Now, while this might sound like an easy prediction, we also include in it debates that are rather absent in the literature. These include objectivity in ML, emergence, replicability, underdetermination, and the—possibility of a—demarcation problem in the context of ML. We could also raise the stakes a bit higher and mention that ML seems to be configuring novel, possibly never seen before, modes of scientific inquiry; that ML seems to depart from syntactic and semantic accounts of models and theories, possibly requiring an account of its own; and why the computation—as opposed to mere modes of inference—and representation of the temporal dynamic of ML these systems displace—or threatens to displace—humans from every the center of production of knowledge. We hope the reader recognizes Humphreys' direct influence on this shortlist.

In what follows, we break down each part to its main goals, as well as provide a brief description of each chapter's main interest.

Epistemic Opacity

The opacity of algorithms is at the heart of virtually every issue pertaining to ML. Moral philosophers, political philosophers, and epistemologists all seem to have opacity in mind when dealing with their arguments. Interestingly, the concept is still very much up in the air, and its philosophical use sometimes lies on purely intuitive grounds. This is partly because, we conjure, opacity works as the motivating starting point for other, perhaps more interesting philosophical issues that presumably do not require any fine-grained definition. Now, this might be an unfortunate argumentative tactic, as the very notion of epistemic opacity is currently under much dispute and far from being intuitive.

This part of the book tries to capture the many complexities lurking behind the concept of epistemic opacity and address the current debates that pertain to it. It also intends to serve as a primer conceptualization for much of the chapters that follow.

The chapter by **Claus Beisbart** provides a solid basis for approaching the complexities of epistemic opacity in ML. In a nutshell, the chapter investigates the specific ways in which ML is considered opaque. This is done by examining statements from scientists and commentators who highlight the significant challenges in achieving certain epistemic goals regarding ML, such as predicting its outcomes. Beisbart then proceeds to contrast these observations with Humphreys' theoretical concept of epistemic opacity. The chapter goes on to explore whether Humphreys' notion fully explains the opacity issues in ML, or whether there is more to it. The chapter concludes with the following thesis: while Humphreys-style opacity applies to some aspects of ML, it is not the sole source of their opacity. Other factors, like the absence of comprehensive higher-level descriptions of ML, for instance, also play a significant role for the epistemic opacity of ML. Beisbart proposes two—seemingly disjoint—approaches to understanding opacity: as an "umbrella term," broadly enough to encompass the difficulties in reaching various achievements in ML, or as a specific issue that accounts for why we struggle with diverse epistemic achievements. The exploration proposed by Beisbart provides a nuanced understanding of the concept of opacity in ML, offering a foundation for further discussion on the topic.

Nico Formanek proposes a turn in the rather standard treatment that opacity has found in the literature. That is, the author neither argues for or against opacity as a *sui generis* foundational concept of specific fields, nor for extending, restricting, or otherwise analyzing the definition. Formanek's core interest, instead, lies in the practical effects that opacity has in the concrete scientific uses of ML. This argumentative maneuver brings the opacity debate to a rather unanticipated new light, one where the historical record furnishes cases where the gradual acceptance and habituation to levels of incompleteness and uncertainty have been central to overcoming practical opacity in past technologies. By understanding this process, the chapter proposes that developers and users of ML can develop a similar habituation. The expectation is that, over time, as users and developers become accustomed to the inherent opacity in ML, they will develop practical strategies to manage it effectively.

Finally, **Jack Casey's** chapter begins with noticing a similarity found in debates between computer-assisted proofs and ML, namely, that the former are taken to be *unsurveyable* methodologies while the latter are *epistemically opaque*. A proof is unsurveyable if it is impossible for a rational agent to understand the proof in its entirety, while an ML is epistemically opaque if it is impossible for a rational agent to know all the relevant epistemic elements of the algorithm. This similarity, Casey argues, has been standardly cast under the same umbrella, that is, computer-assisted proofs and ML are both taken to be epistemically opaque. Nothing new admits the author. However, what has not yet been considered and that constitutes Casey's main case is that concerns about epistemic opaqueness might result in being importantly dissimilar for either methodologies. Casey then builds the argument that epistemic opacity is a matter of degree. To Casey's mind, the level of comfort with a given degree of epistemic opacity is largely determined by one's commitments to scientific (anti-)realism. In this sense, if one holds realist commitments, then one's tolerance for epistemic opacity is lower than holding anti-realist commitments. These ideas, then, bridge the scientific realism and anti-realism debate with computer-assisted proofs and ML.

Justification

When Humphreys presented and discussed (essential) epistemic opacity, he was thinking of a problem of justification of the algorithm and its output. This particular focus is, surprisingly, passed over by many philosophers, who rapidly cash out opacity as a problem pertaining to explanation, understanding, and other epistemic functions. But as Humphreys put it, opacity prevents human agents from "examin[ing] and justify[ing] every element of the computational processes that produce the output of a[n ...] artifacts of computational science" (Humphreys, 2009, 618).

Why is, then, justification important in the context of ML? Justification typically refers to forming beliefs in the right way, or whether beliefs have been formed properly. In this way, a central concept in theories of justification is concerned with what makes a belief rational, credible, or warranted. Claims about having justification for believing that ML's outputs are true, scientifically valid, or represent a fact in the world—you pick your commitment—have multiple repercussions for philosophical studies of ML, both for theoretical and practical philosophy. For starters, it is at the basis of different concepts of knowledge that pervade the literature on ML—philosophical and otherwise. Moreover, justification is also integral to debates about trust and trustworthy ML, debates with a more practical flavor. Interestingly, questions about the justification of ML outputs also seem to be present in the diverse commitments (e.g., factivity conditions) adopted by explanatory AI. Through its chapters, this part of the book laid out the basis for future debates on justificatory ML.

Juan M. Durán motivates the chapter by asking the question, "what is a reliable algorithm?" As Durán intends to answer it, the question is about epistemic justification. The chapter then proceeds in two directions. First, it distinguishes between *internal*-to-the-algorithm and *external*-to-the-algorithmic epistemologies. The former refers to showing "the inner workings of the algorithm," largely driven by diverse transparency methods (e.g., transparency reports, some forms of explanatory AI); the latter describes external factors, metrics, practices, and cultures of research that confer reliability to the algorithm but are not inherent to its internal workings. With this distinction in place, Durán lays out a newer version of his *Computational Reliabilism* (CR), understood as an external-to-the-algorithmic epistemology that builds on the best epistemic efforts for the justification of the algorithm's output. Most of the chapter is geared toward fleshing out CR, its integral parts (i.e., type- and token-reliability indicators), and to what extent an individual is justified in believing algorithms' outputs.

The chapter by **Ramón Alvarado** is engineered around a critical view of CR, as discussed in earlier versions—most prominently, in discussions on computer simulations. Alvarado presents three core objections to CR. First is the "challenge of warrant transmission and reliability-crediting properties." This objection, as admitted by Alvarado, is of a general kind, and not necessarily targeted to CR. At its core, the issue is to know how and to what extent any process, evidence, reason, and the like confer epistemic reliance on an algorithm. The second objection is the so-called "challenge of the indispensability of endogenous features in artifactual reliability." Alvarado's strategy is to question whether reliability indicators, intrinsic to CR, are sufficient to increase the degree of warrant we have to believe in ML outputs. Artifacts, it is argued, are endogenous to the algorithm (i.e., internal to its workings) and thus central to the justification of ML outputs. Finally, there is the "challenge of error-related opacity" which aims at shedding light on an equally important but less discussed epistemic issue related to how a process can fail. After presenting these objections, Alvarado expresses the hope that these observations serve as a starting point to alleviate the limitations previously identified.

Finally, **Carlos Zednik** and **Philippe Verreault-Julien** propose to interlock justification with explainable AI (XAI). They begin by asserting that justification requires decisions to align with relevant norms and that XAI has the potential to uncover the series of reasons influencing ML's outputs, such as identifying spurious correlations and highlighting relevant patterns. By adopting a framework from the philosophy of action, they aim to cover three key issues: first, differentiate between the various types of reasons that may be provided for an ML's outputs; second, assess how these diverse types of reasons contribute to the author's interpretation of justification; and finally to evaluate the capacity of current XAI methods to identify and describe said reasons. The heart of Zednik and Verreault-Julien's chapter is, then, to pinpoint some XAI methods that are suitable for identifying and clearly articulating the reasons that justify ML-driven decisions. The authors also complement their argument by offering insights into how and why some of the methods proposed are limited.

Scientific Explanation (XAI)

Arguably one of the most popular topics in ML nowadays is explanatory AI (XAI). There are a number of good reasons for this. To some, XAI is the key to making ML more transparent, and thus increasing epistemic, cognitive, and morally relevant achievements (e.g., reliability, trustworthiness, credibility, accountability); to others, XAI provides sensitive and valuable answers to *how*—and even *why*—a given ML output was rendered; yet to others, XAI is the path to understanding the behavior of ML, following the footsteps of the much accepted philosophical idea that understanding is the goal of explanation. There are, of course, more reasons for engaging in philosophical discussions on XAI. Is there a (objective) logic of scientific explanation exclusive to ML? Do we admit factivity conditions for XAI? What would the plurality of scientific explanation in ML look like? Is XAI just a few algorithmic techniques that intend to answer how ML outputs are rendered but with no conceptual relation to the philosophical treatment of scientific explanations? Some of these reasons are addressed here, and some of the questions are answered. Under all of these headings, it is undoubted that XAI is central to the philosophy of science just as much as it is to ML in its diverse scientific applications. This part of the book navigates through these choppy waters, searching for precision in the debate, as well as novel interpretations.

Andrés Páez's contribution is an attempt to break with a growing trend of linking models of scientific explanation with ML. To this end, Páez argues that claims that ML is scientifically explainable are either *trivial* or *false*, and that, ultimately, these efforts misrepresent the goals of XAI. They are trivial, the author argues, if an explanation is understood as a set of causes, taken to be either entities or physical states of the computer, that aim to account for a prediction. Claims about ML being scientifically explainable are false, on the other hand, if it is always possible to offer an "explanatory text," a truthful description of the source of the prediction, thereby satisfying the factivity condition on scientific explanation. Páez splits the chapter into two halves. The first half is devoted to justifying the author's claims about the triviality and falseness of scientific explanations in ML. The second half advocates for the use of labels such as "interpretable machine learning" and "understandable AI" instead of subsuming these under the umbrella term "XAI." By doing this, Páez argues, we not only sidestep many of the misconceptions on the goals and purposes of XAI methods that furnish current literature but, more importantly, we recenter our attention on what is achievable, namely, understanding. Páez finalizes proposing a pragmatic account of understanding where conditions of satisfaction for understanding an ML are fleshed out in terms of the agent's success in using the system.

The chapter by **Stefan Buijsman** links XAI with mathematics and mathematical explanation. In this vein, Buijsman proposes to shift the analysis of XAI from scientific explanations—to this author's mind mainly driven by causal explanations—to non-causal explanations as analyzed in the literature on mathematical explanations. Buijsman's key argumentative maneuver consists of taking that ML are mathematical objects. Once this is established—with some caveats, as the author points out—the theoretical apparatus provided by non-causal mathematical explanations follows suit. The chapter also contains a debate on counterfactual and interventionist explanations, today at the center of the many analyses on XAI. Of particular interest to the author are those mathematical equations that are standardly conceived as symmetrical, as opposed to causal relations—recall the canonical objection to the Deductive-Nomological model of explanation. This raises the question of whether counterfactual dependence in the mathematical case is appropriately asymmetrical. The answer to this question, as it should be expected, is complex and with many angles. However, Buijsman favors the idea that the asymmetry of many of the standard explanations found in the ML literature can be accounted for in virtue of concrete implementations of ML.

Finally, **Tim Räz** explores the connections between ML interpretability and—scientific and mathematical—explanations. Through this exploration, the author proposes a pluralistic approach where different notions of interpretability and different notions of—scientific and mathematical—explanations converge. While the former is required for explaining ML phenomena, the author argues, the latter benefits from understanding the nature of explanations. This convergence, however, should be taken as a mutual and prolific way to show their interconnection. In this sense, the chapter introduces perspectives on interpretability and explanation emphasizing a balanced approach between extreme views. It suggests, furthermore, that while a single, homogeneous notion of interpretability is unnecessary, an excessive proliferation of different notions should be avoided. Interpretability should be accepted to be context- and audience-dependent with a psychological dimension, but not entirely subjective nor detached from facts.

Scientific Understanding and Interpretability

We seek to understand scientific theories and hypotheses, the gravitational pull, and the side effects of taking a drug. Philosophers of science have debated the nature of understanding, its scope, and its relation to other cognitive achievements (justification, truth, knowledge, explanation) for a long and very prolific time. It is only natural, one could hyperbolically say, that philosophical studies on ML gravitate toward a philosophical analysis of understanding. Now, the pull seems to exceed mere epistemic work and also include the impact that understanding the behavior and outputs of ML have (or could have) on society at large, economic implications, and a host of ethical and moral considerations. Thus, not only epistemic functions such as explanations are linked to understanding, but also its social, moral, and political implications are the driving force of much of the literature on this central concept. The chapters in this part of the volume are geared toward bridging epistemological concerns in connection with understanding, with moral and societal concerns linked to ML.

Emily Sullivan and **Atoosa Kasirzadeh**'s main goal is to push back on the trend in explainable AI that suggests actionable recourse explanations are the ethical gold standard for AI explanations. Instead, they argue that understanding should be the chief aim of explanations of ML decisions. The strategy begins with laying out the so-called *recourse explanations*, that is, explanations of ML decisions that give recommendations to the end users and data subjects on what actions they could take to get a different model decision in the future. Sullivan and Kasirzadeh refer to this as the "recourse-first approach" to explanation and define it as explanations that provide end users and data subjects with reasons that are actionable and feasible for change. Sullivan and Kasirzadeh argue that the recourse-first approach faces several conceptual pitfalls. Of main interest to these authors are the problems of hidden assumptions, *deflecting responsibility*, and *explanation hacking*. A suitable solution, as proposed by Sullivan and Kasirzadeh, is to advocate for an "understanding-first approach," where users get explanations that aim at understanding the AI decisions. The value of the understanding-first approach is that end users are in the possession of informed self-advocacy against ML decisions and does not face the conceptual pitfalls that we see with the recourse-first approach.

Eva Schmidt's chapter offers a further perspective on the connection between explainability and understanding in ML. In this chapter, Schmidt presents and discusses conditions for understanding through explanations in terms of *stakes*. More precisely, the author argues that whether the belief involved in an arguable case of understanding is sufficiently justified for outright understanding why depends on what is at stake in the particular circumstances considered. This pragmatic perspective is supported by comparing two situations that are identical, except that in one of them, a decision support system (specifically, the well-known COMPAS algorithm) is used in making a harmless decision, whereas in the other, it is used in making a life-determining decision. The COMPAS system has been used to determine the likelihood of defendants to re-offend. A connection with debates on pragmatic encroachment and inductive risk is established to explain how understanding is sensitive to the stakes of a particular situation. Overall, this connection aims to shed light on how non-epistemic values have an influence on whether explainable AI (XAI) methods suffice to generate outright understanding in context.

Scientific Models and Representation

Debates about scientific models and representation seem to be underrepresented in the philosophical literature on ML, thus constituting, to our mind, a non-negligible research gap that needs timely attention. This is particularly the case given the heavy philosophical weight that these concepts pull over others previously discussed such as understanding, explanation, and the formation of beliefs.

Rafael Alvarado and Paul Humphreys made early efforts on representation, linking these concepts back to forms of opacity: with transparent representations "we represent the states of a system in a way that is open to explicit scrutiny, analysis, interpretation, and understanding by humans, and transitions between those states are represented by rules that have similar properties" (Alvarado & Humphreys, 2017, 740). Opaque representations are simply defined as non-transparent representations. While these authors are more interested in debates on the multiple meanings of "Big Data," these initial thoughts furnish a solid basis for the chapters included in this part of the volume.

Now, given the scarcity of these debates in the specialized literature, this part of the book not only attempts to shed light on these authors' positions but it holds the more ambitious aim of setting an agenda on the relevance of these debates. With these intentions in mind, three chapters populate this part of the volume.

Phillip Hintikka Kieval takes note that much of the current philosophical work on the use and applicability of ML in the sciences appeals to one or another form of representation. Under this heading, researchers are inclined to assume that the epistemic value of their deep learning models stems from their capacity to represent the world, preferably to represent it accurately. But Kieval proposes a turn on this growing tendency, one that rescues non-representationalist ideas according to which scientific models are tools for mediating our epistemic activities. With these ideas in mind, the author elaborates on an artifactualist approach for ML in the sciences, which aims to understand their epistemic worth in terms of the material practices of construction and application. More precisely, the argument pivots on the researchers' skills for modeling. Thus considered, the design and construction of deep learning models in science involve iterative processes dependent upon the skill and tacit knowledge of researchers, and which ultimately scaffolds into genuine forms of understanding between the researcher's means of construction and application and their epistemic ends. To Kieval's mind, deep learning models provide diverse epistemic functions, including prediction, exploration, and manipulation. But they do not necessarily do so, unlike standardly believed, through scientific representation.

Timo Freiesleben begins by identifying assumptions underlying the standard narrative about the representational qualities of artificial neural networks (ANN). One central assumption he identifies is that humans rely on concepts—higher-order representations—when engaging in various tasks. For instance, humans perceive images as composed of objects and interpret texts according to grammatical structures. According to Freiesleben, the standard narrative further assumes that ANNs are able to perform similar tasks because they develop similar concepts to those used by humans. The central question Freiesleben poses is whether ANNs actually learn human-like concepts, and if so, whether these concepts are localized in individual units within the network. According to Freiesleben, it is believed among the literature, that by sharing concepts, images, and tasks, ANNs learn something similar to what humans learn—and that is more than just performing a task, but rather conceptual learning. Freiesleben challenges this view, or at least significantly qualifies it. While he acknowledges that ANNs are capable of performing complex prediction tasks and may, in doing so, learn human as well as non-human concepts, he argues that these concepts are not stored in individual units. These findings circle back to the beginning of the chapter, now with a negative twist: the actual forms of representation in ANN differ in important ways from the standard narrative, particularly on issues about the representational qualities of human concepts and ANN.

Finally, in their chapter, **Chiara Manganini** and **Giuseppe Primiero** notice that the concept of validity has typically been spelled out by the literature on deterministic scientific simulations in terms of the relationship between the mathematical model underlying the target system and the computational model used. However, with ML entering the game, the authors argue that this picture needs to be reconsidered. This is because the behavior of said target systems is now being primarily represented by a computational model—as opposed to a mathematical one—that is inherently opaque due to its high complexity. They go on to define three formal relations of probabilistic simulation: Weak Simulation, Weak Bisimulation, and Weak Approximate Simulation and show how these can help us to assess the ability of an ML model to correctly represent real-world phenomena.

Scientific Practice and Scientific Values in ML

The use of ML opens up new paths for scientific inquiries. It is often assumed that their ability to process vast amounts of data at incredible speed grounds their scientific merits. However, questions emerge regarding their impact on diverse scientific endeavors. This is due to the limited understanding of how ML produces knowledge and how they are possibly displacing humans from the center of scientific inquiry. The goal of this part of the book is twofold. First, it aims to inquire into the ways in which the design and use of ML generally affect scientific inquiry. This is done by critically questioning whether scientific pursuits based on ML uphold current accepted epistemic and conceptual standards. Moreover, this part of the book focuses on what can be said about scientific debates that involve systems about which there is often limited knowledge. Second, since ML is used in socially salient fields to carry out predictive tasks (e.g., in judicial systems and in the public health domain), the generalizations that allow predictions are not immune to philosophical concerns such as inductive risk. When assessing this risk, value-laden considerations necessarily play a relevant role. More precisely, the evaluation of whether a certain risk is acceptable in a particular situation is not free from epistemic, societal, and moral considerations. This part of the volume thus brings to the fore discussions regarding the role of values in ML-mediated scientific contexts and the extent to which inductive risk is associated with ML's design, use, and output.

In their chapter, **Sahra A. Styger, Marianne de Heer Kloots, Oskar van der Wal, and Federica Russo** address a topic at the heart of epistemic opacity, explainability, and algorithmic bias, which the authors deem crucial to scientific practice with ML, namely, the question of whether these systems are epistemically displacing humans from the process of scientific inquiry. Thus, the authors take as a starting point this canonical epistemological problem identified by Humphreys in computational science. The authors address this question from the perspective of the philosophy of techno-science and empirical research carried out utilizing ML in neuroscience as an example. They base their analysis on the consideration of ML design practices, thus broadening the discussion beyond mere ML outputs. In doing so, they show that humans are very much involved in the first crucial steps of the design and development of ML and are thus not displaced; they rather uphold a central stage in their role as epistemic agents. A key contribution of this chapter is that humans are not passively undergoing the changing modes of scientific inquiry brought about by ML but are instead actively involved in the choices connected to, among others, the degree of automation these systems (should) acquire. Moreover, through a case of ML in neuroscience, the authors show that human displacement in scientific inquiries is not an all-or-nothing occurrence but rather comes in degrees, depending on the level of system automation. This take on the issue of knowledge displacement opens the possibility for humans to acquire new roles in the process of scientific inquiry.

Milou Jansen and Koray Karaca's chapter examines values and inductive risk in ML, with a special focus on ML used for optimizing decision-making processes in the public domain. Starting from the idea that the evaluation of ML's inductive risk is a necessarily value-laden procedure, the authors highlight the mutual influence of epistemic and societal value judgments. To show the relevance of their considerations, a case study pertaining to an ML is analyzed, which is used to increase the livability of Amsterdam's urban environment. More precisely, the ML is used to classify citizens' reports of malfunctions in public spaces (e.g., a streetlight not working or similar issues) and allocate these to the correct department within the city government of Amsterdam. To this end, the authors first define what inductive risk amounts to in relation to the different ML modeling stages (e.g., methodology choice, data gathering, etc.). Afterward, they advance the claim that the acceptance of an ML also implies the acceptance of the uncertainties and risks connected to it. Therefore, the evaluation leading to the acceptance—or not—of a certain ML should entail epistemic assessments as well as ethical and societal ones. While epistemic values play significant and confirmative roles in the process of knowledge production, the authors claim that this also holds true for non-epistemic values in ML modeling. Indeed, while relevant social factors are often implicit in the evaluation of ML in terms of model performance and outcome, the study by Jansen and Karaca also shows that these factors play a crucial role and should acquire a central place in ML's assessment procedures.

Emanuele Ratti's chapter takes a further perspective on the problem of inductive risk. Differently from other chapters, the focus here is not on the analysis of the background assumptions that lead to the acceptability—or not—of uncertainty and risk with ML. Rather, the author brings attention to the assumption of the so-called *Uniformity of Nature* (UoN) as the central component for ML inductive strategies. This amounts to the fact that inferences of future observations are warranted based on past observations under the assumption that there is, indeed, a uniformity between past and future. By considering the use of ML in biology, the author argues that the UoN assumption is to be considered widely unproblematic in the natural sciences. However, this is not the case when ML is used in different domains that pertain to the prediction of human behavior. For example, the author takes note of these difficulties in political science. So, while UoN can be seen as an unproblematic epistemic norm in the natural sciences, the same cannot be said when it is applied to contexts in which human agency plays a role. In said contexts, the author maintains, a UoN assumption can be ethically controversial. Through the analysis of the often-discussed case of the COMPAS algorithm, it is shown that a defendant's future behavior is predicted based on the previous conduct of individuals sharing similar salient characteristics (e.g., socio-economic status, educational background). Thus understood, the author indicates, a UoN assumption is in place. It is then claimed that assuming UoN in cases similar to COMPAS can be detrimental to human agency. Ratti argues that this is because the evaluation of individual agents' future behavior merely in terms of what they (or similar subjects) have done in the past might constrain an agent's freedom of action.

ML in the Particular Sciences

The use of ML is quickly gaining dominance in diverse fields of science. This part of the book takes a closer look at five promising fields where the use of ML is the driving force of its progress and recent success: astronomy, neuroscience, climate science, particle physics at the Large Hadron Collider, and public health. The chapters included in this part provide novel philosophical challenges related to ML in these particular scientific domains, ranging from their practical use to the acceptance of ML results by the respective communities. This includes a broad range of (classic) epistemological, methodological, and conceptual issues, such as understanding, representation, and scientific progress.

Helen Meskhidze's chapter focuses on ML applications in astrophysics. The chapter considers novel approaches in the so-called "physics-informed machine learning," PIML for short. Approaches in PIML aim to include physical laws and domain-specific knowledge in ML, with the aim of rendering more accurate predictions and increasing these systems' interpretability. Through the consideration of two case studies pertaining to astrophysical shocks and gravitational fields around small bodies, the role of PIML methods on the overall transparency and interpretability of the ML is assessed. In the first case study (i.e., astrophysical shocks), it is argued that there is no noteworthy increase in interpretability through PIML (also due to the specificity of the ML's goal). However, in the second (i.e., gravitational fields around small bodies), it is maintained that PIML shows more promising results in terms of transparency and interpretability compared to standard neural networks. The main argument hinges on the fact that the possibility of adjusting PIML methods contributes to the transparency and, thus, the interpretability of the system.

Florian Boge and Henk de Regt's chapter addresses challenges related to the use of ML, in particular Deep Neural Networks (DNN), in the field of particle physics. This chapter focuses on issues pertaining to scientific understanding, which emerge from two relevant scientific enterprises related to discovery. The first is connected to the possibility of DNNs discovering novel concepts. However, since concepts are an essential part of theories that, in turn, enable understanding, limitations in assessing how ML generates concepts pose considerable constraints to our theoretical understanding. The second issue pertains to DNNs' recognition of anomalous phenomena (i.e., phenomena that deviate from accepted theory) that would otherwise remain inaccessible to humans. Also in the case of anomalous phenomena, noteworthy challenges to understanding are likely to emerge. This is because the discovery of novel phenomena without a theoretical basis would not be well understood. The authors critically address both points and conclude that hybrid approaches that combine ML and theory are realistic in the context of ML-supported discovery of phenomena in particle physics.

The chapter by **Barnaby Crook and Lena Kästner** advances considerations related to the increasing use of Artificial Neural Networks (ANNs) as models of cognitive and perceptual systems in neuroscience. The authors refer to arguments advanced in the literature defending the claim that the high level of predictive accuracy afforded by ANNs cannot contribute to the goal of understanding biological neural systems. Crook and Kästner argue against this often-defended claim and maintain that the use of complex and epistemically opaque ANNs need not hamper our understanding of the human brain. In supporting this argument, a distinction is made between two forms of understanding. The first is model understanding, which is challenging to achieve due to the ANNs' complexity. The second is what the authors call *model-induced understanding* of phenomena. Through the latter, Crook and Kästner aim to show that scientific understanding *can* be achieved by complementing findings from standard neuroscientific research with the insights gained through ANNs and *post-hoc* interpretability methods. Combining these different epistemic perspectives in an iterative re-engineering process can, according to the authors, serve as a base for neuroscientists to understand and explain neuroscientific phenomena.

Greg Lusk's chapter offers an analysis of the role of Neural Networks (NNs) in climate science. Starting from the consideration of ML applications that have started to gain ground in this scientific field already in the 1990s and are thus currently standardly adopted, the chapter aims to inquire into the novelty of NN methods in climate science and how these might substantially affect their epistemology. To this end, the chapter considers how NNs impact central epistemic functions such as understanding, representational transparency, and explanation in the philosophy of climate science. Furthermore, the chapter considers how NNs can complement systems that have a considerable theory-driven component and whether data-driven

NNs could eventually substitute theory-based methods of climate projection. It is further argued that purely NN-based climate projections are unlikely to be successful. This is the case because there is a lack of observational data on climate timescales. Moreover, inferences made from historical data are not necessarily warranted, as climate change implies that the future climate will exceed what can be derived from historical data patterns. This also represents a considerable issue that supports the idea that it is unlikely that theory-informed climate modeling will be discarded in favor of purely NN-based models.

Thomas Grote and Oliver Buchholz's chapter focuses on the deployment of ML in the context of public health. The authors explore why, unlike the increasing use of ML in clinical medicine, the implementation of ML in public health remains limited. One identified reason is the gap between predictions and interventions. Particularly in domains that are subject to considerable change, such as in the context of epidemiology, predictions based on historical data are often considered not sufficiently robust or informative to shape future interventions in public health policy. However, contra skeptic positions, the authors argue that even purely predictive ML can be successfully integrated into public health. They do so through the consideration of a case study from health economics. This study analyzes how ML can spot inefficiencies in the allocation of medical testing by investigating physicians' tendencies to generally overtest low-risk patients while under-testing high-risk ones in diagnosing heart attacks. Drawing on this case, the authors analyze how predictive models can support our understanding of inefficient medical testing and facilitate the improvement of public health outcomes. Grote and Bucholz also highlight some limitations of ML, particularly related to the fact that MLs are most efficient for narrowly defined tasks for which appropriate intervention can be known in advance.

References

Alvarado, R., & Humphreys, P. (2017). Big data, thick mediation, and representational opacity. *New Literary History*, *48*(4), 729–749.

Humphreys, P. (2004). *Extending ourselves: Computational science, empiricism, and scientific method*. Oxford University Press.

Humphreys, P. (2009). The philosophical novelty of computer simulation methods. *Synthese*, *169*, 615–626.

Part I
Epistemic Opacity

Chapter 1
In Which Ways Is Machine Learning Opaque?

Claus Beisbart

Abstract Machine learning models are often said to be opaque: They are supposed to lack transparency and are compared to black boxes. But in which ways are ML models opaque? This chapter aims to help answer this question. To establish a basis for my investigation, I first collect a few statements about the opacity of machine learning (ML) from working scientists and authors who comment on ML. The content of the statements can be summarized by saying that scientists face difficulties in reaching specific epistemic achievements regarding ML, e.g., they struggle to predict their outcomes accurately. This empirical basis is contrasted with a theoretical account, namely the well-known notion of epistemic opacity proposed by Paul Humphreys. My research question then is to what extent epistemic opacity à la Humphreys applies to ML and to what extent it can account for the problems that researchers discuss under the label "opacity of ML". As a heuristic, an analogy with diseases proves helpful: The question is whether Humphreys-style opacity provides a good diagnosis for the symptoms discussed under the heading "opacity of ML". My answer is that epistemic opacity à la Humphreys does apply to some forms of ML and that this is a partial, but not the only, cause for what people describe as opacity. There are other partial causes, for example, that researchers currently lack informative higher-level descriptions of ML models. This leaves us with two options to clarify the idea of opacity. We can either take "opacity" to be an umbrella term for the difficulties in reaching various achievements regarding ML. Alternatively, we can reserve the term for a more specific problem that stands out as a crucial factor in explaining why we struggle with these achievements.

C. Beisbart (✉)
Institut für Philosophie, Universität Bern, Bern, Switzerland
e-mail: claus.beisbart@unibe.ch

J. M. Durán, G. Pozzi (eds.), *Philosophy of Science for Machine Learning*,
Synthese Library 527, https://doi.org/10.1007/978-3-032-03083-2_1

1.1 Introduction

It is now common lore that machine learning (ML) is opaque: ML models seem to be intransparent or black boxes, as it is often put, and this is supposed to constitute a problem. For instance, Burrell (2016, 1) considers "the issue of opacity as a problem for socially consequential mechanisms of classification and ranking". But what exactly does it mean to say that ML is opaque? In which ways is ML opaque? And why exactly is this a problem? This chapter aims to help answer these questions and to clarify the opacity of ML.

In a sense, the opacity of ML is a phenomenon in the real world: ML applications have a feature that people refer to as "opacity". From this perspective, an inquiry about the opacity of ML can be expected to describe and explain the phenomenon. However, the phenomenon is introduced using a term, viz., the word "opacity", and is conceptualized using a concept with this name. Accordingly, we only make progress in understanding the phenomenon if we have a clear grasp of the concept. For this reason, this chapter will focus on the concept of opacity, as applied to ML or computer-based methods (see below for further qualifications). This kind of focus on a concept is typical of philosophical inquiry, which specializes in conceptual work. Still, clarifying the concept of opacity is only fruitful if ML itself and related phenomena are taken into account. Therefore, in this chapter, I will also look at ML itself and its features. The concept of opacity is obviously related to the notions of explainability and interpretability. However, in this chapter, I will not consider the latter notions.

My investigation of the opacity of ML will draw on two sources. One source is talk about opacity, mainly in the sciences, but also in everyday life (Sect. 1.2). I draw on this source because an explication of opacity should be faithful to how people think and talk about opacity (see Carnap, 1950/1962, Chap. 1). It will turn out that the opacity of ML is often articulated by saying that humans lack specific abilities or achievements regarding ML, or have at least difficulties in reaching them. I will thus analyze important achievements that matter to the opacity of ML. My second source is theoretical work by philosopher Paul Humphreys. He has proposed a notion of epistemic opacity that has been extensively discussed in the philosophy of science, but so far, with a focus on computer simulation rather than ML. I will thus introduce Humphreys's account of epistemic opacity in Sect. 1.3. The question then is whether Humphreys-style opacity applies to ML and whether it covers the concerns that people express using the term "opacity". I will answer this question in Sect. 1.4. In Sect. 1.5, I will draw my conclusions.

Before I start, let me mention a challenge for my project. ML comprises various kinds of methods, models, or algorithms. They have in common that their crucial assumptions or rules are learned from data rather than fixed in advance (Alpaydin, 2020, Sect. 1.1). However, they differ in other respects, and this may lead to differences regarding their opacity. There is thus the danger of overgeneralizing claims about opacity. In this chapter, I cannot do justice to all kinds of ML. However, I will try to keep in mind that there are several types of machine learning, and I will consider neural networks, decision trees, and clustering algorithms as examples of

ML. These examples have not been selected because they are particularly opaque, but rather to represent the plurality of ML approaches.

Let me briefly introduce the examples. A *neural network* is a model of interconnected neurons. Typically, information from an input layer of neurons can be propagated through the network to an output layer. The interconnections between the neurons are described using parameterized functions. The parameters are optimized using labeled training data such that the output layer reproduces known labels for a range of inputs. Here, the labels may, e.g., classify masses from mammograms as benign or malignant. After some training and validation, the network is applied to new inputs to infer labels for previously unlabeled data (see, e.g., Samala et al., 2017).

Like neural networks, *decision trees* are optimized to reproduce labels, e.g., classifications, for training data. The crucial idea is to organize the data into a hierarchical structure of types, subtypes, and so on. New data, then, are classified through a series of decisions that determine to which type, subtype, etc., the data belong (see, e.g., Loh, 2011 for an overview, and Carliles et al., 2010 for an application in astrophysics).

Finally, *clustering algorithms* identify clusters in a space in which objects are represented as points. No prior information on cluster membership is provided, so we are talking about *unsupervised* learning. Roughly, objects that are close to each other in some sense are grouped into the same cluster (see e.g., Madhulatha, 2012 for a review). If the underlying space is real space, then the clusters are agglomerations of objects, e.g., clusters of galaxies. If the space is a feature space, by contrast, the clusters represent types of objects, e.g., certain types of stars. There are several methods of clustering, such as k-means clustering, where the definition of exactly k clusters is intended to minimize the intra-cluster spread of objects (see, e.g., Jain, 2010). This method was, for example, used in astronomy to classify stellar spectra (Garcia-Dias et al., 2018).

ML differs from other computer-based methods, notably from the method of computer simulation. In a computer simulation, a digital computer is used to obtain (approximate) solutions to model equations that scientists have come up with. Since the model equations refer to the temporal behavior of a system, the computer effectively obtains a series of state descriptions when solving the equations. The outputs of computer simulations are often visualized using animations (see Humphreys, 2004, 110 for a characterization of computer simulations; see also Durán, 2020). Since Humphreys-style opacity was first proposed in the context of computer simulations, while this paper is concerned with ML, my investigation will also shed light on the question of how ML differs from computer simulation.

1.2 Some Talk About the Opacity of ML

To start our inquiry, it is useful to examine how people talk about the opacity of ML. I have collected a few statements on the opacity of ML by conducting a Google Scholar search using the words "machine learning opaque". I have also searched for what might count as a definition of the opacity of ML. I have bracketed philosophical works at this point to obtain an independent perspective.

Opacity is commonly associated with black boxes and is contrasted with transparency (e.g., Lipton, 2018, 12; Woods, 2018; Pearl, 2020). In his oft-quoted paper, Lipton (2018, 12) defines transparency as "some sense of understanding the mechanism by which the model works." Following Belle and Papantonis (2021) define transparency as "a human-level understanding of the inner workings of the model" (3). Opacity may then be characterized as a lack of understanding of the working of a ML model. This characterization fits quite well with Wood's (2018, 148) complaint that in "artificial intelligence algorithms [. . .] the relationships between the variables used to derive their predictions are obscure." Woods adds to this that the algorithms "often fail to provide insight to the underlying systems," where he seems to refer to real-world target systems studied with ML (Woods, 2018, 128).

Sociologist Burrell (2016, 1) gives the characterization of opacity a slightly different twist:

> [Some algorithms] are opaque in the sense that if one is a recipient of the output of the algorithm (the classification decision), rarely does one have any concrete sense of how or why a particular classification has been arrived at from inputs. Additionally, the inputs themselves may be entirely unknown or known only partially.

Here, the focus is on the inputs and outputs. I assume that the term "inputs" refers specifically to the types of inputs used to train a model, rather than to a single input that triggers a classification or prediction. After all, the former input (the training data) is much more comprehensive than the latter, and thus more difficult to know. Also, many ML models (and ML models are the focus of Burrell's paper) are characterized by the fact that they rely on training data.

The output is also at the center of what Michalsky (1986, 8) calls the "predictive opacity of self-changing systems": "Predicting the behavior of machines that can learn inductively is considerably more difficult than predicting the behavior of machines without such an ability." I take it that the behavior manifests in the outputs that are produced in response to inputs; it is thus difficult to predict the outputs of a ML program.

In a short statement titled "The limitations of opaque learning machines," Pearl (2020, 15) explains the importance of transparency as follows:

> Once you unleash it on large data, deep learning has its own dynamics, it does its own repair and its own optimization, and it gives you the right results most of the time. But when it doesn't, you don't have a clue about what went wrong and what should be fixed. In particular, you do not know if the fault is in the program, in the method, or because things have changed in the environment. We should be aiming at a different kind of transparency.

The focus of this passage is on scientists' abilities to work with a ML model.

These quotations provide some insight into what working scientists and commentators mean by the opacity of ML. To summarize the content of the statements, we may say that they mention things that humans do not do in connection with ML models:

(a) Know the training data (Burrell, 2016);
(b) Know relations among variables (Wood, 2018), understand the workings of the model (Belle & Papantonis, 2021 following Lipton, 2018);
(c) Predict the behavior/the outputs of the models (Michalsky, 1986);
(d) Have a sense of/understand why an outcome was produced (Burrell, 2016);
(e) Pin down mistakes if results are wrong (Pearl, 2020);
(f) Understand the system that is investigated (Wood, 2018);

To simplify terminology, I will say that the items on the list represent (epistemic) achievements. The first four achievements concern the training data (a), the model (b), and the output (c-d), respectively. The other achievements (e-f) are related to working with the model, particularly concerning the target system (f). Opacity then means that humans fall short of these achievements. Often, it is added that humans cannot (easily) reach these achievements.[1]

The achievements most often mentioned are kinds of knowledge and understanding. The other achievements are related to inferences. For instance, to predict an output, scientists must infer it. Now, understanding is often spelled out in terms of abilities, at least some of which are inferential (see, e.g., Grimm, 2006; Hills, 2016; see also Baumberger et al., 2016). Accordingly, we can subsume all the achievements on the list under a broad notion of understanding and the notion of knowledge.

Some of the statements talk about "a sense (of understanding)" (Burrell, 2016, Lipton, 2018). It is not entirely clear what is meant here—for instance, is Lipton talking about understanding in a particular sense of the word, or about what people call "sense of understanding"—the feeling that you have understood something? In what follows, I will bracket a merely subjective feeling of understanding because it may be deceptive and should not be confused with understanding itself (Trout, 2002).

Of course, the above list of quotations is based upon a tiny sample and cannot be expected to be exhaustive. Still, the achievements mentioned seem pretty typical for what is discussed under the heading of opacity. What is typical, too, is that the lack of achievements is seen as a problem, at least prima facie.

What people mean by "opacity" concerning ML is consistent with the meaning of "opacity" in ordinary language. There, the term "opaque" has two meanings, viz. "(of glass, liquid, etc.) not clear enough to see through or allow light through" and "(of speech or writing) difficult to understand; not clear" (Oxford Learner's Dictionary, n.d.)

Some of the above problems, articulated under the label "opacity of ML," consist of a lack of understanding; therefore, opacity is here meant in the second sense

[1] There is a general ambiguity when it comes to what people can (easily) do. A large fraction of the literature focuses on what humans can do without any help or by hand (e.g. Tymoczko, 1979). Dissenting voices suggest that it is more relevant what humans can do using instruments, particularly computers (e.g. Teller, 1980). In this paper, I too focus on what humans can do without any help.

listed in the dictionary. Other problems refer to a lack of knowledge. This lack of knowledge may either be an aspect of a lack of understanding, which brings us back to the second meaning of the term. Alternatively, we may say that "opaque" is used in the first sense, but in a metaphorical way, where seeing is meant to stand for knowing. ML models are then opaque either because their inner workings are difficult to "see" (b) or because they do not let us "see target systems through them" (cf. point f above).

1.3 Paul Humphreys's Notion of Epistemic Opacity

Can we provide a more systematic account of the opacity of ML? To answer this question, it is helpful to draw on a second source, viz., existing philosophical work on opacity. This section aims to explain the concept of epistemic opacity, as developed by philosopher Paul Humphreys. This notion of opacity has been extensively discussed in the context of computer simulations. In the next section, I will then apply this notion to machine learning (ML). (Note, by the way, that something like the opacity of computer-based methods has been noted before Humphreys; for instance, Moor, 1985 speaks of invisibility and of "*invisible complex calculation*", 274).

Humphreys (2009, 618) defines epistemic opacity as follows:

> a process is epistemically opaque relative to a cognitive agent X at time t just in case X does not know at t all of the epistemically relevant elements of the process.

He calls a process "essentially epistemically opaque" if the required knowledge is impossible for the agent due to their nature (ibid.).

In what follows, I will reserve the term "epistemic opacity" to epistemic opacity as defined by Humphreys (although I do not think that the qualification "epistemic" is beneficial here because even the ordinary language notion of opacity is related to knowledge, understanding, and the like). To simplify the presentation, I will occasionally omit the qualification "epistemic" in this section. Note, too, that Humphreys and a coauthor have defined a different notion of opacity, viz., representational opacity (Alvarado & Humphreys, 2017; see also Facchini & Termine, 2022 for different kinds of opacity). In this chapter, I do not consider representational opacity much because it applies to representations, and it can be debated to what extent, and in which sense, ML models represent (see Chap. 12 by Kieval in this volume).

Humphreys's definition of epistemic opacity is valuable and helpful in many respects. First, what may be opaque according to the definition is a process. In computer simulations, the process under investigation is the so-called core simulation, i.e., the process that leads from the input to the output (Humphreys, 2004, 147, see also 110 and 127). However, other processes may also qualify as opaque. This is advantageous because my goal in this paper is to apply the notion of opacity to ML, and when a ML model is run on a computer, this forms a process too.

Second, according to Humphreys's definition, opacity consists in a lack of knowledge. This is in agreement with one of the ordinary language meanings of "opacity".

Third, what is unknown is the entirety of the steps in the process under consideration, not just a single step. It is at least possible to know about an arbitrary step in a simulation. So-called debuggers focus on such steps. The problem behind epistemic opacity is instead that there are too many steps. There are several reasons why computer simulations contain a vast number of computational steps. High resolution requires more time steps and possibly a smaller grid size, and thus more calculations. Also, many computer simulations trace very complex models with a large number of parameters. So, ultimately, for Humphreys, opacity is a problem of large numbers and the corresponding high complexity. It is interesting in this respect that Humphreys (2004, 148) notes some computational processes to have irreducible complexity, which means that their complexity is unavoidable: Roughly, there is no shorter way to provide complete information about the process than listing all of its steps (Humphreys, 2004, 148 and Humphreys, 2020, 16 mentions high speed as another source of opacity, but I doubt that this is a separate issue).

Fourth, the definition stresses that the entirety of the *relevant* steps is unknown. In this way, Humphreys avoids the objection that opacity is not a problem because knowledge of all elements is futile: some steps may not matter after all. Humphreys grants this; as he explains, neglect of certain details is often advantageous because understanding is facilitated (Humphreys, 2004, 148). Accordingly, the steps in the process do not need to be individuated in too fine a manner. Still, even if the steps are individuated in a reasonable way, the entirety of the steps is not, or cannot be, known if opacity is present.

Fifth, epistemic opacity is relativized to an agent and to a time. Accordingly, what may be opaque to one agent may not be opaque to another agent, or the same agent at a later time. Humphreys uses an analogy to illustrate this. In a proof of a mathematical theorem, a step may be relevant for one agent but trivial for another (Humphreys, 2009, 618). The relativity of opacity is a plus because it helps to avoid over-generalization.

Sixth and finally, by defining *essential* epistemic opacity, Humphreys effectively introduces a distinction between two kinds of opacity: In a weak sense, opacity is a lack of knowledge of the relevant details. In a stronger sense, opacity means that such knowledge is impossible. An option that falls between these extremes is that the relevant knowledge is difficult for humans to attain. In what follows, I will focus on this middle path, both for epistemic opacity and opacity of ML (see Beisbart, 2021 for a similar move). In practice, it often doesn't really matter whether we talk about a difficulty or an inability.

Although Humphreys's definition of opacity contains several valuable insights, it also raises some questions. Fortunately, they have largely been answered in the literature.

A first question is how we can talk about opacity sans phrase if the definition relativizes opacity to an agent (and to a time for non-essential opacity). An obvious answer is that opacity sans phrase is opacity for every human agent at all times (cf.

Boge & Grünke, forthcoming; they use essential opacity and quantify over agents to obtain what they call fundamental opacity).

Second, it may be asked what exactly is meant by *knowing the epistemically relevant elements.* What are the elements, and to what kind of things are they relevant? Further, what sort of knowledge is missing—knowledge that there are the steps, how they have been finished, or whether they were correct? Humphreys seems to leave these questions deliberately open and only gives the following examples of relevant elements: "justification for moving from one step in the process to another, the representational correctness of semantic elements of the process, and the defeasibility of the process" (Humphreys, 2020, 15). So, one way to think about epistemic opacity is to refrain from specifying the missing knowledge and to think of the definition as a kind of template that can be applied to discuss various types of ignorance. This aligns well with what Zednik (2021) does when classifying what various stakeholders are interested in knowing about ML methods. Later in this chapter, I will argue that the resulting take on opacity is an interesting option, but that Humphreys's definition is not ideal in this respect.

An alternative option is to spell out in more detail what sort of knowledge or epistemic access is missing. Durán and Formanek (2018) take this approach (see also Durán, Chap. 4 in this volume). As indicated in Humphreys's first example mentioned in the last paragraph (the other examples are less clear), they consider the process to be a justification, e.g., an argument that leads from premises to a conclusion. One can thus ask whether the justification is correct (e.g., a valid argument). Opacity, then, is defined in terms of the difficulty of checking, or the inability to check, that the justification is correct. Similarly, the process can be viewed as a calculation, and we may ask whether the calculation is correct. Opacity would then be a related difficulty or the inability to check the correctness of the calculation.

Taken in either of these ways, opacity is assimilated to unsurveyability, as introduced by Tymoczko (1979). For Tymoczko, the proof of the Four Color Theorem by Appel and Haken is unsurveyable. To complete their proof, Appel and Haken wanted to check that each configuration from a large, so-called unavoidable set is reducible in a mathematical sense. This check requires too many steps to be completed by a human person in a reasonable time and was thus delegated to a computer. What the computer did is not surveyable because humans cannot check by hand whether the result obtained by the computer is correct. Computer simulations are unsurveyable in a similar way because humans cannot check the correctness of the calculations by hand.

If epistemic opacity is essentially unsurveyability, how significant a problem is it? As Durán and Formanek (2018) argue, researchers can be justified in believing the results of simulations, even though these results are opaque. Durán and Formanek explain this by introducing computational reliabilism (see also Chap. 4 by Durán and Chap. 5 by Alvarado in this volume). Independent of this position, the central point is as follows: A step-by-step check of the computer simulation is not necessary to check the correctness of the calculations or the validity of the argument. There are alternative methods, often called methods of verification, that at least make a case that the justification/calculation is correct (see Rider, 2019). For instance, scientists compare the output of a computer simulation with analytically known benchmark solutions or examine whether the outputs converge to a value by sequentially refining the grid resolution.

To summarize this section, according to Humphreys, a process is epistemically opaque to an agent if, and only if, the agent has difficulties knowing the epistemically relevant elements. We can either leave it to the discussion of specific examples to clarify what exactly knowledge of epistemically relevant elements entails. Alternatively, we can answer this question by further developing the definition. A natural way of doing this is to refer to the difficulty in checking the process as justification or calculation: The process is then epistemically opaque if it is unsurveyable, i.e., if there are too many steps to be checked. In the machine learning literature, this is sometimes called lack of simulatability (Lipton, 2018, 12–13). There is no doubt that computer simulations are epistemically opaque, either in the broad sense or in the sense of unsurveyability.

1.4 The Opacity of ML and Epistemic Opacity

We now have material from two sources on hand—first, a couple of statements that articulate problems about ML by using the term "opacity", and, second, the idea of epistemic opacity, as defined by Humphreys. It is thus time to establish a connection between them or to let the two sources coalesce. There is a natural way of doing this: Epistemic opacity à la Humphreys is used to articulate a sense in which computer simulations are opaque. In principle, it has a broader scope because its definition can be applied to every process. ML involves processes too, notably the process of training and the process in which the ML model is applied to new input (our focus will be on the latter). We may thus hope that, in one of its interpretations, Humphreys-style opacity applies to ML as well, and that the resulting epistemic opacity of ML encompasses what people mean when they talk about the opacity of ML. In the remainder of this paper, I will investigate whether this is the case.[2]

There is no doubt that Humphreys-style opacity applies to many ML models. Typical ML applications involve a vast number of computations. This is true both for the training phase and the application of the trained model. The reason is that many ML models involve a large number of variables and parameters. For a recent example, consider the work done by Samala et al. (2018). They tried to reduce the number of parameters in a deep convolutional network trained for classifying

[2] Zednik (2021) too applies Humphreys's notion of opacity to ML. But, as indicated above, he is working with a very broad notion of epistemically relevant elements and concentrates on knowledge. So what he does, basically, is to specify the knowledge that several types of stakeholders are interested in, but typically lack, when it comes to ML. My take on opacity is broader because I do not restrict myself to knowledge. Note however, that my understanding of *epistemic* opacity is narrower than his since I stress the aspect of complexity.

medical images. Even their reduced model contains 19′084 neurons. There are more than 20 million parameters. For one classification task, more than 6 million addition and multiplication operations are required. Accordingly, it is impossible for a human to know or to check each computational step. Thus, ML applications are unsurveyable, and they are epistemically opaque in the broader sense if the computational steps of the computer are epistemically relevant.

But how is this epistemic opacity related to the problems that people articulate using the label "opacity", i.e., to the lack of achievements (a-f) above? Epistemic opacity is undoubtedly in the same ballpark as the problems that scientists discuss regarding ML. The reason is that epistemic opacity is a lack of knowledge, and the achievements that are lacking regarding ML include knowledge as well or are closely related to it. It is also common to epistemic opacity and opacity attributed to ML that they can come in various strengths: knowledge or related achievements can be absent, difficult, or even impossible. As said before, in this chapter, I concentrate on medium strength (difficulty), but for the sake of simplicity, I will often also talk of the lack of some achievements.

Despite the commonalities, the collected statements about ML cover a broader range of problems than can be captured using Humphreys's epistemic opacity. For instance, Humphreys's definition doesn't cover a lack of knowledge of the training data (point a above). There is also a more general mismatch between Humphreys's account of opacity and the statements collected in Sect. 1.2: While some of the latter focus on understanding, e.g. the understanding why some outputs have arisen, Humphreys's definition does not mention understanding (although Humphreys slips into talk about understanding when commenting on opacity; e.g. Humphreys, 2004, 148).

Still, this need not be the end of the story for an attempt to reduce the opacity of ML to epistemic opacity. The statements collected above, one may say, mention various phenomena for which "opacity" suggests itself as an umbrella term. Humphreys's definition of epistemic opacity, by contrast, has the ambition to dig deeper and to identify a kind of feature that explains other problems. That Humphreys does indeed try to give a general diagnosis is suggested by the following statement (Humphreys, 2009, 618–619):

> Many, perhaps all, of the features that are special to simulations are a result of this inability of human cognitive abilities to know and understand the details of the computational process.

Admittedly, this passage refers only to computer simulations, and, when commenting on the opacity of ML, Humphreys (2020) uses a broader notion of opacity. However, it is still a legitimate question how the problems expressed using the term "opacity" relate to his epistemic opacity. It is interesting in this respect that, when Boge (2022) and Zednik (2021) discuss the opacity of ML, they do so within the confines of epistemic opacity à la Humphreys.

To illustrate the central idea here, we may compare the opacity of ML to a disease. The problems discussed under the label "opacity" (i.e., that humans lack the achievements listed above) form various symptoms. Humphreys's definition may then provide a diagnosis of the symptoms or explain them.

Of course, tokens of diseases do not have a single cause. For example, when Sam experiences certain symptoms, say fever and coughing, this may go back to a genetic predisposition for a particular disease, to his habits plus a trigger such as an infection. Similarly, if a ML model is opaque, this may stem from a complicated interplay of causes. Still, in either case, there may be a main cause or a main contribution to the explanation. If epistemic opacity à la Humphreys is the main explanation of why scientists lack the achievements mentioned above, then we can say that the opacity of ML largely reduces to epistemic opacity.

To determine whether this is indeed the case, I will now analyze whether epistemic opacity is the main explanation for why humans struggle with reaching the achievements mentioned above. Since (f) is somehow special because it refers to the target system, I bracket it and concentrate on (a-e). In simple cases, our difficulty in reaching a certain achievement may be an instance of epistemic opacity. In other cases, epistemic opacity may at least be an essential part of the explanation of why we struggle with the achievement. To be fair, it wouldn't be much of a problem if opacity à la Humphreys weren't central to the explanation of why humans struggle with one or the other achievement. After a closer investigation, some symptoms may be traced back to different diseases. Introducing an etiological (cause-based) definition of a disease often has the effect that alleged cases of the disease no longer fall under the disease category. Still, the cause mentioned in the definition should account for a majority of the symptoms or the most important ones. The question now is whether this requirement is fulfilled and whether Humpreys's notion of opacity can explain the most important symptoms.

(a) Know the training data

Scientists typically do not know the inputs that were used for training a decision tree or a neural network, at least not in detail. The reason is that the training data is too massive to be known. For instance, Samala et al. (2018) utilize a network that was initially trained on 1.2 million images; subsequently, additional images were incorporated into the training. Things are different with clustering algorithms. Properly speaking, the latter do not involve training data, and the data used to build clusters can often be visualized in a way that allows humans to obtain an overview of the data. I will thus concentrate on ML models that are trained.

If humans lack knowledge of the training data, this cannot be explained in terms of epistemic opacity as defined by Humphreys. The training data cannot instantiate the latter because it is not a process. Furthermore, the training data is prior to all processes running during the phases of machine learning (both the training process and the process of applying a trained ML model). Thus, it would be strange to explain lack of knowledge of the training data via lack of knowledge of the process (although it is true that some processes, in particular those of training, involve the training data, so knowledge of the training data may be obtained from knowledge of those processes; still, this is a very unnatural way to learn about the training data). Accordingly, Humphreys-style opacity is not a diagnosis of the first problem, not even a partial one. The background is that epistemic opacity was first introduced with an eye to computer simulations, which do not undergo training using data.

(b) Know relations among variables/understand the workings of the model

Consider now knowledge of the relationships between the variables of the model (which I now assume to be the learned model). For both decision trees and networks (but not for clustering algorithms), scientists do not know the relationships between the variables of the models. For instance, for a typical neural network, scientists do not know the exact functions that mediate the interactions between neurons. These functions have been "learned", i.e., obtained using an optimization procedure. Each single function can, in principle, be known, but scientists cannot keep their entirety in mind.

Admittedly, scientists tend to have some general knowledge of the architecture of their decision trees and neural networks. For instance, they may know the number of layers of neurons or the shape of the functions connecting the neurons. However, scientists do not know the precise model that the machine has constructed. Contrast this with other models used in the sciences, e.g., that of a double pendulum or models from fluid dynamics. These models can be written down using a few equations, e.g., the Navier-Stokes equations. The relationships between the variables are all known and relatively simple. There are only very few parameters in the model. Accordingly, sometimes, analytical results about the model are known. Network models are different. Since they have extremely many parameters, they form a very broad class of models that is more difficult to overview (note though that there are now ML models that take physical knowledge into account. In these models, more is known about the relationships between variables; see e.g. Cuomo et al., 2022 for a review).[3]

How is all this connected to epistemic opacity as defined by Humphreys? Relationships between variables are not elements of a process, since elements of something are its proper parts, whereas a relationship between variables is not a proper part of a process. Thus, we are not talking of an instance of epistemic opacity (things are different if one opts for an extremely broad understanding of "element", assimilating "element" to "aspect"). Furthermore, if the process of applying a ML model to new input has many unknown elements, this is the consequence of unknown relationships between variables, and not the other way round. Thus, epistemic opacity is not the explanation of the problem, not even in part, but rather a "downstream" consequence of the fact that the relationships among the variables are unknown.

[3] Admittedly, there are some scientific models that are closer to ML models than the scientific models I have considered in this paragraph. For instance, the models on which computer simulations are built can become very complicated. Further, many computer simulations contain parameters that are calibrated to data of the target system. Accordingly, the resulting computer simulations are based upon some learning too, the details of which may not be understood by the scientists. Still, the variables used in computer simulations allow for an interpretation in terms of the target system, which is more difficult with much ML.

Turn now to the problem that the model or its working is not *understood.* This point applies to both neural networks and decision trees. There are at least two facets to the problem. First, as we have just seen, the model is not known in that the relationships between its variables are not known, so how can one understand the workings of a model about which one does not know much? Still, it may be objected that the model is, in principle, accessible (unless the program code is secret): scientists may learn how the neurons of an artificial network are connected or which decision rules a decision tree has learned. However, there is a second problem: Even if scientists knew the relationships among the variables, the model or its workings would not be intelligible to them. To argue for this point, we can draw on work by de Regt and Dieks (2005). They call a theory intelligible to a scientist if she can anticipate the consequences of the theory without doing exact calculations (p. 151). This account of intelligibility can also be applied to models. Many neural networks and decision trees are not intelligible in this sense because scientists lack a "feeling" for the model: they cannot infer what the model implies without doing the precise calculations (here and elsewhere, "inference" is meant in a comprehensive sense; it may not feel like going through an argument). They might do so if they had a sense of the properties to which a network model is most sensitive; however, scientists often cannot identify the features to which the model is most sensitive. Additionally, they lack sufficient experience with the model to make reasonable predictions about its behavior for specific inputs. (As this example shows, the most significant consequences of the model refer to the outputs. Since I will get to the prediction of outputs in the discussion of the next achievement, I will not elaborate on this issue here.)

This lack of understanding is quite far from Humphreys-style opacity. First, the achievement that humans lack in this regard does not refer to a process, but rather to features and consequences of a model. Second, the central problem is not a lack of knowledge, but rather the difficulty of running, or the inability to run, certain inferences. Finally, the problem is not that there are too many steps (epistemically relevant elements) in some calculations, because the definition of intelligibility excludes going through many steps. A crucial part of the problem is rather that there is no higher-level description that enables a more intuitive grip on the model (but see Buckner, 2019 for attempts to characterize the workings of neural networks at a higher level). Additionally, at least for today, scientists lack experience with these models, which means they lack implicit expert knowledge of what the model implies.

(c) Predict the behavior/the outputs of the models

Predicting the output of a ML model is sometimes possible. For instance, if a neural network or a decision tree has the task of classifying masses from mammograms, and a new image is shown to experts, they can determine with high probability whether the mass is benign or not. Assuming the ML model works well, they can thus predict the outcome of the ML model. The outcomes of clustering algorithms can also often be predicted, at least in a rough outline. For instance, in a visual representation of a galaxy catalogue, the human eye can detect clusters of galaxies.

Still, there are cases in which humans are not good at predicting the outputs of ML models, for instance, when ML models are used to predict the weather (see e.g. Lam et al., 2023 and Price et al., 2025). It is also challenging to predict the output of ML models when the input falls outside the range of the training data used (this is the issue of generalizability; see, e.g., Zhang et al., 2017 and Zhang et al., 2021).

Why is it difficult to predict the outcomes of some ML models? Part of the reason is epistemic opacity à la Humphreys. If scientists knew all epistemically relevant elements of the process run on the computer (more precisely, if they knew in advance the outcomes of all computations done by the computer), they would be able to predict the outcome. However, there are too many computational steps, making this knowledge difficult, if not impossible, as implied by epistemic opacity.

Now, scientists do not need to know all the steps in the calculations performed by the computer to make predictions about the outputs. As indicated, there can, in principle, be alternative routes to predicting the outputs. The problem, however, is that, frequently, all these routes are closed, often for reasons that do not stem from epistemic opacity. Therefore, epistemic opacity is not the main cause of our difficulty in predicting outcomes. Let me explain.

As for a first route, scientists may, in principle, know the ML model and calculate the output of the model themselves. In practice, this is not possible for two reasons. First, as we have seen (point b), scientists do not know the model. As indicated above, this point may, with some goodwill, be taken to instantiate epistemic opacity, namely, if we allow that the relations between the variables are the epistemically relevant elements. But there is a second problem. Even if the scientists knew the model, they would have a hard time calculating its solution for the given input. The reason is that the model is extremely complicated. So, the problem is not only that some kind of knowledge is missing, but also that humans cannot perform many calculations in a short time.

Now, at least some aspects of an output may be predicted without a detailed knowledge of a model. For instance, in many models in physics, e.g., in hydrodynamics, energy is known to be conserved. For other models, scientists know that the behavior is composed of certain oscillations. This allows scientists to predict certain aspects of the model output if they can assume that the computer program correctly evaluates the model.

This second alternative route to prediction, however, is also closed for many ML models. The reason is that ML models are not sufficiently well studied and well understood, as we have shown under point b. Whereas many features of the behavior of, say, the Ising model are known, this is not (yet) the case with ML models.

A third alternative route for predicting outputs or certain aspects thereof would begin with the training data. For instance, it may be foreseeable that a classifier classifies a certain mass as benign because the mass is very similar to other masses that are labeled as benign in the training data. However, such a prediction is only possible based on knowledge of the training data, which is often too massive to be known (see above). Furthermore, the prediction may prove incorrect in some cases. So-called adversarial examples are cases in point (Szegedy et al., 2013; Kurakin et al., 2018). These are based upon inputs that are correctly classified. The adversarial examples change the inputs in a way that is barely perceptible to humans, but leads to a different machine classification. Thus, for adversarial examples, the machine outcomes differ from what humans predict.

To summarize our findings on predicting the behavior of outputs: If ML models make detailed predictions, then it is difficult for humans to anticipate the predictions. Part of the reason is epistemic opacity: If scientists knew all epistemically relevant elements in advance, they could make a prediction. Still, there are additional important reasons why the outputs of ML models are difficult to predict: ML models are not the kind of models that humans can currently work with (see Chap. 12 by Kieval and Chap. 13 by Freiesleben in this volume; this point is also captured using the notion of representational opacity by Alvarado & Humphreys, 2017). Additionally, it is practically impossible to predict the output based on the training data.

(d) Have a sense of/understand why an outcome was produced

For neural networks and decision trees, scientists often do not understand why a particular output was produced, or why it has certain features (this kind of problem is central for the account of opacity proposed by Beisbart, 2021). It's different again for clustering algorithms, which is why I will not address them here.

But what exactly is the problem for neural networks and decision trees? According to a recent account of understanding why something is the case (Hills, 2016), understanding why specific outputs have been produced would imply a cluster of abilities, including the abilities to explain why the outputs have arisen, and to draw inferences about slightly modified inputs and outputs. The second ability would mean that scientists can answer "what-if" questions for counterfactual conditions (this ability is mentioned explicitly by Pearl, 2020; in what follows, I neglect inferences back from perturbed outputs to inputs).

In some sense, scientists know the crucial parts of an explanation of the output: the input that was used to produce the output, and the computer program or the model. But this knowledge is very shallow and does not suffice for understanding why a specific output has arisen. For one thing, the scientists do not really know the model. They can point to "this" model, but they do not know it in detail (point b). For another thing, scientists cannot run counterfactual inferences about slightly perturbed inputs and thus fall short of an ability that is central to understanding according to Hills (in this volume, Chap. 8 by Buijsman also stresses the importance of counterfactuals for explaining and understanding ML models). This inability to run inferences about slightly perturbed inputs is related to the inability to predict in detail the outcomes of runs of the model: Clearly, if scientists could predict outcomes of the model, they also could run the required inferences about perturbed inputs. Accordingly, humans struggle with making inferences about counterfactual inputs partly because they cannot predict outputs based on actual inputs (note, however, that there may be other possibilities for inferring outcomes for perturbed inputs, e.g., by tracing the effects of perturbations). As we have seen above (point c), there are various reasons for this latter inability; epistemic opacity is only one of them, and it does not stand out in this context. On a more general note, reasoning about counterfactual scenarios moves further away from epistemic opacity, as this reasoning does not refer to an actual process anymore, as it figures in Humphreys's definition.

(e) Pin down mistakes if the results are wrong

The results of an ML model are incorrect, for instance, if it misclassifies benign masses as malignant or if its forecasts turn out to be wrong. A single incorrect result isn't too much of a problem because ML models are only intended to provide the correct result with high probability. Still, if the fraction of wrong results is too high or if an important class of inputs is misclassified, this is a problem for a ML model.

Suppose, then, that a row of results of a ML model are wrong. According to Pearl (2020), pinpointing the source of the error is challenging. Pearl mentions three possibilities: either there is a bug in the program, the model is incorrect, or the new inputs differ markedly from the data used to train the model. Additionally, a hardware failure may have led to the mistake. What scientists face here is a degeneracy, an underdetermination problem, or an instance of holism (see Lenhard & Winsberg, 2010 for a similar point regarding computer simulations). There are several possible explanations of what has gone wrong, and scientists struggle to determine the correct explanation.

Pearl is correct that such degeneracy is common; his point applies to all kinds of machine learning models (and also to computer simulations; Lenhard & Winsberg, 2010). Still, the degeneracy can sometimes be broken. For instance, scientists may know, or discover, that the new inputs are significantly different from the training data. A precondition for this is that the training data is known. As we have seen before (point a.), the training data is often not known in detail. Still, sometimes a rough characterization of the training data is sufficient to help scientists see that new inputs differ markedly from the training data.

If scientists knew the model and understood its workings, then they might be able to tell whether the problem derives from the model. However, as we have seen before, scientists often do not know the model. Moreover, they lack an understanding of its workings. In particular, they often do not understand to which kinds of features from the input the model is really sensitive. We may put this again by saying that they lack a higher-level description of what the model does.

If the incorrect results are due to a new type of input data or a bug in the program, this can be determined by examining the computational process. At this point, epistemic opacity à la Humphreys becomes relevant: Since scientists do not know, or cannot check, all epistemically relevant steps in the computational process, they cannot find out where a hardware failure or a bug has led to the wrong result.

Note, though, that epistemic opacity is, once more, not the only cause of the problem. What is an issue, too, is a lack of knowing the training data, and a lack of understanding the model, which is not only a matter of epistemic opacity.[4]

---

[4] As mentioned before, according to Humphreys (2020, 15), one example of epistemically relevant elements is "the representational correctness of semantic elements of the process". This example seems to gesture at a possible representational inaccuracy of the conceptual model. If epistemic opacity is supposed to cover the examination of accuracy, it is of course more relevant to the identification of mistakes. Still, representational inaccuracy is not naturally discussed at the level of the process that is running when the ML model is evaluated.

To summarize this section: For some of the problems that working scientists summarize under the label "opacity", Humphreys-style opacity is part of the diagnosis. If scientists knew all epistemically relevant elements, they would be in a better position to predict the outputs of ML models, explain them, and understand them. However, knowing the epistemically relevant elements is not sufficient for all achievements on the list; for instance, running certain inferences requires more than just knowing about the process by which the ML model is numerically evaluated. Nor is lack of epistemic opacity always necessary for avoiding the difficulties that humans encounter regarding ML. For instance, the outputs of ML models may, in principle, be inferred from a detailed knowledge of the training data. That such an inference is not possible in practice is not a matter of epistemic opacity à la Humphreys. In sum, then, epistemic opacity is neither sufficient for all problems that people summarize under the label "opacity of ML," nor is it necessary for each problem of this kind.

A significant problem with many ML models is not only that they are extremely complicated in their details, but also that scientists lack good higher-level descriptions of such models. This differs from some computer simulations that are also epistemically opaque. Although such simulations also involve extremely complex computational processes, they are built upon domain-specific conceptual models, the core of which are sometimes a few laws of nature. It is thus easier to characterize those simulations at a higher level of description. Accordingly, scientists can describe their simulations in terms of physical, biological objects, and assumptions. For instance, they can say that individuals from a population interact in these specific ways. In fact, all computer simulations have an interpretation in terms of a target system; accordingly, domain knowledge about the relevant type of system can be applied to the simulation. By contrast, scientists who work with ML models cannot describe the model's working using their domain knowledge (although domain knowledge about networks may help them understand the model).

This result aligns with recent findings by Boge (2022). Boge argues that DNNs are opaque in two dimensions, which he calls h-opacity and w-opacity, respectively. While h-opacity is supposed to be continuous with epistemic opacity of simulations, w-opacity is distinctive of ML. Here, w-opacity roughly is a lack of knowledge or understanding of what the ML model learns. In my terms, he is referring to the lack of a higher-level description of the ML model. But I think it is not helpful when Boge tries to mold w-opacity in the form of Humphreys-style opacity. It stretches things too much if we say that the higher-level features, the lack of which is decisive for w-opacity, are elements of a process. This is because elements of a process are parts of it rather than its features.

1.5 Conclusions

Where does all this leave us? To conclude, I will expand on the analogy between the opacity of ML and diseases.

Working scientists discuss a broad range of problems related to ML under the heading of "opacity." These problems primarily consist of a lack of epistemic achievements on the part of humans concerning ML, or difficulties, if not outright inabilities, to attain these achievements. If we want to systematize the problems a bit, we can group them into clusters or draw some distinctions. For instance, some of the achievements refer to the target systems, while others do not.

In an analogous way, people discuss a broad range of symptoms affecting various kinds of people under headings like "cold". The symptoms include coughing, a sore throat, fever, etc. If we want to systematize the symptoms a bit, we can group them into clusters or draw some distinctions.

Whenever a method is found to be opaque for a particular person in some sense, this is often due to an interplay of various causes, some more proximal and some more distal. In the same way, whenever a person has a cold in some sense, this goes back to an interplay of various causes. To provide a more systematic account of opacity or colds, we may ask how the different symptoms of opacity or colds relate to one another and how they are explained. Maybe some symptoms cause others, or there are common causes for symptoms. In the case of ML, it would be attractive to explain the symptoms using features of ML methods themselves.

Concerning opacity, the symptoms consist of a lack of specific achievements, or related difficulties, if not an inability to reach these achievements. The relevant achievements are clearly interconnected. For instance, explanatory understanding requires the ability to draw inferences about slightly modified inputs, which is clearly given if scientists can predict outputs. Further, if we try to explain why humans lack the achievements, some features of ML arise again and again, e.g., the large number of computational steps or the lack of an informative higher-level description. Still, a neat theory of opacity is unlikely. One reason is that our difficulties in achieving specific tasks have multiple causes. This is because tasks related to ML can be accomplished in different ways. For instance, in principle, there are several ways to explain the outputs of ML models. The deeper reason is that ML models can be described at different levels, e.g., as performing computations, as evaluating a model, or as projecting examples from the training set, etc. (an analogous point is made by Barberousse et al., 2009 regarding computer simulations). Accordingly, we may explain the outputs in terms of computations, in terms of a model, or in terms of an inductive inference from the training data.—For similar (but not exactly the same) reasons it is unlikely that we come up with a neat theory of what people take to be colds (well, if colds are not confused with influenza, they have in common that they go back to viruses; still, many different viruses can produce a cold, and a general theory on who can be how affected by which type of cold is unlikely).

This leaves us with two options for clarifying the concept of opacity. We can either remain at the level of symptoms and try to summarize them using a definition. Alternatively, we can select one or two important causes and refer to them (and possibly their consequences) as opacity.

If we choose the first option, it will be challenging to provide a clear-cut clarification of opacity with necessary and sufficient conditions. Human achievements and related difficulties or inabilities to reach them are not the sorts of things that allow for a systematic account. Additionally, people may disagree on when precisely the lack of an achievement constitutes a problem, and different types of stakeholders perceive different kinds of problems as more or less severe (Zednik, 2021). It is more likely that "opacity" is an umbrella term for all types of problems that are similar or loosely connected (a similar point has been made for many concepts; see Murphy, 2002, Chap. 2, for an overview).

Perhaps, epistemic opacity, as defined by Humphreys, was originally intended to capture the use of "opacity" as an umbrella term for various interrelated symptoms. This is suggested by the fact that Humphreys doesn't define what knowledge of the relevant elements is. Still, if opacity à la Humphreys is meant to encompass all kinds of epistemic problems that agents face regarding ML, it doesn't seem ideally suited for this purpose. One reason is that it focuses on knowledge, while there are other achievements that one may want to attain regarding ML, such as understanding or the ability to run certain inferences. That such achievements are difficult to obtain should also be covered by an account of opacity. A second problem with Humphreys's definition is the focus on a process and its elements. Some problems with the epistemic access to ML models are not naturally described as relating to the process in which a computer simulation is run. For instance, if scientists don't know the training data or the relations between the variables, this is not ignorance about the process. Humphreys's emphasis on the elements of the process leads readers to the high number of computational steps. This is undoubtedly a significant problem, but it is not the only one.

If we nevertheless want to explicate what opacity of ML is in a broad sense, the notion of understanding suggests itself as a better alternative, precisely because it is pretty vague. Recent research in philosophy has tried to say a bit more about understanding, and we can become clearer by focusing on what some philosophers call "objectual understanding" (see Beisbart & Räz, 2022 for this proposal). The latter is an understanding of some object or topic, and it comprises a grasp of the connections among various parts of the object/topic (see, e.g., Kvanvig, 2003, 192). A precondition for this is that we know some of the parts, and grasping the connections then manifests in various abilities, particularly in inferential skills. The opacity of a ML model may thus be defined using the difficulty for humans in understanding the model, its training, its behavior, and its target (a similar definition for the opacity of computer simulations is also possible).

The other option is to identify opacity with a specific cause that is responsible for a range of symptoms. This parallels etiological definitions of diseases, where the latter are identified and individuated using their primary cause (see, e.g., Broadbent, 2009 for discussion). An advantage is that such a definition is more specific. At least under a natural interpretation, Humphreys-style opacity focuses on one cause, viz., computational complexity. I have argued, however, that such a definition does not do justice to various symptoms that have been discussed under the label "opacity". The vast number of steps is important, but if it only forecloses some routes to achieving certain tasks, we cannot call it the main cause for the symptoms (a similar problem is also discussed for diseases, see e.g., Broadbent, 2009).

All in all, then, we face a choice between a broad account of opacity that covers various kinds of difficulties with ML models and a narrower account that tries to identify a specific cause of the symptoms. We confront similar choices in medicine, where a disease or a group of diseases can be identified using either symptoms or causes.

Since human difficulties in reaching certain achievements change over time and depend on various factors, such as the tools available and the stock of knowledge that has already been obtained, opacity is contingent upon several circumstances. This makes a general theory of opacity even more unlikely. But it also leaves hope that at least some symptoms associated with opacity can be overcome.

Acknowledgement I thank Juan M. Durán, Nico Formanek, Layla Huber, and Tim Räz for very valuable comments on an earlier version of this manuscript.

References

Alpaydin, E. (2020). *Introduction to machine learning* (4th ed.). The MIT Press.

Alvarado, R., & Humphreys, P. (2017). Big data, thick mediation, and representational opacity. *New Literary History, 48*, 729–749.

Barberousse, A., Franceschelli, S., & Imbert, C. (2009). Computer simulations as experiments. *Synthese, 169*, 557–574.

Baumberger, C., Beisbart, C., & Brun, G. (2016). What is understanding? An overview of recent debates in epistemology and philosophy of science. In S. Grimm, C. Baumberger, & S. Ammon (Eds.), *Explaining understanding. New perspectives from epistemology and philosophy of science* (pp. 1–34). Routledge.

Beisbart, C. (2021). Opacity thought through: On the intransparency of computer simulations. *Synthese, 199*, 11643–11666. https://doi.org/10.1007/s11229-021-03305-2

Beisbart, C., & Räz, T. (2022). Philosophy of science at sea: Clarifying the interpretability of machine learning. *Philosophy Compass, 17*(6), e12830. https://doi.org/10.1111/phc3.12830

Belle, V., & Papantonis, I. (2021). Principles and practice of explainable machine learning. *Frontiers in Big Data, 39*, 688969.

Boge, F. J. (2022). Two dimensions of opacity and the deep learning predicament. *Minds & Machines, 32*, 43–75. https://doi.org/10.1007/s11023-021-09569-4

Boge, F. J., & Grünke, P. (forthcoming). Computer simulations, machine learning and the Laplacean demon: Opacity in the case of high energy physics. In A. Kaminski, M. Resch, & P. Gehring (Eds.), *The Science and Art of Simulation II*. Springer.

Broadbent, A. (2009). Causation and models of disease in epidemiology. *Studies in History and Philosophy of Science Part C: Studies in History and Philosophy of Biological and Biomedical Sciences, 40*(4), 302–311.

Buckner, C. (2019). Deep learning: A philosophical introduction. *Philosophy Compass, 14*(10), e12625. https://doi.org/10.1111/phc3.12625

Burrell, J. (2016). How the machine 'thinks': Understanding opacity in machine learning algorithms. *Big Data & Society, 3*(1), 2053951715622512.

Carliles, S., et al. (2010). Random forests for photometric redshifts. *The Astrophysical Journal, 712*(1), 511–515. https://doi.org/10.1088/0004-637X/712/1/511

Carnap, R. (1950/1962). *Logical foundations of probability* (2nd ed). University of Chicago Press. (first edition 1950).

Chang, S., Cohen, T., & Ostdiek, B. (2018). What is the machine learning? *Physical Review D, 97*, 5.

Cuomo, S., et al. (2022). Scientific machine learning through physics–informed neural networks: Where we are and what's next. *Journal of Scientific Computing, 92*(3), 88.

De Regt, H. W., & Dieks, D. (2005). A contextual approach to scientific understanding. *Synthese, 144*, 137–170.

Durán, J. M. (2020). What is a simulation model? *Minds and Machines, 30*(3), 301–323.

Durán, J. M., & Formanek, N. (2018). Grounds for trust: Essential epistemic opacity and computational reliabilism. *Minds and Machines, 28*(4), 645–666.

Facchini, A., & Termine, A. (2022). Towards a taxonomy for the opacity of AI systems. In V. C. Müller (Ed.), *Philosophy and theory of artificial intelligence 2021. PTAI 2021. Studies in applied philosophy, epistemology and rational ethics* (Vol. 63). Springer. https://doi.org/10.1007/978-3-031-09153-7_7

Garcia-Dias, R., et al. (2018). Machine learning in APOGEE-unsupervised spectral classification with K-means. *Astronomy & Astrophysics, 612*, A98.

Grimm, S. R. (2006). Is understanding a species of knowledge? *British Journal of Science, 57*, 515–535.

Hills, A. (2016). Understanding why. *Noûs, 50*, 661–688. https://doi.org/10.1111/nous.12092

Humphreys, P. (2004). *Extending ourselves: Computational science, empiricism, and scientific method*. Oxford University Press.

Humphreys, P. (2009). The philosophical novelty of computer simulation methods. *Synthese, 169*, 615–626.

Humphreys, P. (2020). Why automated science should be cautiously welcomed. In M. Bertolaso & F. Sterpetti (Eds.), *A critical reflection on automated science. Human perspectives in health sciences and technology* (Vol. 1). Springer. https://doi.org/10.1007/978-3-030-25001-0_2

Jain, A. K. (2010). Data clustering: 50 years beyond K-means. *Pattern Recognition Letters, 31*(8), 651–666. https://doi.org/10.1016/J.PATREC.2009.09.011

Kurakin, A., Goodfellow, I. J., & Bengio, S. (2018). Adversarial examples in the physical world. In *Artificial intelligence safety and security* (pp. 99–112). Chapman and Hall/CRC. https://arxiv.org/abs/1607.02533

Kvanvig, J. (2003). *The value of knowledge and the pursuit of understanding*. Cambridge University Press.

Lam, R., Sanchez-Gonzalez, A., Willson, M., Wirnsberger, P., Fortunato, M., Alet, F., et al. (2023). Learning skillful medium-range global weather forecasting. *Science, 382*(6677), 1416–1421.

Lenhard, J., & Winsberg, E. (2010). Holism, entrenchment, and the future of climate model pluralism. *Studies in History and Philosophy of Science Part B: Studies in History and Philosophy of Modern Physics, 41*(3), 253–262.

Lipton, Z. C. (2018). The mythos of model interpretability. *Queue, 16*(3), 31–57. Preprint: https://arxiv.org/abs/1606.03490

Loh, W.-Y. (2011). Classification and regression trees. *Wiley Interdisciplinary Reviews: Data Mining and Knowledge Discovery, 1*(1), 14–23.

Madhulatha, T. S. (2012). An overview on clustering methods. *IOSR Journal of Engineering, 2*(4), 719–772. https://arxiv.org/abs/1205.1117

Michalsky, R. S. (1986). Understanding the nature of learning: Issues and research directions. In R. S. Michalsky, J. G. Carbonell, & T. M. Mitchell (Eds.), *Machine learning: An artificial intelligence approach, Volume II, 2, 1* (pp. 3–26).

Moor, J. H. (1985). What is computer ethics? *Metaphilosophy, 16*(4), 266–275.

Murphy, G. L. (2002). *The big book of concepts*. MIT Press.

Oxford Learner's Dictionary, Online Resource. (n.d.). Accessed May 19, 2023, from https://www.oxfordlearnersdictionaries.com/

Pearl, J. (2020). The limitations of opaque learning machines. In J. Brockman (Ed.), *Possible minds: 25 ways of looking at AI* (pp. 13–19). Penguin Press.

Price, I., Sanchez-Gonzalez, A., Alet, F., et al. (2025). Probabilistic weather forecasting with machine learning. *Nature, 637*, 84–90. https://doi.org/10.1038/s41586-024-08252-9

Rider, W. J. (2019). The foundations of verification in modeling and simulation. In C. Beisbart & N. Saam (Eds.), *Computer simulation validation. Simulation foundations, methods and applications*. Springer. https://doi.org/10.1007/978-3-319-70766-2_11

Samala, R. K., Chan, H.-P., Hadjiiski, L. M., Helvie, M. A., Cha, K. H., & Richter, C. H. (2017). Multi-task transfer learning deep convolutional neural network: Application to computer-aided diagnosis of breast cancer on mammograms. *Physics in Medical Biology, 62*, 8894–8908.

Samala, R. K., et al. (2018). Evolutionary pruning of transfer learned deep convolutional neural network for breast cancer diagnosis in digital breast tomosynthesis. *Physics in Medicine & Biology, 63*(9), 095005.

Szegedy, C., et al. (2013). Intriguing properties of neural networks. arXiv preprint. https://arxiv.org/abs/1312.6199

Teller, P. (1980). Computer proof. *The Journal of Philosophy, 77*(12), 797–803. https://doi.org/10.2307/2025805

Trout, J. D. (2002). Scientific explanation and the sense of understanding. *Philosophy of Science, 69*, 212–233.

Tymoczko, T. (1979). The four-color problem and its philosophical significance. *The Journal of Philosophy, 76*(2), 57–83.

Wood, D. A. (2018). A transparent open-box learning network provides insight to complex systems and a performance benchmark for more-opaque machine learning algorithms. *Advances in Geo-Energy Research, 2*(2), 148–162.

Zednik, C. (2021). Solving the black box problem: A normative framework for explainable artificial intelligence. *Philosophy and Technology, 34*(2), 265–288.

Zhang, C., Bengio, S., Hardt, M., Recht, B., & Vinyals, O. (2017). Understanding deep learning requires rethinking generalization. https://arxiv.org/abs/1611.03530

Zhang, C., Bengio, S., Hardt, M., Recht, B., & Vinyals, O. (2021). Understanding deep learning (still) requires rethinking generalization. *Communications of the ACM, 64*(3), 107–115.

Claus Beisbart holds a PhD in physics and another one in philosophy, both from the Ludwig Maximilian University of Munich. Since 2012, he has been a professor of philosophy of science at the University of Bern. His areas of expertise include the epistemology of AI and reflective equilibrium. From 2015 to 2020, he was co-editor-in-chief of the Journal for General Philosophy of Science.

Chapter 2
How I Stopped Worrying and Learned to Love Opacity

Nico Formánek

Abstract This chapter claims that two different problems are lurking behind epistemic opacity. One is philosophical, related to skepticism and very hard to solve. The other one is practical, has been solved for different technologies in the past, but for ML the jury is still out. Philosophical treatments have understandably focused on coming up with proposals how to solve the philosophical problem of opacity. This chapter traces how the practical problem has been solved for methods that predated ML and suggests that habituation to an agreed upon level of residual opacity plays a central role. We therefore should expect that habituation to opacity in ML is also possible.

Keywords Epistemic opacity · Philosophy of science · Computer intensive methods · Skepticism · Pragmatism · Habituation

2.1 Introduction

Opacity has been seen as a problem for computational sciences at least since Moor's *What is computer ethics?* (Moor, 1985). He called it "invisibility factor" (Moor, 1985, p. 272) and used it as a distinguishing characteristic for the field of computer ethics. Later, Humphreys used the concept similarly to argue for the novelty of the field philosophy of computer simulation (Humphreys, 2009). Recently, Boge (2022) and Duede (2022) claimed that ML techniques have their own distinct version of opacity thereby giving rise to the field of philosophy of ML. These arguments are all similar in that they take opacity to be a novel foundational problem of computer intensive methods, which implicitly justifies the creation of a new subfield of philosophy. A different strand of research has sought to precisify the concept of opacity. Humphrey's definition of epistemic opacity is central to that debate. Later

N. Formánek (✉)
Philosophy of Computational Sciences, HLRS, Universität Stuttgart, Stuttgart, Germany
e-mail: nico.formanek@hlrs.de

J. M. Durán, G. Pozzi (eds.), *Philosophy of Science for Machine Learning*, Synthese Library 527, https://doi.org/10.1007/978-3-032-03083-2_2

contributions have variedly extended or criticized his definition (see Saam, 2017; Kaminski et al., 2018; Durán & Formanek, 2018; Alvarado, 2021; Creel, 2020; Zednik, 2021; Beisbart, 2021).

My aim in this chapter is neither to add to these debates, nor will I argue for or against opacity as a sui generis foundational concept of this or that field. I also will not engage in extending or restricting its definition. I am interested in the practical effects of opacity in computer intensive methods in science, especially machine learning methods.

Without doubt opacity has become a widely discussed problem even in computer science as can be seen for example from Lipton's review (Lipton, 2018). In fact one can view the entire subfield of explainable AI (XAI) or interpretable machine learning (IML) as a reaction to the epistemic challenge of opacity (Watson, 2022). Clearly there are different strategies to cope with opacity and it might be tempting to classify them according to how they deal with it. Many strategies will try to reduce opacity, either by establishing mathematical error bounds (e.g. statistical learning theory or algorithmic information theory) or by approximating opaque models with hopefully less opaque ones (as is done in most XAI or IML techniques). Other strategies are inductive in nature, they accept a certain level of opacity if the error rates are satisfactory in practice. I will discuss one example, namely Monte Carlo methods, below. Such strategies don't give us any insight into the model in question, they just allow us to use it without worrying too much.

I will also discuss one strategy, if one can call it that, which has been overlooked so far in the philosophical literature (maybe because it is so obvious): simple habituation to opacity. This strategy, I will argue, is actually the most important and useful one in coping with problems arising from opacity. It trumps every other strategy in effectiveness and efficiency. I will show how habituation as a strategy to cope with opacity works in several computational and non-computational contexts. These cases will be telescopes, Monte Carlo simulations and computer aided proofs. In all these cases I will argue that habituation and not some more complex epistemic strategy was the main reason for the final absorption of the method into the scientific toolkit. I claim that we will observe the same dynamic in the case of ML techniques.

This is the practical side of the opacity problem in computer intensive science. There is also a philosophical side to it, which is not as easily resolved because the habituation strategy doesn't work there. The distinction between the practical and the philosophical sides of opacity shows that their respective solutions must be very different. Both these aspects have been discussed under the umbrella of epistemic opacity in the literature and sometimes I will talk about them as if they are separate problems. This makes sense insofar I claim that one of them can be solved and the other cannot, but as they stem from a similar worry at their inception it is only after some time that they can be separated cleanly.

Habituation does not mean that opacity is a practical problem only as long as it hampers the adoption of the method in question. Extremely opaque methods will perhaps never be adopted. But if opacity has been reduced to a tolerable level, habituation comes into play and sooner than later it won't be a live problem anymore.

2.2 Varieties of Opacity

Philosophers have tried to bring out different aspects of opacity by giving various definitions of the concept. Humphreys for example thought knowledge of the relevant steps of a computational process is necessary for eliminating opacity (Humphreys, 2009), while Durán and Formánek thought surveyability is enough (Durán & Formanek, 2018). Beisbart extends the definition of opacity by additionally requiring understanding (Beisbart, 2021)—see also Chap. 1 by Beisbart in this volume.

Unfortunately there is a lot of equivocation involved even in articles claiming only to study the epistemic part of opacity. If a company is hiding an algorithm for business purposes (e.g. Sullivan's example in Sullivan, 2023) that is not another layer of epistemic opacity. It is merely accidental. I will therefore impose the following restriction on the concept: *Opacity is only an epistemic problem insofar as its resolution depends necessarily on epistemic resources.* That is, if opacity can be overcome by e.g. sharing or open-sourcing code it is not epistemic enough to be considered here. In Burrell's terminology (Burrell, 2016) such lack of public access would be *type one opacity*, while *type three opacity* would be her concept closest to epistemic opacity.

There are lots of other subtle distinctions within epistemic opacity. Most are concerned with the sources of opacity. Kaminski et al. (2018) distinguish between mathematical, social and technical opacity. Boge (2022) and Duede (2022) think the sources of opacity in ML are different from the ones in computer simulation. These distinctions do not matter for the argument I am going to present. Important is the shared characteristic of them being instances of epistemic opacity. I will therefore adopt the following minimal definition: *Epistemic opacity is the type of opacity that prevents understanding of aspects of a process.* This definition is somewhat similar to the spirit of Beisbart's analysis of opacity (Beisbart, 2021) although it is much more vague. This is intentional as I want to capture a general feature that is present in all previous discussions of opacity. Namely epistemic worry.

There are two different kinds of problem sets attached to epistemic worry—one practical, one philosophical. Solutions to these two problem sets differ. This is what I will argue in the next section.

2.3 Epistemic Opacity as a Philosophical Problem

It follows from the minimal definition of opacity given above that it undermines understanding. It blocks the extension of knowledge through understanding and worse, it prevents us from purposefully intervening on the method or target system—in short it is a major barrier for scientific progress. Presumably we also lose control of the errors of opaque methods by not being able to quantify them. The philosophical problem of opacity is therefore a skeptical problem which leads us to doubt the validity of results derived from any opaque method. This connection to skepticism is a first indication that opacity as discussed in the philosophy literature should be expected to have the same type of solution as other philosophical problems—namely a philosophical one. Rather than entering the thicket of defining philosophical (and practical) problems, let me just remark this: Philosophical problems have a certain appeal to philosophers, they tend to come up naturally while looking philosophically at things. Following a distinction by William James they might be called live problems. Obviously if one doesn't look at things philosophically such problems can be dead. One can force oneself to consider them, but it doesn't come naturally. Some live philosophical problems are dead practical problems, they have no bearing on (scientific) practice. Examples are the problem of universals, the problem of induction and the question of realism. This does not mean that they have to stay dead practical problems, things could change in practice. But at the moment we have to concede that they are not considered routinely as such. Obviously many practical problems are philosophically dead, they are of no interest to the philosopher. It is also possible that problems are philosophically and practically live, so philosophy and practice don't exclude each other.

Probably Moor was the first to point at opacity as a philosophical problem in the context of computing. In his famous paper on computer ethics he tells us that a "mark of a basic problem in computer ethics is one in which computer technology is essentially involved" (Moor, 1985, p. 267) and one such essential involvement is "invisible complex calculation". He elucidates on why it is invisible and essentially involved: "Computers today are capable of enormous calculations beyond human comprehension. Even if a program is understood, it does not follow that the calculations based on that program are understood. Computers today perform and certainly supercomputers in the future will perform calculations which are too complex for human inspection and understanding." (Moor, 1985, p. 274). I quote this passage in full detail because succeeding considerations of opacity use exactly the same argumentative strategy if they want to establish an essential involvement of an opaque method. Generally the structure of such arguments is as follows:

1. The opacity of the method in question is established (generally by pointing to human calculatory limitations).
2. This in turn blocks any sensible human access, understanding, justification (can be replaced with an epistemic term of choice) of the target system or method.
3. The results of said opaque methods are relied on.
4. Therefore opacity is **essentially involved** in any relied upon result from said methods. They cannot be had without opacity.
5. Optional step: Methods with essentially involved opacity are to be explained with novel philosophical approaches

Note that the essential involvement depends on opacity blocking epistemic access to the method and the actual reliance on its results. If we didn't rely on them, opacity would pose no problem, similarly relying on transparently obtained results is deemed unproblematic.

Examples of this argumentative strategy of viewing opacity as a philosophical problem can be found in Humphreys who uses it to argue for the novelty of computer simulation methods (Humphreys, 2009); Burrell (2016) as well as Kaminski and Glass (2019) use it to argue for the novelty of ML methods. Durán and Formánek employ it to motivate computational reliabilism (Durán & Formanek, 2018; Duran, 2023), while Boge claims that opacity in ML can result in "unprecedented gaps between discovery and understanding" (Boge, 2022, p. 70) and Duede to say the least uses it to motivate seeing ML as a novel kind of science (Duede, 2022).

It should be noted that in all these arguments opacity is a means to justify the significance of a philosophical point (e.g. novelty or uniqueness of method, change of epistemology). This reinforces my point that opacity as analyzed by philosophers is a philosophical problem. The solutions offered have been similarly philosophical, they have proposed general and abstract strategies to cope with opacity as much as general and abstract frameworks to dissolve opacity. If opacity is accepted as irreducible it is only a means to license a philosophical shift elsewhere. I also don't think that such shifts in philosophical understanding have any bearing on scientific practice. We don't believe anymore that philosophy of science should prescribe the norms of science, so similarly we shouldn't expect epistemology to tell us how to deal with opacity practically.

2.4 Epistemic Opacity as a Practical Problem

One would expect opacity to pose a huge obstacle to scientific and software engineering practice. If you don't understand the code you are trying to improve, you surely have a problem. You won't be able to debug it. If you don't understand the inner workings of your scientific instrument you don't know when it will fail. It is therefore very instructive to take a look at how practical problems with opacity have been dealt with historically. To this end I'll describe three cases where opacity was a concern for scientific practice and where in the end a residual amount of opacity was considered acceptable.

2.4.1 Telescopes

Telescopes provide, not only because of their literal transparency, a case in point. Their theory was unknown and not understood even for centuries after their inception and successful deployment. The physical processes limiting their optical performance were only gradually uncovered. The general story goes as follows (a much more detailed version of this fascinating story can be found in the first chapter of Alvarado's *Simulating Science* (Alvarado, 2023)): When Galileo started using simple telescopes in the 17th century the physical theory of geometric optics was unknown, the magnifying properties could not be explained and even worse artifacts could not be reliably distinguished. Ingenious tests were envisioned which aimed to increase the trust in the instrument but they didn't help to establish an optical theory at all. That came much later! Walther states that, "[t]he theory of perfect optical instruments was, in its major features, completed during the second half of the 19th century." (Walther, 1989, p. 416). He is referring here to Abbé's theory of optics which was developed in the context of microcopes (see Mappes & Dienerowitz, 2024). Actually Hacking made a similar point about microcopes not being understood before Abbé (seeing Hacking, 1981, p. 2). In any case, what these tests did was to convince scientists and the general public that, despite all unknowns, telescopes are legitimate scientific instruments. They were used and improved despite their opacity. Thereby pointing to the irrelevance of the philosophical problem in practical contexts. And of course they successfully delivered scientific results.

2.4.2 Monte Carlo Methods

Monte Carlo methods (short: Monte Carlos) are another case, being a numerical method for computer simulations, where opacity should be expected to be a practical problem. And in a sense the problem in Monte Carlos is worse than in telescopes—even though we have a full mathematical understanding of them. Because they are random sampling procedures they produce random error and this error needs to be estimated. As full theoretical estimates are computationally impossible one reverts to approximation. But approximations always come with a price which the following quote by Sokal illustrates: "However, at present we have very little understanding of the conditions under which this windowing algorithm may produce biased estimates of τ_{int} or of its own error bars." (Sokal, 1997, p. 16). It is noteworthy that this quote doesn't stem from a time where Monte Carlos were new and adding to the trouble the problem wasn't resolved anytime later. The same problem is still present in today's Monte Carlos as it cannot mathematically be solved. As I have argued elsewhere (Formánek, 2025), the impossibility to prove general convergence properties of stochastic methods such as Monte Carlos is due to the problem of induction. In fact it affects any inductive method (see Howson, 2000). One might think that such inherent opacity might hamper the adoption and development of any method. But this was not the case in Monte Carlos. This is best seen by a concluding comment in Borrelli's study on early Monte Carlos: "As often the case with physicists, the practical success of this method in helping locate new resonances in the data quickly swept away any doubts on the validity of the hypotheses, and Monte Carlo event generators like FAKE soon came to be seen as unproblematic computational tools." (Borrelli, 2019, p. 509). A clearer statement of a practical solution of opacity is indeed hard to come by. Practical success overrides skeptical worries.

2.4.3 *Computer Aided Proofs*

Interestingly the first computer aided proofs triggered a debate in mathematics (for a review see MacKenzie, 2001) which bore close resemblance in terms and content to the much later discussion of the philosophical problem of opacity. But the practical problem of opacity in mathematics was actually conceived differently. It stemmed from the problem that mathematicians lost trust in their own process of proof. Lenhard calls this Voevodsky's problem after the mathematician who famously dedicated himself to a large computer aided formalization programme after he discovered several errors in proofs he made (Lenhard, 2022). Opacity comes through lack of accessibility of the human generated proof, "[w]hen the pollution by errors cannot be cleaned through more time and energy because there is a lack of accessibility, a prohibitive complexity." (Lenhard, 2022, p. 8) The problem of opacity in mathematical proofs might be the practical problem that is closest related to the philosophical problem, but still the mathematics community has decided to attack it through practical computation (for example with the help of interactive proof assistants) rather than through philosophical reflection. This shows again that the problem of opacity is treated differently in practice.

In conclusion one can safely say that the following observation by Greif holds for opacity in general and not only in computer science: "[In computer science,] epistemic opacity is mostly perceived as an engineering problem that requires engineering solutions, but is considered acceptable in practical terms to some extent." (Greif, 2022, p. 112).

2.5 Solutions to the Philosophical and Practical Problem of Opacity

As the cases of telescopes, Monte Carlos and computer aided proofs show the philosophical problem and the practical problem of opacity are not straightforwardly related. Different proposals to cope with opacity exist for both cases. Similar to the problem of induction, the philosophical problem of opacity doesn't go away by pointing out specific cases about how knowledge was obtained while relying on opaque methods. Proposed solutions demand quite severe changes of epistemology, like the redefinition of justification along externalist lines. The existence of the practical problem of opacity can also not be used to argue that a philosophical problem exists, because the practical problem is always solved in practice. Analogously it is clear that a solution of the philosophical problem would not in turn be a solution to the practical problem. The problems which trigger practical opacity, like lack of error control in Monte Carlos or accessibility of proofs, still exist, despite any shift to reliabilist epistemology or the like (see Chap. 4 by Durán and Chap. 5 by Alvarado in this volume). Solutions to the practical problem are of a completely different type and as I will suggest have something to do with habituation.

2.5.1 Getting Used to Opacity

The main practical strategy to cope with opacity is to acknowledge it and proceed as if it didn't exist.[1] This is what I call habituation to opacity. That habituation has played a role in coping with opacity is evident from the cases I discussed. In the case of telescopes—although not immediately—they were used, relied upon and trusted long before they were completely understood. I venture to say that we would still use them even if we hadn't discovered a satisfactory theoretical description of optical instruments. Such a development is best explained firstly by the relative lack of relevance of the practical problem of opacity for the aim of using telescopes and secondly by getting used to a residual level of opacity. In Monte Carlos this process is even more evident because we possess a full mathematical theory of them and still have to uncontrolledly approximate their errors. This is explicitly accepted in the literature and despite that science goes on employing them to great success. Habituation to opacity again explains this behavior. That habituation is a powerful strategy not only to cope with opacity can be seen from many different examples. The interpretation of quantum mechanics once was a hotly discussed topic in physics, but it now receded into the background and is at best a footnote in quantum mechanics textbooks. Physicists have not solved that problem, they have gotten used to it instead. And this was possible because it didn't have much practical impact on their research programmes. Transferring this process to opacity we can get an idea of the dynamics of habituation as it plays out in the historical process. First there will be a lively debate in the special sciences about a perceived foundational problem opacity introduces—which might even spill over to philosophy of science (where it then continues to linger long after science proper has habituated. Because it is of philosophical appeal!). This debate either gets gradually pushed out of focus by more practical concerns or it turns out that opacity really is a practical obstacle to progress. If science habituates to the foundational problem of opacity, the problem has never been "truly" solved, it just remains unquestioned. But if opacity poses a live problem to science, it has to be reduced or the method in question will be abandoned. We therefore should expect mature scientific methods to exhibit a level of opacity that was deemed satisfactory by (historical) scientific consensus. A specifically philosophical worry about the process of habituation in science might be that it is irrational. Science could habituate to anything. I don't

[1] For what its worth, this is also the strategy to deal with another famous philosophical problem—Induction. As Howson laconically remarks: "Yet no uncontroversially definitive answer has ever been forthcoming to Hume, and most people have followed his own example of behaving as though nothing had happened." (Howson, 2000, p. 1).

think that "anything goes" in scientific practice, but philosophers have to get used to the thought that science's guardrails are not something they can impose. We also should expect developing scientific methods in general to have a higher level of opacity. But we must be careful to make predictions about the possible adoption of a method on the basis of its opacity.

2.6 Opacity in ML

ML methods have entered the scene as opacity monsters. Artificial neural nets especially suffer from opacity, as we don't have a mathematical theory of their errors yet (although attempts are made: see e.g. Roberts et al., 2022). But even if we had such a theory it is a good guess as any that, like in the Monte Carlo case, opacity will still be present. But will it hamper scientific progress? Will it be a live problem? At least in the moment opacity is perceived as a practical problem (Lipton, 2018; Watson, 2022) and sought to be reduced. It is unclear if the currently fashionable methods from XAI can reduce opacity below the required threshold for habituation to kick in. It is also unclear where this threshold is located. A theoretical consideration suggests that XAI methods either merely propagate opacity to a different level[2] or that they generate new opacity by being coarse grained approximations. A similar point was made by Durán et al. (2021, p. 331) with respect to transparency increasing methods—which might aptly be called "transparency regress". What seems clear though is that opacity doesn't pose a problem for development, selection and training of ML models. Furthermore, if empirical scaling laws hold generally, then generalization errors are bottlenecked just by compute, dataset and model size (cf. Kaplan et al., 2020). This is the allure of scaling. Thus opacity might only continue to be practically relevant in ML if Moore's law levels off. In this case, should the opacity of artificial neural nets not be reduced to a practically accepted level, then they will be abandoned and replaced. This is because only then their opacity will prevent further progress. It is hard to speculate which method could be come to replace them. For example, support vector machines have a much better developed mathematical theory, but they have been pretty much left for dead in the wake of neural nets' success. The practical trade-off is not made along the lines of more and less opacity, but along the lines of practical success. Opacity might sometimes have something to do with it, but often it does not.

In philosophy the situation presents itself very differently. Opacity of ML methods is seen as a fundamental problem hindering scientific progress and challenging traditional modes of knowledge production. ML methods, it is argued, exhibit a novel category of opacity not seen in other computational processes like simulations

[2] LIME for example shifts the selection of explanatory relevant features to superpixelization methods, which are in themselves opaque (Glass & Formánek, 2025).

(e.g. Duede, 2022 or Boge, 2022). Echoing Humphreys' argument for a novel epistemology for computer simulations, an epistemology for ML, which accounts for this novel type of opacity, is promptly called for. These novelty accounts don't say anything about how much opacity is actually present in specific methods nor do they tell us that we must reduce it. Nonetheless some philosophers have felt the call to urge practitioners to prefer transparent models (Sullivan, 2022) or to reduce opacity (Creel, 2020)—so far without much success. Neural nets still reign supreme. The general philosophical stance is that opacity is something to worry about, something which requires specific theoretical interventions to be dealt with. But this just shows how disconnected the philosophical problem of opacity is from the practical one. To end this section with a more positive take on the prospects of philosophy, I do think that a study of conceptual shifts in science under opacity is fruitful and could make explicit which level of opacity is considered acceptable. But this is only possible at the end of a scientific development—as for example has been done for the concept of intelligibility under computer simulations by Lenhard (2019). Attempting such an analysis for ML seems a little premature.

2.7 Conclusion

Like the problem of induction I suspect opacity to be a problem which can never be solved in a philosophically satisfying way. As a little brother of skepticism it will stay (a)live. But in scientific practice that doesn't matter, all that matters is that the practical problem of opacity fades to irrelevance. Judging from history one should not expect that philosophy can be any help in that process.

Acknowledgments I wish to thank Claus Beisbart, Juan Durán and Johannes Lenhard for very helpful comments on a draft of this chapter.

References

Alvarado, R. (2021). *Explaining epistemic opacity*. Retrieved March 14, 2024, from https://philsci-archive.pitt.edu/19384/

Alvarado, R. (2023). *Simulating science: Computer simulations as scientific instruments* (Vol. 479). Springer International Publishing. https://doi.org/10.1007/978-3-031-38647-3

Beisbart, C. (2021). Opacity thought through: On the intransparency of computer simulations. *Synthese*. https://doi.org/10.1007/s11229-021-03305-2

Boge, F. J. (2022). Two dimensions of opacity and the deep learning predicament. *Minds and Machines*, *32*(1), 43–75. https://doi.org/10.1007/s11023-021-09569-4

Borrelli, A. (2019). Program FAKE: Monte Carlo event generators as tools of theory in early high energy physics. *NTM Zeitschrift für Geschichte der Wissenschaften, Technik und Medizin*, *27*(4), 479–514. https://doi.org/10.1007/s00048-019-00223-w

Burrell, J. (2016). How the machine 'thinks': Understanding opacity in machine learning algorithms. *Big Data & Society*, *3*(1), 205395171562251. https://doi.org/10.1177/2053951715622512

Creel, K. A. (2020). Transparency in complex computational systems. *Philosophy of Science*, *87*(4), 568–589. https://doi.org/10.1086/709729

Duede, E. (2022). Instruments, agents, and artificial intelligence: Novel epistemic categories of reliability. *Synthese*, *200*(6), 491. https://doi.org/10.1007/s11229-022-03975-6

Durán, J. M. (2023). *Machine learning, justification, and computational reliabilism*. Retrieved March 13, 2024, from https://philarchive.org/rec/DURMLJ

Durán, J. M., & Formanek, N. (2018). Grounds for trust: Essential epistemic opacity and computational reliabilism. *Minds and Machines*, *28*(4), 645–666. https://doi.org/10.1007/s11023-018-9481-6

Durán, J. M., & Jongsma, K. R. (2021). Who is afraid of black box algorithms? On the epistemological and ethical basis of trust in medical AI. *Journal of Medical Ethics*, *47*(5), 329–335. https://doi.org/10.1136/medethics-2020-106820

Formáanek, N. (2025, to appear). Inductive assumptions in ML - Always required, often ignored. In M. Pantsar & A. Olteanu (Eds.), *Philosophy of artificial intelligence: Optimistic and pessimistic views*. Springer.

Glass, C. W., & Formánek, N. (2025). A minimalistic definition of XAI explanations. In F. Al Machot, M. T. Horsch, & S. Scholze (Eds.), *Designing the conceptual landscape for a XAIR validation infrastructure*. Springer Nature Switzerland. https://doi.org/10.1007/978-3-031-89274-5_11

Greif, H. (2022). Analogue models and universal machines. Paradigms of epistemic transparency in artificial intelligence. *Minds and Machines*, *32*(1), 111–133. https://doi.org/10.1007/s11023-022-09596-9

Hacking, I. (1981). Do we see through a microscope? *Pacific Philosophical Quarterly*, *62*(4), 305–322. https://doi.org/10.1111/j.1468-0114.1981.tb00070.x

Howson, C. (2000, November 2). *Hume's problem: Induction and the justification of belief* (1st ed.). Oxford University Press. https://doi.org/10.1093/0198250371.001.0001

Humphreys, P. (2009). The philosophical novelty of computer simulation methods. *Synthese*, *169*(3), 615–626. https://doi.org/10.1007/s11229-008-9435-2

Kaminski, A., & Glass, C. (2019). Das Lernen der Maschinen. In K. Liggieri & O. Müller (Eds.), *Mensch-Maschine-Interaktion* (pp. 128–133). J. B. Metzler. https://doi.org/10.1007/978-3-476-05604-7

Kaminski, A., Resch, M., & Küster, U. (2018). Mathematische Opazität. Über Rechtfertigung und Reproduzierbarkeit in der Computersimulation. In A. Friedrich, P. Gehring, C. Hubig, A. Kaminski, & A. Nordmann (Eds.), *Arbeit und Spiel* (pp. 253–278). Nomos Verlagsgesellschaft mbH & Co. KG. https://doi.org/10.5771/9783845285429-253

Kaplan, J., McCandlish, S., Henighan, T., Brown, T. B., Chess, B., Child, R., Gray, S., Radford, A., Wu, J., & Amodei, D. (2020, January 22). *Scaling laws for neural language models*. arXiv: 2001.08361 [cs, stat]. https://doi.org/10.48550/arXiv.2001.08361

Lenhard, J. (2019). *Calculated surprises: A philosophy of computer simulation*. Oxford University Press.

Lenhard, J. (2022). Proof, semiotics, and the computer: On the relevance and limitation of thought experiment in mathematics. *Axiomathes*. https://doi.org/10.1007/s10516-022-09626-2

Lipton, Z. C. (2018). The mythos of model interpretability: In machine learning, the concept of interpretability is both important and slippery. *Queue*, *16*(3), 31–57. https://doi.org/10.1145/3236386.3241340

MacKenzie, D. A. (2001). *Mechanizing proof: Computing, risk, and trust*. MIT Press.

Mappes, T., & Dienerowitz, M. (2024). Abbes Auflösung. *Physik Journal*, *23*(1), 25–30. Retrieved January 29, 2024, from https://pro-physik.de/zeitschriften/download/21871

Moor, J. H. (1985). What is computer ethics? *Metaphilosophy*, *16*(4), 266–275. https://doi.org/10.1111/j.1467-9973.1985.tb00173.x

Roberts, D. A., Yaida, S., & Hanin, B. (2022, May 26). *The principles of deep learning theory*. https://doi.org/10.1017/9781009023405

Saam, N. J. (2017). What is a computer simulation? A review of a passionate debate. *Journal for General Philosophy of Science*, *48*(2), 293–309. https://doi.org/10.1007/s10838-016-9354-8

Sokal, A. (1997). Monte Carlo methods in statistical mechanics: Foundations and new algorithms. In C. DeWitt-Morette, P. Cartier, & A. Folacci (Eds.), *Functional integration* (pp. 131–192, Vol. 361). Springer US. https://doi.org/10.1007/978-1-4899-0319-8_6

Sullivan, E. (2022). Understanding from machine learning models. *The British Journal for the Philosophy of Science*, *73*(1), 109–133. https://doi.org/10.1093/bjps/axz035

Sullivan, E. (2023). How values shape the machine learning opacity problem. In I. Lawler, K. Khalifa, & E. Shech (Eds.), *Scientific understanding and representation; Modeling in the physical sciences* (pp. 306–322). Routledge. https://doi.org/10.4324/9781003202905-27

Walther, A. (1989). Irreducible aberrations of a lens used for a range of magnifications. *Journal of the Optical Society of America A*, *6*(3), 415–422. https://doi.org/10.1364/JOSAA.6.000415

Watson, D. S. (2022). Conceptual challenges for interpretable machine learning. *Synthese*, *200*(2), 65. https://doi.org/10.1007/s11229-022-03485-5

Zednik, C. (2021). Solving the black box problem: A normative framework for explainable artificial intelligence. *Philosophy & Technology*, *34*(2), 265–288. https://doi.org/10.1007/s13347-019-00382-7

Nico Formánek studied physics and philosophy at TU Darmstadt. He now is a philosopher at the High Performance Computing Center (HLRS) at Uni Stuttgart, where he runs the Philosophy of Computational Sciences group. His main interest lies in the impact of computing methods on science. He has published on the philosophy of computer simulations, computer-based inductive methods and epistemology of ML.

Chapter 3
Epistemic Opacity and Scientific Realism and Anti-Realism

Jack Casey

Abstract Objections to computer-assisted proofs are remarkably similar to those expressed with regards to machine-learning models; both usually being centered on the epistemic opacity of either methodology. In this chapter, I explore how these objections differ, arguing that their differences indicate that epistemic opacity is a matter of degree. I then argue that one's level of comfort with a given degree of epistemic opacity will be, in large part, determined by one's position with regards to scientific realism. If one is a realist, I argue, one's tolerance for epistemic opacity will be lower. This brings to light the bearing of the well-established debate between scientific realism and anti-realism into contemporary discussions over the acceptability of epistemic opacity in machine-learning.

Keywords Epistemic opacity · Machine-learning models · Computer-assisted proofs · Scientific realism · Scientific anti-realism

3.1 Introduction

The first computer assisted proof of a theorem was published by Appel and Haken in 1979. Their proof of the four-color theorem initiated a revolution in mathematics, and the use of computer-assisted methods of proof is now commonplace in contemporary mathematics. The proof also initiated a philosophical controversy that still rages today, centered mainly on the question of whether computer assisted proofs should be regarded as having the same status as conventional proofs in mathematics. More recently, superficially similar epistemic concerns have been expressed with regards to knowledge gained through the modelling of natural systems by machine learning algorithms. For that reason, then, it's common to see a discussion of the epistemic worries generated by computer-assisted proofs

J. Casey (✉)
Leverhulme Centre for the Future of Intelligence, University of Cambridge, Cambridge, UK
e-mail: jc2605@cam.ac.uk

J. M. Durán, G. Pozzi (eds.), *Philosophy of Science for Machine Learning*,
Synthese Library 527, https://doi.org/10.1007/978-3-032-03083-2_3

feature alongside a discussion of the epistemic worries generated by the employment of computational resources in natural science—even if this is often accompanied by tacit acknowledgement that we might be dealing with very different concerns (Humphreys, 2009: 617; Durán & Formanek, 2018: 649; see also Chap. 4 by Durán, 2025, in this volume).

Arguments against the equal acceptance of computer-assisted proofs in mathematics often center on the apparent *unsurveyability* of computer-assisted proofs. Roughly, a proof is surveyable if its possible (even if only in principle) for a rational agent to understand the proof in its entirety, and thereby come to *know* the conclusion. Tymoczko names surveyability as one of three necessary characteristics of proofs (1979: 59) and traces the unsurveyability of (some) computer-assisted proofs as being the crucial difference between computer-assisted, and traditional proofs (1979: 73). It's for this reason, primarily, that computer-assisted proofs are usually regarded with either suspicion, or as having a lesser status in comparison with traditional proofs (Brown, 1999; Tymoczko, 1979; Kitcher, 1983; also see McEvoy, 2008; Parshina, 2023).

Similarly, in the case of machine learning models, epistemic concerns center on the *black box* nature of some machine learning algorithms (see Durán & Jongsma, 2021). The common complaint with regards to the lack of opacity in machine learning contexts amounts to a complaint that machine learning algorithms are also unsurveyable—being either too complex, or too large, for a single (or even multiple) person(s) to work through the entirety of their architecture.

My contention in this chapter is twofold; firstly, that the epistemic concerns around computer-assisted proofs in mathematics, and the employment of machine learning algorithms (MLAs) in natural science both stem from the fact that both processes are *epistemically opaque*. This is not a novel idea (Humphreys, 2009). What has not been considered, however, is how those concerns are importantly dissimilar, and what this tells us about epistemic opacity. In this chapter, I will argue that, though indeed both stem from concerns over epistemic opacity, the different conclusions that are drawn on the basis of their being epistemically opaque indicates that opacity is a matter of degree.

Secondly, after arguing that epistemic opacity is a matter of degree, I then argue that one's acceptable degree of epistemic opacity will be (at least partly) determined by one's commitment to either scientific realism or scientific anti-realism. Realists, I argue, will have a low tolerance for epistemic opacity. Anti-realists, on the other hand, I argue will be much more comfortable with epistemically opaque processes.

Dialectically, this is notable. It's natural to think that the arguments against the implementation of computer-assisted proofs on epistemic grounds might similarly serve those positioned against the employment of machine learning models in natural science. This chapter then will go some way to explaining why those potential lines of criticism are not translatable into the latter context. Furthermore, it opens up room for the reassessment of the scientific realism/anti-realism debate in a novel context.

3.2 Computer-Assisted Proofs

The first computer-assisted proof of a theorem was the proof of the four-color theorem (4CT) in 1977 by Appel and Haken. A small literature bloomed in the wake of Tymoczko's paper concerning the philosophical implications of the proof (1979).[1]

The four-color problem is as follows: is it possible, using only four colors, to color any given map such that no two adjacent areas of the map are the same color? The problem was first posed by Guthrie to De Morgan in 1852 (Parshina, 2023: 108; MacKenzie, 1999: 103). It stood unsolved for more than 100 years—with a brief period in between in which the theorem was believed to be proved, before a mistake was recognized — before Appel and Haken published their computer-assisted proof in 1977.[2]

The proof begins by translating the problem from one concerning maps to one concerning *graphs*. Regions on a map become *vertices*—points on the graph. That a border exists between two regions are indicated by *edges*—lines joining one vertex to another. The number of edges meeting at any vertex is called the *vertex degree*. Specifically, we're considering *planar graphs*—graphs on which edges cannot intersect, other than at vertices. In essence, the proof consists of reducing all possible combinations of edges and vertices to a finite set of configurations and demonstrating that all possible configurations are such that all vertices can be colored differently from any other vertices to which they are joined by an edge, using only 4 distinct colors in total.

Using Euler's formula, it's demonstrable that the minimal counterexample (i.e., the most difficult instance to prove 4-colorable) to 4CT would be found after a graph undergoes *triangulation* (Tymoczko, 1979: 64). Any region of the graph bounded by four or more edges will have at least two non-adjacent vertices on the boundary. We can join these vertices with a new edge, not intersecting any other edges. Doing this until all we are left with is three-sided regions *triangulates* the graph. Doing so only makes it more difficult to color the graph using only 4 colors, since it only further restricts possible colorings. If 4CT holds for triangulated graphs, it holds generally. The strategy then is to identify the minimal counterexample, assume that it is not four-colorable, and prove that it is (thereby proving that any other configuration is also four-colorable). It's trivial to demonstrate that vertices of degree 3 are four-colorable—given three edges, it's clearly possible to color each vertex a different color. It's also not too difficult to prove the same for vertices of degree 4 using the *Kempe chain argument*. The proof for 5 is much more difficult.

[1] See Parshina (2023) for an overview of the debate.

[2] To avoid unnecessary complexity, I will include technical details of the proof only to the extent that they are philosophically relevant for the discussion at hand. Those interested should consult Haken and Appel's original proof (1977), and Tymoczko's overview (1979: 63–69). I here take my description of the proof from Tymoczko's overview.

The computer-assisted portion of Haken and Appel's proof appears here, then. Following work by Birkhoff (1913), they first identified a finite set of configurations of vertices and edges and demonstrated that any planar graph *must* contain one of these configurations, calling this set of configurations *unavoidable.* Secondly, they then demonstrated that these configurations were *reducible*—a configuration being reducible if the 4-coloring of any graph containing that graph as a subgraph is entailed by the 4-colorability of any graph with fewer vertices. The route for providing a proof is then clear: if it can be demonstrated that presence of these subgraphs indicates 4-colorability, and it can also be demonstrated all graphs contain at least one of these configurations as subgraphs, it indicates that no counterexample to 4CT is possible.

Generating the *unavoidable* set of possible configurations was assisted by computer. Though the set is large—there were 1834 configurations in the unavoidable set in 1977—after generating the set, there are not so many configurations as for it to be unsurveyable.[3] Assessing the reducibility of the set, however, *is* too complex a task for a single human being. To check the reducibility of one configuration alone, approximately 500,000 logical operations are performed (Parshina, 2023: 109). As a result, Appel and Haken had a computer fulfil the task (using a methodology called *discharging*), with the computations taking over 1200 h (Tymoczko, 1979: 68).

The resultant proof then is well understood from a theoretical standpoint. As Tymoczko says, the proof is far from being a simple brute force proof—the generation of the unavoidable set, and the method for testing them, rests upon 'novel and sophisticated theory developed by the authors' (Tymoczko, 1979: 68). Nonetheless, it's far beyond a human being's capabilities to assess each configuration and see whether it is indeed reducible; and it's the fact that it cannot undergo such an assessment that generated skepticism with regards to its status as a *proof,* in the traditional sense.

I'll here focus primarily on those objections that stem from the *unsurveyability* of 4CT (and, by extension, the objection to the classification of all computer-assisted unsurveyable proofs as proofs simpliciter).[4] The first to focus on the unsurveyability of 4CT as being the reason it stands apart from conventional proofs was Tymoczko (1979). Tymoczko's argument is that one major characteristic of proofs is that are *surveyable.* That is, proofs must be 'comprehended by mathematicians' to count as proofs (Tymoczko, 1979: 59). He very firmly places a requirement on the possibility that a proof can be 'looked over, reviewed, verified by a rational agent', in order to count as being a proof in the traditional sense (Tymoczko, 1979: 59). Tymoczko credits surveyability as being the source of the *aprioricity* of mathematical theorems. In the case of Appel and Haken's proof, there is very clearly a step within the proof to which we, as rational agents, are necessarily ignorant.

[3] Though the set has subsequently been reduced significantly, first by Appel and Haken themselves, and then by Robertson et al. (1996), the computational task of assessing reducibility still stands way beyond the capabilities of a single human being.

[4] Other forms of objection have been made. See Parshina (2023) for a classification.

There is intuitive pull to the idea that there's very clearly a difference between a proof in which part of the proof is computer-assisted, and one on which we are fully acquainted with the chain of reasoning. Tymoczko argues that computer-assisted proofs differ in so far as the justification for our believing the computer-assisted step is necessarily grounded in empirical knowledge. For example, the justificatory power of the step relies on our understanding of the physics that describes the way computers function, for example. We only trust this step because of our knowledge of physics that undergirds our understanding of how computers function generally, and what that computer is doing in this specific case. All this knowledge, Tymoczko notes, is necessarily *a posteriori*. On this basis, Tymoczko argues that Appel and Haken's proof should not be considered as standing alongside traditional proofs, which are typically justified through *apriori* means alone.[5] Nonetheless, Tymoczko takes this development in stride; he suggests that, though 4CT might require us to reevaluate the status of mathematics as a purely *apriori* endeavor, this is not necessarily a concerning development (Tymoczko, 1979: 80). Nonetheless, he maintains that the fact we are ignorant of a step in the proof process of 4CT suggests that this proof is of an importantly different kind, and that it is of a different kind is directly attributed to the necessity of our ignorance of part of the justificatory process, and our reliance on empirical knowledge to serve as justification for our belief that this part does indeed stand in the relevant relations of logical inference, even if we can't directly see that it does so.

Brown (1999) shares similar concerns. He bases his skepticism over the status of Appel and Haken's proof in the *fallibility* of computers. Similarly to Tymoczko, his contention is that the source of fallibility in computer-assisted proofs is found in empirical science. We have hypotheses over how computers function, and those hypotheses may be wrong (even if we consider that possibility remote). The portions of unsurveyable proofs whose justifications stand on our understanding of how computers function inherit this uncertainty. While we might be very confident of our understanding of how computers function, we can never get to the certainty inherent in *apriori* justifications; our confidence in the reliability of the unsurveyable portions of proofs can never extend beyond that afforded to well-confirmed scientific theories, which undergird our understanding of how computers function. The fact that there is *some* possibility that we are mistaken, means the status of 4CT as *proof simpliciter* cannot be maintained, according to Brown. Brown differs from Tymoczko, in that he is not as sanguine towards the prospect of the quasi-empirical turn in mathematics apparently demanded by the development of computer-assisted proofs.

[5] Some have pushed back against this claim, notably Burge (1998).

3.3 Machine Learning Algorithms, and Epistemic Opacity

Superficially similar epistemic concerns have been raised with regards to machine learning algorithms. Before we begin, I'll give a general overview of how machine learning algorithms function; technical details are omitted—the level of granularity required to see the philosophical implications for our purposes is not high. Typically, machine learning algorithms function by first ingesting training data, to which they attenuate their internal model—adjusting their internal parameters according to that data. Dependent on the degree to which they are predictively successful given recognition of relevant features—with their success being judged externally — they further adjust their internal parameters, either strengthening or weakening the weight they place on particular features as indicators for whatever predictive output they gave, based on whether or not that particular weighting led to a successful prediction or not, respectively. Through this process, they are able to 'learn' autonomously, in so far as they can come to recognize relevant features of input data on which to make predictions, through the weightings of their internal parameters, without those parameters being set by an external agent.

Machine learning algorithms have now been deployed in numerous areas of empirical enquiry and are increasingly complex. Given sufficient complexity, it becomes impossible to follow the decision procedures internal to them. That is, it is impossible to tell *why* they made the prediction they did. Algorithms with sufficient complexity to preclude understanding of them are often described as being *black-boxes.*[6] To use one oft-cited example to illustrate this issue, consider the example of the employment of a machine learning algorithm to detect skin cancers (Esteva, et al., 2017). The algorithm is trained on labelled dataset consisting of 129,540 images of skin lesions. After this training, it is then tasked with classifying novel images according to the type of skin lesion found in the image. Whilst the study suggests the model outperformed physicians involved in terms of correct classification, no justification for *why* it gave the classification that it did is available. There are saliency maps, but they offer nothing greater than an understanding of which pixel in the image the model took to be important; we lack any ability to understand *why* that pixel was important. As Durán and Jongsma note, there is intuitive pull to the idea that it's objectionable for us to be reliant on a tool for cancer diagnosis, when we are unaware of *how* the diagnosis was made, in so far as the models complexity means we're unable to identify which feature the model identified as being the reason for a particular diagnosis (Durán & Jongsma, 2021: 330). It is often on the basis that MLAs are increasingly used in highly impactful contexts such as these that objections to their inherent complexity are often raised.

Objections to the use of MLAs in empirical science, then, usually rely upon this lack of access to the internal workings of black-box algorithms. As the thought usually goes, in so far as the model is too complex to comprehend, we lack some

[6] From this point on, I'll refer to black-box MLAs as simply 'MLAs'.

corollary of a chain of reasoning on which to understand the basis on which the prediction was made, and as a result we should be suspicious of that prediction. If a model picks out a particular pixel as salient, we want to know *why* that pixel is salient, but the complexity of many models precludes us attaining such an understanding.[7]

This class of objections, then, look remarkably similar to those expressed with regards to computer-assisted proofs. We can understand how the two are related by casting those concerns in terms of Humphreys' notion of *epistemic opacity*:

> [A] process is epistemically opaque relative to a cognitive agent X at time t just in case X does not know at t all of the epistemically relevant elements of the process. (Humphreys, 2009: 618)

In the case of 4CT, there is a step within the proof which is epistemically opaque. That is, in affirming the *reducibility* of configurations in the unavoidable set, we are reliant on a process to which we're not directly epistemically acquainted. The computer-assisted part of the proof is *epistemically opaque*. Similarly, in the case of machine learning models, the reason for discomfort in their inclusion in a clinical setting is that there is an epistemically relevant part of the process (namely, the inference made from an image to a diagnosis of cancer) to which we are necessarily ignorant given the complexity of the model. All this is to say that, in both cases, the source of the epistemic concerns that have been registered seem plausibly to stem from the epistemic opacity present in both cases. Furthermore, that opacity seems particularly stubborn; in both cases, the opacity stems from the limitations of our cognitive resources as human beings.[8]

Much of this has been said before. What has not been discussed is the difference between the two cases. Here I wish to draw attention to the difference in the conclusions that are usually drawn given epistemic opacity in either case. In the case of computer-assisted proofs, on the whole, the consensus is *not* that, given epistemic opacity at some stage in a computer-assisted proof, the proof should be rejected wholesale. Rather, the more limited recommendation is usually made—as we saw with Tymoczko — that computer-assisted proofs should have a lesser status than conventional proofs, or that mathematics reliant on computer-assisted proofs must be recognized as quasi-empirical (Lam, 1990). That is, the truth of the result is not really at issue in computer-assisted proofs.[9] Commentators often go to great lengths

[7] Much work has gone into the task of making MLAs *explainable*, or *interpretable*, where a system is interpretable if its 'operations can be understood by a human, either through introspection or through a produced explanation' (Biran & Cotton, 2017: 8; also see Beisbart & Räz, 2021: 1). Miller goes some way in demonstrating the amount of work required at the level of the conceptual foundations of the field, in so far as the notion of *explainability* requires much greater clarification (Miller, 2019). Durán and Jongsma argue more forcefully that XAI simply removes the problem by one step (2021: 330).

[8] In contrast with more mundane forms of epistemic opacity, such as when a proprietary model is used, precluding us from gaining an understanding of its inner workings (see Burrell, 2016).

[9] Note here, if 4CT is indeed true, for Tymoczko, it is a *necessary a posteriori* truth. The possibility for error only allows for 4CT to be *necessarily false*, or, alternatively, for our apparent justification

to express that their incredulity stems from the mere *possibility* that computer-assisted proofs *could* be mistaken, not that they think that they actually are. As we saw, most arguments against computer-assisted proofs are often predicated on the fact that relying on computer-assisted proofs means we're then partially reliant on those physical theories that undergird our understanding of how those computers function, which introduces an unacceptable level of jeopardy, but only by virtue of the fact there is a possibility our best physical theories might turn out to be false. Those arguments only run when one holds that *no* jeopardy should be (or is) the natural state of mathematical proofs.

A much stronger conclusion is usually drawn in the case of MLAs in empirical science; in many cases, skepticism regarding the truth of the result is expressed. Essentially: we should not trust these processes because we don't have an understanding of how they work, and as a result they could *actually be wrong*. The sense of possibility here is a different one. Rather than some vanishingly small chance that our best physical theories turn out to be mistaken, objectors believe that some erroneous correlation might actually be made, and our inability to survey those models means we cannot detect it. In the case of machine learning algorithms, a far stronger recommendation is usually made. In so far as they're employed to recognize patterns in the world, skepticism arises as to whether a pattern has indeed been properly recognized. The epistemic issue is much more immediate. Indeed, the objection that the employment of machine learning algorithms makes the knowledge derived from them *a posteriori,* given epistemic opacity, has no force in this context—all knowledge gained via *any* empirical means is already *a posteriori* to begin with.

What does this indicate, then? First of all, that a process is epistemically opaque does not entail skepticism of the truth of claim that is justified on the basis of that opaque process. If it did, then we would see far greater skepticism with regards to the truth of 4CT, rather than mere skepticism with regards to its status as a proof. Moreover, that this stronger form of skepticism *is* expressed with regards to MLAs indicates that sufficient epistemic opacity *can* entail skepticism with regards to the truth of prediction.

One way to account for this difference is that epistemic opacity comes in degrees.[10] In case of the computer-assisted proofs such as 4CT, the epistemic opacity is fairly low. We have a good theoretical understanding of how the proof functions, it's just that we can't sort through the vast number of configurations ourselves. Nonetheless, the computer-assisted portion of the proof is epistemically opaque *enough* that it warrants a downgrade in the proof's status, given the high threshold established by the fact mathematics is usually taken to be a purely *apriori*

of 4CT to be misguided—that is, 4CT might be true, but the justification for believing it to be true might be missing, for example if there was an undetected error in the unsurveyable portion of the proof (Tymoczko, 1979: 77, also see Kripke (1980)).

[10] San Pedro (2020) also argues that epistemic opacity is a matter of degree. Durán and Formanek (2018) also talk of 'degrees of opacity'.

discipline. In the case of machine learning algorithms, the epistemic opacity is often high enough to warrant suspicion of the truth of the prediction—we lack a clear theoretical basis on which the function of the algorithm stands. It's no surprise that recommendations for reducing opacity often proceed along lines suggesting that we make them coherent with existing theoretical foundations (see Rudin, 2019). That epistemic opacity is a matter of degree is in line with other suggestions, elsewhere, too. Sullivan, for example, argues that fully-fledged transparency in a model isn't a desirable goal. Some details don't matter (2022a, 2022b: 1). In so far as *fully-fledged* transparency isn't a desirable goal, though *some* is, this implies that epistemic opacity is not an absolute matter.

Indeed, we can consider what would be the case if we raised the degree of epistemic opacity in the case of computer-assisted proofs to something closer to the degree of epistemic opacity we see in machine learning algorithms—Tymoczko consider just such a case, and perhaps unsurprisingly, the skepticism of the result would be much more severe (Tymoczko, 1979: 72). The case he considers is one concerning a genius mathematician, Simon, whose ability fair outstrips his peers. Simon begins by proving many new theorems by traditional methods. Eventually, given his ability, his proofs become too long to comprehend. He begins to handwave to portions of proof that are too extensive with the simple phrase 'Simon says'. Tymoczko's claim is that the logic of 'Simon says' is the same as the appeal to computer in a computer-assisted proof; both being, in essence, appeals to authority (Tymoczko, 1979: 72). Nonetheless, there's an intuitive pull to the idea that computer-assisted proofs are on a far safer epistemic footing than appeals to authority such as 'Simon says'. Viewing this issue through the lens of epistemic opacity, though both are indeed formally similar, we can account for this difference in so far as there is much greater degree of epistemic opacity in Simon's case. And it is because of this greater degree of opacity that we are rightly more skeptical of cases like Simon. Indeed, even if it turned out that Simon was universally correct, and the 4CT proof turned out to be false, we would still suggest that we were right to be more suspicious of Simon.

3.4 Epistemic Opacity, Realism and Anti-Realism

What this tells us then is that epistemic opacity is a matter of degree. Processes of knowledge production can be epistemically opaque to some particular degree. Given greater opacity in a process, a greater degree of skepticism in the truth of the claim made on the basis of that process is warranted.

Two questions naturally arise, then: how do we gauge the level of epistemic opacity in a process, and what determines what level of epistemic opacity is acceptable? My suspicion is that both have highly complex answers. The acceptable level of opacity will probably be determined by a variety of context-dependent factors. For one, it seems likely that it will partly be determined by what one wants to do with the knowledge. If we're considering a low-stakes situation, the threshold might be high. If I'm using the algorithm to determine whether or not a person has cancer, the threshold might be much lower. Relatedly, the availability of extant, alternative explanations: if we're considering phenomena whose dynamics we are largely ignorant of, then perhaps a higher degree of opacity will be palatable. If we lack a theoretical basis for how a particular disease functions, for example, but we have an opaque model which is apparently predictively successful, we are more likely to be comfortable with its employment, even given a degree of epistemic opacity we might otherwise deem intolerable. In relation to computer-assisted proofs: though a computer-assisted proof might not be palatable as a proof itself, its existence might be enough to encourage mathematicians that the aim of finding a conventional proof for the same theorem is worth dedicating their time to. Similarly, measuring opacity seems possible—my argument partly rests on the idea that we can at least recognize relative differences in cases.[11]

For now, I want to sidestep both of these rather thorny issues. These foregoing considerations seem to be largely instrumental in nature. I will here focus on whether there any *intrinsic* reasons that affect the acceptable degree of epistemic opacity. Primarily, what I want to consider is whether there are any background philosophical positions that will act to generally determine one's acceptable level of epistemic opacity. One such belief which I take to act as a significant determinant of one's general acceptance of epistemic opacity is one's position with regards to the debate between scientific realism and anti-realism.

If one's position with regards to scientific realism and anti-realism are as impactful as I will argue, this is useful to consider as (i) given that these are fundamental, background philosophical beliefs, these will apply generally, across most, if not all, contexts mentioned. Given that epistemic opacity appears likely to be highly context-dependent, this might give us some firmer footing on which to tackle this much more complicated question, and (ii) dependent on which one of these views is correct, this might help us establish general normative standards with regards to what *should* be the acceptable of epistemic opacity. That is, if it should turn out that subsequent argumentation strongly favors one particular position over the other, then it might be that we *should* be more lenient (or stringent) *generally*, with regards to what we consider to be acceptable levels of epistemic opacity.

To begin, scientific realism and anti-realism are usually framed as positions on the correct epistemic attitude to adopt with regards to scientific theories (Kukla, 1998: ch. 1; Psillos, 1999; Chakravartty, 2017: 1.2). Realism is typically marked by a positive attitude to the existence of entities whose existence is entailed by scientific theories. The reason for this positive attitude is usually ascribed to the realist commitment to the *literal* or *approximate truth* of scientific theories. For example, if our best physical theories talk of *electrons*, the realist suggests we should believe that electrons exist, in just the same sense in which we readily commit to the existence of non-theoretical objects we're directly acquainted with (tables, chairs,

[11] San Pedro (2020) considers the relation between epistemic opacity and actual scientific modelling and simulation practices.

etc.). The primary argument for the position, termed the *no-miracles argument,* is that the best explanation for the predictive success of our best scientific theories is that what they say is *true*—not ascribing truth to them makes that predictive success *miraculous* (Putnam, 1975: 73). In so far as they are true, what our best theories refer to must exist for them to be true.[12]

Anti-realism, on the other hand, is usually marked by either a skeptical, or, alternatively, a dismissive attitude to questions regarding the truth of our best scientific theories. Taking their cue from the so-called *pessimistic meta-induction*, or the recognized historical tendency for our best scientific theories to be supplanted by novel, incommensurable theories over time, anti-realists recommend that we take a more conservative attitude to question of the truth of our best scientific theories.[13] Having been proven wrong in the past, anti-realists are skeptical of the claim that predictive success can underwrite a belief that those theories are true. To explain what differentiates our best scientific theories from unsuccessful ones, they typically point to their predictive success alone, or to the coherence of a theory with observable statements to which they are more comfortable committing to the truth of. Van Fraassen's view of science as 'a jungle red in tooth and claw' is often appealed to (Van Fraassen, 1980: 40); the success of our best theories is not miraculous, they're just the only ones that have managed to survive (see Wray, 2010).

Though strictly speaking realism is usually cast, as mentioned, as a position with regards to the correct epistemic attitude to our best scientific theories, there are underlying motivations for the position that find their expression in that explicit claim. Laudan, for example, describes realism as follows:

> At its core, realism is a normative doctrine about what the aims or values of science ought to be. Specifically, the realist maintains that the goal of science is to find ever truer theories about the natural world. But the modern-day realist typically conjoins to this axiological thesis a descriptive one: to wit, that the history of science, especially in recent times, can best be understood as an exemplification of the programmatic ideas of realism ... [A]lthough the normative claim of the realist is logically independent of the descriptive claim, the former is epistemically parasitic on the latter. If it should turn out that the descriptive claim is false ... then serious doubts will be raised about its normative counterpart. (Laudan, 1984, 106)

According to Laudan's characterization, the realist motivation is deeper than one merely concerning the correct epistemic attitude to our best scientific theories—rather, it is a position with regards to the aims of science itself. The aim of science *ought* to be the identification of truths concerning the natural world. That is, realists take it to be an aim of science to accurately describe reality. This motivation then finds its expression in the commitment to the literal truth of theories. As such, they

---

[12] I omit some detail with regards to exact formulations of realism here, simply because it has no impact on the argument. Some realists, for example, commit to the *semantic thesis* of realism (namely, that our best theories are literally (approximately) true, whilst reserving from committing to the mind-independent existence of that which they describe (see Chakravartty, 2017, 1.2; 2007: ch. 6).

[13] See Wray (2015) for an overview of the various formulations of the argument.

look likely to be uncomfortable with the utilization of more epistemically opaque methods in scientific enquiry. If, as the realist suggests, a core aim of science is descriptive accuracy, mere predictive success is not enough. In so far as the realist is committed to a conception of science as having a core aim of furnishing us with truth, methodologies which offer mere predictive success, and not an interpretable, accurate description will fail to meet this standard. All this is to say, those of a realist bent who subscribe to a view of science as having as a core aim an accurate description of the world will likely be uncomfortable with epistemic processes with high epistemic opacity, for those processes offer no such descriptions. Understood as a normative claim, then, realism will hold that epistemically opaque processes, however predictively successful, are in violation of the dictum that science *should* furnish us with an accurate description of reality. Therefore, their tolerance for epistemic opacity will necessarily be lower.

This conception of realism as concerning the *aims* of science is not universally shared, however (Chakravarty, 2017: 1.1). And as Lauden notes, the normative claim is logically independent of the descriptive one. Some realists will not frame their position in terms of the *aims* of science and will restrict themselves to a view of realism as merely requiring a commitment to the truth of our best scientific theories. Doing so, perhaps they might find epistemically opaque processes palatable in the following way: we can transpose that position into the current context, substituting interpretable theories of natural science to which they would commit to the truth of for the epistemically opaque models internal to black-box MLAs. It seems possible for the realist to maintain that the only reasonable explanation for the predictive success of those models is that they accurately represent the target phenomena. As such, they might be comfortable committing to the accuracy of those models, even when the model itself is epistemically opaque.

Whether or not the realist *can* maintain their position coherently, however, is not the question. This requirement to commit to the *truth* of epistemically opaque models is quite onerous. Commitment to realism entails a commitment to the truth of models which necessarily lay outside of our epistemic reach. It's one thing to commit to the truth of understandable theories, which sit embedded in a wider framework of theoretical understanding—with our trust in those theories being at least partially dependent on them cohering with other, empirically verified theories. It's quite another to be asked to commit to the truth of something to which we are not acquainted, purely on the basis of predictive success. As such, though the realist *can* coherently commit to the truth of epistemically opaque models, they might not want to. That is, given that they commit to the (approximate) truth of predictively successful models, such a high commitment might mean their threshold for epistemic opacity is lower. If we are to commit to the truth of MLA models, then realists might prudently reserve this commitment to processes with less epistemic opacity.

If one considers this situation from an anti-realist perspective, the difference is stark. While MLAs might stand beyond our ken, if the sole metric by which scientific theories are to be judged is their predictive success as the anti-realist suggests, then nothing is standing in the way of extending the status of 'science'—

whatever that might amount to — to models whose inner workings are beyond our cognitive capabilities. There seems to be no principled reason to require that theories be interpretable to us; indeed, the position of the anti-realist is that we neither commit to the truth nor the falsity of our best scientific theories. As such, that a model has a high degree of epistemic opacity proves to be no issue. Take, for instance, Van Fraassen's *constructive empiricism*. As long as predictions of the model coheres with observation of observables (Van Fraassen, 1980), where *observables* refers to those entities we can observe unmediated by the use of theory—chairs and tables being observable, electrons (in so far as our observation of them requires the use of technology, with the understanding of how that technology functions being embedded in theoretical knowledge) being unobservable—then the model is *empirically adequate* (in comparison with *true*, according to the realist). The epistemic opacity of the model is simply of no concern—if it works, it works. In so far as anti-realism makes no demand that we commit to the truth of the models, our threshold for epistemic opacity should be much higher—if the model turns out to be false—which, given higher degrees of epistemic opacity, is more likely —, that's fine. We were never committed to its truth in the first place.

That being said, though the situation described seems to fit nicely with the underlying motivations of the anti-realism position overall, the particularities of actual expressions of anti-realism might preclude them being entirely comfortable with highly epistemically opaque processes. For example, Van Fraassen's constructive empiricism ascribes empirical adequacy to *theories*. MLAs do not produce theories to which we can ascribe empirical adequacy. This, however, is perhaps unsurprising, given that these positions were developed in a context in which predictive success only ever emerged as a result of the employment of theories. There seems no principled reason empirical adequacy cannot be extended to MLA developed models, as well. If not true to the letter, it appears true to the *spirit* of anti-realism.

It seems likely, then, that one's comfort or discomfort with epistemic opacity will be, in large part, determined by one's position with regards to scientific realism and anti-realism. In so far as realism demands us to commit to the truth of predictively successful models, a lower threshold for epistemic opacity is likely. In so far as anti-realism ascribes the lesser status of empirical adequacy to models, just in case the predictions of those models agree with facts concerning observables, there's more scope for epistemically risky models, and therefore a higher tolerance for epistemic opacity.

Though these are technical positions, they concern very general considerations that most will have opinions on (even if such opinions have not been reflected upon). They concern a basic conception of what science is, and what its aims are. That is, if one considers science as being in the business of increasing understanding, rather than merely being a methodology for the construction of technology, one is more likely to be aligned with the realism. Those of a more pragmatic stripe, who are dismissive of the notion that any deep understanding of nature is to be gleaned using science—typified, perhaps, by Mermin's famous instruction to simply 'shut up and calculate', concerning discussion regarding the correct interpretation of quantum mechanics (Mermin, 1989: 9)[14] — are likely to feel more at home in the anti-realist camp. As such, though they might not cast them in terms of realism and anti-realism, most will have preformed opinions that will act as general determinants of the acceptability of epistemic opacity. Philosophical clarification of this issue—ignoring the inherent unlikelihood of such an event—might then act to give general normative guidelines with regards to epistemic opacity. Should argumentation strongly favor anti-realism, for example, then it might be that we have good reason to dismiss our concerns regarding the epistemic opacity of MLAs; don't worry that we don't have any understanding of how they work, that worry is simply a hangover of realism—all that matters is they work. This underlying tension between realism and anti-realism can be seen as expressed in other disagreements. For example, over what models should do; as to whether they should be assessed solely according to their purpose (Parker, 2020), along anti-realist lines, or whether there is a more objective feature of correspondence between model and phenomenon (underlying something like 'link uncertainty', for Sullivan (2022a)), as the realist would have it. The recent focus on *understanding*, and the question of whether or not models furnish us with understanding, can too be seen as being driven by underlying realist commitments (see Sullivan, 2022b; de Regt, 2017; Strevens, 2013). Understanding might not even be a desideratum for anti-realists, outside of the instrumental benefits it offers. Up to now, that this may be a purely partisan issue hasn't really been recognized; the presumption that models *should* provide understanding in the first place looks to be fundamentally grounded in unspoken realist commitments.

3.5 Concluding Remarks

There remain questions about the coherency of a realist demand for lower epistemic opacity. While it does indeed seem prudent to reduce epistemic opacity, given one's commitment to the literal truth of a model, if the primary argument for taking theories to be literally true in the first place is the predictive success of the theory, it seems difficult to explain how one might coherently demand lower epistemic opacity. Either predictive success is indicative of truth, or it isn't. As such, this problem seems like it might require a restatement of some realist positions. That, or they extend the umbrella of the no-miracles argument to epistemically opaque models, in which case, the position seems to be at odds with its underlying normative character.

This is an interesting development, in so far as it promises to reinvigorate the debate between realism and anti-realism, transposing the dispute from one concerning the correct epistemic attitude to theories, to a much wider dispute, one which appears much closer to the heart of their motivating principles—whether we

[14] Often misattributed to Feynman (Mermin, 2004).

should conceive as science as being in the business of furnishing us with truths about our world, or have a much more limited aim, in being predictively successful. Of course, from the perspective of the anti-realist in this dispute, this more 'limited aim' is simply an entailment of the increased predictive power technology offers us, which simultaneously demands our displacement from the center of epistemic considerations.

References

Appel, K., & Haken, W. (1977). Solution of the four color map problem. *Scientific American, 237*(4), 108–121.

Beisbart, C., & Räz, T. (2021). Philosophy of science at sea: Clarifying the interpretability of machine learning. *Philosophy Compass, 17*(6), 1–11.

Biran, O., & Cotton, C. (2017). Explanation and justification in machine learning: A survey. In *IJCAI-17 workshop on explainable AI (XAI)* (Vol. 8, pp. 8–13).

Birkhoff, G. (1913). The reducibility of maps. *American Journal of Mathematics, 35*(2), 114–128.

Brown, J. R. (1999). *Philosophy of mathematics: An introduction to a world of proofs and pictures*. Routledge.

Burge, T. (1998). Computer proof, apriori knowledge, and other minds: The sixth philosophical perspectives lecture. *Philosophical Perspectives, 12. Language, mind, and ontology*, 1–37.

Burrell, J. (2016). How the machine 'thinks': Understanding opacity in machine learning algorithms. *Big Data and Society, 3*(1), 205395171562251.

Chakravartty, A. (2007). *A metaphysics for scientific realism: Knowing the unobservable*. Cambridge University Press.

Chakravartty, A. (2017). Scientific realism. In E. N. Zalta (Ed.), *The Stanford Encyclopedia of philosophy*. https://plato.stanford.edu/archives/sum2017/entries/scientific-realism/

De Regt, H. W. (2017). *Understanding scientific Understanding*. Oxford University Press.

Durán, J. M. (2025). Beyond transparency: Computational reliabilism as an externalist epistemology for algorithms. In J. M. Durán & G. Pozzi (Eds.), *Philosophy of science for machine learning: Core issues and new perspectives*. Synthese Library, Springer.

Durán, J. M., & Formanek, N. (2018). Grounds for trust: Essential epistemic opacity and computational reliabilism. *Minds and Machines, 28*(4), 645–666.

Durán, J. M., & Jongsma, K. R. (2021). Who is afraid of black box algorithms? On the epistemological and ethical basis of trust in medical AI. *Journal of Medical Ethics., 47*(5), 329–335.

Esteva, A., Kuprel, B., Novoa, R., et al. (2017). Dermatologist-level classification of skin cancer with deep neural networks. *Nature, 542*, 115–118. https://doi.org/10.1038/nature21056

Humphreys, P. (2009). The philosophical novelty of computer simulation methods. *Synthese, 169*(3), 615–626.

Kitcher, P. (1983). *The nature of mathematical knowledge*. Oxford University Press.

Kripke, S. A. (1980). In D. Byrne & M. Kölbel (Eds.), *Naming and necessity: Lectures given to the Princeton University philosophy colloquium*. Harvard University Press.

Kukla, A. (1998). *Studies in scientific realism*. Oxford University Press.

Lam, C. W. H. (1990). How reliable is a computer-based proof. *The Mathematical Intelligencer, 12*(1), 8–12. https://doi.org/10.1007/bf03023977

Laudan, L. (1984). *Science and values: The aims of science and their role in scientific debate*. University of California Press.

MacKenzie, D. (1999). Slaying the kraken: The sociohistory of a mathematical proof. *Social Studies of Science, 29*(1), 7–60. https://doi.org/10.1177/030631299029001002

McEvoy, M. (2008). The epistemological status of computer-assisted proofs. *Philosophia Mathematica, 16*(3), 374–387.

Mermin, N. D. (1989). What's wrong with this pillow? *Physics Today, 42*(4), 9.
Mermin, N. D. (2004). Could Feynman have said this? *Physics Today, 57*(5), 10–11.
Miller, T. (2019). Explanation in artificial intelligence: Insights from the social sciences. *Artificial Intelligence, 267*, 1–38.
Parker, W. S. (2020). Model evaluation: An adequacy-for-purpose view. *Philosophy of Science, 87*(3), 457–477.
Parshina, K. (2023). Philosophical assumptions behind the rejection of computer-based proofs. *Kriterion—Journal of Philosophy, 37*(2–4), 105–122.
Psillos, S. (1999). *Scientific realism: How science tracks truth*. Routledge.
Putnam, H. (1975). *Mathematics, matter and method*. Cambridge University Press.
Robertson, N., Sanders, D. P., Seymour, P., & Thomas, R. (1996). A new proof of the four-color theorem. *Electronic Research Announcements of the American Mathematical Society, 2*(1), 17–25.
Rudin, C. (2019). Stop explaining black box machine learning models for high stakes decisions and use interpretable models instead. *Nature Machine Intelligence, 1*, 206–215. https://doi.org/10.1038/s42256-019-0048-x
San Pedro, I. (2020). Degrees of epistemic opacity. Manuscript.
Strevens, M. (2013). No Understanding without explanation. *Studies in History and Philosophy of Science Part A, 44*(3), 510–515. https://doi.org/10.1016/j.shpsa.2012.12.005
Sullivan, E. (2022a). Understanding from machine learning models. *The British Journal for the Philosophy of Science, 73*(1), 109–133. https://doi.org/10.1093/bjps/axz035
Sullivan, E. (2022b). How values shape the machine learning opacity problem. In Lawler, Khalifa, & Shech (Eds.), *Scientific understanding and representation* (pp. 306–322). Routledge.
Tymoczko, T. (1979). The four-color problem and its philosophical significance. *Journal of Philosophy, 76*, 57–82.
van Fraassen, B. C. (1980). *The scientific image*. Oxford University Press.
Wray, K. B. (2010). Selection and predictive success. *Erkenntnis, 72*(3), 365–377. https://doi.org/10.1007/s10670-009-9206-6
Wray, K. B. (2015). Pessimistic inductions: Four varieties. *International Studies in the Philosophy of Science, 29*(1), 61–73. https://doi.org/10.1080/02698595.2015.1071551

Jack Casey is a Teaching Associate at the Leverhulme Centre for the Future of Intelligence, University of Cambridge, and a College Research Associate of Clare College, Cambridge. He works primarily on philosophy of science, with a focus on the metaphysics of science. He finished his PhD in 2019, under the supervision of David Liggins and Helen Beebee, with a thesis entitled Grounding and the Laws of Nature. His work has appeared in the Journal of the American Philosophical Association.

Part II
Justification

Chapter 4
Beyond Transparency: Computational Reliabilism as an Externalist Epistemology of Algorithms

Juan M. Durán

Abstract This chapter examines the epistemology of algorithms, framing the discussion as a question of epistemic justification. Current approaches emphasize algorithmic transparency, which involves elucidating internal mechanisms—such as functions and variables—and demonstrating how (or that) these compute outputs. Thus, the mode of justification through transparency is contingent on what can be shown about the algorithm and, in this sense, is *internal* to the algorithm. In contrast, I propose an *externalist* epistemology of algorithms called *computational reliabilism* (CR). While I have previously developed CR in the context of computer simulations (Durán, Explaining simulated phenomena: A defense of the epistemic power of computer simulations, 2013; Durán, Computer simulations in science and engineering. Concepts - practices - perspectives. Springer, 2018; Durán, Formanek, Minds and Machines 28(4), 645–666, 2018), this chapter extends the framework to a broader range of algorithms used across scientific disciplines, particularly in machine learning and deep neural networks. At its core, CR posits that an algorithm's output is justified if it is generated by a reliable algorithm, where reliability is determined by reliability indicators. These indicators arise from formal methods, algorithmic metrics, expert competencies, research cultures, and other scientific practices. The chapter's primary objectives are to delineate the foundations of CR, explain its operational mechanisms, and outline its potential as an externalist epistemology of algorithms.

4.1 Introduction

The use of algorithms for scientific purposes is delivering remarkable results. A couple of examples will suffice to illustrate this. In molecular biology, AlphaFold can predict protein structures with atomic accuracy in cases where no similar struc-

J. M. Durán (✉)
Department of Values, Technology and Innovation, Faculty of Technology, Policy and Management, Delft University of Technology, Delft, The Netherlands
e-mail: j.m.duran@tudelft.nl

J. M. Durán, G. Pozzi (eds.), *Philosophy of Science for Machine Learning*, Synthese Library 527, https://doi.org/10.1007/978-3-032-03083-2_4

tures are known (Jumper et al., 2021). In medicine, BenevolentAI has combined structured and unstructured biomedical data sources to identify rheumatoid arthritis drugs like *baricitinib* as therapeutic for COVID-19 symptoms (Medeiros, 2021). In an increasingly number of cases, algorithms have successfully extended the class of tractable chemistry, biology, physics, and medicine, broadening the range of modeling and experimental capabilities available to researchers.

Yet, unlike other methods, the algorithm's scientific merits cannot be easily determined by association with a body of scientific knowledge, by adequacy to empirical data, or by diverse theoretical constructs—such as explanation and observation. This is for a variety of reasons. Algorithms are epistemically and methodologically opaque,[1] making it difficult to associate a given algorithm and its output with the general scientific canon. Likewise, empirical phenomena are often temporarily, spatially, or cognitively inaccessible for validation of the algorithm, potentially casting doubts over any representational value of these systems.

When confronted with these issues, philosophers and computer scientists gravitate towards *transparency*, an umbrella term capturing diverse methods linking the internal mechanisms and properties of algorithms to their outputs (Creel, 2020; Wachter et al., 2018; Ribeiro et al., 2016). To see how transparency works, consider BenevolentAI. At its core, this algorithm is a search engine that combines structured and unstructured biomedical data sources, drug industry data, and automated retrieval of information from diverse scientific research papers. The data is curated and standardized via data analysis and data fabric. It is then fed into knowledge graphs that structure the data into relationships between diseases, genes, and diverse drugs (Smith et al., 2021). Richardson led the team that used BenevolentAI to identify rheumatoid arthritis drugs—notably *baricitinib*—as suitable therapeutics for COVID-19 symptoms (Richardson et al., 2020). To justify Richardson's belief that the output has scientific value, proponents of transparency would emphasize showing how baricitinib is computed through procedures that integrate biomedical data, instantiate key variables, invoke function calls that identify structural relationships within the algorithm, employ relevant conditional statements, and execute other algorithmic operations. Another way to make BenevolentAI transparent is via a *knowledge graph* (Richardson et al., 2020). This visualizes how baricitinib inhibits AAK1 (associated with interrupting the COVID-19 virus' passage into cells) and JAK 1/2 (critical for signal transduction pathways), and how baricitinib binds with GAK (known to decrease certain viruses' infectiousness). This knowledge graph also provides reasons to consider drugs like fedratinib, sunitinib, and erlotinib as less effective and, depending on the case, unsafe. For instance, it is shown how these drugs only inhibit AAK1 and neither decrease the chances of cell infection

[1] The philosophical literature on opacity—particularly in the context of computational systems—is too extensive to cite exhaustively. Key contributions include Humphreys (2004), Humphreys (2009), Burrell (2016), Alvarado and Humphreys (2017), Durán and Formanek (2018), Beisbart (2021), Boge (2022), Beisbart (2025).

(by binding with GAK) nor inhibit cytokine signaling (by inhibiting JAK 1/2) (Richardson et al., 2020).

Are Richardson and his team justified in believing that baricitinib is a medically valid outcome for the issue at hand? What reasons do they have to discard other drugs as either less effective or unsafe? What supports their claim that BenevolentAI is a reliable system for the intended purposes? These are questions about the epistemic reliance on algorithms and the justification of their outputs. To a great extent, transparency provides answers to these questions. This paper, however, is an effort to provide an alternative answer, one that does not depend on methods aimed at the transparency of the algorithm. More specifically, this paper lays the groundwork for *computational reliabilism* (CR), a reliabilist epistemology centered on algorithms, aimed at justifying their outputs.

As the name suggests, CR borrows bits and pieces from epistemological reliabilism, notably Goldman's process reliabilism (Goldman, 1979, 2012). However, the version of CR I develop here draws from, but also expands on, my previous work on computer reliabilism for computer simulations (Durán, 2013, 2018; Durán & Formanek, 2018; Durán & Jongsma, 2021). With these ideas in mind, this chapter is divided as follows. Section 4.2 presents and discusses two epistemologies of algorithms: one that is internal to the algorithm (e.g., transparency) and one that is external to the algorithm (i.e., CR). As expected, these epistemologies have different modes of justification, which are exemplified in Sect. 4.2.1. The example provided is only intended to motivate CR. In Sect. 4.3, I lay the groundwork for *computational reliabilism* (CR). Here, I present three types of *reliability indicators* (type-RIs) that credit reliability to algorithms. These are (1) type$_1$-RI technical performance of algorithms (Sect. 4.3.1.1), (2) type$_2$-RI computer-based scientific practice (Sect. 4.3.1.2), and (3) type$_3$-RI social construction of reliability (Sect. 4.3.1.3). In Sect. 4.4, I briefly take stock of my findings and suggest further lines of investigation that substantiate the merits of CR. In gist, this article invites us to reflect on a crucial but often overlooked question: under what conditions are researchers justified in believing the algorithms' outputs? My answer is that reliability comes through myriad methods, practices, and processes at diverse stages of specification, coding, use, and maintenance of the algorithms.

4.2 Internalist and Externalist Epistemologies for Algorithms

A central motivation for seeking justification is that algorithms are often epistemically opaque. This concept has two distinct but related interpretations. The first interpretation addresses how algorithms involve multiple complex elements (functions, variables, decisions, data, etc.) in their specification, coding, execution, and maintenance. This means that little can usually be said about how these algorithms cluster data, which criteria are used for creating categories, and overall why algorithms behave the way they do. This interpretation is captured in the epithet 'black-box' algorithms as a way to express how far removed algorithms sometimes are from human insight. The second interpretation sets the focus on our limited capacities to say something meaningful about the output of an algorithm (Humphreys, 2009, p. 649). That is, no human being (or group of human beings) can know which functions, variables, decisions, data, etc. are relevant to a given output.

Whereas the first interpretation focuses on the algorithm as an opaque method, the second highlights our cognitive, epistemic, and other limitations in making knowledge claims about the algorithm's output. Both interpretations, I believe, can be cast as human agents lacking proper justification for the algorithm's output. Here justification is taken to be epistemic, and understood in the general sense of a belief being formed in the proper manner. Thus, either because the algorithm is a black-box or because human agents are cognitively limited, there is no basis for claims about the proper formation of a belief about whether the algorithm's output is true, has scientific value, or represents a fact in the world (Humphreys, 2020).[2]

Under this heading, transparency emerges as a promising epistemology of algorithms. It begins with uncovering the algorithm's internal mechanisms and properties—such as its functions, variables, and implemented metrics—and proceeds by linking these to its outputs. Justification, then, is understood as the provision of reasons or supporting evidence by the internal workings of the algorithm for believing that the output is true, has scientific value, or accurately represents a fact about the world (Creel, 2020; Zednik, 2021; Burrell, 2016).[3] Recall from the introduction that BenevolentAI employs *knowledge graphs* to visualize how the algorithm prioritizes baricitinib over alternative drugs. On a transparency-based account, access to the functioning of the knowledge graph justifies the belief that baricitinib is an appropriate candidate for combating COVID-19.

Another example of a transparency-conferring mechanism is LIME: a general algorithm that explains the predictions of any classifier by locally approximating

[2] For simplicity and continuity with the literature on the tripartite view of knowledge, I distinguish between the belief that the algorithm's output $\bar{o}$ is true—or false—scientifically valid—or sound—or that it represents—or misrepresents—from *S* having a propositional attitude toward that belief. In the former case, I do not intend to defend either a realist or an anti-realist position. I take this to be an important debate that philosophers interested in algorithms must eventually confront but this is not the place for it—see Chap. 3 by Casey in this volume. Thus, for *S* to believe that the algorithm's output $\bar{o}$ is "true" is for *S* to have a propositional attitude toward the claim that $\bar{o}$ is true, has scientific value, represents a fact in the world. Following Elgin (2017), the latter sense of truth is not understood as (absolute) correspondence with reality, but rather as sufficient adequacy for the purposes at hand—an adequacy that enables practical engagement, scientific progress, and understanding (see also Parker, 2020). I am grateful to Jack Casey for his close reading on this attempt to clarify the issue. Likewise, the claim that a human agent *S* holds the belief that $\bar{o}$ is "true" constitutes a distinct and higher-order belief. In this respect, it is important to note that beliefs, in this sense, can be occurrent or dispositional, subject to revision or revocation, and that their acceptance may vary in degrees of strength and confidence. Beliefs are also taken to be historically situated, incremental, and perspectival (Massimi & McCoy, 2020).

[3] Like others–for example, Räz and Beisbart (2022)—I take at least some of Sullivan's work to be internalist, despite its strong references to externalist approaches (e.g., Sullivan, 2022). It is less controversial, however, that Sullivan endorses an evidence-based epistemology.

it with an interpretable model. Formally, LIME computes a model $g \in G$, where G is a class of potentially interpretable models (e.g., linear models, decision trees, falling rule lists). In practice, if an algorithm predicts that a patient has the flu, LIME can highlight the symptoms in the patient's history responsible for the prediction. 'Sneeze' and 'headache', for example, are key variables used by LIME. Indeed, they are flagged as net contributors to the flu prediction. In contrast, 'no fatigue' is a variable used as evidence against the prediction (Ribeiro et al., 2016). Let us note in passing that many forms of explanatory AI (XAI) provide a rich source for transparency, as they often involve tracking back the *path-dependency* of the algorithm that relates a given function (or set of functions), variables, etc., to its output (Durán, 2021).

Thus understood, transparency purports justification as an *internal to the algorithm* matter. That is, the justification of our beliefs that the output is true depends exclusively on some form of surveying the inner workings of the algorithm. To put this idea more or less formally,

Definition 1 A human agent *S* is justified in believing the algorithm's output $\bar{o}$ just in case: (a) it is shown, directly or indirectly, the algorithmic path-dependency to $\bar{o}$; and (b) *S* has reasons or supporting to believe that the path-dependency to $\bar{o}$ are the case.[4]

Opposing this view is computational reliabilism (CR), presented here as an *external to the algorithm* epistemology. It consists in identifying formal methods, algorithmic metrics, expert competencies, and research cultures that constitute our best epistemic and normative efforts to specify, code, use, and maintain reliable algorithms. I refer to these as *reliability indicators* (RIs), which, as I will explain in Sect. 4.3, can be divided into *types* and *tokens*.

By construction, then, CR does not depend on revealing the internal mechanisms and properties of the algorithm. Instead, it relies on reliability indicators that are external to it, including the algorithm's socio-technical dimensions. For this reason, CR aspires to be an epistemology of *algorithmic systems*, rather than one limited to program-scripts or program-processes—that is, to algorithms in the narrow sense (Eden, 2007).[5] For simplicity, however, I will occasionally speak of algorithms. A provisional working definition might be stated as follows:

Definition 2 A human agent *S* is justified in believing the algorithm's output $\bar{o}$ if and only if $\bar{o}$ was generated by a reliable algorithmic system. A reliable algorithmic

[4] There is a more nuanced—and arguably more scientifically accurate—approach to transparency, namely, *contextual transparency*. The underlying idea is that transparency does not arise solely from the inner workings of the algorithm, but also from "knowledge of the environmental patterns and regularities that are being tracked and of the abstract representational structures that are tracking them" (Zednik, 2021, p. 268). A similar view is expressed in Burrell (2016). In this form of contextual transparency, reasons or supporting evidence are both internal and external to the algorithm.

[5] Thanks are due to Emma-Jane Spencer for pressing me to clarify the scope of CR.

system is one that produces outputs $\bar{o}$ that are true—scientifically valid, represent—most of the time. To achieve this, the algorithm must have been specified, coded, implemented, and maintained in accordance with a range of reliability indicators.

While I will dedicate a large portion of this chapter to the characterization of type and token reliability indicators, a preliminary conclusion can be drawn now: we can say that transparency and CR have different justificatory modes. According to the former, we have justification by having access to the inner workings of the algorithm. According to the latter, we have justification by identifying methods (formal and otherwise), metrics, expert competencies, cultures of research, and the like external to the algorithm that make up our best epistemic and normative efforts to increase the algorithm's reliability.

As a final attempt to illustrate these two justificatory modes, let me briefly present and discuss an example of an algorithm that classifies individual suspects as {criminal; non-criminal} based on their facial traits. The idea is to show that transparency may justify the belief that a given suspect is a criminal—or not—whereas CR flags the algorithm as unreliable and therefore lacking the epistemic merit required for justification. It goes without saying that this example is only meant to contrast these two justificatory modes. No conclusions regarding the relative value of these two epistemologies are intended to be drawn from this.

4.2.1 Merchants of Unreliability

In 2016, computer scientists Wu and Zhang developed a Convolutional Neural Network (CNN) that analyzed over 1850 ID photos and classified them as {criminal; non-criminal}.[6] About 1120 of these photos were of people with no criminal convictions, and the remaining were of people who were either wanted for crimes or convicted of crimes. The CNN's operation was simple. It picks out facial traits (e.g., distance between the eyes, length and curvature of the mouth) and classifies each photo as {criminal} or {non-criminal.}. No other concept or category was operational. Despite this—or perhaps because of this—the predictive accuracy measured using the Area Under the Receiver Operator Characteristic Curve (AUC-ROC) was very impressive: Wu and Zhang measured 0.9540 accuracy in the classifications. This means that the CNN was able to successfully classify faces of individuals as being {criminal} or {non-criminal} approximately 95% of the time (Durán et al., 2024; Wu and Zhang, 2016).

To further validate their algorithm and rule out that such a high predictive accuracy resulted from overfitting, Wu and Zhang retrained the CNN on a dataset where the labels 'criminal' and 'non-criminal' were assigned randomly as negative and positive instances with equal probability. For the retraining case, the CNN failed

[6] It is worth noticing that Wu and Zhang's use of photos from actual people, in contrasts with other approaches that use synthetically generated photos (Turk & Pentland, 1991; Blanz & Vetter, 1999).

to distinguish between the two categories, plummeting the average classification's accuracy to 48%, with a false negative rate of about 51%, and the false positive rate close to 50%. Wu and Zhang also accounted for unbalanced datasets, choice of photos (light, angle, over and under exposure, clothing, etc.), and other issues pertaining to accuracy. To most algorithmic standards, these results speak in favor of a reliable CNN capable of consistently classifying the photos in question.

Wu and Zhang naturally defend the scientific merits of their algorithm. To their mind, as to many, high predictive accuracy means that the algorithm's outputs represent, have scientific value, are true, etc. It is thus no coincidence that they confidently announce the "law of normality for faces of non-criminals" (Alston, 1995; Wu & Zhang, 2016). But high predictive accuracy is no standard for claims about scientific value or truth. One could argue that while the Ptolemaic model exhibited high predictive accuracy in its measurements, the model fundamentally misconceived and misconstrued planetary motion. Additionally, it is well known that high accuracy can be manufactured by carefully selecting input data and calibrating the algorithm's variables and functions to achieve a desired degree of predictability. For example, finding optimal values for hyperparameters (number of hidden layers, batch size, choice of activation function, etc.) is fundamental for having faster convergence, high accuracy, and overall better results. Now, algorithms allow for multiple optimal hyperparameter configurations, depending on the dataset, the algorithm's intended purpose, and the specific task (Morales-Hernández et al., 2022). Furthermore, optimal configurations for one algorithm do not typically translate to others, making them incompatible in many different ways (van Rijn & Hutter, 2018). As a result, selecting optimal values for hyperparameters, along with the best configuration for a given algorithm, is largely a matter of human decision. Without further provisions in place, such as ensuring compliance with scientific standards, professional integrity, and standardized measurements for the optimality of hyperparameters, predictive accuracy can (relatively easily) be manufactured.

Against all epistemic and moral odds, Wu and Zhang insist that high predictive accuracy confers scientific value on their CNN and their outputs. To further defend this, they retrained the parameters of every layer in the CNN while also modifying the architecture (Hao & Stray, 2019; Wu & Zhang, 2017). As a result, the output maintained the same level of accuracy relative to previous executions of the algorithm. In fact, the CNN correctly picks out specific facial attributes from photos, and then classifies them into the appropriate category ca. 95% of the time.[7] But again, taking high predictive accuracy as an indication of the reliability of automated inference on criminality algorithms is problematic. It confounds justification of a technically correct output with the justification required for believing that output.[8]

[7] Despite these efforts, the system's high accuracy remains questionable. While there is no evidence of output manipulation, one can't help but wonder whether the system would maintain the same level of accuracy when faced with a larger and more diverse datasets.

[8] What is operating here is the distinction between *output accuracy*, which is concerned with the correctness of the final computed outputs, and *procedural accuracy*, which is concerned with the execution of steps and adherence to methods.

Wu and Zhang have no justification for believing that someone is a criminal based on facial traits alone. This is the case regardless of how accurate their algorithm is at picking out and classifying photos.

In this context, transparency does not seem to be of much help for justification. When Wu and Zhang try to justify their outputs on high predictive accuracy, they look at what their AUC-ROC values are telling them. This means that specific inner functions and properties of the CNN responsible for the output will support the justification. But justifying the CNN's output using the same functions used to compute them is epistemically circular and inadmissible for the proper formation of beliefs. Rather, these functions only speaks of the algorithm's robutness, and only in a very limited way. It follows that Wu and Zhang can pin down the functions and properties of their CNN that account for the high predictive accuracy, but at no point can they use those functions and properties alone for claims about justification. In other words, transparency here does not help to distinguish what we are compelled to believe from what cements that belief (Durán & Jongsma, 2021).[9]

4.3 Computational Reliabilism (CR)

Claims about justification find a home in CR, a branch of process reliabilism in which a subject *S* is justified in believing output $\bar{o}$ if the algorithmic system is reliable (see Definition 4.2 on page 60) (Goldman, 2012; Durán et al., 2024). An algorithmic system is reliable when it generates $\bar{o}$ that are true—have scientific merit, represent— rather than false—lack scientific merit, misrepresent—in most cases. Importantly, reliability here is not merely a matter of track record, but concerns the algorithm's propensity to generate true $\bar{o}$ across an appropriate range of cases. The debate in epistemology over the most suitable form of reliabilism is extensive and cannot be addressed here. Suffice it to say that I favor *propensity reliabilism* over Goldman's *frequentist reliabilism*, aligning with Alston, who writes: "A reliable [method] is one that *would* usually deliver favorable results over an appropriate range of cases *if and when* they occur" (Alston, 1995).[10]

I will not pursue this point further, as the distinction between relative frequency and propensity reliabilism is unproblematic for the purposes of this chapter. The question, rather, is how to confer reliability to an algorithmic system. To this end, CR employs *reliability indicators* (RIs) as markers of methodological,

[9] As suggested earlier, transparency is a broad concept that admits different interpretations. A partisan of post-hoc explanatory AI, for instance, could argue that the algorithm was not taking into account scientifically salient aspects of the pictures. By means of this, one could in principle identify high-level features that refer to domain knowledge (e.g., a list of criminality-based characteristics) and thus have reasons or supporting evidence for claims about justification.

[10] Humphreys defended a version of Goldman's process reliabilism (Humphreys, xxxx, 2020). For a recent critique of reliabilism in the context of algorithms, see Duede (2022) and Chap. 5 by Alvarado in this volume.

cognitive, social, normative, and epistemological competence (see also Russo et al., 2024; Pozzi & Durán, 2025). RIs are algorithm-related methods, metrics, practices, domain-specific knowledge, and related competencies that either possess a reliability-conferring property or function as means for conferring reliability—I do not wish to be taken as endorsing a strong metaphysical commitment on this point. Although I will not discuss the nature of these or the mechanisms by which they operate, a simple example should suffice to illustrate that it is not of a 'spooky' kind. Rather, it is quite familiar to philosophers of science (Kitcher, 1993). Consider the microscope. Its reliability is typically attributed to several interdependent factors: the application of optical laws in its design and calibration; the effective observation of entities, which also depends on the observer's prior knowledge; and the presence of a scientific community equipped with the background expertise to assess and adjudicate observational claims. The instrument's proper functioning, the researcher's methodological competence, and the community's epistemic norms together confer reliability on the use of the microscope.

The RIs I now discuss play analogous reliability-conferring roles, insofar as they consist in accessible scientific practices, methodological standards, research cultures, disciplinary debates, and other activities that are, to varying degrees, epistemically grounded and scientifically meritorious. Nothing spooky about that. The real challenge, rather, is to be as precise as possible in identifying RIs for specific cases. Here is where this chapter falls short. However, this is for a good reason. Recall that my only pretense with this chapter is to lay down the groundwork for an externalist epistemology of algorithms, and therefore my treatment of CR will be very general. The reader interested in concrete applications of CR to different domains is cordially invited to read Durán and Jongsma (2021) for cases on medicine and healthcare, Durán et al. (2024) for forensic science, and Humphreys (2020) for computer vision.

4.3.1 Reliability Indicators

For conceptual clarity, I distinguish between *type*-RIs and *token*-RIs. While the former refers to a unique category of indicators, the latter refers to an individual occurrence for that category. With this distinction in mind, the following type- and token-RIs are at the heart of CR:

- *Type*$_1$*-RI—Technical performance of algorithms* focuses on the specification, coding, execution, maintenance, and other technical features that contribute to the performance of the algorithm (e.g., high accuracy and low rate of errors, but also tolerance to domain change, repurposability, reusability, modularity, etc.). In this sense, typical cases of token$_1$-RI include practices and protocols for collecting, curating, storing, distributing, and analyzing data; the use of out-of-distribution data and data augmentation, parametrizations; benchmarking; choice of architecture; treatment of algorithmic kludges (Clark, 1987; Lenhard &

Winsberg, 2010); recasting (Durán, 2020); error treatment, and other techniques pertaining to achieving the desired performance of algorithms. Within this $type_1$-RI, one might also include an account of the employment of these practices, metrics, and methodologies, along with the specific circumstances under which the algorithm is specified and coded.
- *$Type_2$-RI—Computer-based scientific practice* focuses on securing algorithmic-based scientific research. It results from the operationalization and implementation of scientific concepts, causal structures, models and theories, laws and law-like principles, taxonomies, but also scientific metaphors and intuitions, values (epistemic and otherwise), idealizations, abstractions, and representations. This type-RI intends to capture the degree to which scientific units of analysis are implemented and operationalized into the algorithm. The selection and justification of domain knowledge are equally crucial in enhancing an algorithm's reliability. Let us note that the viability and success in doing so largely depend on algorithmic-related decisions, such as programming language choice, the use of formal techniques like verification methods (Fetzer, 1998), and the utilization of sub-modeling and multi-modeling (Durán, 2020).
- *$Type_3$-RI—Social construction of reliability* focuses on broader goals related to the acceptance—or rejection—of algorithms and their outputs by diverse communities (e.g., scientific, academic, general public), the realization of intended values and goals, and the overall assessment of an algorithm's scientific merits. Much of this occurs through $token_3$-RI, where debates about the replication of results, checks on the coherence of $\bar{o}$ with the established body of scientific beliefs and scientists' theoretical and other commitments, and other forms of intellectual exchange take place, embedded, so to speak, in diverse research cultures.[11]

Under this heading, CR is understood as a family of reliability-eliciting algorithmic-related indicators capable of crediting an algorithm as a reliable belief-forming method. It is important to note that by accepting a reliabilist epistemology, one also accepts the propensity likelihood that governs the reliability of a process. This translates into acknowledging that algorithms can occasionally be inefficient, contain errors, be unsuitable for specific purposes, misrepresent, and compute incorrect results. If failures perpetuate over time, the relative propensity governing CR shifts, rendering the algorithm ultimately unreliable.[12]

Furthermore, proponents of CR take note of human cognitive limitations in accessing some token-RIs, which conditions the claims about the reliability of

---

[11] To rephrase what Douglas (2004) persuasively argued: scientific and computational practices, as represented in $type_1$- and $type_2$-RI, together with the social processes associated with them, as represented in $type_3$-RI, are neither reducible to one another nor entirely independent.

[12] Humphreys notices that it is not only the high frequency, but also the quality of outputs that affects the reliability of an algorithm. He calls this problem *statistically insignificant but serious errors* (SIS-Errors) (Humphreys, 2020). My attempt to circumvent SIS-Errors can be found in Durán (2025).

an algorithm. They also need to accommodate the fact that token-RIs are neither absolute nor universally applicable. Not all token-RIs are credited, relevant, and applicable under the same criteria, nor does the same token-RI equally apply to all algorithms. CR is thus understood as perspectival, provisional, and subject to corrections, with no particular token-RI considered to have an all-or-nothing reliability-conferring property.

Thus understood, token-RIs come in degrees. The degree to which one token-RI is more relevant than another, or contributes to the overall reliability of the algorithm will depend on the context in which the algorithm is specified, coded, used, and maintained. It will depend on the epistemic and non-epistemic values and goals at stake. It will also depend on the culture of specifying, coding, maintaining, and using the algorithm of a given community (MacKenzie, 2005; Sundberg, 2010). In this sense, no individual (set of) token-RI can guarantee the reliability of all algorithms. Furthermore, even under the assumption that some token-RI is suitable for a given algorithm, this does not ensure that our reliability claims are eternally warranted. Old token-RIs can lose their appeal as new ones come to light. For these reasons, this chapter holds no pretensions to claim the completeness of the various type- and token-RIs presented here. Further arguments could be given on the need for additional type- or token-RIs not discussed here, or that some indicators are somewhat misplaced, or that some others need replacement. None of this is to say, however, that there are no stable type- and token-RIs that apply across many reliable algorithms. In fact, most of the token-RIs discussed next maintain, to my mind, a permanence in time despite changes and fine-tuning that occur with new technological and scientific developments.

Finally, I recognize that CR may not be readily accepted by everyone. As a reliabilist epistemology, one might feel that it still needs to address a few concerns. For starters, there are issues pertaining to the relevance and availability of type- and token-RIs, potential conflicts emerging among token-RIs, and their precedence, order, and weight. Unfortunately, these issues will not find a complete answer here. To my mind, it is the richness and urgency of this problem that requires putting into practice demands for an account at least as complex as the one presented here. In this sense, CR does not provide, nor intend to provide, absolute assurances. Instead, CR aims to highlight that our best epistemic efforts can be geared towards the reliability of algorithms. Little more can be expected given the fallibility and limitations of human cognition. Taking note of these caveats, I now discuss a few types- and token-RIs in more details.

4.3.1.1 Type$_1$-RI: Technical Performance of Algorithms

In earlier versions of CR (Durán, 2013, 2018; Durán & Formanek, 2018), RIs mainly focus on the specification, implementation, tractability, and overall performance of algorithms. For instance, the first three token$_1$-RI discussed in Durán and Formanek (2018) put forward defining criteria for assessing the utility value of algorithms and their outputs *qua* computational methods. Consider *validation* procedures as an example.[13] In automated diagnosis, algorithms are used for patient prognosis. One way to increase our confidence in the output is to compare the disease progression as indicated by the algorithm with clinical data from prior patients that share the same endotype or phenotype (Myszczynska et al., 2020). This practice validates the synthetic data computed by the algorithm with empirical data collected via diverse scientific methods (e.g., observation, experimentation, intervention, measurement, and others). The utility value of the algorithm is then considered appropriate if validation standards are satisfactory.

In this respect, subjecting algorithms' outputs to validation methods increases—or reduces—our confidence in the reliability of an algorithm, as it is a good indication of the algorithm's accuracy and margin of error. Validation methods also give a fair sense of the capacity to generalize the algorithm from the training data to new, undiscovered data. From a scientific perspective, validating algorithms also contributes to the rigor and reproducibility of research, ensuring that findings are based on sound methods.

Now, it should be expected that validation methods encompass a variety of techniques and methods.[14] As such, they are not all appropriate for the same goals. This means that a given validation techniques cannot be simply applied to different algorithms without prior critical discussion. There must be agreement on how suitable a given validation technique is for the algorithm and data in question, as well as the purposed goals and tasks (Lorscheid et al., 2012; Fagiolo et al., 2007). This is an often overlooked aspect of the social dimension of engineering the performance of algorithms. In Durán and Formanek (2018), we argued for a *history of (un)successful implementations* that affords this interpretation. The idea is simple and intuitive: good practices with visible success—such as high accuracy, low margin of error, ease of implementation, and formal verification—tend to endure over time, while less successful practices tend to be eradicated.

To illustrate this $token_1$-RI a bit further, consider *design prototyping*, a subfield of software engineering that assists developers in assessing alternative design strategies and deciding which is best for a particular goal. Since there are no standard methods for choosing the best strategy, researchers need to compare the requirements of the algorithm with various design approaches to evaluate which one possesses the best characteristics for fulfilling the intended objectives. The example I used in previous publications is a computer simulation involving networking. For this, there are different topologies: ring, star, tree, and mesh. In order to pick the most suitable one, diverse performance characteristics need to be evaluated to see which topology is better at meeting performance goals and constraints (Pfleeger & Atlee, 2009, Chapter 5).

The same point can be made with an example closer to machine learning. Take the case of BenevolentAI presented earlier, which utilizes *Best First Search* (BFS), a

[13] It is important to recognize that various forms of verification and validation exist, each conferring various degrees of reliability (Oberkampf & Roy, 2010).

[14] See Chap. 14 by Manganini and Primiero in this volume.

search algorithm highly successful for navigating graphs and trees. The primary goal of BFS is to find the most promising path to a target node based on a given heuristic. In this respect, BFS has proven to be extremely effective for searching suitable drugs within BenevolentAI's knowledge graph (Segler et al., 2018, p. 604). Classified under $type_1$-RI *history of (un)successful implementations*, BFS contributes to the reliability of BenevolentAI and the justification of its outputs.

Likewise, past failures must be, and typically are, avoided by competent programmers. The history of computing is littered with cases of failed software that changed specification and coding practices. Therac-25 is one tragic case (Leveson & Turner, 1993). As reported, the algorithm used by Therac-25 was not thoroughly validated, and the testing process was insufficient to catch critical bugs that led to radiation overdoses. Furthermore, there was poor error handling and reporting in the software. Error messages were often cryptic, and operators were not adequately trained to understand and respond to them. Finally, there were no redundancy safety mechanisms that could ensure that software failures do not result in such catastrophic outcomes. From the perspective of CR, these all amount to diverse indicators of the unreliability of the algorithm used in Therac-25.

4.3.1.2 $Type_2$-RI: Computer-Based Scientific Practice

Assessing the technical performance of algorithms facilitates justification in terms of increasing accuracy, predictive power, low error rates, tolerance to domain change, and the ability to multi-purpose algorithms and data, among other factors. However, this assessment is silent on the adequacy of algorithms for scientific purposes. A reliabilist epistemology must offer standards by which the algorithm used in a scientific context can be warranted to a greater or lesser degree.

Let me illustrate these ideas with a familiar example. Wu and Zhang's automatic facial recognition system exemplifies how accuracy alone does not exhaust the reliability of an algorithm. As mentioned in Sect. 4.2.1, the AUC-ROC measured 0.9540 predictive accuracy for their CNN. Such tremendous results cemented these researchers' confidence in the scientific merits of the algorithm. However, as discussed, there is no basis for such optimism. Criminality is a socially constructed concept that depends on diverse and sometimes contradictory interpretations of the socio-economic basis of criminality, psychological studies of criminals, and laws that determine when and to what degree someone is considered a criminal. Without reference to some of these concepts and frameworks, scientists are licensed to hold that the prediction—however accurate—lacks the merits required for legitimate scientific claims.

Under CR, the reliability of algorithms is not assessed based on high predictive accuracy. Science involves more than just measuring and classifying algorithmic outputs. In this respect, I believe that algorithms cannot and should not operate in isolation from the broader context of scientific undertakings. We need to delve not only into standard non-algorithmic scientific practice, but also into a form of scientific practice that evolves with and heavily depends on algorithms. The example that immediately comes to mind is AlphaFold 2 (Jumper et al., 2021), but the use of BenevolentAI in discovering drugs to combat COVID-19 also illustrates the scientific significance of algorithms. $Type_2$-RI is an attempt to capture the family of token-RI connected to a larger body of scientific theories, beliefs, and practices within which algorithms are specified, coded, utilized, and maintained. In what follows, I lay out two obvious candidates.

Expert Knowledge

Expert knowledge is an umbrella term that covers the myriad of background education, knowledge, activities, training, virtues, and skills of researchers that bring to bear a broad range of talents to the specification, coding, use, and maintenance of algorithms in scientific contexts. Understood as a reliability indicator, *expert knowledge* reports on the many ways in which scientific expertise, technical expertise, and general competencies can be implemented into an algorithm.

To best understand this indicator, we must look at its various functions. For starters, it puts forward the algorithm's competencies, scope, and theoretical assumptions as conceptualized by the researchers involved in the specification, coding, maintenance, and execution of the algorithm. It also accounts for the ability to describe a target system and its conditions for adequacy (e.g., to be applicable in a specific domain, to be representative of a particular condition, to be context-sensitive, to be repurposed). Expert knowledge covers social practices tailored to the development of algorithms, aptitudes to anticipate their merits intelligibly, and abilities of agents to manipulate them. For instance, setting up the variety of initial conditions, datasets, parameters and hyper-parameters (epochs, batch size, number of neurons, number of layers, dropout rate, etc.), all of which are complex yet critical for the performance and scientific merits of the algorithm. Consider determining which parameter to prioritize as a reliability indicator. Their selection and optimization are not trivial and yet fundamental for the general performance of algorithms (convergence of results, accuracy, overall performance) (Hutter et al., 2014). van Rijn and Hutter have conducted an informative experiment to show that the final performance metrics for deep learning models vary according to how different researchers select and optimize algorithmic parameters and instantiations (van Rijn & Hutter, 2018). Thus understood, experts contribute to the overall reliability of an algorithm by specifying relevant internal data-types, structures, relations, operations, and the like. They also credit reliability (or might identify instances of unreliability) by their pick and choose of datasets, parameters, and other variables.

As a reliability indicator, expert knowledge also attempts to accommodate the complexities of algorithms through the division of cognitive labor. Rather than being developed in isolation, algorithms involve a myriad of direct and indirect stakeholders (e.g., software engineers, physicians and chemists—in the case of BenevolentAI—, biologists—in the case of AlphaFold—, and psychologists and legal officers—in what should have been the case for Wu and Zhang). A core team specifies and codes algorithms utilizing ready-made computer modules others have coded. They employ measuring techniques others have designed, constructed, and calibrated. They analyze data using mathematical and statistical techniques others have validated. They make use of mathematical and computational methods others have devised and tested. There is no development of algorithms in solitude. Teams with diverse cognitive strengths and talents collectively collaborate in a variety of ways. Hence, the success or failure of algorithms is tailored to this collective knowledge, just as much as it depends on individual competencies. What one team member overlooks, another might notice. What one team member forgets, another might foresee. What one team member does not know how to solve, another might be able to teach. Thus diversified, the range of achievable solutions is far greater than what is available in atomized practices.

Interestingly, the role and value of experts are being increasingly recognized in philosophical studies on algorithms. Ratti and Graves (2022) argued that documenting developers' motives and the code and specification of an ML are indicators of the reliability of the system. Newman (2016) has argued along similar lines with respect to computer simulations. Newman considers the entire practice of software engineering to be at stake, from test plans to selecting programming languages and modeling tools, including configuration management.

While I am sympathetic to these ideas, my interpretation of expert knowledge is somewhat broader. It includes technical personnel with no training in software development, practices that exceed software engineering standards, and accommodates the possibility that complete documentation of an algorithm is not always available.

In practice, non-technical personnel are intimately involved in algorithmic development (e.g., physicians and chemists in the case of BenevolentAI, and biologists and chemists in the case of AlphaFold), despite having little to no idea how key features of the system are specified and implemented. Their expertise is, however, crucial for the assessment of the reliability of the algorithm, and thus must be considered. Typically, their role is to inform, supervise, and sometimes even test the specification and coding of algorithms. But of course, these roles and interactions vary among cultures of research.

In connection with this, local practices and vernacular terminology often exceed what is captured by software engineering standards. Consider for instance how algorithms might naturalize or 'fossilize' concepts. Once a concept is coded into the system, it is universally and indistinguishably applied across large and heterogeneous databases with varying degrees of success. Take the concept of 'health' as a case in point. One interpretation takes statistical measures and standards of normal biological measurements of someone's body as the baseline for whether they are healthy. This concept of 'health' can be relatively straightforwardly implemented on an algorithm. However, the same concept also allows interpretations tailored to the diverse values of an individual or a community (Richman, 2004). If a community considers blood transfusion to be harmful, they will treat any members of the community who have received a blood transfusion as unhealthy (Richman & Budson, 2000). Implementing a cogent definition of health is no trivial matter.[15]

Lastly, anyone who has written a piece of code knows all too well that not every line is documented. And even if algorithms were exhaustively documented, this would not guarantee an understanding of the code and its various functions. Thorough documentation—when it occurs—and well-intentioned software engineering may still fall short of capturing the methodological and epistemological competencies embedded in algorithmic systems. We still need to highlight the subtle interpretations, gentle disagreements, and non-verbal practices that permeate computational and scientific work, and that ultimately find their way into the algorithm.

Let me finish by noticing that this reliability indicator brings about another important aspect of algorithmic systems, namely, that they might only be *locally* reliable. The idiosyncrasies attached to documenting, specifying, coding, executing, and maintaining algorithms might make them only reliable in one context but not necessarily in another. This is, I believe, at the root of IBM's Watson for Oncology's difficulties of implementation in South Korea and Denmark, despite its success in the US market (Vulsteke et al., 2018; Emani et al., 2022). Notoriously, Watson for Oncology was capable of analyzing large amounts of data and multiple variables, rendering accurate diagnoses and treatments for cancer patients in the US. But while IBM presents Watson for Oncology as offering more objective medical decisions and more accurate diagnoses than actual oncologists (Swetlitz, 2016), it has been reported that many of these claims have been aggrandized (Ross & Swetlitz, 2017, 2018). When implemented in South Korea and Denmark, only a fraction of the outputs computed by the algorithm matched—or closely matched—the local clinician's best diagnosis (Hamilton et al., 2019).

Knowledge-Based Integration

Scientific results do not come in discrete bits, nor are the objects of scientific inquiry independently sanctioned. Instead, scientific theory and practice constitute a web of mutually supportive claims and commitments that are reached after complex negotiations in complex socio-economic and political environments. However, many studies utilizing algorithms portray a sanitized image of scientific research, where there is privileged access to structured data, undisputed model implementation, and meaningful representations of the world.

Wu and Zhang, for example, state that the quality of their databases and the methods implemented for data analysis prevent "the garbage of human biases from creeping in" (Durán et al., 2024; Wu & Zhang, 2017). Given "race, gender and age, the faces of [the] general law-abiding public have a greater degree of resemblance

[15] Another, perhaps more thoroughly studied example is the implementation of the notion of fairness in algorithmic systems (Narayanan, 2018; Hao & Stray, 2019).

compared with the faces of criminals" (Durán et al., 2024; Wu & Zhang, 2017). It is, however, doubtful whether Wu and Zhang's CNN has any scientific merits. One reason (to add to those previously mentioned) is that Wu and Zhang's CNN is largely disconnected from accepted bodies of scientific knowledge. More precisely, the categories their CNN purports to use (i.e., {criminal} and {non-criminal}) are posited in isolation from established evidence, models of criminal psychology, social studies on crime, and the relevant theories on criminality. Thus, the CNN's outputs are based solely on picking out facial traits from selected photos, rather than being premised on a larger body of knowledge implemented in the algorithm.

I will call approaches that conceive of algorithms as disconnected from the larger body of scientific knowledge *just a bunch of data analysis* (JBDA). By doing this, I intend to emphasize that an algorithm performing mere data analysis, but disconnected from concepts, theories, law-like principles, hypotheses, and other scientific units of analysis, is unlikely to merit scientific credentials, regardless of its predictive accuracy. To my mind, JBDA ignores the 'bigger picture' of knowledge integration, interpretation, and operationalization into algorithms. In fact, I consider JBDA as misleadingly portraying algorithms as an objective, unambiguous, and scientifically grounded examination of data that generates scientifically meaningful outputs. Nothing could be further from the truth. Wu and Zhang's CNN approach is an archetypal JBDA, as it depicts scientific practice as granular, consisting of discrete pieces of information, separately secured and individually sanctioned. To these authors' minds, "like most technologies, machine learning is neutral" (Durán et al., 2024; Wu & Zhang, 2017). JBDA approaches advocate a form of scientific practice that is non-perspectival, socially disinterested, and impartial (i.e., epistemically and normatively neutral[16]), and disembodied from a larger corpus of scientific knowledge.

Are there instances where JBDA approaches are scientifically intelligible? I believe so. As suggested, there are indeed cases where mere data analysis renders valuable scientific insight about a subject matter. But for such cases, one needs to provide further justification that relates JBDA with a larger corpus of knowledge. A plausible interpretation of an account of scientific practice with algorithms capable of accommodating cases of JBDA takes the bulk of scientific knowledge and practices in the field under study as background knowledge and as affording sufficient grounds to underwrite particular claims made with the algorithm.[17] To briefly illustrate this idea, as this point encroaches on issues discussed under $type_3$-RI (see Sect. 4.3.1.3), consider BenevolentAI*, an algorithm whose working principles are JBDA. Suppose that BenevolentAI* puts forward baricitinib* as a

[16] For a critical view on this perspective in the context of algorithms, (see Pozzi & Durán, 2025).

[17] Meskhidze makes a similar claim in the context of ML applications in astrophysics. The chapter discusses "physics-informed machine learning," where physical laws and domain-specific knowledge are embedded within the algorithm and are crucial to its performance. From the standpoint of an epistemology of algorithms, my approach diverges in that Meskhidze prioritizes fostering transparency and interpretability, in contrast to externalist accounts such as reliabilism (see Chap. 18 by Meskhidze in this volume).

drug with high chances of combating COVID-19 symptoms. Would researchers be justified in believing baricitinib*? Surely not at face value, but only once it is embedded in a larger body of knowledge about COVID-19 and after some clinical trials show its scientific worth. This interpretation, however, shifts the reliability of the algorithmic system to a human agent capable of re-interpreting the outputs in light of background knowledge. Examined in isolation, algorithmic systems whose working principles are solely based on JBDA can sustain claims about single justification, but they are less likely to be *reliable* in any sense of the term.

What the examples of BenevolentAI and BenevolentAI* show, I believe, is that we might still be justified under JBDA-like algorithms if (a) the algorithm—as a whole or as constituent parts—implements scientific models, theories, principles, categories, and/or other elements purposed in our corpus of scientific knowledge, and/or (b) its outputs are later assessed by the relevant community and within a corpus of scientific knowledge (more on this in Sect. 4.3.1.3.)

Let me further illustrate this reliability indicator. Take again BenevolentAI, which utilizes information gathered from scientific research papers, structured and unstructured biomedical data, and drug and pharmaceutical industry data. BenevolentAI also implements knowledge graphs that structure data into causal relationships between known diseases, genes, environmental factors, and approved drugs (Smith et al., 2021). Furthermore, BenevolentAI aligns with auxiliary assumptions, theories, and structures of drug molecular profiles, as well as mechanisms integral to the process of damaging healthy cells and tissues. It also incorporates theories about genetics, medical studies of disease, and biological models relating genes to drug effects. Through this knowledge-based integration, outputs generated by BenevolentAI are better justified than those by BenevolentAI*. Indeed, the JBDA-like version does not implement any accepted model or concept into its algorithm, does not operationalize knowledge graphs, and does not represent mechanisms integral to knowledge about damaging healthy cells and tissues. It is the theoretical rigor, along with a history of (un)successful implementations (Durán & Formanek, 2018), domain and expert competence, and possibly some skilled insight that leverages the reliability of BenevolentAI over and above BenevolentAI*.[18] To put the same idea rather bluntly, the lack of such token$_2$ reliability indicators places BenevolentAI* at a justificatory disadvantage compared to BenevolentAI.

4.3.1.3 Type$_3$-RI: Social Construction of Reliability

The performance of algorithms (type$_1$-RI) is undoubtedly required for justification. But while a necessary condition, it is certainly not sufficient. The paradigmatic example is Wu and Zhang's CNN. This algorithm leverages a subset of RI$_1$ (most prominently, validation) but makes claims about an alleged law of facial recognition that is difficult to justify. Expert and knowledge integration (type$_2$-RI) are also

[18] Thanks go to Emanuele Ratti for pressing on clarifying this point.

fundamental to the reliability of algorithms. But again, necessary but not sufficient. BenevolentAI furnishes a good example. The algorithm is robust and built on a solid scientific basis. However, baricitinib was later flagged as counter-prescribed for immunocompromised patients. So, what is missing? To my mind, the reliability of algorithms must also be assessed within social processes that aim to achieve standards of scientific value and thresholds for acceptance of $\bar{o}$. Let me put the same idea in different form. The performance of algorithms along with expert knowledge and knowledge-based integration observe that the relevant scientific structures and processes, commitments and categories, entities and concepts are—to an acceptable degree–correctly implemented into, and computed by the algorithm. But the justification of $\bar{o}$ requires something else. We need to further seek for the consistency of $\bar{o}$ with a larger corpus of knowledge. In this way, we observe that the output aligns with accepted scientific commitments and satisfies standards of quality, evidential support, and relevance, along with other key scientific qualifications.

Let me quickly illustrate how this reliability indicator would work using the story behind BenevolentAI. After announcing barinitricib, diverse groups within the scientific community began to debate the benefits—and dangers—of this drug. A major concern that emerged was that the mechanisms of action of baricitinib would block JAK-STAT signaling pathways (mainly via JAK1 and JAK2), thus impairing interferon-mediated antiviral responses (Favalli, 2020, p. 1013). Blocking interferon would allow attack by other viruses (e.g., herpes zoster and herpes simplex) which in some cases may be more harmful than COVID-19. Favalli (2020) later reported on the potential harms of administering baricitinib to some patients, most importantly immunodeficient patients. As a response, Richardson and colleagues accepted the conditions under which BenevolentAI was a reliable indicator—and reasonably use this debate and incorporate new functions into the algorithm.

As a reliability indicator, this scientific debate regulated on how much evidence was required to accept BenevolentAI's outputs. It draw thresholds of which errors and artifacts can be tolerated, and to what extent. It also determined which assumptions are fit for purpose. Commitments to reliable algorithms are commitments to a network of scientific methodologies, standards, and traditions expressed through scientific debate. As Elgin points out, this network enables scientists to build on each other's work. They can be confident that justified outputs have the epistemic value their discipline prescribes (Elgin, 1996, p. 77). Of course, disputes and disagreements among community members are to be expected. There may be conflicts over values, methods, and what constitutes acceptable evidence. Take again Favalli's concerns about administering baricitinib to a specific group of patients. Whereas Richardson largely agreed with Favalli's concerns, research on the drug continued. Further laboratory testing confirmed Richardson's beliefs.

The social formation and justification of beliefs is a complex enterprise—one that is not always successful. It involves situated background assumptions and perspectives, scientific inquiry permeated by contextual values and interests, and a myriad of concepts implemented in algorithms that are socially constructed, historically contingent, and often idiosyncratic to particular disciplines. Take again variations in the definition of 'health' and 'disease' (Boorse, 2011; Sisti & Caplan, 2017). Each concept operates under a myriad of cultural, political, economic, and moral values. Caruana et al. (2015) discuss a neural network specified to predict pneumonia risk scores in patients and their readmission to hospital. Caruana et al. find that asthmatic patients are at low risk and thus less likely to require hospitalization than other patients (with chronic lung disease, for instance). Now, while these findings are statistically accurate—as evidenced by the presence of $type_1$-RI and $type_2$-RI indicators—the system is nonetheless perceived as unreliable, owing to physicians' divergent background assumptions about what a predictive algorithm ought to provide. According to Theunissen and Browning (2022), physicians often presume that the algorithm predicts outcomes based on a shared baseline of care, rather than adapting to differential care relative to patients' backgrounds. As a result, the algorithm's outputs may be called into question due to oversimplifications in one or more of its embedded assumptions. This further underscores that neither $type_1$-RI nor $type_2$-RI is individually—or jointly—sufficient for establishing the reliability of algorithmic systems.

CR makes an effort to foster belief-forming methods that accommodate social interventions, scientific scrutiny, and inter-domain justifications. Beliefs are justified in relation to a network of interconnected scientific beliefs. Yet, we cannot expect it to be recurrently successful. Contingent values and interests within the scientific community also find their way into the use and perception of the outputs of algorithms. Just like many other scientific methodologies, entrenching the reliability of algorithms requires a delicate balancing act. We bring together our best technical knowledge, theories, methodologies, and social skills. We do so in an attempt to justify believing that the output of a given algorithmic system can be scientifically valuable, represent, or even true.

4.4 Final Thoughts

I have presented CR as a reliabilist epistemology for the justification of algorithms' outputs. I also presented and discussed diverse types and token reliability indicators that form the basis of reliable algorithmic systems. In sum, I set out to defend the following claim: we have—or increasingly have—justification for believing algorithms' outputs when they are generated by a reliable belief-forming method. To be reliable here means that the algorithm is specified, coded, used, and maintained utilizing specific, tailor-made reliability indicators designed for the purpose at hand.

Admittedly, CR construes justification as inherently provisional. As discussed, reliability indicators might change over time and even be replaced. They might also be hard to measure or resolve conflicts. But I believe this is part of the self-critical and self-correcting endeavors that we find in scientific research. It might also be the best epistemic effort we can offer given a context and our limited resources. These activities constitute our best knowledge, metrics, methodologies, and practices, even if subjected to further scrutiny and revision.

I also listed a few issues with CR that I was unable to address but which might be considered conditional to its acceptability. I can accept that. But CR is a step in the right direction, if not as a more adequate epistemology of algorithms, at least as an alternative to internalist epistemologies. In this regard, the chapter has achieved its goal of setting up an externalist epistemology of algorithms, a goal that should be appreciated in its own right.

Acknowledgments This paper has been in the making for quite some time, and many people deserve acknowledgment for their thoughtful comments and suggestions. Let me begin with my co-editor: thank you, Giorgia, for your many close readings of this chapter. I am also grateful to Jack Casey for engaging discussions on these topics and for helping me avoid several missteps.

Many others have contributed in various ways: Federica Russo, Emanuele Ratti, Edoardo Datteri, Giuseppe Primiero, Viola Schiaffonati, Rawad El Skaf, Manuel Barrantes, Karin Jongsma, Andrea Ferrario, Nico Formanek, Charles Rathkopf, Emma-Jane Spencer, and Atocha Aliseda. Some of them believe in CR, some do not, but all have been supportive and encouraging of my ideas. Naturally, all wrongs—and all rights—are mine.

Finally, thanks go to Kass and Diego. To Kass, for endlessly listening and sharing thoughts as I structured and shaped CR. I am sure I bored you on a cosmic scale—thank you for still listening, under all conditions. And to Diego, for teaching me one absolute truth: that everything is silly.

References

Alston, W. P. (1995). How to think about reliability. *Philosophical Topics*, *23*(1), 1–29.

Alvarado, R., & Humphreys, P. (2017). Big data, thick mediation, and representational opacity. *New Literary History*, *48*(4), 729–749. https://doi.org/10.1353/nlh.2017.0037

Beisbart, C. (2021). Opacity thought through: On the intransparency of computer simulations. *Synthese*, *199*, 11643–11666.

Beisbart, C. (2025). In which ways is machine learning opaque? In Durán, J. M., & Pozzi, G. (Eds.), *Philosophy of science for machine learning: Core issues and new perspectives*. Synthese Library. Springer.

Blanz, V., & Vetter, T. (1999). A morphable model for the synthesis of 3D faces. In *Proceedings of the 26th Annual Conference on Computer Graphics and Interactive Techniques* (pp. 187–194). ACM Press/Addison-Wesley Publishing Co.

Boge, F. J. (2022). Two dimensions of opacity and the deep learning predicament. *Minds and Machines*, *32*(1), 43–75.

Boorse, C. (2011). Concepts of health and disease. In Gifford, F. (Ed.), *Handbook of the philosophy of science* (pp. 13–64). Elsevier.

Burrell, J. (2016). How the machine 'thinks': Understanding opacity in machine learning algorithms. *Big Data & Society, 3*(1). https://doi.org/10.1177/2053951715622512

Caruana, R., Lou, Y., Gehrke, J., Koch, P., Sturm, M., & Elhadad, N. (2015). Intelligible models for HealthCare: Predicting pneumonia risk and hospital 30-day readmission. In *Proceedings of the 21th ACM SIGKDD International Conference on Knowledge Discovery and Data Mining* (pp. 1721–1730). Association for Computing Machinery. https://doi.org/10.1145/2783258.2788613

Clark, A. (1987). The kludge in the machine. *Mind and Language*, *2*(4), 277–300.

Creel, K. A. (2020). Transparency in complex computational systems. *Philosophy of Science*, *87*(4), 568–589. https://doi.org/10.1086/709729

Douglas, H. (2004). The irreducible complexity of objectivity. *Synthese*, *138*, 453–473.

Duede, E. (2022). Instruments, agents, and artificial intelligence: novel epistemic categories of reliability. *Synthese*, *200*(6), 491.

Durán, J. M. (2013). *Explaining simulated phenomena: A defense of the epistemic power of computer simulations* (PhD thesis). Institut für Philosophie, Universität Stuttgart. Retrieved from https://elib.uni-stuttgart.de/bitstream/11682/5409/1/Thesis_Duran.pdf
Durán, J. M. (2018). *Computer simulations in science and engineering. Concepts - practices - perspectives*. Springer.
Durán, J. M. (2020). What is a simulation model? *Minds and Machines*, *30*, 301–323.
Durán, J. M. (2021). Dissecting scientific explanation in AI (sXAI): A case for medicine and healthcare. *Artificial Intelligence*, *297*, 103498. https://doi.org/10.1016/j.artint.2021.103498
Durán, J. M. (2025). In defense of reliabilist epistemology of algorithms. *European Journal for Philosophy of Science*, *15*, 37.
Durán, J. M., & Formanek, N. (2018). Grounds for trust: Essential epistemic opacity and computational reliabilism. *Minds and Machines*, *28*(4), 645–666.
Durán, J. M., & Jongsma, K. R. (2021). Who is afraid of black box algorithms? On the epistemological and ethical basis of trust in medical AI. *Journal of Medical Ethics*, *47*(5), 329–335. https://doi.org/10.1136/medethics-2020-106820
Durán, J. M., van der Vloed, D., Ruifrok, A., & Ypma, R. J. F. (2024). From understanding to justifying: Computational reliabilism for AI-based forensic evidence evaluation. *Forensic Science International: Synergy*, *9*, 100554. ISSN:2589-871X. Available at: https://www.825sciencedirect.com/science/article/pii/S2589871X24001013
Eden, A. H. (2007). Three paradigms of computer science. *Minds and Machines*, *17*, 135–167.
Elgin, C. Z. (1996). *Considered judgement*. Princeton University Press.
Elgin, C. (2017). *True enough*. MIT Press.
Emani, S., Rui, A., Rocha, H. A. L., Rizvi, R. F., Juacaba, S. F., Jackson, G. P., & Bates, D. W. (2022). Physicians' perceptions of and satisfaction with artificial intelligence in cancer treatment: A clinical decision support system experience and implications for low-middle-income countries. *JMIR Cancer*, *8*(2), e31461. https://doi.org/10.2196/31461
Fagiolo, G., Moneta, A., & Windrum, P. (2007). A critical guide to empirical validation of agent-based models in economics: Methodologies, procedures, and open problems. *Computational Economics*, *30*, 195–226.
Favalli, E. G., Biggioggero, M., Maioli, G., & Caporali, R. (2020). Baricitinib for COVID-19: a suitable treatment? *The Lancet*, *20*, 1012–1013.
Fetzer, J. H. (1998). Program verification: The very idea. *Communications of the ACM*, *37*(9), 1048–1063.
Goldman, A. (1979). What is justified belief? *The Justification of Belief*, *105*(9), 1–23.
Goldman, A. I. (2012). *Reliabilism and contemporary epistemology*. Oxford University Press.
Hamilton, J. G., Genoff Garzon, M., Westerman, J. S., Shuk, E., Hay, J. L., Walters, C., Elkin, E., Bertelsen, C., Cho, J., Daly, B., Gucalp, A. (2019). "A tool, not a crutch": patient perspectives about IBM Watson for oncology trained by Memorial Sloan Kettering. *Journal of Oncology Practice*, *15*(4), e277–e288.
Hao, K., & Stray, J. (2019). *Can you make AI fairer than a judge? Play our courtroom algorithm game*. MIT Technology Review. Retrieved June 9, 2025, from https://www.technologyreview.com/2019/10/17/75285/ai-fairer-than-judge-criminal-risk-assessment-algorithm/
Humphreys, P. W. (2004). *Extending ourselves: Computational science, empiricism, and scientific method*. Oxford University Press.
Humphreys, P. W. (2009). The philosophical novelty of computer simulation methods. *Synthese*, 169(3), 615–626.
Humphreys, P. (2020). *Neural Nets: Why reliabilism is an inappropriate epistemology for them*, YouTube video, 13 October 2020. Retrieved August 15, 2024, from https://www.youtube.com/watch?v=2VFPXbrCqzM&t=830s
Hutter, F., Hoos, H., & Leyton-Brown, K. (2014). An efficient approach for assessing hyperparameter importance. In E. P. Xing & T. Jebara (Eds.), *Proceedings of the 31st International Conference on Machine Learning* (Vol. 32(1), pp. 754–762). PMLR.
Jumper, J., Evans, R., Pritzel, A., Green, T., Figurnov, M., Ronneberger, O., Tunyasuvunakool, K., Bates, R., Žídek, A., Potapenko, A., Bridgland, A., Meyer, C., Kohl, S., Ballard, A., Cowie, A.,

Romera-Paredes, B., Nikolov, S., Ain, R., Adler, J., ..., Hassabis, D. (2021). Highly accurate protein structure prediction with AlphaFold. *Nature, 596*, 583–589.
Kitcher, P. (1993). *The advancement of science: Science without legend, objectivity without illusions*. Oxford University Press.
Lenhard, J., & Winsberg, E. (2010). Holism, entrenchment, and the future of climate model pluralism. *Studies in the History and Philosophy of Modern Physics, 41*, 252–262.
Leveson, N. G., & Turner, C. S. (1993). An investigation of the Therac-25 accidents. *Computer, 26*(7), 18–41.
Lorscheid, I., Heine, B.-O., & Meyer, M. (2012). Opening the 'black box' of simulations: increased transparency and effective communication through the systematic design of experiments. *Computational and Mathematical Organization Theory, 18*, 22–62.
MacKenzie, D. (2005). Computing and the cultures of proving. *Philosophical Transactions: Mathematical, Physical and Engineering Sciences, 363*(1835), 2335–2350.
Massimi, M., & McCoy, C. D. (Eds.). (2020). *Understanding perspectivism. Scientific challenges and methodological prospects*. Routledge.
Medeiros, J. (2021). How tech is changing healthcare. From rapid development and rollout of the Covid-19 vaccines to the science of isolation, machine-learning-enabled gene editing and digitised medicine. *Wired*. Retrieved from https://www.wired.co.uk/article/future-health-trends
Morales-Hernández, A., Van Nieuwenhuyse, I., & Rojas González, S. (2022). A survey on multi-objective hyperparameter optimization algorithms for machine learning. *Artificial Intelligence Review*. https://doi.org/10.1007/s10462-022-10359-2
Myszczynska, M., Ojamies, P., Lacoste, A., Neil, D., Saffari, A., Mead, R., Hautbergue, G., Holbrook, J., & Ferraiuolo, L. (2020). Applications of machine learning to diagnosis and treatment of neurodegenerative diseases. *Nature Reviews Neurology, 16*, 440–456. https://doi.org/10.1038/s41582-020-0377-8
Narayanan, A. (2018). Translation tutorial: 21 Fairness definitions and their politics. In *Proc. Conf. Fairness, Accountability, and Transparency (FAT*)*, New York, NY, USA. Tutorial presentation.
Newman, J. (2016). Epistemic opacity, confirmation holism and technical debt: Computer simulation in the light of empirical software engineering. In F. Gadducci & M. Tavosanis (Eds.), *History and Philosophy of Computing. HaPoC 2015. IFIP Advances in Information and Communication Technology* (Vol. 487, pp. 256–272). Springer.
Oberkampf, W. L., & Roy, C. J. (2010). *Verification and validation in scientific computing*. Cambridge University Press.
Parker, W. S. (2020). Model evaluation: An adequacy-for-purpose view. *Philosophy of Science, 87*, 457–467.
Humphreys, P. *Epistemic opacity and epistemic inaccessibility*, unpublished manuscript. Available at: https:V/wordpress.its.virginia.edu/Paul_Humphreys_Home_Page/files/2016/02/epistemic-opacity-and-epistemic-inaccessibility.pdf
Pfleeger, S. L., & Atlee, J. M. (2009). *Software engineering: Theory and practice* (4th ed.). Pearson.
Pozzi, G., & Durán, J. M. (2025). From ethics to epistemology and back again: Informativeness and epistemic injustice in explanatory medical machine learning. *AI & Society, 40*, 299–310. https://doi.org/10.1007/s00146-024-01875-6
Ratti, E., & Graves, M. (2022). Explainable machine learning practices: Opening another black box for reliable medical AI. *AI and Ethics*. https://doi.org/10.1007/s43681-022-00141-z
Räz, T., & Beisbart, C. (2022). The importance of understanding deep learning. *Erkenntnis*. https://doi.org/10.1007/s10670-022-00605-y
Ribeiro, M. T., Singh, S., & Guestrin, C. (2016). "Why should I trust you?": Explaining the predictions of any classifier. In *KDD '16: Proceedings of the 22nd ACM SIGKDD International Conference on Knowledge Discovery and Data Mining* (pp. 1135–1144).
Richardson, P., Griffin, I., Tucker, C., Smith, D., Oechsle, O., Phelan, A., Rawling, M., Savory, E., & Stebbing, J. (2020). Baricitinib as potential treatment for 2019-nCoV acute respiratory disease. *The Lancet, 395*(10223), e30–e31. https://doi.org/10.1016/S0140-6736(20)30304-4

Richman, K. A. (2004). *Ethics and the metaphysics of medicine. Reflections on health and beneficence*. MIT Press.
Richman, K. A., & Budson, A. E. (2000). Health of organisms and health of persons: An embedded instrumentalist approach. *Theoretical Medicine and Bioethics, 21*(4), 339–352.
Ross, C., & Swetlitz, I. (2017). IBM pitched its Watson supercomputer as a revolution in cancer care. It's nowhere close. *Statnews*. Retrieved from https://www.statnews.com/2017/09/05/watson-ibm-cancer/
Ross, C., & Swetlitz, I. (2018). IBM's Watson supercomputer recommended 'unsafe and incorrect' cancer treatments, internal documents show. *Statnews*. Retrieved from https://www.statnews.com/wp-content/uploads/2018/09/IBMs-Watson-recommended-unsafe-and-incorrect-cancer-treatments-STAT.pdf
Russo, F., Schliesser, E., & Wagemans, J. (2024). Connecting ethics and epistemology of AI. *AI & Society, 39*(4), 1585–1603.
Segler, M. H. S., Preuss, M., & Waller, M. P. (2018). Planning chemical syntheses with deep neural networks and symbolic AI. *Nature, 555*(7698), 604–610. https://doi.org/10.1038/nature25978.
Sisti, D., & Caplan, A. L. (2017). The concept of disease. In Solomon, M., Simon, J. R., & Kincaid, H. (Eds.), *The Routledge companion to philosophy of medicine* (pp. 5–15). Routledge.
Smith, D. P., Oechsle, O., Rawling, M. J., Savory, E., Lacoste, A. M. B., & Richardson, P. J. (2021). Expert-augmented computational drug repurposing identified baricitinib as a treatment for COVID-19. *Frontiers in Pharmacology, 12*, 709856. https://doi.org/10.3389/fphar.2021.709856
Sullivan, E. (2022). Understanding from machine learning models. *British Journal for the Philosophy of Science, 73*(1), 109–133.
Sundberg, M. (2010). Cultures of simulations vs. cultures of calculations? The development of simulation practices in meteorology and astrophysics. *Studies in History and Philosophy of Modern Physics, 41*, 273–281.
Swetlitz, I. (2016). Watson goes to Asia: Hospitals use supercomputer for cancer treatment. *Statnews*. Retrieved from https://www.statnews.com/2016/08/19/ibm-watson-cancer-asia/
Theunissen, M., & Browning, J. (2022). Putting explainable AI in context: Institutional explanations for medical AI. *Ethics and Information Technology, 24*(2), 23.
Turk, M., & Pentland, A. (1991). Eigenfaces for recognition. *Journal of Cognitive Neuroscience, 3*(1), 71–86.
van Rijn, J. N., & Hutter, F. (2018). Hyperparameter importance across datasets. In *Proceedings of the 24th ACM SIGKDD International Conference on Knowledge Discovery & Data Mining* (pp. 2367–2376). ACM.
Vulsteke, C., Ortega Arevalo, M., Mouton, C., Stam, K., Goethals, R., Ameye, F., Populaire, C., Peeters, M., & Verdonck, P. (2018). Artificial intelligence for the oncologist: Hype, hubris, or reality? *Belgian Journal of Medical Oncology, 12*(7), 330–333.
Wachter, S., Mittelstadt, B., & Russell, C. (2018). Counterfactual explanations without opening the black box: Automated decisions and the GDPR. *Harvard Journal of Law and Technology, 31*(2), 841–887.
Wu, X., & Zhang, X. (2016). Automated inference on criminality using face images. arXiv: 1611.04135v1. https://arxiv.org/pdf/1611.04135v1
Wu, X., & Zhang, X. (2017). *Responses to critiques on machine learning of criminality perceptions* (Addendum of arXiv:1611.04135). arXiv:1611.04135v3. https://doi.org/10.48550/arXiv.1611.04135
Zednik, C. (2021). Solving the black box problem: A normative framework for explainable artificial intelligence. *Philosophy & Technology, 34*(2), 265–288.

Juan M. Durán is an Assistant Professor in the Faculty of Technology, Policy and Management at TU Delft. His research focuses on epistemology and the philosophy of science in relation to computer-based scientific and engineering practices, including computer simulations, artificial intelligence, and Big Data. He also actively investigates the ethical dimensions of AI technologies and their impact on society.

In 2019, Durán received the Herbert Simon Award from the International Association for Computing and Philosophy (IACAP) in recognition of his outstanding early-career contributions. The award honors scholars whose original research is transforming the dialogue at the intersection of computing and philosophy.

Chapter 5
Challenges for Computational Reliabilism in AI and Other Computational Methods

Ramón Alvarado

Abstract Computational reliabilism has been recently deployed to justify our reliance and trust in computational technologies such as machine learning methods in artificial intelligence. Roughly, these deployments can be understood as seeking to (a) respond to or circumvent the challenges related to epistemic opacity in computational methods, and in doing so, (b) warrant or justify our beliefs regarding the reliability of computational processes and their results; and hence, (c) to reassure us of the possibility of trust in computational methods, practices and artifacts even if these are insurmountably opaque. This chapter aims to elucidate three major challenges to computational reliabilism that have a bearing on its viability *both* as a general epistemological framework capable of dealing with the advent of computational methods, *and* as a pragmatic epistemic resolution to the justification problems related to the adoption of opaque computational methods. These challenges are the following:

1. The challenge of warrant transmission and reliability-crediting properties,
2. The challenge of the indispensability of endogenous features in artifactual reliability, and
3. The challenge of error-related opacity

5.1 Introduction

The concept of computational reliabilism, originally coined in the context of epistemological problems related to computer simulations (Durán & Formanek, 2018), has been recently deployed to justify our reliance and trust in many other computational technologies, including machine learning methods in artificial intelligence such as deep neural networks (Durán & Jongsma, 2021). Roughly, these deployments can be understood as seeking to serve several— and often

R. Alvarado (✉)
Department of Philosophy, University of Oregon, Eugene, OR, USA
e-mail: ralvarad@uoregon.edu

J. M. Durán, G. Pozzi (eds.), *Philosophy of Science for Machine Learning*,
Synthese Library 527, https://doi.org/10.1007/978-3-032-03083-2_5

interrelated— purposes under the umbrella of a unified epistemological framework adept to account for a justified reliance on computational practices, methods and devices. In particular, an overarching hope of those championing such a framework is that computational reliabilism can:

- Respond to or circumvent the challenges related to epistemic opacity in computational methods, and in doing so,
- warrant or justify our beliefs regarding the reliability of computational processes and their results; and hence,
- to reassure us of the possibility of trust in computational methods, practices and artifacts even if these are insurmountably opaque.[1]

In this chapter I aim to elucidate what I deem to be three major challenges to computational reliabilism. I deem these challenges to have a bearing on its viability *both* as a general epistemological framework capable of dealing with the advent of computational methods—particularly in scientific inquiry— *and* as a pragmatic epistemic resolution to the justification problems related to the adoption of opaque computational methods, both of which are often cited as motivations for its adoption (Durán & Formanek, 2018; Durán & Jongsma, 2021). In particular, I focus on the following three challenges:

1. The challenge of warrant transmission and reliability-crediting properties
2. The challenge of the indispensability of endogenous features in artifactual reliability, and
3. The challenge of error-related opacity

5.2 Computational Reliabilism as a Kind of Reliabilism

Computational reliabilism— a term coined by Durán and Formanek (2018)— is the view that in the context of computational methods such as AI systems researchers are justified in believing or trusting the results yielded by such methods "*because* there is a reliable process (i.e. the algorithm) that yields, *most of the time*, trustworthy results." (Durán & Jongsma, 2021 p.332 italics mine)[2] According

[1] That reliability and trustworthiness are not the same thing is well known and well-documented in the literature. Nevertheless, that points 2 and 3 are distinct is not always clear or explicitly stated. However, that they are clearly distinct issues is evidenced by the fact that trust can be had in things whose reliability is questionable and by the fact that otherwise reliable methods and practices are sometimes not trusted. The relationship between epistemic efforts such as explainability, interpretability and trustworthiness, is a non-trivial issue that is the subject of important ongoing debates, some of which, despite their important insights, may be beyond the scope of this chapter.

[2] Other versions of reliabilism related to computational systems expand the foci of the processes in question beyond just the algorithms and suggest that in contexts such as AI, a sociotechnical understanding of such processes is required. Although Durán and Formanek's original framing of computational reliabilism is directed towards algorithmic processes, there is a sense in which the

to Durán and Formanek, computational reliabilism borrows closely from Alvin Goldman's epistemological reliabilism. In particular, it borrows from *process reliabilism* which suggests that a given inference/assertion can be deemed reliable if it is the product of a reliable process (2011). How such a process can be deemed reliable varies, but strictly speaking, the details do not matter that much: a reliable assertion need not include details about the reliable processes that produced it, the reliability in question is that of the assertion and not that of the processes by which it was produced, or so it is argued. Following from this, a process can be reliable even if the reasons *why* it is reliable are not accessible to an agent (Comesaña, 2010 p.571). In the context of perceptual claims, for example, the reliability of the process must be treated as somewhat of a brute fact. That is, we must simply accept that our perceptual system yields true beliefs more often than not, even if we do not know how or why it does so or how and why we know so.

Although at times Durán and Formanek "heartily endorse" views of reliability that suggest that scientists are justified in trusting the results of their computational methods simply because they "trust the assumptions upon which they are built" (2018, p.652), as suggested by Beisbart (2017) and others, a closer look at their views suggests that computational reliabilism— rightly— forgoes these kinds of assumptions (ibid.). Unlike conventional reliabilism, computational reliabilism does not need to presuppose the reliability of the methods from which the results are obtained. Rather, it is suggested, computational reliabilism takes in consideration 'reliability indicators' as "markers of methodological and epistemological competence of the computer, algorithms and social processes involved in the formation of beliefs." (Durán, 2023). Accordingly, Durán and Formanek suggest that computational reliabilism requires a *retrospective reliability chain* "that conditions the sources that attribute reliability to [computational methods] to be reliable in and by themselves." Furthermore, such sources, they accept, "must be *shown* to be reliable." (2018, p.656 italics mine).

As we shall see in the next section, this last point seeks to distinguish the type of reliability chain referred to by Durán and Formanek somewhat from a mere appeal to a chain of epistemic entitlements—i.e., non-evidentiary epistemic warrants used to justify ordinary knowledge claims and widely believed to be epistemologically acceptable (if not necessary) in everyday epistemic endeavors (Graham, 2012; Symons & Alvarado, 2019). By contrast, according to Durán and Formanek, the way in which computational methods, or their constitutive algorithms, are deemed reliable is through the consideration of factors—external to the algorithms themselves— that function as reliability indicators. These include "identifying methods (formal or otherwise), metrics, expert competencies, cultures of research, and the like that make up for our best epistemic and normative efforts that might increase the degree of warrant we have to believe the outputs of

considerations are broadened in later work (See Durán and Jongsma, 2021 and Durán, 2025) to include both technical details and social practices.

ML systems" (Duran, 2025). Reliability indicators for machine learning methods, according to Durán, include the following:

- Technical robustness of algorithms,
- Computer-Based scientific practices, and
- A social construction of reliability[3]

Hence, according to this view, we can rely on the results of ML in virtue of the fact that they are the product of robust algorithms, which in turn come from trustworthy computer science practices. In addition, we should also trust them because the reliability of a system is ultimately a contextual and socially negotiated concept that includes the input of a diverse group of experts, practitioners, and users. These methods, metrics, practices and the processes that include them produce reliable results more often than when such methods, metrics and practices are absent.

The "most of the time" or "more often than not" clauses above make sense of Durán's suggestion that, at its core, computational reliabilism is meant to be a frequentist approach to reliability (Durán & Formanek, 2018; Durán & Jongsma, 2021; Durán, 2025). Accordingly, the reliability of a process must be understood as a tendency to produce "high proportion of true beliefs relative to false ones." (ibid., p.653)[4] For instance, Durán and Formanek explicitly state that computational reliabilism is predicated on whether or not "the probability that the next set of results of a *reliable* [computational method] is trustworthy is greater than the probability that the next set of results is trustworthy given that the first set was produced by an unreliable process by mere luck." (2018, p.654) Trusting the results of computational methods therefore, according to this view, "depends on a chain of reliable processes that, in the end, allow researchers to be justified in believing the results" (p. 655) Where this chain ends, however, is simply left unanswered (ibid., fn.6, p.655).

Still, as mentioned above, advocates of computational reliabilism deem the framework as capable of responding to the challenge of *essential epistemic opacity* in computational methods — that is, the kind of opacity that is relative either to an agent's or a process' nature rather than relative to an agent's contingent epistemic limitations (Humphreys, 2009; Alvarado, 2021). Because of this, perhaps it is pertinent to spend a few lines clarifying certain aspects of such opacity and distinguishing it from conventional understandings of the concept as it relates to computational methods (see Burrell, 2016; Duede, 2022).[5]

When it comes to epistemic opacity two important distinctions need to be made. As briefly mentioned in the paragraph above, views such as Humphreys' suggest that

[3] While verification and validation techniques as well as expert knowledge are still discussed in the context of machine learning, Durán considers them as belonging to the categories of technical robustness and the social construction of reliability respectively.

[4] This is something echoed in Ferrario's (2023) approach to their own, highly formalized, version of reliabilism.

[5] See the work of Beisbart and that of Formanek in this volume for alternative takes on the issue of opacity.

essential epistemic opacity occurs in virtue of an agent's nature—e.g., its life-span, its perceptive apparatus, how much they can know or surveil, etc. In other words, the reason why some or all the epistemically relevant elements of a system, process or device are opaque has to do with the essential limitations of an agent. This makes it so that this kind of opacity is insurmountable for agents of some particular natures. It is precisely in this way that it differs from other, weaker and more general instances of opacity whose occurrence is relative to a specific agent, *at a specific time.* This also implies, as the name suggests, that changing the nature/essence of an agent amounts to changing the agent. And so, the distinction and its strength hold.

As we will later see, however (see footnote 17, p.12), and as I and others have shown (Alvarado, 2021; Saam, 2017), sometimes, particularly in the context of computational methods, this latter kind of opacity can also arise in virtue of the features of a system and not just in virtue of any agent-relative features. Some systems are opaque in virtue of their own properties, their complexity, irreducibility, etc. As such, the causes of such opacity are what I call agent-independent—i.e., the opacity in these cases arises from factors that are independent of any limitations on the part of the agent (Alvarado, 2021). Furthermore, because no agential resources can correct such instances, these latter instances also qualify as yet another kind of opacity: agent-neutral opacity— i.e., the kind of opacity for which any epistemic agent would face the same challenges (Alvarado, 2023a, 2023b). It is important to note that these two latter notions of opacity— agent-independent and agent-neutral— are not just distinct from the version of general epistemic opacity that is relative to an agent's contextual and contingent epistemic circumstances, but are also distinct from each other and imply non-trivially distinct things.[6]

In short, 'Relative' and 'essential' opacities are kinds of epistemic opacity, both of which could be agent-relative. While 'relative' opacity is necessarily so, essential opacity is not. This is because it can arise in virtue of elements that have little or nothing to do with an agent. What the latter three kinds of opacity— essential, agent-neutral and agent-independent— have in common however, is that they are for all intents and purposes insurmountable. Furthermore, according to some (Humphreys, 2009; Burrell, 2016; Alvarado & Humphreys, 2017; Horner & Symons, 2014; Symons & Alvarado, 2019; Alvarado, 2020, 2021), these are the kinds of opacity to which computational methods are particularly and unprecedentedly prone to.

Because of this serious challenge, finding an epistemological framework capable of circumventing these obstacles seems highly desirable. Whether or not computational reliabilism is such a framework, as we will see, is not immediately obvious.

[6] For a thorough discussion of the distinctions and the interplay between these kinds of opacity, see Alvarado's discussion of Nagel's agent-neutral reasons (2023a, 2023b, p.136) and Alvarado (2021).

5.3 The Challenges

In this section, I provide an overview of three challenges that I believe have a negative effect on the viability of computational reliablism as an epistemological framework that can appropriately capture the novel and/or unique epistemic issues associated with the use of computational methods. This is particularly the case in formal epistemic contexts such as scientific inquiry or in safety-critical contexts. In both of these contexts, endogenous features— such as the nature and source of error— of a system are simply indispensable reliability indicators and computational reliabilism seems unable to account for them. As I will show, these challenges also elucidate the limitations of computational reliabilism even when it is framed solely as a pragmatic epistemic resolution to the justification problems related to the adoption of opaque computational methods. As a reminder to the reader the challenges are the following:

1. The challenge of warrant transmission and reliability-crediting properties
2. The challenge of the indispensability of endogenous features in artifactual reliability, and
3. The challenge of error-related opacity

5.3.1 The Challenge of Warrant Transmission and Reliability-Crediting Properties

Although argued for elsewhere by Symons and Alvarado (2019), it may prove useful to begin by briefly noting that one of the serious challenges for computational reliabilism as an epistemic framework is that it fails to account for the epistemic challenges related to warrant transmission, particularly as related to sanctioning novel computational technologies. As we saw in the previous section, the success of computational reliabilism strongly hinges on what Durán and Formanek call 'reliability indicators.' According to them these include "algorithmic-related methods and practices" that have "*reliability-conferring*" properties (Durán, 2025, p.7).

According to these views, some of these practices can confer their reliability on to a process, a device that comes from such process, or the results of such a process. There is, however, a problem with the concept of reliability-conferring indicators and that is that there is no clear sense in which they can *confer* such reliability. While there is something to the intuition that in everyday epistemic practices— and in those practices without much at stake (epistemically or otherwise)— the provenance of a result or a device is taken to be a sort of indicator that confers reasons for trust, in epistemically rigorous contexts and in those contexts in which trust comes at a high cost, this is not the case. One can easily imagine that anyone involved at Los Alamos research laboratory that accepted the reliability or the results of an equation simply on the basis that it came from Einstein would be held as epistemically and socially irresponsible: both the equations and their results must have been checked before being incorporated into the workflow. Similarly, any General that accepted Oppenheimer's assertions simply on the basis that they came either from him, his team, from Los Alamos laboratory or any of its associated practices would be forgoing due diligence. It was Oppenheimer, yes, but also the reliable methods and the tests and the peer review (even if internal to the project's secrecy) that made it more acceptable for anyone to go along. And yet, notice that the reasons why the characters involved in this scenario would go along are still not reasons that actually —technically speaking— confer any degree of reliability or trustworthiness to the device itself. Tests are involved and inferences strengthened: the device itself must provide such warrants. This is particularly the case if the thing being conferred reliability to in virtue of established methods and processes is not itself a-yet-established method or technology. For example, the best practices by the best experts in the best of settings behind the invention of a never-before-seen technology are not enough to ensure that such technology is itself reliable. As I have argued for elsewhere (Alvarado, 2023a, p.72), even in the case of technological devices that are now considered paradigmatic scientific instruments such as the classic optical telescope, neither the name and reputation of Galileo or Kepler alone sufficed to establish the new device as a scientifically reliable instrument. What is more, neither Galileo's assertions that his *theoretical* approach resulted in a better instrument than the original, nor Kepler's initial efforts to drum up consensus about the validity of Galileo's instrument amongst the scholarly community proved to be sufficient warrant for its serious adoption (Alvarado, 2023a, 2023b). Rather, extensive testing as well as criteria for what constituted proper testing had to be designed and developed for telescopes to undergo before they were sanctioned as scientifically reliable instruments (Van Helden, 1994). These tests did not just involve the interested parties, e.g., astronomers, lens crafters and mathematicians. Rather, these tests included calligraphists to create simpler fonts so that they could not be easily guessed at a distance, papermakers to avoid indentations that could cause shadows on the printed material, standardization of materials and design specification, as well as competitions amongst instrument makers against these standardizations to ensure fair and sound benchmarking. All of which were competing against one aim: what could epistemically undermine our inferential warrants towards the results derived from the telescope? More to the point, what could potentially diminish the credibility of assertions made with this instrument about the nature of Saturn's—at the time still unelucidated—side protrusions?

The difficulties regarding the transferability of epistemic warrants, or the lack thereof, from methods, practices and people to technical artifacts, become clearer in epistemic contexts in which epistemic endeavors that require hard rigor depend on one another, where the scale of resources required demand significant efforts in preparedness and constraint, and in contexts that could cause significant harms to others. Deeming the atomic bomb as a reliable weapon both in its power and its limitations, for example, solely in virtue of the fact that it was made with the best available methods and by the best physicists at the time, without probing the artifact itself to produce its own epistemic warrants, is clearly an inadequate epistemic strategy. As demonstrated by Symons and I (2019), epistemic warrants that justify reliance or trust on one method or on a group of experts are not simply transferable to the technical artifacts built with such a method or by such group of people. The notion that either existing reliable methods or human expertise can simply "*credit* reliability" forward to computational methods (or any technical artifact for that matter) as Durán suggests— while intuitive in less epistemically costly settings—is thus not immediately obvious.

Even in the case of highly restrained epistemic endeavors that require technological mediators, ensuring that such warrants transfer from one process to another requires that hard-to-attain epistemic conditions hold. Consider arguments that appeal to analogies between epistemic practices in computer-assisted mathematical proofs and those in computer simulations (Barberousse & Vorms, 2014). One condition that these appeals must assume that holds in both settings is that the methods involved are transparent conveyers of epistemic content— that is, that the process of manipulating epistemic content that is supported by one kind of warrant, say an a priori epistemic warrant like the ones supporting mathematical inferences, did not introduce elements into the process that required the support of another kind of epistemic warrant, namely a posteriori epistemic warrants such as those provided by empirical evidence (Burge, 1993, 1998). The analogy between computer-assisted mathematical proofs and computer simulations, as elucidated by Symons and I, was already a dubious stretch given the distinct practices and distinct epistemic norms involved in each endeavor. As noted by Winsberg (2010) and others,[7] computational methods such as computer simulations, involve a vast and non-trivially motley set of experts, practices and methods (Durán, 2020). These practices and methods are far removed from the rigorously cautious process of mathematical proofs. Transparent conveying of epistemic content is simply not something that can be easily achieved in such settings. The fact that we cannot simply trust the mathematics involved in such processes (which require an a priori warrant) but that we have to also trust practices and equipment (which require their own a posteriori warrants) demonstrates the non-transparency of the conveyers at play. Hence, epistemic warrants fail to simply transfer from one step to the next and each part of the process must muster its own through distinct means.

It is worth noting that even in more ordinary epistemic contexts, the transfer of epistemic warrants is epistemically non-trivial. Take for example the problem of *epistemic alchemy* (McGlynn, 2014) in which a seemingly warranted proposition is derived from a poorly warranted one. In the case of epistemic entitlements, this would imply deriving a purportedly justified claim from claims supported by non-evidentiary warrants. While it may very well be the case that in ordinary epistemic practices this may not be too much of a problem (ibid.), it is not clear that this practice or that epistemic entitlements themselves, such as the ones relied upon by

[7] See Winsberg and Alvarado's (2025) Stanford Encyclopedia of Philosophy entry on *Computer Simulation in Science* for the most comprehensive discussion of these and other considerations regarding the epistemological status of computer simulations in inquiry.

Burge and others (see Barberousse & Vorms, 2014), are adequate in the context of scientific inquiry to start with. Rather, extensive sanctioning processes that are in fact reactionarily independent from— i.e., critically engaging with and seeking to overcome the limitations of— existing methods or current expertise consensus have to be designed and implemented for novel technologies to be appropriately sanctioned for scientific use (Alvarado, 2023a).

This points to the fact that making the case that warrant transmission occurs, if it occurs at all between practices, methods, experts and their artifactual products, takes a non-trivial philosophical effort. Unfortunately, other than noting that these reliability-conferring properties are not to be understood as 'spooky' (Durán, 2023), this effort is all but absent in accounts of computational reliabilism to date. Whatever version of computational reliabilism that strongly hinges on the notion of reliability-conferring, or reliability-crediting properties will have to seriously contend with the issues of warrant transmission. Until then, the future promissory note of existing computational reliabilism accounts seems to be of too high a denomination to just go unchecked, signaling a genuine and serious challenge to the viability of computational reliabilism as an epistemic framework.

To be fair however, this is a major epistemological problem for most accounts of reliability. Hence, in what follows I will take it as a given that there may be such things as reliability indicators. As we will see, even if this is the case, major problems emerge for computational reliabilism for in the case of the reliability of artifacts, these indicators happen to be found elsewhere from where reliabilist accounts conventionally look for them.

5.3.2 The Challenge of the Indispensability of Endogenous Features in Artifactual Reliability

As we saw above, in reliabilism, whether something is reliable or not does not depend on an agent's internal evidentiary threshold for justification to deem it so, as evidentialist views suggest (Goldman, 2011). Rather, under views of this type, the reliability of a process can be determined solely by considering exogenous elements of a process. In the case involving conventional epistemic agents, such as ourselves, and our claims to knowledge, reliabilism is supposed to give us a theoretical framework that makes sense of the intuition that we seem perfectly capable of justifiably relying on either our own perceptive systems or on the testimony of others without necessarily having to invoke exhaustive mechanistic details about how either of them functions.[8]

[8] On a more common sensical sense, this safeguards the intuition that one does not have to know *that* one knows in order to know something and that one does not need to know *how* one knows something in order to know something, e.g., knowing that something was seen does not entail/require knowing how eyes, brains, nervous systems, etc., work.

As we saw in Sect. 5.2 above, under views like these, the degree of reliability of a process depends on its proclivity, or lack thereof, to produce truthful outcomes and not on any internal properties of the process itself or on whether or not agents find them sufficiently justified to subscribe to them (Goldman, 2011; Goldman & Beddor, 2021). Process reliabilism extends this seemingly virtuous omission of conventional reliabilism to considerations about results or assertions that require or come from a procedure carried out by an agent or set of agents. Importantly, process reliabilism points towards exterior—often environmental—factors that can make or break an epistemically conducive setting. There is no point under this type of views, for example, in citing an agent's good eyesight as a reliability indicator on a severely foggy day. It is the foggy day, or the lack thereof, that determines whether or not we have a reliable sighting; famously, seeing a farm from afar in a county littered with fake farm facades as tourist attractions is not sufficient to conclude that a farm has been sighted. Rather, a process by which a piece of knowledge could be relayed or come to be known under these circumstances has to be reliable and the factors that deem it so are mainly external, exogenous to the agent in question, their inner workings or those of the process itself. While pointing in the right direction, these examples fail to show that what reliabilism was really trying to keep at bay was the internalist notion that an agent's inner sense of justification was all that sufficed for them to claim knowledge. Reliabilism, in other word, is an externalist epistemic framework.[9]

Durán and Formanek (2018) try to safeguard this intuition of reliabilism in *computational* reliabilism. They suggest, as noted above, that computational reliabilism uses "reliability indicators (RIs) as markers of methodological and epistemological competence of the computer, algorithm, and social processes involved in the formation of beliefs." Importantly, according to Durán and Formanek (*ibid.*), these indicators, which "can be understood as algorithmic-related methods and practices" (Durán, 2023), are nevertheless to be considered "exogenous to the algorithm." This is in large part what makes computational reliabilism a reliabilist epistemological account in the first place.[10] *And*, it is also what allows it to be positioned, as we saw in previous sections, as a plausible solution to the challenges posed by the severe opacity many say to be characteristic of computational methods (Humphreys, 2009), particularly machine learning and other AI technologies (Burrell, 2016; Alvarado, 2020, 2021, 2022a, 2022b). In other words, one of the main appeals (if not *the* main appeal) of computational reliablism vis-à-vis computational methods is that it is not supposed to need to take into consideration the inner working of a process, its endogenous features, in order to provide justification of our reliance on its

[9] This will become important at the end of this chapter when we briefly discuss the fact that computational reliabilism, as championed by Duran and coauthors, seems to be more of an evidentialist non-externalist epistemic account and not an externalist reliabilism after all.

[10] It must be noted that Durán's insistence that the notion of reliability must be acknowledged as a social construct risks taking computational reliabilism more in the direction of evidentialism, a view which invokes an epistemic agent's internal evidential thresholds as relevant determinants in the assessment of claims.

results. Hence, the opacity— in its many varieties and however severe— of the inner workings of a system, device or process is supposed to be a non-challenge. This virtuous omission of computational reliabilism is what substantiates the claims in the literature that nobody really is, and nobody should be, afraid of black boxes (Durán & Jongsma, 2021), even —or particularly— if lives are at stake (London, 2019).

Yet, a few issues for this view immediately emerge. First, the strong external/internal distinction that conventional reliabilism applies to considerations involving naturally-occurring epistemic agents, such as us, as well as to considerations regarding our surrounding environment, may not necessarily apply to considerations involving artifacts, or to processes generally understood as artifactual,[11] or even to their products: be these propositions, results, computations, etc.[12] This is because technical artifacts are constituted by features without which they could not carry out their designed or adopted central function and which emerge in virtue of material or design properties that can only be seen as endogenous— i.e., arising from within. At the same time, their designed/assigned tasks also determine what these essential arrangements and materiality will and can be. How they were made, what they were made of, and how they do what they do in virtue of how they were made, what they were made of and what they were made for, are indispensable considerations to fully capture the nature of artifacts (Kroes & Meijers, 2002; Symons, 2010) and hence for understanding their functional capacities and limitations (Alvarado, 2023b). Understanding their capacities and limitations is, in turn, essential for reliability assessments and for properly-grounded trust allocation (Simon, 2020; Alvarado, 2022a, 2022b).[13] As endogenous features, however, computational reliabilism is not supposed to need them to assess the reliability of a process, system, technology or their results.[14] Yet ignoring them

[11] Science in general and methodology in particular can be thought of as artifactual processes of inquiry: like technical artifacts, scientific methodology is designed and implemented with an aim, intention and hence a function.

[12] There may be something to such an externalist/internalist demarcation in cases in which agents such as ourselves are involved—e.g., perhaps due to phenomenological, biological or cognitive features that clearly separate us from our non-organic surroundings. There is us, the cohesive organism with reflective cognitive capacities, and there are certain *external* conditions that ought to be met, processes to be carried out, in order for someone to be in a position where they can be said to know something (see the barn example above). That this demarcation—as fuzzy as it may be— can be meaningfully made at all in the case of technical artifacts is not immediately obvious.

[13] This is further in alignment with some developments in the epistemology of trust in science and technology which (rightly) suggest that in order for trust to be appropriately allocated, or 'grounded', the reasons that justify such trust must align with actual properties, namely capacities, of the trustee (Buechner et al., 2013; Simon, 2020; Oreskes, 2021, p.55; Alvarado, 2022b, 2023b).

[14] Recently, in personal correspondence, Formanek suggested to me that computational reliabilism was supposed to be a halfway house between externalist and internalist theories of knowledge and that as such, it was compatible with my assessment that a purely externalist position would not do. This is unfortunately not something reflected in their earlier published account. And yet, it would be interesting to see how Formanek's more recent interpretation accommodates Duran's more recent characterization that reliability indicators related to methods and practices could

seems at best epistemically questionable, at worst epistemically irresponsible (Winsberg & Harvard, 2022). It is true that we often allocate trust without proper warrant. This could amount to a mistake or perhaps even a necessity in cases in which there is insufficient information or time to gather it. Doing so knowingly, however, and particularly in contexts in which one's aim is assessing the reliability of a method or device for future or continued use, simply violates basic norms of epistemic hygiene, particularly those of scientific inquiry (Symons & Alvarado, 2019; Alvarado, 2023a, 2023b, p.138).

Consider a medical doctor telling you that a given pharmaceutical cures acne 95% of the time but could kill the other 5%. Also, imagine that when asked about what information there is about the 5%, their reply is that there is not any more information beyond the rate and the extent of the risk: we know it fails *this* often, and we know that failure often implies death. Consider further that they tell you that they deem this substance to be very reliable because at least they know both the rate *and* the extent of failure. This is already a problematic scenario which would require a healthy dose of skepticism or informed consent from the part of a patient. It is also an unlikely scenario, so let us add some more detail to it and make the doctor in question a bit more epistemically responsible. Rather than simply assuming the reliability of the process by which the medicine was brought about, the doctor cites some reliability indicators similar to those in computational reliabilism to justify their reliance on such a medicine. That is, rather than simply appealing to an epistemic entitlement and telling you that you should trust them, they insist that some endogenous factors should serve as reliability indicators to you as well. They tell us that not only does the medicine yield reliable results most of the time, but that the processes by which this medicine has been designed, developed and deployed are also reliable because they represent "methods (formal or otherwise), metrics, expert competencies, cultures of research, and the like that make up for our best epistemic and normative efforts that might increase the degree of warrant we have to believe in the results of [such medicine]." The pill, they argue, was developed in a world-renowned laboratory, it was produced with the best equipment and sanctioned by the most credible people, etc. Imagine further that if you asked about how the most credible people sanctioned the pill, the answer is that you do not really need to know but if you insist and must know, they did it by testing it on roughly a hundred random people. Five of which died and ninety-five who got better. Once you take all these details into consideration, contemplate whether or not the question of whether this substance *is* reliable has been answered. In other words, can we accept the information above and still ask "yes, but is this substance

nevertheless be consistently accepted as "exogenous to the algorithm" (2023). Or the following opening statement in Duran's most recent preprint: "CR is a computationally-inspired off-shoot of process reliabilism that does not require showing the inner workings of an algorithm. CR credits reliability to machine learning by identifying reliability indicators external to the algorithm (validation methods, knowledge-based integration, etc.). Thus, we have justification for believing the output of machine learning when we have identified the appropriate reliability indicators." (2023).

reliable?", without being conceptually confused? That is, if someone provided all the information described above, could a fully rational and capable agent still ask—without risking inconsistency—whether the pharmaceutical is reliable or not? Because if it is possible to consistently ask that question even after having received the above-mentioned information, then it means that the information above does not answer the question about the reliability of the pharmaceutical. At the very least, we can say, *something else other than* the information that we have so far received must determine whether or not the pharmaceutical is reliable or not. We can go back to the atomic bomb example and the cases of warrant transmission discussed in the previous section for reference: the answer is yes, the reliability of such substance has not yet been determined.[15]

The argumentative element of the example above comes from the fact that the question—even if Durán's considerations are true—is not already answered and asking it continues to make sense. In other words, even if a technology comes from methods (formal or otherwise), metrics, expert competencies, cultures of research, and the like that make up for our best epistemic and normative efforts that are alleged to increase the degree of warrant we have to believe in the results of such technology, whether or not that technology itself, both in kind or in token, is *in fact* reliable, remains an open question (Symons & Alvarado, 2019; Alvarado, 2021, 2023a). This signals at the very least that these exogenous details are not sufficient to determine the reliability of such an artifact, and/or that there must be something else, perhaps something more, that does so.[16]

Here is yet another example that illustrates this point. Consider the recent announcement by Google's GraphCast researchers that GraphCast has "significantly outperformed conventional weather forecasting methods in predicting global weather conditions up to 10 days in advance." (Edwards, 2023) Some may think that knowing that this system outperforms conventional weather prediction tools 90% of the time (Lam et al., 2022) could suffice to deem it as reliable or even as *more* reliable than the conventional tools it is being compared to. Nevertheless, notice that in such a context, since it is the kind of computational methods that is likely to be relied upon in order to make safety-critical decisions, knowing the

[15] It is in cases like these that a confusion may emerge between questions about reliability and trustworthiness. What the doctor is alluding in this case may be reasons for trust. They are citing trust indicators but not necessarily reliability indicators. Although this is of course problematic, as we saw in the previous section, for the sake of argument we can say that you may have reasons other than the actual/factual reliability of the substance to trust it: the reliability of others, social pressure, lack of alternatives, etc. As pointed by Dretske (2000), these are simply not epistemic reasons, so even if they may be adequate for non-epistemic contexts or technology (e.g., in medicine), they are simply inadequate for epistemic trust in epistemic contexts (Alvarado, 2023b).

[16] Note here that the argumentative strategy does not need for us to specify what this other thing is. Rather, what has been demonstrated is that something, whatever, other than the suggested elements must be at play. What these other determinant elements are is a separate question and one for which a full other paper must be written. For now, it suffices to point towards the next challenge in the following section, namely: that certain endogenous features must be taken into account and that computational reliabilism, as fleshed out so far in the literature, cannot account for.

source and nature of error— that is, why and to what extent it fails—becomes a non-trivial, indispensable consideration. The number of resources required for weather related interventions are such that decision-makers must be able to both prepare for and constraint their deployment with accuracy and precision. Just knowing how often a system like this fails is not enough to assess its reliability. Further, note that in such cases, even just knowing how bad the system fails is not enough. It may prove irrelevant, for example, that the benchmark evaluation of one method over the other ranks as more accurate in average by 1 degree Celsius in normal weather, if once every month it is off by 10 degrees in the case of unpredictably weather events, which is when these systems are the most needed/useful.[17] What would help deem this system reliable is knowing how and why it does better, where it can fail and why it can fail, and what about its inner workings makes it vulnerable to unreliability, just like in the case of the telescope described earlier. Such inquiry may, of course, involve an operationalization of what 'better' means that may include standardization for benchmarking and comparison purposes, the methodology and rationalization for the evaluation criteria, which could be deem as exogenous to the instrument itself. But also, it would have to involve knowledge of the kinds of sensors the instrument has, the kinds of properties such sensors are designed to capture, as well as the way in which such sensors were designed to capture such a property. In other words, knowing about its endogenous properties.

Consider one last example to this point. It was noted during the recent COVID-19 pandemic that certain oximetry instruments, particularly certain non-invasive optical techniques to measure arterial oxygen saturation, were inaccurately measuring the oxygen levels of certain patients: namely those with a darker skin. The prevalence and visible consequences during such a crisis made the pattern more easily accessible. We know, thanks to such a dense data space during the pandemic, that it is likely to fail when used in darker skins. But notice how knowing certain endogenous features of the oximeter can help understand the nature and source of the inaccuracies and hence, enrich our reliability assessment in a way that merely recording rate of success/failure and cases of success/failure does not. While the golden standard to measure oxygen in blood is to take a careful measurement of gas molecules in blood, pulse oximetry works mainly by measuring changes in light absorbance in the arterioles over time (Al-Halawani et al., 2023). Normally, these devices use specific equations that account for the way light tends to disperse in general coupled with other equations that posit the way light may disperse in tissue. However, for most intents and purposes, these equations typically forgo considerations regarding relative variations in the distribution of fundamental elements of tissue. That is, while it takes into consideration the chemical makeup of tissue, it does not take into consideration whether some tissue has more of something than of other thing. This makes it so that the formal methods associated with the calculation of oxygen in a patient's blood through this method do not account for

[17] Thanks to Mats Krüger Svensson whose comment on the news aggregator website HackerNews inspired this example.

the fact that melanin—an element that has a causal relationship to the dispersion of light—is higher in some tissues with distinct skin tones than in others. Hence, it is not calibrated to non-trivial differences that affect its main metric: namely the dispersion of light through tissue. Notice, the richness of insight regarding both the reliability of the artifact but also the parameters within which such reliability can be assessed that this further knowledge grants our inquiry. As we will see below, knowing the nature and source of error are not only more enriching to reliability assessments but rather seem indispensable for them both in theory and in practice.

Importantly, considering these aspects of a technical artifact is not only key in life-critical settings, but also in knowledge-critical processes—those processes in which acquiring, building on, or expanding genuine knowledge is the essential central task.[18] In such contexts, questions such as 'why/how/where did the error happen?' become epistemically relevant even when the technology functions correctly "most of the time." Particularly if we want to *appropriately*, that is justifiably, rely on it.[19] In a sense, this is the one thing that sections in scientific papers devoted to methodology seek to elucidate: that the methods used were crafted in such a way that those considerations that could directly undermine the validity of their results have been sufficiently addressed and hopefully assuaged. Importantly, when addressing the possibility of error in a technical artifact in these kinds of contexts, as we saw in the examples above, this does not simply mean gathering a rate of error, or even quantifying a magnitude of error when it does occur. Rather, it means having a general understanding of the nature and the source of such error such that we can speak to it, track it, and hopefully address it in future iterations of the process in question. These features, again, seem like centrally endogenous and not merely exogeneous features of both the technical artifact and the process by which it was made.

[18] In some foundational scientific projects, for example, in which the aim is to simply know more about a phenomenon in order so that the underlying mechanisms of other higher-level phenomena could be elucidated and hence understood, simply knowing that a process yields consistent results the majority of the time does not suffice. In this context the task is to know more so that we can know more. Hence, if we fail at identifying the elements that could undermine our capacity to know more, then we are failing at trying to know more and our ability to know more about the higher-level phenomena is on flimsy epistemic grounds.

[19] The emphasis on appropriate justification for an epistemic technology is important here because, as a practical matter, we can and do rely on things for which we have no proper justification. We do this in certain instances of crises, but also in most of our everyday activities. That we can do so, and that we do so, however, does not do much to the plight that in some circumstances doing otherwise is what should be done. In contrast with Durán and Formanek, I find little use in accommodating existing scientific practices in a normative inquiry. The reasons why we could or should rely on an artifact are not appropriately informed by the reasons why people actually or currently do so. This is something acknowledged as far back as the times of the early iterations of the telescope and it essentially constitutes the many epistemic developments usually ascribed to Baconian scientific methodology.

Hence, when it comes to understanding the capacities and limitations of a technical artifact, elements such as its materiality, material arrangement, structural integrity and hence possible points of failure are indispensable features to consider, e.g., a hammer made of a single piece of cast iron will be able to perform distinct tasks in distinct settings from a hammer made of plastic or even a hammer made of more than one part. In the case of computational methods, hardware and software architectures, the way code functions are timed, the kinds of errors these architectures and specific programs are prone to, and the kinds of languages and/or techniques deployed to perform a function must be considered. These material, structural, or functional elements can be categorized as constitutive of—or at least coextensively defined with— "algorithmic-related methods and practices," yet whether these elements are exogenous or not to the algorithm itself, external or internal to a process or results whose reliability is under consideration, is not entirely clear.

And yet, whether these features are endogenous or exogenous, they must be accounted for. As briefly noted above, in the case of certain artifacts — computational or not— namely those used in both safety-critical settings, like medicine, and those used in foundational epistemic endeavors, such as science, endogenous features, such as the nature and source of failure points, become even more indispensable. This is not unique to computational methods, there is a reason why precision instruments and not less expensive alternatives are the norm in laboratories: because it is epistemologically relevant to consider endogenous properties of an artifact to deem it epistemically reliable (i.e., they serve epistemological values such as standardization, reproducibility, objectivity, etc.). Similarly, there is a reason why medical equipment is expected (and often forced) to undergo rigorous vetting and sanctioning processes before it is introduced into medical practice: because it is both medically and morally relevant to do so. Yes, it is an epistemic good to make sure that we can rely in the results of our instruments as we chose a path forward in a patient's treatment, and yes, we have an epistemic responsibility not to take the reliability of systems that can affect people's lives or well-being lightly (Harvard & Winsberg, 2021; Winsberg & Harvard, 2022), but importantly, we are also morally required, in settings such as medicine or civil engineering, to safeguard people from error that may undermine the function of such systems. Doing so requires us to not neglect properties and features endogenous to both the artifact and the processes by which it was built. If this is the case then, as we will see, either computational reliabilism cannot account for such features and it is therefore an inadequate epistemological framework to deal with artifacts of any sort but particularly computational artifacts; or, computational reliabilism can and hence must account for such endogenous features, in which case it must directly address the challenge of epistemic opacity of a process' endogenous features and not merely circumvent it.

5.3.3 *The Problem of Error-Related Essential Epistemic Opacity*

In this section I aim to show that what is at stake in instances of epistemic opacity is not merely the inaccessibility to essential features of how a process *works*. This has been a central misunderstanding within the literature. Rather, what is at stake is the opacity of how such a process may *fail*. The inability of computational reliabilism to take endogenous features into consideration makes this challenge a serious problem for the viability of the framework as a solution to the epistemic opacity of computational methods. However, as we will see, the insurmountable essential epistemic opacity of certain kinds of error in computational methods such as machine learning— particularly neural networks, which are at the center of more recent developments in large language models and generative pre-trained transformers—makes it so that the challenge is even more severe than mere conceptual inadequacy on the part of the epistemological framework.

In the face of the challenges posed by the *essential epistemic opacity* of computational methods—elucidated first by Humphreys (2009) and later expanded by others (Burrell, 2016; Alvarado & Humphreys, 2017; Alvarado, 2020, 2021)—philosophers have come up with argumentative strategies that make sense of the fact that inquiry, even the inquiry that makes use of opaque methodology, can nevertheless be successful.[20] Some of these strategies aim to undermine the severity, both conceptually and practically, of the challenge of epistemic opacity. San Pedro (2024), for example, suggests that epistemic opacity can be contextually assessed as a matter of degrees and that certain mitigating procedures can ease its impact.[21] Similarly, after a thorough assessment of different kinds of sources for opacity in machine learning technologies, Jenna Burrell (2016) offers a series of best-practice remedies to assuage it. Others, mainly practitioners, simply deny that essential epistemic opacity is the case and offer evidence that suggests to them ways to make certain methods—for example, in machine learning— 'interpretable.' Still others,

[20] I take it that the reader is sufficiently acquainted with the nuance surrounding the concept of epistemic opacity, particularly as it relates to its kinds, its sources, and their respective strengths and implications. For the most thorough review of the concept of opacity, its sources, and its implications to date, see Alvarado, 2021. Hence, here I will only focus on the essential epistemic opacity particular to computational artifacts and not on the general sense of opacity first identified by Humphreys (2004) and later expanded on by others to include social and contingent sources of opacity: e.g., scientific infrastructure, technical literacy (Hubig & Kaminski, 2017), state and corporate secrecy (Burrell, 2016), natural resources, etc. Elsewhere (2020, 2021, 2023a), heavily borrowing from the work of Symons and Horner (2014) and expanding on Humphreys' (2009) views, I have argued that in the context of computational artifacts, essential epistemic opacity is uniquely— even if not exclusively or exhaustively— at play. This is particularly the case in AI/ML systems, whose architecture and analytic dimensionality further exacerbates the 'too many everything problem" (Alvarado, 2020).

[21] This, of course, ignores or neglects the fact that the concept of essential epistemic opacity implies that it is *impossible* for agents of a limited epistemic nature to overcome it. This is a topic I address elsewhere (Humphreys, 2009; Alvarado, 2021).

accept that essential opacity is the case but deny that it makes a significant difference to the aims of inquiry (Durán & Jongsma, 2021; Duede, 2022). Formanek (see Chap. 2 by Formanek in this volume), for example, suggests that the best solution to opacity in artificial neural networks is to simply get habituated to it. With time, he argues, we will either forget about it or find ways around it. In any case, his argument goes, opacity simply does not matter for practical purposes.[22]

All of these argumentative strategies acknowledge that those seriously concerned with epistemic opacity have a certain commitment to the value of what is taken to be its opposite: transparency. And this is, to a certain extent, true. Transparency, however, can mean different things in different contexts, e.g., access to source code, surveyable or interpretable practices and components, explainable processes, etc. What is often assumed is that some level of access to 'how something works' or 'how something produces results' is what is at play. Hence, as a way of undermining the challenge of opacity, some of these strategies aim to undermine the value of such a commitment to transparency. One may argue, for example, that it is simply unsound to think that every single part of a process ought to be maximally accessible (intelligible, surveyable, etc.,) to an agent in order for them to successfully rely on it. If this is the idea of transparency, one may argue, then we go back to thinking about opacity in such general terms that it ultimately trivializes the concept in at least two ways: (a) everything is opaque to everyone at one level or another, and if so, (b) then opacity does not seem to pose any meaningful challenge after all since life and inquiry goes on successfully despite its ubiquity.

Additionally, it can be pointed out that while it is easy to understand *some* level of transparency of these methods as a desirable epistemic good, it is still possible to undermine the normative import of this good by diminishing the normative weight of this transparency: i.e., one could always argue that while it is *good* to have transparent methods, this transparency is by no means an epistemic necessity nor an obligation. It is well known, for instance, that transparency is not *sufficient* for explainability; someone could easily further argue that its *necessity* is equally questionable.[23] Even more devastatingly, it could be argued that the kind of maximal

[22] Although some philosophers walk a fine line between the last two strategies, computational reliabilism is often championed as promising a way forward in line with the latter argumentative strategy: i.e., even in the case of essential epistemic opacity, so this strategy supposes, we can be justified in our reliance on certain computational methods and hence also in our trust on their results. Duede (2022), for example justifies our reliance on deep learning techniques in science by attempting to contextualize its use in broader context of discovery. This is, as is often the case, yet another pragmatic defense for the utility of such an artifact. As I have argued here and elsewhere, if scientific inquiry is intrinsically an epistemic endeavor and not just merely a problem-solving one, this pragmatic approach is simply inadequate and can be reduced to an attempt to justify practices because 'practitioners practice them' and have happened to gain some utility from them, and not as I see it a critical epistemological inquiry.

[23] Full transparency of a complex and large source code, for example, in many instances of interest would not yield anything resembling interpretability or explainability. In other words, one can have access to all the details of a system without necessarily understanding them. Hence, transparency is not sufficient for understanding how something works.

transparency alluded to above could, in fact, get in the way of any actual epistemic good. A barrage of source code, for example, could obscure the functionality of a piece of software to someone investigating what it is supposed to do.

One way to counter this last point is to simply say that not every single aspect of a process of a system *can/must* be relevant. We must discern between those things that ought to be accessible to an agent and those that do not need to be. Consider the following, according to Humphreys, a computational system, process or device is *essentially* epistemically opaque iff

> it is impossible, given the nature of X, for X to know all of the epistemically relevant elements of the process. (2009)

Researchers of opacity have either taken this definition at face value and focused on the strength of the claim, i.e., the impossibility clause, or, as we saw above, on what the term 'all' implies. However, rather than focusing on the 'all' clause in the definition above, one can also focus on what Humphreys may have meant by the 'epistemically relevant' aspects of a process and try to see what those are. In a formal proof—be it in mathematics, geometry or logic— for example, one may find that *every* inferential step towards a result and hence every premise that informs such steps is an epistemically relevant element of the process. Durán and Formanek (2018) rightly point out, however, that in the context of computational methods, it is not immediately obvious which are the epistemically relevant elements of a process: is it every single line of code? Elements related to the overall aims and purpose of a software system? Or do the complete functions of underlying components such as compilers, circuitry, etc., count as epistemically relevant? Regardless of these difficulties, however, advocates of computational reliabilism claim to circumvent them by pointing out that transparency is simply not needed, and that having it may not do any good in the contexts where philosophers conventionally demand it. Although computational reliabilism does not simply assume the reliability of the methods in question, it can still make this latter claim by arguing, as we briefly saw, that there are some reliability indicators to help us ground our confidence and trust.

All of these views are partially correct: yes, the value of transparency is often implied in those that see opacity as a genuine epistemic challenge; yes, the value of this transparency is not immediately obvious, particularly since it seems that full transparency of all the elements of a process is neither sufficient nor necessary for explanatory endeavors; furthermore, even if such transparency is an epistemic good, in many circumstances of interest, the value of these goods may be outweighed when considered against exceptional ethical, prudential and/or practical urgencies. It could even be said, without much controversy, that it is true that prioritizing epistemic concerns in certain circumstances of ethical, prudential or practical relevance may prove frivolous, irresponsible or even reprehensible.

However, as we saw in the previous section, while we do not need a list of all the epistemically relevant elements of the technology being deployed, we definitely *need some very specific ones*, namely the elements related to error. As I have noted elsewhere "While the contents of an exhaustive list of epistemically relevant elements of a system may be debatable, that error and error assessment must be included is less controversial." (2022) For example, while we may be able to forgo knowing exactly how an Aspirin *works* when we deploy it —as suggested by London (2019)— what a medical professional cannot forgo when deciding on its use, or the use of any pharmaceutical technology, is having a good idea of *how and why it could fail* and not just the rate at which it fails.

While knowing that a given medical practice could fail (where failing implies the loss of life or the loss of quality therein) can help an individual calculate the risks they are willing to take on their own life or those of others, or help an institution to calculate the mortuary and legal resources they will need when deploying a given technology, thinking that these considerations suffice in such a context is at best limited and at worst seriously misguided. A medical practitioner deploying these pharmaceuticals should know not just how often they fail, but more or less when they can fail. This is more often than not informed by knowing why they may fail, which is in turned informed in part by the chemical substances *in* the pharmaceutical, their relative ratio to one another and even the order in which they were mixed. Hence, they also need to know why and how it can fail. In other words, they need to have an idea of the sources and the nature of possible errors.

Simply stated, when it comes to technical artifacts in formal epistemic contexts such as rigorous scientific inquiry some things cannot *not* be in a list of relevant epistemic elements of *any* process: these include the nature and source of its possible errors.

Importantly, this is different from requiring a system and its inner workings to be fully transparent, or from requiring a full explanation of all the epistemically relevant steps a system takes to achieve a function. While the reliability indicators associated with computational methods such as machine learning suggested by Durán include "methods (formal or otherwise), metrics, expert competencies, cultures of research, and the like that make up for our best epistemic and normative efforts that might increase the degree of warrant we have to believe the outputs of ML systems," (2023, p.5) and while there is a sense in which the complete epistemically relevant elements of a given process remain elusive, there is an undeniable sense, as noted above, in which in certain settings knowing something about the ways in which something could fail is an indispensable element of *any* list of reliability indicators.

Hence, so much is clear: the heart of the problem of epistemic opacity is not necessarily not knowing the details of how something works. The real problem is access to the nature and source of error in order to run an adequate reliability assessment.

This is something that is acknowledged by — and, contrary to what Formanek argues (see Chap. 2 by Formanek in this volume), justifiably troubles— both philosophers *and* practitioners of machine learning. As Judea Pearl succinctly puts it (and as Beisbart cites in this volume, see Chap. 1):

> Once you unleash it on large data, deep learning has its own dynamics, it does its own repair and its own optimization, and it gives you the right results most of the time. But when it doesn't, you don't have a clue about what went wrong and what should be fixed. In

> particular, you do not know if the fault is in the program, in the method, or because things have changed in the environment.

As can be gathered from this quote, error is at the center of what is worrisome about the opacity inherent to machine learning methods such as deep learning. Hence, error is an epistemically relevant element of any reliability assessment of such technology.

The problem for computational reliabilism is thus twofold. The first issue is that they must, as we saw in the previous section, concede that at least some of the internal details of a process or device qua process or device *must* be known in order to appropriately rely on it. Note how this is in sharp contrast to the view that computational reliabilism can rely only on elements exogenous to the process/device itself to ascribe reliability. If this is so, then the main appealing feature of computational reliabilism—namely that it can ensure an appropriate epistemic framework even without appeals to transparency or without caring for opaque elements of a process—seems to be no longer. The second problem is, that if they do concede to the points above, then they must face the challenge of *essential* epistemic opacity concerning the nature and source of error in most computational systems of interest, and particularly those involving machine learning.

In particular, these views have failed to acknowledge the severity of the challenges posed by the irreversibility of probabilistic, stochastic models whose results are multiply realizable and overdetermined (Symons & Boschetti, 2013). They also neglect to consider the catastrophic/strange nature of error in ML technologies such as deep neural networks (Alvarado, 2022a, 2022b). Both of these are instances in which tracking the source of error and the nature of such error is simply unachievable. In the first case, the fact that a computational system may take several distinct paths to reach its results even in the case of identical input, makes it so that determining which path was actually taken is not possible (Symons & Boschetti, 2013). If this is the case and there is an error, tracking where the error happened is simply not achievable. This is a particularly severe challenge in the case of computational technologies such as deep neural networks and transformer models which require a significant number of layers of analysis.

In the second case, the kinds of error that these computational methods are prone to, are simply too extreme and unpredictable. This is exemplified by issues such as hallucinations in generative AI (Ji et al., 2023; Lee, 2023), but also in the case of adversarial attacks to neural networks in which a random positioning of a single pixel could alter the analysis of an image. Importantly, in recent research similar results to these adversarial attacks have been found to emerge even in non-modified images that involve complex textures such as those of natural foliage or intricate fence designs. Elsewhere (Alvarado, 2022a, 2022b) I note the following:

> There are at least thousands of naturally occurring examples of images that contain information that works just as an artificially induced adversarial attack would and that can mislead well-known image classifiers. Furthermore, it has been shown that even well-established defense strategies that ensure some widespread image classifiers remain resilient versus some artificial adversarial attacks are nevertheless unable to defend against naturally occurring examples.

As further noted there, these naturally occurring adversarial examples can include simple elements such as "inclement weather conditions and obscured objects, and it can also include objects that are anomalous." (Hendrycks et al., 2019) This means that the neural network does not need to be intentionally targeted to produce unexpected error. Rather, features intrinsic to natural images can make neural networks yield unpredictable errors. Although this signals that we can at least recognize certain patterns that elicit this kind of error, the fact that non-manipulated images of random scenes and objects can cause this kind of error means that there can be an infinite number of patterns out there that can do the same in a myriad of fields: telescope imagery, x-rays, etc. Non-surprisingly, this kind of issue can even be found in cases of text-based analysis (Zhang et al., 2020). If this is so, then any kind of machine learning data analysis may be prone to such a problem. Interestingly, these errors are "different "bugs" from traditional software" (Sun et al., 2018) Moreover, this is not the only way in which errors of this type in these types of systems are opaque. For example, due to their highly distributed computing, neural networks also fail in what is called a 'graceful' manner with minute deviations from weight to weight and layer to layer (Alvarado, 2022a, 2022b). This makes it so that such errors carry on distributed in the very many weights of deep neural networks and the automated optimization changes that these systems undergo in their many layers of analysis. And lastly, the arrangement of these networks structures is such that this error is not easily detected either internally by the network or externally by those supervising it. Thus, these errors are delivered with the same degree of confidence as truthful results.

This last point illustrates, once more, the essential epistemic opacity of computational methods and the severity of the challenge it represents for epistemic frameworks seeking to justify our reliance in such methods, particularly when they are used in context where reliability is crucial. The section before this one showed the indispensability of endogenous features of processes and devices to the reliability assessments of artifacts. This discussion also showed the further indispensability to reliability assessments of the nature and source of error, particularly in knowledge-critical and safety critical contexts. In this section, we saw that it is precisely these two elements, the source and nature of error that matters most when we talk about epistemic opacity. The last point in the paragraph above illustrates that such a challenge in the context of computational methods such as deep neural networks may very well be, as Humphreys (2009) once suggested of other computational methods, *essentially* insurmountable and thus, a genuine challenge to computational reliabilism. Importantly, the essentiality of such an opacity need not arise from internal features of an epistemic agent such as its nature, as Humphreys suggested. Rather, it can be traced to endogenous features of the system in question and as such be agent-neutral and agent-independent. That is, they arise in virtue of features that have little or nothing to do with an agent's epistemic limitations (hence, agent-independent) and it is the kind of opacity that would apply to any given agent with sufficiently similar epistemic limitations (agent-neutral) (Alvarado, 2021).

5.4 Conclusion

In this chapter I have presented three distinct yet interrelated challenges to computational reliabilism and its viability as an epistemic framework that can provide justificatory grounds for our reliance in novel and opaque computational technologies, particularly those related to novel AI methodology such as machine learning. In particular, I argued that the challenges related to warrant transmission, the indispensability of endogenous features such as the source and nature of error in reliability assessments, and the ultimate opacity of the latter, as exemplified by the irreversibility of such systems and the intractable nature of errors related to adversarial disruptions, represent a serious problem for the viability of computational reliabilism. With regards to the second of these challenges, I also showed that it is the same virtues for which the epistemic framework is lauded that make it incapable of accounting for such endogenous features as the source and nature of error in an artifact. With regards to the third challenge, I also showed that, contrary to conventional interpretations in the literature, what is at play in the challenge of epistemic opacity is not just any kind of general transparency regarding how something works, but rather the access—or lack thereof— to the nature and source of error. In doing so, I hope to have also shown that these challenges are not only serious but maybe even insurmountable for such an epistemic framework as computational reliablism as currently articulated in the literature. Nevertheless, I do not aim for this analysis to constitute a final devastating critique of computational reliabilism. My hope is rather that, along with other recent efforts (Smart et al., 2020; Ferrario, 2023), these observations serve as a starting point to rescue it from its present inadequacy.

References

Alvarado, R. (2020). Opacity Artificial Intelligence, machine learning, big data and democratic processes. In K. Macnish & J. Galliott (Eds.), *Chapter in Big data and democracy*. Edinburgh University Press.

Alvarado, R. (2021). Explaining epistemic opacity. (Preprint). http://philsci-archive.pitt.edu/id/eprint/19384

Alvarado, R. (2022a). What kind of trust does AI deserve, if any? *AI and Ethics, 3*(4), 1–15.

Alvarado, R. (2022b). Should we replace radiologists with deep learning? Pigeons, error and trust in medical AI. *Bioethics, 36*(2), 121–133.

Alvarado, R. (2023a). *Simulating science: Computer simulations as Scientific instruments* (Vol. 479). Springer Nature.

Alvarado, R. (2023b). AI as an epistemic technology. *Science and Engineering Ethics, 29*(5), 32.

Alvarado, R., & Humphreys, P. (2017). Big data, thick mediation, and representational opacity. *New Literary History, 48*(4), 729–749.

Al-Halawani, R., Charlton, P. H., Qassem, M., & Kyriacou, P. A. (2023). A review of the effect of skin pigmentation on pulse oximeter accuracy. *Physiological Measurement, 44*(5), 05TR01.

Barberousse, A., & Vorms, M. (2014). About the warrants of computer-based empirical knowledge. *Synthese, 191*(15), 3595–3620

Beisbart, C. (2017). Advancing knowledge through computer simulations? A socratic exercise. In *The science and art of simulation I: Exploring-understanding-knowing* (pp. 153–174). Springer International Publishing.
Beisbart, C. (in this volume). In which ways is machine learning opaque? In J. M. Durán & G. Pozzi (Eds.), *Philosophy of science for machine learning: Core issues and new perspectives*. Synthese Library, Springer.
Buechner, J., Simon, J., & Tavani, H. T. (2013). Re-thinking trust and trustworthiness in digital environments. In *11th Computer Ethics: Philosophical Enquiry (CEPE 2013)* (pp. 1–15).
Burge, T. (1993). Content preservation. *The Philosophical Review, 102*(4), 457–488.
Burge, T. (1998). Computer proof, apriori knowledge, and other minds: The sixth philosophical perspectives lecture. *Philosophical Perspectives, 12*, 1–37.
Burrell, J. (2016). How the machine 'thinks': Understanding opacity in machine learning algorithms. *Big Data & Society, 3*(1), 2053951715622512.
Comesaña, J. (2010). Evidentialist reliabilism. *Noûs, 44*(4), 571–600. http://www.jstor.org/stable/40959693
Dretske, F. (2000). Entitlement: Epistemic rights without epistemic duties? *Philosophy and Phenomenological Research, 60*(3), 591–606.
Duede, E. (2022). Deep learning opacity in scientific discovery. *Philosophy of Science, 90*(5), 1–13.
Durán, J. M. (2020). What is a simulation model? *Minds and Machines, 30*(3), 301–323.
Durán, J. M. (2022). Models, explanation, representation, and the philosophy of computer simulations. In *Philosophy of computing: Themes from IACAP 2019* (pp. 221–249). Springer International Publishing.
Durán, J. M. (2023). Machine learning, justification, and computational reliabilism. Preprint. https://philpapers.org/archive/DURMLJ.pdf
Duran, J. M. (2025). Beyond transparency: Computational reliabilism as an externalist epistemology of algorithms. *arXiv preprint arXiv:2502.20402*.
Durán, J. M., & Formanek, N. (2018). Grounds for trust: Essential epistemic opacity and computational reliabilism. *Minds and Machines, 28*(4), 645–666.
Durán, J. M., & Jongsma, K. R. (2021). Who is afraid of black box algorithms? On the epistemological and ethical basis of trust in medical AI. *Journal of Medical Ethics, 47*(5), 329–335.
Edwards, B. (2023). AI outperforms conventional weather forecasting for the first time: Google study. https://arstechnica.com/science/2023/11/ai-outperforms-conventional-weather-forecasting-for-the-first-time-google-study/
Ferrario, A. (2023). Justifying our credences in the trustworthiness of AI systems: A reliabilistic approach. *Available at SSRN 4524678*.
Formanek, N. (in this volume). How I stopped worrying and learned to love opacity. In J. M. Durán & G. Pozzi (Eds.), *Philosophy of science for machine learning: Core issues and new perspectives*. Synthese Library, Springer.
Goldman, A. I. (2011). Toward a synthesis of reliabilism and evidentialism?: Or: evidentialism's troubles, reliabilism's rescue package. In *Evidentialism and its discontents* (pp. 254–280).
Goldman, A., & Beddor, B. (2021). Reliabilist epistemology. In E. N. Zalta (Ed.), *The Stanford encyclopedia of philosophy (Summer 2021 edition)*. https://plato.stanford.edu/archives/sum2021/entries/reliabilism/
Graham, P. J. (2012). Epistemic entitlement. *Noûs, 46*(3), 449–482.
Hendrycks, D., Zhao, K., Basart, S., Steinhardt, J., & Song, D. (2019). Natural adversarial examples. *arXiv preprint arXiv:1907.07174*.
Humphreys, P. (2004). *Extending ourselves: Computational science, empiricism, and scientific method*. Oxford University Press.
Horner, J., & Symons, J. (2014). Reply to Angius and Primiero on software intensive science. *Philosophy & Technology, 27*(3), 491–494.

Humphreys, P. (2009). The philosophical novelty of computer simulation methods. *Synthese, 169*(3), 615–626.

Ji, Z., Lee, N., Frieske, R., Yu, T., Su, D., Xu, Y., et al. (2023). Survey of hallucination in natural language generation. *ACM Computing Surveys, 55*(12), 1–38.

Harvard, S., & Winsberg, E. (2021). Causal inference, moral intuition, and modeling in a pandemic. *Philosophy of Medicine, 2*(2), 1–10.

Hubig, C., & Kaminski, A. (2017). Outlines of a pragmatic theory of truth and error in computer simulation. In *The science and art of simulation I: Exploring-understanding-knowing* (pp. 121–136). Cham: Springer International Publishing.

Kroes, P., & Meijers, A. (2002). The Dual Nature of Technical Artifacts-presentation of a new research programme.

Lam, R., Sanchez-Gonzalez, A., Willson, M., Wirnsberger, P., Fortunato, M., Pritzel, A., et al. (2022). GraphCast: Learning skillful medium-range global weather forecasting. *arXiv preprint arXiv:2212.12794*.

Lee, M. (2023). A mathematical investigation of hallucination and creativity in GPT models. *Mathematics (2227–7390), 11*(10), 2320.

London, A. J. (2019). Artificial intelligence and black-box medical decisions: accuracy versus explainability. *Hastings Center Report, 49*(1), 15–21.

McGlynn, A. (2014). On epistemic alchemy. In *Scepticism and perceptual justification* (pp. 173–189). Oxford University Press.

Oreskes, N. (2021). *Why trust science?* (Vol. 54). Princeton University Press.

Saam, N. J. (2017). Understanding social science simulations: Distinguishing two categories of simulations. In *The Science and Art of Simulation I: Exploring-Understanding-Knowing* (pp. 67–84). Cham: Springer International Publishing.

San Pedro, I. (2024). Degrees of epistemic opacity. *Teorema: Revista Internacional De Filosofía, 43*(2), 5–21.

Simon, J. (Ed.). (2020). *The Routledge handbook of trust and philosophy*. Routledge.

Smart, A., James, L., Hutchinson, B., Wu, S., & Vallor, S. (2020, February). Why reliabilism is not enough: Epistemic and moral justification in machine learning. In *Proceedings of the AAAI/ACM Conference on AI, ethics, and society* (pp. 372–377).

Sun, Y., Huang, X., Kroening, D., Sharp, J., Hill, M., & Ashmore, R. (2018). Testing deep neural networks. *arXiv preprint arXiv:1803.04792*.

Symons, J. (2010). The individuality of artifacts and organisms. *History and Philosophy of the Life Sciences, 32*, 233–246.

Symons, J., & Alvarado, R. (2019). Epistemic entitlements and the practice of computer simulation. *Minds and Machines, 29*(1), 37–60.

Symons, J., & Boschetti, F. (2013). How computational models predict the behavior of complex systems. *Foundations of Science, 18*, 809–821.

Symons, J., & Horner, J. (2014). Software intensive science. *Philosophy & Technology, 27*(3), 461–477.

Van Helden, A. (1994). Telescopes and authority from Galileo to Cassini. *Osiris, 9*, 8–29.

Winsberg, E. (2010). *Science in the age of computer simulation*. University of Chicago Press.

Winsberg, E., & Alvarado, R. (2025). Computer simulations in science. In E. N. Zalta & U. Nodelman (Eds.), *The Stanford encyclopedia of philosophy (Forthcoming, Fall 2025 Edition)*. https://plato.stanford.edu/archives/win2022/entries/simulations-science/

Winsberg, E., & Harvard, S. (2022). Purposes and duties in scientific modelling. *Journal of Epidemiology and Community Health, 76*(5), 512–517.

Zhang, W. E., Sheng, Q. Z., Alhazmi, A., & Li, C. (2020). Adversarial attacks on deep-learning models in natural language processing: A survey. *ACM Transactions on Intelligent Systems and Technology (TIST), 11*(3), 1–41.

Ramón Alvarado is a philosopher of science and computational methods at the University of Oregon's philosophy department. His work focuses on the epistemology of computer simulations and machine learning in science as well as on the ethics of artificial intelligence technologies as used in science and society. He is the Editor-In-Chief of the journal *Global Philosophy* for Springer Nature as well as the President of the International Association of Computing and Philosophy (IACAP) and of the Association for Philosophy and Computing (APC).

Chapter 6
Can XAI Justify?

Carlos Zednik and Philippe Verreault-Julien

Abstract AI systems are increasingly being used to drive high-stakes decision-making in domains such as medical diagnosis, credit approval, and human resources. Ensuring that AI-driven decisions are properly justified is crucial. Justification implies that decisions align with relevant norms. However, the *opacity* of many AI systems hinders this kind of justification since it makes it difficult to determine which features influence these systems' outputs and whether they have learned to track spurious correlations rather than relevant patterns. Although explainable AI (XAI) offers potential solutions by uncovering the *reasons* for AI systems' outputs, its role in justification remains underexplored. This chapter adopts a framework from philosophy of action to differentiate between the different types of reasons that may be given for an AI system's outputs, assess how these types of reasons contribute to justification, and evaluate the capacity of current XAI methods to identify and describe such reasons. Overall, while the chapter shows how XAI methods could be used to identify reasons with which to justify AI-driven decisions, it also highlights some of these methods' current limitations.

Keywords Xai · Justification · Reasons · Motivation · Explanation · Opacity

6.1 Introduction

AI systems facilitate decision-making in domains as varied as medical diagnosis, credit approval, and human resources. In many of these domains, the stakes are high: the decisions being made can profoundly impact human well-being. For

Both authors have contributed equally.

C. Zednik (✉) · P. Verreault-Julien
Eindhoven University of Technology, Eindhoven, Netherlands
e-mail: c.a.zednik@tue.nl; philippe@pvjulien.net

J. M. Durán, G. Pozzi (eds.), *Philosophy of Science for Machine Learning*, Synthese Library 527, https://doi.org/10.1007/978-3-032-03083-2_6

this reason, it is important that AI-driven decisions[1] are *justified*. Roughly, for a decision to be justified means that it demonstrably accords with relevant norms (Malgieri & Pasquale, 2024). In medicine, for example, diagnoses should conform with received medical knowledge; in banking, creditworthiness should be assessed based on financially relevant attributes such as income as opposed to irrelevant ones such as hair color; in human resources, hiring decisions should be fair rather than discriminatory and be motivated by business reasons rather than by nepotism.[2]

Justification is important in these domains because it allows for decisions to be adequately contextualized, evaluated, and if necessary, challenged (e.g. Baum et al., 2022; Durán & Formanek, 2018; Ferrario & Loi, 2022; Symons & Alvarado, 2022; Vredenburgh, 2022; Widdau & Zednik, 2022). If skin cancer is diagnosed on the basis of skin discolorations that are known to be medically irrelevant, then this may be a reason to mistrust the accuracy of the diagnosis. If creditworthiness is assessed on the basis of relevant features such as income, this can be used by a credit applicant to effect different assessments in the future (e.g., by securing a higher-paying job before re-applying). If a hiring decision is known to violate principles of non-discrimination, this may be grounds to challenge the decision and possibly even seek legal retribution. For all of these reasons and more, decision-makers in high-stakes domains should be able to provide information to demonstrate that AI-driven decisions are justified.

Although justification is important, the fact that many state-of-the-art AI systems are *opaque* threatens the ability to justify the decisions they inform (e.g. Chap. 1 by Beisbart and Chap. 2 by Formanek, in this volume; Durán & Formanek, 2018; Humphreys, 2009; Krishnan, 2020; Maclure, 2021). Because of their opacity, it often remains unknown which features these systems actually consider when generating a particular output, whether they have learned to track features that can be considered problematic, or how to best effect a different outcome in the future. In addition to causing other harms such as reducing end-user trust and posing an obstacle for effective debugging, opacity therefore makes AI-driven decisions hard to justify.

Insofar as opacity is a problem, *explainable AI* (XAI) might be considered a solution (Doshi-Velez & Kim, 2017; Zednik, 2021; cf. Durán, 2021). Indeed, a variety of sophisticated analytic methods can be used to uncover the features being considered by a particular AI system, the representations being learned, and the inputs that determine particular outputs, among other information. Nevertheless, questions arise about whether the insights garnered from these methods suffice for

[1] The term 'AI-driven decision' is here used as a catch-all term for any kind of action that is significantly influenced or determined by an AI system's outputs. Paradigmatically, this may be a decision to, e.g., grant a loan or hire a candidate, but it may also include an action-guiding classification or prediction. Notably, although in many real-world applications such decisions are made through complex and interactive processes that involve both humans and machines, for simplicity of argument here these complexities will be ignored.

[2] See also Chap. 10 by Sullivan and Kasirzadeh in this volume for a related discussion of 'core' reasons.

the purposes of justifying. Can explainable AI contribute to the justification of AI-driven decisions? If so, how? Moreover, what, if any, are the limitations?

In what follows, we adopt a framework from the philosophy of action to distinguish between the different kinds of *reasons* that one might give for an AI system's behavior. Then, we analyze the extent to which these kinds of reasons figure in justification, and evaluate the extent to which current methods from explainable AI can be thought to identify and describe such reasons. In general, although we propose that XAI methods can sometimes provide relevant insights, they are also limited in important ways. Most notably, in addition to technical limitations such as a lack of robustness and interpretability, explainable AI cannot itself provide insight into the kinds of normative reasons that are ultimately needed for justification. Therefore, although explainability is often helpful, it cannot be considered generally sufficient for justification.

6.2 Reason & Justification

6.2.1 Justifying Human Actions

When people act, we often ask for the reasons that favor their actions. We want to gain insight into why people act, what motivates their actions, and whether they have good reasons for doing as they do. Consider an example discussed by Alvarez (2017): Othello's murder of his wife, Desdemona. Iago, Othello's ensign, manipulates circumstances and people to make Othello believe that his wife has been unfaithful. This leads Othello to a state of intense jealousy and anger. Othello confronts Desdemona in their bedchamber, refusing to believe her protests of innocence. In a state of emotional turmoil, he smothers her, only to learn afterwards that Desdemona has indeed been faithful all along, and that Iago had deceived him.

What are the reasons for Othello's murderous action? According to Alvarez, what motivates Othello–the *motivating reason*–is the putative fact that Desdemona was unfaithful. Motivating reasons are reasons that, for the agent, favor actions, and for which the agent acts.[3]

Notably, in this case, the putative fact that motivates Othello's action cannot also explain the action. Since Desdemona has not actually been unfaithful, it is merely a putative fact, rather than an actual fact. As a result, what explains Othello's murderous action–the *explanatory reason*–must instead be a psychological fact:

[3] In the following, we broadly follow Alvarez's (2017) externalist treatment of reasons for action, but want to remain agnostic with respect to the ontology of reasons, and in particular, whether motivating reasons are only facts about mental states (psychologism) or whether they can also include external objects, states, and events (non-psychologism).

Othello's having a belief that Desdemona has been unfaithful. This belief is what makes a difference to whether or not Othello acts in the way he acts.[4]

Two noteworthy clarifications about motivating and explanatory reasons. First, there is often a close representational connection between the psychological fact that explains an action and the (putative) fact that motivates it. In the current example, the belief that explains Othello's murder represents the (putative) fact that Desdemona has been unfaithful. Second, not all explanatory reasons are psychological facts. Although Othello's false belief is the most immediate or proximate explanation of his action, Iago's deceptive scheme is also explanatorily relevant. Indeed, the latter explains the former, which in turn explains the action. The etiology—viz. the causal history—of beliefs and other difference-making psychological states may thus provide explanatory reasons that supplement the relevant psychological facts.

Of course, we are not only interested in understanding why Othello kills Desdemona, but also in determining whether he is justified in doing so. *Normative reasons* favor an action, but from a perspective external to the agent. Although there is widespread agreement that normative reasons are facts (e.g. Raz, 2009), for a reason to be considered normative it need not appeal to fundamental moral principles. Rather, it only needs to appeal to relevant rules or standards—more generally, norms—that can be shared with others. To say that a reason normatively counts in favor of an action implies that others could not reasonably reject it (Scanlon, 1998). Since in most social contexts unfaithfulness does not warrant murder, there is no normative reason for the killing. Thus, Othello's action is unjustified.

6.2.2 Justifying AI-Driven Decisions

The tripartite distinction between explanatory, motivating, and normative reasons can, we contend, also be fruitfully applied to AI systems.[5] More generally, the language of reasons can be used to make sense of these systems' outputs and the decisions to which they contribute, with the goal of determining whether these decisions are justified. Indeed, in many contexts it has already become commonplace to request reasons to explain or justify certain AI-driven decisions. Consider, for example, the American Consumer Financial Protection Bureau (CFPB) (2022), which recently published a circular clarifying compliance requirements with the Equal Credit Opportunity Act:

[4] Although we here appeal to a notion of difference-making, we remain agnostic with respect to whether difference-making should be understood in strictly causal terms. There are good reasons to believe that explanatory reasons may also provide non-causal explanations (see e.g. Reutlinger & Saatsi, 2018), but our argument doesn't hinge on the existence of such explanations.

[5] See Baum et al. (2022) and Cappelen and Dever (2021, 162ff.) for a similar suggestion. Notably, it is not necessary to think of AI systems as genuine agents in a philosophically "thick" sense in which they are equated with persons, nor to think of these systems' behaviors as genuine "actions". Nevertheless, these terms will be used here for matters of convenience.

> Whether a creditor is using a sophisticated machine learning algorithm or more conventional methods to evaluate an application, the legal requirement is the same: Creditors must be able to provide applicants against whom adverse action is taken with an accurate statement of reasons.

Similarly, the German Institute for Standardisation (DIN, 2023) recently published a national standard which specifies some of the requirements for an AI system to be considered trustworthy:

> [...] when the stakes are high (as in the case of an AI system used for medical diagnosis), user trust might additionally require knowledge of the reasons for a particular output, prediction, or decision, because this kind of knowledge allows a user to question, challenge, or influence the system's behavior.

Although these and many other documents (e.g., HLEGAI, 2019) make frequent appeals to reasons, the term remains ambiguous. In particular, these documents don't clearly describe the nature of the reasons to be provided, leaving unclear whether these reasons should be motivating, explanatory, and/or normative.

To illustrate the different kinds of reasons that might be given for AI-driven decisions, consider the kinds of financial risk-assessment systems that are targeted by the CFPB and that are used to predict the likelihood that a particular individual will default on a loan. Suppose that Jimmy applies for a loan to buy a car. A financial risk-assessment system predicts his risk of default on the basis of input data such as Jimmy's income, gender, employment status, education history, marital status, and number of dependents, among others. Given this data, the AI system might output the following: "Applicant has a 70% chance of defaulting". On the basis of this prediction, Jimmy's bank denies the application. Having read the CFPB's circular, Jimmy then asks the bank for a statement of reasons. But what kinds of reasons might actually be given?

The bank could provide explanatory reasons. Just as Othello's belief explains the murder, it is the AI system's computational states and processes—the values of internal variables and parameters, mathematically interlinked to systematically transform inputs into outputs—that explain the system's assessment. Indeed, the computational states and processes in such a system can be fruitfully viewed as analogues of the psychological states and processes of human agents.[6] Of course, as in the human case, here too it is possible to supplement the "psychological" facts of the system with etiological facts about the training data, learning algorithm, architecture, and objective function that help explain the values of the system's internal variables and parameters. Ultimately, however, it is the variables and

[6] To be clear, the analogy is epistemic rather than ontological. It need not be assumed that an AI system's computational states and processes are similar physically or even functionally to the psychological states and processes of human agents, but only that they play a similar role, as theoretical posits, in an observer's descriptions, predictions, explanations, and/or justifications of that systems' outputs. This is in line with a widespread but nevertheless contentious habit, within the AI community, of using mentalistic terms to describe AI systems' internal processes, as well as with several classic and contemporary philosophical treatments thereof (e.g. Papagni & Koeszegi, 2021; Zerilli et al., 2019; Zednik, 2021).

parameters themselves–together with the system's inputs–that explain the system's outputs for any particular input.

Although explanatory reasons may help Jimmy better understand the system's behavior, these are not the reasons that regulators typically require, nor the reasons that Jimmy actually wants. Instead, Jimmy wants to know the reasons that, for the system, counted in favor of the assessment. Was it Jimmy's relatively low income that led to the poor evaluation? Was it his high level of student debt? Perhaps, more problematically, was it Jimmy's gender or immigrant background? Even though it is the system's computational states and processes that bring about the system's evaluation, these states and processes are not what motivate the system to make the evaluation. Instead, what motivates the system is the particular set of features or patterns that the system has learned to detect, track, and represent. In particular, these may include the combined feature of having a relatively low income and high student debt, as well as the learned correlation between this combined feature and a high likelihood of default. In order to answer Jimmy's request for reasons, the bank should not describe the computational states or processes that explain the assessment, but rather the features or patterns that motivate it.

Knowledge of motivating reasons is important because it is these reasons that allow Jimmy to ask whether or not the decision to reject his application is justified. More precisely, Jimmy would like to know whether the motivating reasons are appropriate in the sense of according with relevant norms. Do the relevant features or patterns constitute *good* reasons to predict that he has a 70% chance of default? In order to answer this question, Jimmy must evaluate the system's motivating reasons in light of relevant laws, norms, or values. Since Jimmy has an immigrant background, he is concerned that the AI system might have invoked this particular feature–instead of, or in addition to, other features–when producing its output. However, by knowing the system's motivating reasons–i.e. his low income and high student debt–Jimmy is able to determine that the AI-driven decision conforms with the norm of not basing financial decisions on protected attributes such as race, sex, or geographic origin. Put differently, the decision accords with the norm of non-discrimination. Because of this accordance, Jimmy is likely to consider the decision justified.

6.2.3 The Interplay of Reasons

Before proceeding, it is worth reflecting on the subtle differences and interdependencies between motivating, explanatory, and normative reasons. In particular, three aspects are worth highlighting. First, while some reasons are internal to a system, others are external. Explanatory reasons that relate to a system's computational states and processes (or analogously, an agent's psychological states and processes) are internal to the system; they are the *system's* states and processes. However, other explanatory reasons may be external. In particular, reasons that cite the etiology of a system's states and processes are likely to be at least partially external, because they are likely to refer to the environment in which the system has been trained or developed. Motivating reasons are also internal in the sense of being (putative) facts that, *for the system*, count in favor of the action. Nevertheless, these facts will typically be about environments outside of the system which the system has learned to represent and respond to–the training environment, or the environment from which inputs are acquired. Finally, normative reasons are always wholly external to the system (Henin & Le Métayer, 2022) because they pertain to moral, legal, epistemic, or social rules or standards of evaluation in which the system's behavior unfolds.

Secondly, while in most typical cases justification is likely to require consideration of explanatory, motivating and normative reasons, in some cases only normative reasons matter. Indeed, normative reasons are always essential–these are the kinds of reasons that provide the evaluative component that is necessary for making normative assessments. Normative reasons show why some action is right or wrong, e.g. from a legal or ethical perspective. Nevertheless, normative reasons are not always sufficient for justification. In Kantian or (versions of) virtue ethics, for example, motivations determine whether or not an action is done for the right reasons or reflect a person's character and virtues. Helping an elderly person cross the street in the expectation of a reward would thwart its moral value. In contrast, other moral frameworks like utilitarianism downplay the relevance of motivations for normative assessment. Likewise, in the law, motivations tend to be less important in civil law than in criminal law and can be unnecessary to, e.g., establish guilt.

Like some areas of ethics and law, in the case of AI systems it might be argued that the overall reliability of a system is relevant for justification, and that therefore, explanatory or motivating reasons may not be needed (see e.g. Durán & Formanek, 2018; see Chap. 4 by Durán, in this volume). However, here it is important to distinguish epistemic justification from other kinds of justification. Even if we accept that reliability may be sufficient for epistemic justification (which is controversial), it is unlikely to be sufficient for the purpose of moral justification (Durán & Jongsma, 2021; Smart et al., 2020). To bridge that gap between epistemic and moral justification, some (e.g. Maclure, 2021; Smart et al., 2020) emphasize the importance of integrating human oversight around these systems' outputs. In addition to such oversight, however, regulatory frameworks such as the ones identified above place strong requirements concerning the relevance of AI systems' reasons, in addition to their overt behaviors. In practice, knowing reasons plays an important role in justification.

All that being said, how exactly do explanatory or motivating reasons contribute to justification? As suggested previously, justification of AI-driven decisions involves showing that these decisions conform to relevant norms (Henin & Le Métayer, 2022; Malgieri & Pasquale, 2024). As indicated above, in some cases only showing that the system's behavior conforms to a norm might be sufficient (e.g., by ensuring that the outputs are "fair" in some statistical sense). In some other cases, however, we might not only be concerned with the outcomes themselves, but must also confirm that these outcomes are reached for the right motivating reasons. To illustrate the importance of motivation, consider again the case of Othello. Had

Othello's murderous action been motivated by the putative fact that Desdemona posed an immediate threat to his own life and that killing her was his only viable course of action, then there might have been a normative reason for the action: self-defense. In that context, to assess whether a given reason is a *good* reason we first need to know what the motivation is. Likewise for AI systems, it may not always be sufficient to justify an AI system's behavior by appealing to its overall reliability. For instance, a system trained to visually classify images of huskies may be very reliable in a statistical sense. Nevertheless, probing that classifier's motivating reasons may show that, in fact, it classifies huskies on the basis of a whiteness in the image background (Lapuschkin et al., 2019; Ribeiro et al., 2016). Regardless of its statistical reliability, there is a clear sense in which the classifier's outputs are not justified because its decisions are reached for the wrong motivating reasons. Although justification itself ultimately depends on the presence or absence of normative reasons, demonstrating that a particular decision is justified may also require knowledge of motivating reasons. Explanatory reasons, especially external ones, can also help assess whether an AI-driven decision is justified. For example, imagine that in the loan-application case we discover that one reason that explains the decision is the presence of bad labeling or inappropriate sampling in the training data. As a result, the system might have learned a spurious correlation between applicant-features and likelihood of default. Again, the presence of that explanatory reason undermines the justification; the learning process was not reliable.

Thirdly, although normative reasons are often taken to fall under the realm of the moral or the legal, there are other sources of normativity that are potentially relevant for the justification of AI-driven decisions. Since knowledge is often viewed as a constraint on practical reasoning (e.g. Hawthorne & Stanley, 2008; Mueller, 2021; see also Buijsman & Veluwenkamp, 2022), a lack of epistemic reasons qua normative reasons (Sylvan, 2016) may imply that an action is unjustified. Epistemic norms may specify the explanatory or motivating reasons that would be relevant to know for the purpose of justification. In particular, AI behavior based on false motivating reasons may be unjustified. That financial risk assessment should not rely on protected classes is a legal (and probably ethical) norm. However, as the example of the husky-classifier shows, in the context of AI-driven decision-making, epistemic norms can also be relevant. In particular, for a system's motivating reason to be considered good, that reason will typically need to be (approximately) true. To illustrate, consider the motivating reason for Jimmy's loan rejection, viz. that applicants with low income and high student debt are likely to default. One normative reason that counts in favor of the system's decision is that it is reached without appealing to protected attributes. However, this motivating reason is only appropriate if it is also true. Suppose that having a low income and a high student debt doesn't actually make people more likely to default. Indeed, student debt might actually be an indicator of high education and future career prospects. On this supposition, would the decision to reject Jimmy's loan application still be justified? Insofar as there are no legal requirements of accuracy, we may want to say that the prediction is justified in a *legal* sense. However, the prediction is clearly not justified in an *epistemic* sense, because the decision is grounded in a false belief. Epistemic norms also play a role in helping us assess whether the output or its use in a decision was justified.

6.3 Opacity as a Threat to Justification

The previous section was written as if the reasons for an AI-driven decision are always transparent to an external observer. Given knowledge of relevant motivating, explanatory, and normative reasons, it should be possible to determine whether or not a particular AI-driven decision is justified. That said, it is now widely accepted that many state-of-the-art AI systems are *opaque* (e.g. Chap. 1 by Beisbart and Chap. 2 by Formanek, in this volume; Burrell, 2016; Creel, 2020; Lipton, 2018). Opacity negatively affects many different kinds of stakeholders, depending on their unique tasks or purposes with respect to a particular system, as well as on their prior knowledge and expertise (Zednik, 2021). For example, expert developers tasked with debugging an AI system can be negatively affected insofar as opacity prevents them from knowing the nature of a bug or how to fix it. Moreover, non-expert end-users who receive an opaque AI system's outputs will be unlikely to trust these outputs if they are not given clear and comprehensible information about how these outputs were generated. More relevant to present concerns, opacity also affects stakeholders who are tasked with justifying AI-driven decisions, because it threatens their ability to know relevant motivating and explanatory reasons.

6.3.1 Opacity and Motivation

Consider again the kinds of financial risk-assessment systems introduced above. State-of-the-art systems of this kind are typically based on classifiers that have been trained using sophisticated machine learning techniques. These techniques are applied to large repositories of previous applicants' demographic, financial, and behavioral data to identify statistical correlations between, e.g., age, income, and the likelihood of default (World Bank, 2019). The power and utility of these techniques derives from the fact that they regularly uncover predictive features and statistical patterns that were previously unknown even to domain experts. Notably, in addition to being unknown, these features and patterns are often also highly unintuitive and intricate. For the system that is eventually used to process Jimmy's application, for example, although no single demographic or financial attribute in the training data might indicate (from the perspective of a domain expert) a high likelihood of default for applicants with Jimmy's profile, it is nevertheless possible that a very subtle and previously unrecognized combination of attributes is statistically correlated therewith. Moreover, it may be that seemingly irrelevant features such as the number of letters in an applicant's first name, or the color of an applicant's hair, might reliably predict the likelihood of default. However unintuitive or intricate these features and patterns may be, Jimmy's bank is likely to use machine learning to uncover and use them for financial gain.

The fact that machine learning facilitates decisions that ground on such unintuitive and intricate features and patterns has led to gains in efficiency and accuracy in many different domains even beyond finance. At the same time, it is also the source of great risk. Because the relevant features and patterns may be so unintuitive and intricate, they might not be easily described and understood by human observers, regardless of their level of technical expertise or domain knowledge. More precisely, it may be difficult for human observers to *interpret* these patterns using concepts from natural language or from an appropriate technical domain such as finance or medicine. From the perspective of these domains, it can be exceedingly difficult to make sense of what the relevant AI systems are actually doing, why they are doing it, and how they work.

Lack of interpretability undermines the ability to justify AI-driven decisions because it threatens our ability to understand the motivating reasons that favor these decisions. To better understand the link between interpretability and motivating reasons, it is worth delving a bit deeper into what it is that makes an AI system interpretable. Roughly, interpretation involves associating elements of the system with features or patterns in the environment or problem domain that are recognized or understood by a human observer. For example, activity in a particular unit of an artificial neural network might be associated with the presence (or absence) of a particular object or attribute in the input (e.g., low income, or high student debt). In many respects, this kind of interpretation is similar to attributing representational contents to the system's variables, parameters, or architectural elements (Zednik, 2021); a neural unit that is associated with low income represents that particular feature.[7] That said, this kind of attribution of representational content—and thus, interpretation—is difficult if the relevant content is unintuitive and intricate, because a human observer will have difficulty identifying and describing that content in the first place. As the degree of unintuitiveness and intricacy increases, the degree of interpretability decreases.

Recall now that a motivating reason—a reason that favors an action for an agent or a system—can be viewed as the representational content of a particular internal (e.g. psychological or computational) state or process. On this view, interpretation is tantamount to identifying a system's motivating reasons. Accordingly, an *in*ability to interpret is tantamount to an inability to identify the motivating reasons that favor a particular course of action, for a system. The fact that a system may be making decisions about Jimmy's creditworthiness on the basis of a highly unintuitive and intricate feature or pattern that is difficult for a human observer to comprehend suggests that this observer will have great difficulty in knowing the reasons that

[7] 'Representation' is here used in the relatively loose sense of carrying information. For present purposes, no more specific commitments are made (or need be made) about the specific conditions in which informational content may actually be considered representational (cf. e.g. Ramsey, 2007). See also Chap. 12 by Kieval and Chap. 13 by Freiesleben in this volume for discussions about representation in humans versus machines.

motivate the system's decisions. Insofar as knowledge of motivating reasons can facilitate justification, opacity in the sense of a lack of interpretability makes it harder to determine whether or not an AI-driven action is justified.

6.3.2 Opacity and Explanation

There is another way in which state-of-the-art AI systems can be opaque. In addition to not knowing how to meaningfully interpret a system by attributing representational contents, it may be hard for human observers to identify the corresponding representational vehicles. Consider again the fact that financial risk-assessment systems are based on classifiers that are trained using techniques from machine learning. Many of these techniques work by fine-tuning the parameters of high-dimensional nonlinear systems—for example, the connection weights of a deep neural network with a large number of layers and units, or the decision criteria in a random forest consisting of a large number of decision trees of arbitrary depth. Because the number of parameters in such systems is high, and because the interactions between these parameters are nonlinear, it is difficult for a human observer to describe the way in which individual system elements contribute to the system's behavior (Humphreys, 2009). When Jimmy applies for a loan and his application is denied, no bank employee—no matter their level of technical expertise or domain knowledge—is likely to be in a position to know which elements—input variables, intervening variables, or system parameters—actually contribute to the denial, how exactly they contribute, and how these elements might have to be modified in order for the decision to be reversed.

This kind of opacity can be likened to a lack of knowledge about explanatory reasons. Recall that in the case of Othello's murder, the explanatory reason for the action is a psychological state (a belief) that has the motivating reason (the putative fact that Desdemona has been unfaithful) as its representational content. Even if it appears that the financial risk-assessment system motivates its decision on the simple fact that Jimmy has a relatively low income (i.e., even if the system can be interpreted), it may still be difficult to know where in the system that fact is represented, and how (or even if) that fact actually contributes to the final decision to deny. More specifically, insofar as any representational content is likely to be distributed across a large number of variables and parameters—i.e., insofar as the representation is *distributed* (Hinton et al., 1986)—it can be difficult for a human observer to identify and clearly delineate the vehicle of this representational content and to specify that vehicle's causal contribution to the system's output as a whole. Because of this difficulty, it can be difficult to confirm that the relevant representational content—i.e. the motivating reason—is in fact the reason that favored the system's action.

Like a lack of knowledge about motivating reasons, a lack of knowledge about explanatory reasons can therefore be viewed as an obstacle for determining whether or not an AI-driven decision is justified. Lacking knowledge of explanatory reasons increases the uncertainty about a particular system's motivation. Because such uncertainty poses an obstacle to evaluating the system's motivation against a relevant norm, it also interferes with the ability to determine whether or not that system's behavior is justified.

6.4 Uncovering Motivating and Explanatory Reasons

The preceding sections suggest that, although knowledge of motivating and explanatory reasons helps (and may in some cases be necessary) to justify AI-driven decisions, opacity challenges our ability to acquire this kind of knowledge. Can this challenge be met?

The explainable AI (XAI) research program is dedicated to overcoming opacity that results from the increased use of machine learning. Specifically, its aim is to apply statistical, mathematical, or computational methods to meet different stakeholders' epistemic needs (Zednik, 2021; see also Chap. 7 by Páez and Chap. 10 by Sullivan and Kasirzadeh, in this volume).[8] Insofar as opacity is an obstacle to justifying the behavior of AI systems in high-stakes domains, it is worth asking whether any of these methods may be able to overcome opacity in the sense of providing insight into motivating and explanatory reasons. Although this is not the occasion to provide a comprehensive overview of XAI methods, there is reason to believe that some of these methods are well-suited for uncovering motivating and explanatory reasons.

6.4.1 Feature-Importance Measures

Methods from the broad family of *feature-importance measures* are designed to identify features within the input space that are particularly important with respect to a specific output. Insofar as Jimmy's demographic and financial data are entered into the financial risk-assessment system, such methods can be used to determine which particular data points are most important for the decision that his application should be denied. Notably, these methods can be tailored to the type of input data being processed. For domains that center on visual images (e.g. medical image classification), they can be used to construct heatmaps to highlight pixels or pixel-

[8] The application of such methods to opaque AI systems is often termed *post hoc explanation*. An alternative to post hoc explanation has been termed *inherent interpretability* (Rudin, 2019). The aim of this alternative approach is to replace current machine learning techniques that yield opaque systems with alternative techniques that avoid opacity altogether. The latter approach will not be considered here, in part because of lingering doubts about the resultant systems' predictive power, as well as about the legitimacy of claiming that any system can be considered "inherently" interpretable.

regions that are deemed particularly relevant for a classification. In contrast, for domains that center on tabular data (e.g. financial decision-making), they might be used to group or rank the most high-importance table elements.

Because they identify relevant features in the input, feature-importance measures appear well-suited for identifying motivating reasons. The features or patterns that are deemed particularly important for a decision can also be thought of as features or patterns that the system has learned to detect, track, or more generally, represent. Thus for example, if the method of *Shapley Additive Explanation* (Lundberg & Lee, 2017) highly ranks the INCOME and STUDENT_DEBT fields in Jimmy's application, this may be viewed as evidence for claiming that the values in those particular fields favor the denial for the system: Jimmy's income is too low and the amount of student debt is too high.

Beyond motivating reasons, some feature importance measures can also be used to characterize explanatory reasons. For example, *Layerwise Relevance Propagation* (LRP, Montavon et al., 2019) uses the "blame attribution" subroutine from the backpropagation learning algorithm to identify the particular input elements that bear the greatest causal responsibility for a specific output. Thus, LRP provides evidence for the claim that the features being highlighted do not only motivate the system's behavior, but also explain that behavior. That said, it is worth emphasizing that not all feature-importance measures are equally attuned to detecting a system's causally relevant states and processes. In particular, unlike LRP, SHAP highlights input elements that correlate statistically with particular outputs whether or not they also play a causal role in the production of those outputs. Therefore, whereas all feature-importance measures appear well-suited for revealing motivating reasons, only some of them are also likely to reveal explanatory reasons.

6.4.2 Surrogate Modeling Methods

The heterogeneous family of *surrogate modeling methods* aims to develop relatively simple models to replicate a more complex AI system's behavior. For methods such as *Linear Model-Agnostic Explanation* (Ribeiro et al., 2016), the behavior is replicated locally (i.e., for a subsection of the input domain) using a simple linear approximation. In contrast, for methods such as *tree-extraction* (Wu et al., 2018), the behavior is replicated globally (for the input domain as a whole) using a decision tree with relatively few decision criteria.

Like feature-importance measures, surrogate modeling methods can also be thought to provide insight into motivating reasons. That said, whereas feature-importance measures are most useful for identifying features in the input data, surrogate modeling methods can also be used to identify higher-order features that a system has learned to detect without those features being explicitly contained in the input. For example, a decision tree extracted from a financial risk assessment system might have, as one of its decision criteria, a feature that might be interpreted as "high job security". Although this feature might not be explicitly contained in the training or input data, it is possible that this feature nevertheless partitions the training set in a way that might be used for the purposes of assessing financial risk: farm workers and postdoctoral researchers (who typically have limited-term contracts) may be more likely to default on a loan than farm owners and department chairs (who typically have permanent contracts). Insofar as the decision tree with a JOB_SECURITY criterion adequately replicates the original system's behavior, that criterion might be viewed as a motivating reason for that system's decision.

As was the case for feature-importance measures, different surrogate modeling methods vary in the extent to which the motivating reasons being uncovered can also be considered explanatory. In particular, there is no guarantee that the criteria of an extracted decision tree actually correspond to causally relevant latent variables in the AI system being modeled. Therefore, although the motivating reasons identified by such a method might be plausible and illuminating to a human observer, these reasons might not actually explain the system's behavior. That said, current research has recently attempted to improve these methods so that the variables of the surrogate model also track the causally-relevant variables of the system being modeled. In particular, *causal abstraction analysis* (Geiger et al., 2021) yields a low-dimensional causal network whose nodes are aligned with causally-relevant variables (or variable groupings) of a high-dimensional neural network. In that sense, whereas traditional tree-extraction may yield motivating reasons that may or may not also be explanatory, causal abstraction analysis is more likely to yield both.

6.5 Limits of XAI

The previous section shows how different XAI methods may be used to overcome some of the negative effects of opacity by providing insight into motivating and explanatory reasons. Since knowledge of these kinds of reasons facilitates justification, these methods can play an important role in justifying AI-driven decisions in many different domains. That being said, it is important to be wary of some of the characteristic limitations of state-of-the-art explainable AI. Although some of these limitations may already be familiar from a growing body of critical discussions, what remains less well-understood is the particular way in which these limitations relate to justification.

6.5.1 Residual Lack of Interpretability

Whereas feature-importance measures and surrogate models may be able to reveal the kinds of features or patterns that might potentially be referenced by motivating reasons, they do not always do so in a way that actually facilitates justification. In particular, although the features and patterns being revealed may be viewed as contents which the AI system has learned to detect, track, or more generally represent, they may still not be interpretable in the sense of being recognizable to a human observer. Recall that interpretability is tied to an observer's ability to identify and describe elements of the system using concepts from natural language or a relevant technical domain such as medicine or finance. Although several XAI methods are capable of revealing the features being detected, these methods do not guarantee that the features are in fact interpretable in this sense. For example, a heatmap developed using LRP could highlight an amorphous blob or a seemingly random collection of distributed pixels (Rudin, 2019). Similarly, the individual criteria in an extracted decision tree need not be as meaningful or familiar as an applicant's level of job security. Although these features may well play an important role in motivating decisions for the system, it will be difficult for a human observer to actually characterize and understand that motivation.

This residual lack of interpretability–even after the application of XAI methods–is problematic for the purpose of justifying a relevant AI system's behavior. This is because it prevents a human observer from evaluating the system's motivating reasons against relevant norms. Insofar as most such norms are formulated using concepts from natural language or from the technical domain to which these norms pertain, the evaluation of a system's motivating reasons is hindered if those reasons cannot be described using concepts from that same domain. For example, if the decision on Jimmy's loan application actually turns out to be made on the basis of an unintuitive or intricate feature such as the sum total of letters in Jimmy's first name divided by the timestamp value (in milliseconds) at which Jimmy's application was submitted, then it will be difficult to know which norm to use to evaluate the correctness or appropriateness of this feature for the purposes of financial decision-making. Is it appropriate to decide on an individual's creditworthiness on the basis of such an intricate numerical value, given that there is no explicit rule declaring that doing so is *in*appropriate, whether or not this particular feature is reliably correlated with the likelihood of default?

6.5.2 Verification and Validation

A second issue concerns the verification and validation of XAI methods as epistemic tools. Verification (in this particular context) involves ensuring that a given measure or model does in fact do what it is supposed to do: consistently identify motivating or explanatory reasons. In contrast, validation consists in assessing whether the XAI method adequately represents its target: the original AI system. Current XAI methods are still notoriously limited in both respects.

Consider verification first. Even if XAI methods can be used to identify interpretable motivating reasons, it may not always be clear that these reasons are actually motivating the system's actions, as opposed to being mere computational artifacts of either the XAI method being used, or the AI system to which that method is applied. More generally, it may be difficult to verify that a given XAI measure or model does in fact do what it is supposed to do. Verification is a problem because many XAI methods lack *robustness* (Alvarez-Melis & Jaakkola, 2018). For one, small and irrelevant changes in the input have been known to yield significant changes in XAI methods' results. Even more problematic, different runs of the same XAI method can sometimes yield different explanations, and different XAI methods regularly, when applied to one and the same system, disagree. This indicates that XAI methods might not always be doing what they are supposed to be doing, viz. providing consistent explanations across similar inputs and unchanged systems.

Validation is a problem because we often don't know whether the motivating reasons being identified are true of either the training environment or the real world. XAI methods often approximate an AI system's actual behavior in order to provide interpretable representations. But whether those representations are accurate is by and large an open question. The problem is twofold. First, it is often unclear when one can be sure that a particular reason has been adequately described. More generally, there is a lack of widely shared and accepted standards for evaluating the goodness or accuracy of the information that an XAI method delivers. Applying one method as opposed to another often involves making trade-offs between different criteria such as faithfulness or predictive accuracy. For instance, suppose a surrogate model makes similar predictions to the original AI system 95% of the time, but another more accurately represents causally-relevant motivating reasons. Absent evaluation standards for comparing the quality of different methods, it is difficult to know which method is better (though cf. Durán forthcoming for one promising proposal).

The second problem is that, even if we can be confident that a particular method accurately represents the reasons that motivate the system's output, whether those facts are indeed facts or merely putative facts is often difficult to assess.[9] But this is often essential for the purpose of justification. In particular, false motivating reasons would violate the epistemic norm of truth. For the behavior of AI systems to be justified, their motivating reasons should reflect facts about the world.

6.6 Normative Opacity and Normative Discovery

The preceding discussion might be taken to suggest that knowing a system's motivating and explanatory reasons is the main obstacle to justifying its behavior. Furthermore, it can give the impression that XAI cannot play a role in our normative practices. In this section, we would like to push back against these two ideas.

[9] In a different context, Sullivan (2022) discusses this particular validation problem as 'link uncertainty', viz. an uncertainty concerning the empirical validation of a deep neural network and its target phenomena. See also, e.g., Humphreys (2009) for a related discussion of the semantic opacity of computer simulations.

6.6.1 Normative Opacity

Although explainable AI is well-suited—within limitations—for providing insight into a system's motivating and explanatory reasons, justification also requires reference to normative reasons. In order to justify a particular AI-driven action, the system's motivating and explanatory reasons must be evaluated against relevant norms. As discussed previously, however, normative reasons are external to the AI system itself. Normative facts—if there are any such facts—are facts about abstract evaluative principles, rather than facts about the AI system itself. Insofar as the methods being developed in explainable AI are designed to extract information about the system itself rather than about its socio-technical context, these methods will be unable to reveal the normative reasons that are needed to ultimately determine whether the system's behavior is justified.

Current methods from XAI can, despite some limitations, alleviate the lack of interpretability and explainability. They cannot, however, alleviate a lack of normative clarity—or, equivalently, a prevalence of *normative opacity*. Indeed, whereas for some systems, deployed by certain stakeholders in certain contexts, it is relatively clear which norms their actions should be compared against, in other contexts it is not. One issue we are currently facing with using AI systems is the lack of established moral and legal norms: it is unclear what normatively counts in favor of the behavior of AI systems.

Governmental and nongovernmental organizations are responding to this lack of normative clarity by drawing up regulatory frameworks to govern the development and deployment of AI systems. In October 2023, the White House issued an executive order that instructed federal agencies to develop guidelines to assess the risks of AI systems. The EU introduced a legislation draft in April 2021, the *Artificial Intelligence Act* (AI Act), whose aim is to govern the use of AI systems by anyone in the EU. Other states are also developing various governance regimes that will institute legal norms, and national as well as international standardization organizations are engaged in efforts to specify good practices in the development and deployment of AI systems. All these frameworks will provide norms against which to assess the behavior of AI systems. Among others, we can expect upcoming regulations to require impact assessments. The behavior of an AI system may then be justified if, for instance, it doesn't unduly interfere with people's autonomy. Or, a behavior that would be too risky may be prohibited.

But there is normative unclarity also beyond the regulatory domain. In particular, many epistemic norms remain unclear. Although systems for algorithmic medical diagnosis now regularly achieve levels of accuracy that are higher than those of human doctors, in many cases medical knowledge is still insufficiently developed to confirm or disconfirm the correctness of any individual diagnosis. Similarly, although experts in the financial sector may have over a century of expertise with which to evaluate the quality of an AI-assisted decision on a particular loan application, that expertise is far from perfect. In these domains and others, the epistemic norms are unclear. As long as this unclarity persists, it will be difficult to justify AI-driven decisions even if motivating and explanatory reasons can be identified.

As a result, even if XAI methods deliver the relevant explanatory or motivating reasons, we might still not be in a position to evaluate whether the system's behavior is justified. Explanatory or motivating reasons aren't sufficient for the purpose of justification. A lack of normative clarity is also an obstacle. To assess justification, we need to have appropriate normative reasons external to the system. No amount of XAI can deliver those reasons.

6.6.2 Normative Discovery

An unappreciated potential role for XAI is helping in *normative discovery*. In the context of scientific discovery, Zednik and Boelsen (2022) argue that XAI methods can help scientists refine target phenomena, identify experimental starting points for causal inference, or generate possible explanations, thereby aiding them in discovering (even if not quite predicting or explaining) empirical facts. Could XAI play an analogous role in the discovery of norms or normative facts?

According to some views in moral philosophy, an action is morally justified if the normative reasons coincide with the motivating reasons (Markovits, 2010). But if the reasons don't coincide, a lack of justification is not the only possible conclusion; an alternative conclusion is that new norms need to be articulated, or existing norms revised. To illustrate, consider the case of anti-discrimination laws. Wachter (2022) argues that machine learning systems cluster people into new forms of groups. Contrary to traditional groups that are typically formed on the basis of self-identified characteristics, algorithmic groups are based on patterns in the data. For that reason, they can be quite diverse, unconventional, and even obscure. These groups may include attributes like being a Star Wars fan, dog owner, and tennis news reader–but also seemingly obscure attributes such as frequent-mouse-clickers or more-than-100-open-brower-tabs-users.

Although some groups receive the protection of anti-discrimination laws (e.g. women), these laws don't protect these new algorithmic groups. According to Wachter, there is thus a gap between the protection offered to traditional groups and these new algorithmic groups. In turn, "[t]his gap necessitates rethinking and reinterpretation of legal doctrine resulting in the inclusion of those new algorithmic groups under the law" (Wachter, 2022, 8). To address that concern, Wachter proposes that the new algorithmic groups should receive the protection of the law since these harms are similar to those addressed by existing laws. This would ensure the spirit of the law is followed.

Traditionally protected groups often have so-called immutable characteristics, i.e. features that people cannot reasonably change (e.g. sex or race). Wachter introduces the notion of 'artificial immutability', viz. the idea that characteristics assigned by AI to individuals often act de facto immutable traits.

Wachter's concerns are to a large extent independent of any specific explanatory or motivating reasons AI systems may have. In fact, that AI systems are opaque and often uninterpretable counts in favor of reinterpreting the law. But suppose that, at least in some cases, XAI could deliver motivating reasons grounded on belonging to an algorithmic group. What if the reason to deny Jimmy a loan is that he is a banjo-loving, PC-user, high-frequency-mouse-clicking, Diplomacy-player? Although appropriate XAI methods may reveal that the financial risk-assessment system is deciding on his creditworthiness on the basis of Jimmy's membership in just such an algorithmic group, whether or not that decision is in fact justified will require the development or discovery of relevant norms. Insofar as XAI methods play an important role in discovering these kinds of algorithmic groups, they might ultimately inspire the articulation of normative frameworks with which to evaluate such AI systems' outputs.

6.7 Conclusion

AI-driven decisions are used in many different high-stakes domains. Although these decisions merit justification, the opacity of the AI systems involved poses a considerable obstacle. Explainable AI is a research project whose goal is to reduce the systems' opacity. But can the methods from XAI help justify AI-driven decisions?

Justifying AI-driven decisions requires *reasons*. In this chapter, we use the distinction between explanatory, motivating, and normative reasons to spell out the way in which different reasons contribute to justification. Moreover, the distinction allows us to clarify the role of XAI for that purpose. While normative reasons are essential, they stem from norms external to the AI systems themselves, and are for this reason beyond the purview of Explainable AI. What XAI can potentially do, however, is to provide insight into a system's explanatory and motivating reasons, and in this sense, contribute to justification.

That being said, current XAI methods, although useful, still face significant hurdles. Although many of these methods help to identify features and patterns that can be viewed as a system's motivations, these features and patterns may remain uninterpretable. Moreover, it is not always clear whether the motivations identified through the use of explainable AI are the system's actual motivations and are thus responsible for the outputs. Finally, and perhaps most significantly, normative opacity poses an obstacle that XAI cannot overcome. For this reason, although XAI can be viewed as a promising research program that can sometimes help, it is by no means guaranteed to yield the insights that are needed to justify AI-driven decisions.

Acknowledgements The authors would like to acknowledge Ramón Alvarado, Juan M. Durán, and researchers in the TU/e Philosophy & Ethics group for helpful feedback on earlier drafts of this chapter, as well as audiences at the European Philosophy of Science Association and Science & Art of Simulation conferences.

References

Alvarez, M. (2017, Winter). Reasons for action: Justification, motivation, explanation. In E. N. Zalta (Ed.), *The Stanford encyclopedia of philosophy*. Metaphysics Research Lab, Stanford University.

Alvarez-Melis, D., & Jaakkola, T. S. (2018). On the robustness of interpretability methods. arXiv. https://doi.org/10.48550/arXiv.1806.08049.

Baum, K., Mantel, S., Schmidt, E., & Speith, T. (2022). From responsibility to reason-giving explainable artificial intelligence. *Philosophy & Technology, 35*(1), 12. https://doi.org/10.1007/s13347-022-00510-w

Beisbart, C. (in this volume). In which ways is machine learning opaque? In J. M. Durán & G. Pozzi (Eds.), *Philosophy of science for machine learning: Core issues and new perspectives*. Synthese Library, Springer.

Buijsman, S., & Veluwenkamp, H. (2022). Spotting when algorithms are wrong. *Minds and Machines, 33*(4), 541–562. https://doi.org/10.1007/s11023-022-09591-0

Burrell, J. (2016). How the machine 'thinks': Understanding opacity in machine learning algorithms. *Big Data & Society, 3*(1), 2053951715622512. https://doi.org/10.1177/2053951715622512

Cappelen, H., & Dever, J. (2021). *Making AI intelligible: Philosophical foundations*. Oxford University Press.

Consumer Financial Protection Bureau. (2022). Consumer financial protection circular 2022-03: Adverse action notification requirements in connection with credit decisions based on complex algorithms.

Creel, K. A. (2020). Transparency in complex computational systems. *Philosophy of Science, 87*(4), 568–589. https://doi.org/10.1086/709729

DIN Deutsches Institut für Normung. (2023). DIN SPEC 92001-3.

Doshi-Velez, F., & Kim, B. (2017). Towards a rigorous science of interpretable machine learning. arXiv:1702.08608 [Cs, Stat], March. https://arxiv.org/abs/1702.08608.

Durán, J. M. (2021). Dissecting scientific explanation in AI (sXAI): A case for medicine and healthcare. *Artificial Intelligence, 297*(August), 103498. https://doi.org/10.1016/j.artint.2021.103498

Durán, J. M. (Forthcoming). Causal-manipulationist approaches to explaining machine learning. In P. Illari & F. Russo (Eds.), *The Routledge handbook of causality and causal methods*. Routledge.

Durán, J. M. (in this volume). Beyond transparency: Computational reliabilism as an externalist epistemology for algorithms. In J. M. Durán & G. Pozzi (Eds.), *Philosophy of science for machine learning: Core issues and new perspectives*. Synthese Library, Springer.

Durán, J. M., & Formanek, N. (2018). Grounds for trust: Essential epistemic opacity and computational reliabilism. *Minds and Machines, 28*(4), 645–666. https://doi.org/10.1007/s11023-018-9481-6

Durán, J. M., & Jongsma, K. R. (2021). Who is afraid of black box algorithms? On the epistemological and ethical basis of trust in medical AI. *Journal of Medical Ethics, 47*(5), 329–335. https://doi.org/10.1136/medethics-2020-106820

Ferrario, A., & Loi, M. (2022). How explainability contributes to trust in AI. In *2022 ACM conference on fairness, accountability, and transparency* (pp. 1457–1466). ACM. https://doi.org/10.1145/3531146.3533202

Formanek, N. (in this volume). How I stopped worrying and learned to love opacity. In J. M. Durán & G. Pozzi (Eds.), *Philosophy of science for machine learning: Core issues and new perspectives*. Synthese Library, Springer.

Freiesleben, T. (in this volume). Artificial neural nets and the representation of human concepts. In J. M. Durán & G. Pozzi (Eds.), *Philosophy of science for machine learning: Core issues and new perspectives*. Synthese Library, Springer.

Geiger, A., Lu, H., Icard, T., & Potts, C. (2021). Causal abstractions of neural networks. arXiv. https://arxiv.org/abs/2106.02997

Hawthorne, J., & Stanley, J. (2008). Knowledge and action. *The Journal of Philosophy, 105*(10), 571–590. https://www.jstor.org/stable/20620129

Henin, C., & Le Métayer, D. (2022). Beyond explainability: Justifiability and contestability of algorithmic decision systems. *AI & SOCIETY, 37*(4), 1397–1410. https://doi.org/10.1007/s00146-021-01251-8

High-Level Expert Group on Artificial Intelligence. (2019). *Ethics guidelines for trustworthy AI*. European Commission.

Hinton, G. E., McClelland, J. L., & Rumelhart, D. E. (1986). Distributed representations. In D. E. Rumelhart & J. L. McClelland (Eds.), *Parallel distributed processing: Explorations in the microstructure of cognition: Volume 1: Foundations* (pp. 77–109). MIT Press.

Humphreys, P. (2009). The philosophical novelty of computer simulation methods. *Synthese, 169*(3), 615–626.

Kieval, P. H. (in this volume). Representation learning without representationalism. A non-representationalist account of deep learning models in scientific practice. In J. M. Durán & G. Pozzi (Eds.), *Philosophy of science for machine learning: Core issues and new perspectives*. Synthese Library. Springer.

Krishnan, M. (2020). Against interpretability: A critical examination of the interpretability problem in machine learning. *Philosophy & Technology, 33*(3), 487–502. https://doi.org/10.1007/s13347-019-00372-9

Lapuschkin, S., Wäldchen, S., Binder, A., Montavon, G., Samek, W., & Müller, K.-R. (2019). Unmasking clever Hans predictors and assessing what machines really learn. *Nature Communications, 10*(1), 1096. https://doi.org/10.1038/s41467-019-08987-4

Lipton, Z. C. (2018). The mythos of model interpretability: In machine learning, the concept of interpretability is both important and slippery. *Queue, 16*(3), 31–57. https://doi.org/10.1145/3236386.3241340

Lundberg, S. M., & Lee, S.-I. (2017). A unified approach to interpreting model predictions. In *Advances in neural information processing systems* (Vol. 30). Curran Associates, Inc.

Maclure, J. (2021). AI, explainability and public reason: The argument from the limitations of the human mind. *Minds and Machines, 31*(3), 421–438. https://doi.org/10.1007/s11023-021-09570-x

Malgieri, G., & Pasquale, F. (2024). Licensing high-risk artificial intelligence: Toward ex ante justification for a disruptive technology. *Computer Law & Security Review, 52*(April), 105899. https://doi.org/10.1016/j.clsr.2023.105899

Markovits, J. (2010). Acting for the right reasons. *The Philosophical Review, 119*(2), 201–242. https://doi.org/10.1215/00318108-2009-037

Montavon, G., Binder, A., Lapuschkin, S., Samek, W., & Müller, K.-R. (2019). Layer-wise relevance propagation: An overview. In W. Samek, G. Montavon, A. Vedaldi, L. K. Hansen, & K.-R. Müller (Eds.), *Explainable AI: Interpreting, explaining and visualizing deep learning* (Lecture notes in computer science) (pp. 193–209). Springer International Publishing. https://doi.org/10.1007/978-3-030-28954-6_10

Mueller, A. (2021). The knowledge norm of apt practical reasoning. *Synthese, 199*(1), 5395–5414. https://doi.org/10.1007/s11229-021-03030-w

Páez, A. (in this volume). Axe the X in XAI: A plea for understandable AI. In J. M. Durán & G. Pozzi (Eds.), *Philosophy of science for machine learning: Core issues and new perspectives*. Synthese Library. Springer.

Papagni, G., & Koeszegi, S. (2021). A pragmatic approach to the intentional stance semantic, empirical and ethical considerations for the design of artificial agents. *Minds and Machines, 31*(4), 505–534. https://doi.org/10.1007/s11023-021-09567-6

Ramsey, W. M. (2007). *Representation reconsidered*. Cambridge University Press.

Raz, J. (2009). Reasons: Explanatory and normative. In C. Sandis (Ed.), *New essays on the explanation of action* (pp. 184–202). Palgrave Macmillan. https://doi.org/10.1057/9780230582972_11

Reutlinger, A., & Saatsi, J. (Eds.). (2018). *Explanation beyond causation: Philosophical perspectives on non-causal explanations*. Oxford University Press.

Ribeiro, M. T., Singh, S., & Guestrin, C. (2016). "Why should i trust you?": Explaining the predictions of any classifier. In *Proceedings of the 22nd ACM SIGKDD international conference on knowledge discovery and data mining, 1135–1144. KDD '16*. Association for Computing Machinery. https://doi.org/10.1145/2939672.2939778

Rudin, C. (2019). Stop explaining black box machine learning models for high stakes decisions and use interpretable models instead. *Nature Machine Intelligence, 1*(5), 206–215. https://doi.org/10.1038/s42256-019-0048-x

Scanlon, T. M. (1998). *What we owe to each other*. Harvard University Press.

Smart, A., James, L., Hutchinson, B., Wu, S., & Vallor, S. (2020). Why reliabilism is not enough: Epistemic and moral justification in machine learning. In *Proceedings of the AAAI/ACM conference on AI, ethics, and Society* (pp. 372–377). Association for Computing Machinery.

Sullivan, E. (2022). Understanding from machine learning models. *The British Journal for the Philosophy of Science, 73*(1), 109–133. https://doi.org/10.1093/bjps/axz035

Sullivan, E., & Kasirzadeh, S. (in this volume). Explanation hacking: The perils of algorithmic recourse. In J. M. Durán & G. Pozzi (Eds.), *Philosophy of science for machine learning: Core issues and new perspectives*. Synthese Library. Springer.

Sylvan, K. (2016). Epistemic reasons I: Normativity. *Philosophy Compass, 11*(7), 364–376. https://doi.org/10.1111/phc3.12327

Symons, J., & Alvarado, R. (2022). Epistemic injustice and data science technologies. *Synthese, 200*(2), 87. https://doi.org/10.1007/s11229-022-03631-z

Vredenburgh, K. (2022). The right to explanation*. *Journal of Political Philosophy, 30*(2), 209–229. https://doi.org/10.1111/jopp.12262

Wachter, S. (2022). The theory of artificial immutability: Protecting algorithmic groups under anti-discrimination law. *Tulane Law Review, 97*, 149.

Widdau, C. S., & Zednik, C. (2022). Opake Systeme künstlicher Intelligenz Und Das Problem Der Verantwortungslücke. In *Algorithmen Und Autonomie: Interdisziplinäre Perspektiven Auf Das Verhältnis von Selbstbestimmung Und Datenpraktiken* (p. 121).

World Bank Group. (2019). *Credit scoring approaches guidelines*. World Bank Group.

Wu, M., Hughes, M., Parbhoo, S., Zazzi, M., Roth, V., & Doshi-Velez, F. (2018). Beyond sparsity: Tree regularization of deep models for interpretability. In *Proceedings of the AAAI conference on artificial intelligence* (Vol. 32, No. 1). https://doi.org/10.1609/aaai.v32i1.11501

Zednik, C. (2021). Solving the black box problem: A normative framework for explainable artificial intelligence. *Philosophy & Technology, 34*(2), 265–288. https://doi.org/10.1007/s13347-019-00382-7

Zednik, C., & Boelsen, H. (2022). Scientific exploration and explainable artificial intelligence. *Minds and Machines, 32*(1), 219–239. https://doi.org/10.1007/s11023-021-09583-6

Zerilli, J., Knott, A., Maclaurin, J., & Gavaghan, C. (2019). Transparency in algorithmic and human decision-making: Is there a double standard? *Philosophy & Technology, 32*(4), 661–683. https://doi.org/10.1007/s13347-018-0330-6

Carlos Zednik is Assistant Professor for Philosophy of AI at Eindhoven University of Technology. In his research, he investigates methodological principles of cognitive psychology and neuroscience, as well as methods, norms, and applications of explainable artificial intelligence. Beyond research, he contributes to international AI standardization efforts as well as policy developments to promote the use of AI in education.

Philippe Verreault-Julien is a postdoctoral researcher at Eindhoven University of Technology, where he works on the ethics, epistemology, governance, and safety of artificial intelligence systems.

Part III
Scientific Explanation (XAI)

Chapter 7
Axe the X in XAI: A Plea for Understandable AI

Andrés Páez

Abstract In a recent paper, Erasmus et al. (2021, *Philosophy & Technology, 34*(4), 833–862) defend the idea that the ambiguity of the term "explanation" in explainable AI (XAI) can be solved by adopting any of four different extant accounts of explanation in the philosophy of science: the Deductive Nomological, Inductive Statistical, Causal Mechanical, and New Mechanist models. In this chapter, I show that the authors' claim that these accounts can be applied to deep neural networks as they would to any natural phenomenon is mistaken. I also provide a more general argument as to why the notion of explainability as it is currently used in the XAI literature bears little resemblance to the traditional concept of scientific explanation. It would be more fruitful to use the label "understandable AI" to avoid the confusion that surrounds the goal and purposes of XAI. In the second half of the chapter, I argue for a pragmatic conception of understanding that is better suited to play the central role attributed to explanation in XAI. Following De Regt (2017, *Understanding scientific understanding*. Oxford University Press) and Kuorikoski and Ylikoski (2015, *Synthese, 192*, 3817–3837), the conditions of satisfaction for understanding an ML system are fleshed out in terms of an agent's success in using the system.

Keywords Explainable AI · Scientific explanation · Scientific understanding · Deep neural networks · Factivity in XAI · Inferential conception of understanding

To appear in: Durán, J. M., & Pozzi, G. (Eds.), *Philosophy of science for machine learning: Core issues and new perspectives*. Synthese Library. Cham: Springer.

A. Páez (✉)
Department of Philosophy, Universidad de los Andes, Bogotá, Colombia
e-mail: apaez@uniandes.edu.co

J. M. Durán, G. Pozzi (eds.), *Philosophy of Science for Machine Learning*,
Synthese Library 527, https://doi.org/10.1007/978-3-032-03083-2_7

7.1 Introduction

The concept of explanation has a long and variegated history in the philosophy of science. Starting with Hempel and Oppenheimer's (1948) logic-based approach, at least a dozen different analyses of the concept have been proposed.[1] Given this rich theoretical repository, some philosophers have argued that the solution to the problem of specifying a meaning for "explanation" in the context of artificial intelligence (AI) is to adapt an extant account of scientific explanation to machine learning (ML) in general, and to deep neural networks (DNNs) in particular. Erasmus et al. (2021) offer the most developed account of this strategy. They examine four different accounts of explanation in the philosophy of science: the Deductive Nomological, Inductive Statistical, Causal Mechanical, and New Mechanist models. Their claim is that any of them is applicable to DNNs as it would to any scientific phenomenon. This claim derives from a more general principle that they call "the indefeasibility thesis" about explanation (2021, 840). The thesis states that explanations are invariant with respect to the complexity of both the explanans and the explanandum. There is no threshold of complexity beyond which a phenomenon becomes unexplainable. Therefore, despite their complexity, DNNs are scientifically explainable.

In this chapter, I argue that the thesis that opaque ML systems are scientifically explainable is either trivial or false, and that it misrepresents the goals of explainable AI (XAI). It is trivial if an explanation is simply understood as the set of causes, entities or states that physically or computationally produce a prediction; not the linguistic or mathematical *description* of the known elements in the set, but the elements themselves, known or unknown. It is false if the claim is that it is always possible to offer an "explanatory text," a *truthful* description of the source of the prediction, thereby satisfying the factivity condition on scientific explanations (Páez, 2009, 2019). Most of the first half of this chapter will be devoted to justifying the second claim. Now, if there are explanatory gaps in machine learning, and more specifically, if the predictions of DNNs cannot be scientifically explained, then the goal of *explainable* AI thus formulated will be unattainable. We should not insist on using a concept that cannot perform its desired function. Instead, I will advocate for the use of labels such as "interpretable machine learning" (Watson & Floridi, 2021) or "understandable AI" to avoid the confusion that surrounds the goal and purposes of XAI methods. In the second part of the chapter, I argue that understanding is a success concept that is better suited to play the role often attributed to explanation in XAI. Following De Regt (2017) and Kuorikoski and Ylikoski (2015), the conditions of satisfaction for understanding an ML system are fleshed out in terms of an agent's success in using the system.

---

[1] See Woodward and Ross (2021) for a survey of the earliest models of explanation, roughly until 1990, and Ross and Woodward (2023) for more recent causal approaches.

7.2 Clarifying the Explanandum

Before we can tackle the question of explanation in the context of machine learning, it is essential to clarify what it is that we are trying to explain, that is, to clarify the explanandum. There are two possible explananda in ML. From the point of view of local *post-hoc* interpretability methods, the explanandum is some specific prediction churned out by a trained model. These methods usually try to pinpoint input features that make a difference in the prediction. Local interpretations include counterfactual probes (Mothilal et al., 2020; Wachter et al., 2018), and different types of perturbation-based methods such as LIME (Ribeiro et al., 2016), Grad-CAM (Ancona et al., 2019; Selvaraju et al., 2017), SHAP (Lundberg & Lee, 2017), TCAV (Kim et al., 2018), among many others (see Ivanovs et al., 2021 for a survey).

From the point of view of *global* interpretation methods, the explanandum is the entire model, more specifically, the non-linear function performed by the trained model. Global interpretations generally take the form of surrogate models. The most widely used classes of surrogate models are linear or gradient-based approximations, decision rules, and decision trees (Frosst & Hinton, 2017; Wu et al., 2018). Some surrogate models are generated through knowledge distillation techniques (Bastani et al., 2017; Jung et al., 2017; Kim et al., 2022; Tan et al., 2018). Given a model f, the goal is to generate an interpretable model M such that $M(x) \approx f(x)$. A different approach is to use example-based methods. The idea is to select subsets of the dataset to explain the behavior of ML models or to make explicit the underlying data distribution. This approach only works when the data is structured and can be represented in a human understandable way. They include similar examples or factuals (Schoenborn et al., 2021), influential instances (Koh & Liang, 2017; Koh et al., 2019) and prototypes (Kim et al., 2016), among others. These methods yield results that are more easily understood.

Most theories of explanation in the philosophy of science have taken individual facts or particulars, such as things or events, as their explanandum. Thus, the explanandum of a scientific explanation is a true statement that expresses a fact or that states that the things or events to be explained exist or occur. In that respect, they resemble *post hoc* local interpretability methods. Kitcher's (1989) unificationist model of explanation uses a different approach. Scientific explanation is a matter of providing a unified account of a range of different phenomena. Erasmus and Brunet (2022) discuss Kitcher's unificationist model in the context of ML, but their argument is parasitic on their defense of the applicability of Deductive-Nomological explanations to DNNs. I will therefore skip the specifics of that discussion to focus on the four theories of explanation that seem more readily applicable in the context of ML.

Notice that if we restrict the discussion to these four models of scientific explanation, whose explananda are always individual facts or events, global explanations will be excluded because the target of any global interpretation method in ML is the original model itself, not one of its predictions. As a result, ML models will be scientifically unexplainable. ML models are complex mathematical structures that admit of infinitely many input-output pairs; their abstract nature makes them akin to a law of nature, not to a singular fact or event. This means that the account provided by Erasmus et al. (2021) is incomplete at best because it cannot account for global explanations. The authors could try to amend this shortcoming by adding a model of explanation of laws to the mix, but this approach faces the problem that there is no agreed upon account of the laws of nature in the philosophical literature, let alone an account of the explanation of laws.[2]

It could be argued that although surrogate models do not explain singular predictions, they can be seen as explanatory *schemata*. Plugging in the missing values in a surrogate decision tree, for example, will provide an explanation of a singular prediction.[3] However, the price of taking this way out of the problem is that factivity will have to be sacrificed at the outset because surrogate models are always false, as Cynthia Rudin explains:

> Explanations must be wrong. They cannot have perfect fidelity with respect to the original model. If the explanation was completely faithful to what the original model computes, the explanation would equal the original model, and one would not need the original model in the first place, only the explanation. (2019, 207)

Since Erasmus and colleagues want to defend the factivity of explanation (see below), surrogate models, even when regarded as explanatory schemata, would not qualify as bona fide explanations.[4] The remaining question is whether their account is successful in the explanation of singular predictions.

7.3 Scientific Explanation of ML Systems?

In this section, I will examine whether the four theories of explanation discussed by Erasmus et al. (2021) can play the role attributed to them. I will focus mostly on the way they use Hempel's Deductive-Nomological (D-N) model because discussion of that model will bring out the main elements in play.[5] Once the stakes have been made clear, it will be much easier to dismiss the use of the other three models.

[2] In "Studies in the logic of explanation," Hempel and Oppenheimer famously state that the explanation of laws "presents peculiar problems for which we can offer no solution at present" (1948, 165n33). These problems force them to limit their theory to the explanation of particular events. The situation is no better today than in 1948. I'm grateful to Juan M. Durán for reminding me of this telling passage.

[3] I am grateful to Stefan Buijsman for suggesting this possibility.

[4] Following Elgin (2017) and Potochnik (2017), in Sect. 7.4, I argue that the factivity condition on scientific explanation can be replaced by a more pragmatic approach in which surrogate models only need to be "true enough" of their target models.

[5] For the purpose of this chapter, I will ignore the fact that Hempel's D-N and I-S models are nowadays considered museum pieces due to the well-known objections that have been raised against them. Salmon (1989) offers the best discussion of these objections and counterexamples.

7.3.1 Deductive-Nomological Explanations

I will begin by examining the idea that Hempel's Deductive-Nomological model of scientific explanation can be used to explain the output of a DNN. Hempel's model can be represented by the well-known schema that deduces the explanandum from laws and initial conditions, which jointly constitute the explanans (1965, 336):

$$\frac{\begin{array}{c} L_1, L_2, \ldots, L_\mathrm{n} \\ C_1, C_2, \ldots, C_\mathrm{n} \end{array}}{E}$$

Here L_1, L_2, ..., L_n are general laws, C_1, C_2, ..., C_n are sentences describing the particular facts involved, and E is a sentence describing the explanandum phenomenon. As Hempel is well aware, general laws in science take many forms and even nowadays there is no agreed-upon definition of what a law of nature is. However, to have an explanatory character, Hempel argues, laws should not make reference to particulars. We must require them to be of "essentially general form" (1965, 343). Finally, all the elements in the explanation must be true. This is generally known as the *factivity condition* on scientific explanation.

Before we move on to examine whether D-N explanations can explain the output of a DNN, it is important to understand the theoretical claim that undergirds the very possibility of attempting such a move. As we saw in the Introduction, Erasmus and collaborators argue, essentially, that everything is explainable because the possibility of devising an explanation does not depend on the complexity of the phenomenon:

> The features that make something an explanation turn out to be invariant with respect to the complexity of both the explanans and the explanandum. ... Adding complexity to the explanation of the phenomenon does not entail that the phenomenon is any less explainable. This is not a claim about the quality, superiority, or goodness of a given explanation. Our concern is whether increasing the complexity of a given explanation makes it no longer an explanation. (2021, 840)

Let us examine how the thesis applies to D-N explanations. We will only consider an increase in the complexity of the explanans because in supervised ML the explanandum always remains simple, viz. a prediction or a classification.

How could more information be added to the explanans of a D-N explanation? Erasmus et al. argue that one could, for example, replace one of the law-like statements and substitute it for a set of *n* laws jointly entailing it (here E is a set of empirical conditions):

> This may make the explanans less manageable or more difficult to understand, but we still have a DN explanation. That is, if x follows deductively from L + E, then it also follows from L* = {L_1+L_2+ ... +L_n} + E, where $L_{1\ \ldots\ \mathrm{n}}$ are (complexity-increasing) laws which, taken together, entail L. The expansion of the explanans does not make its laws any less law-like nor its conclusion any less deductively valid. Put another way, the explanandum is no

> less explainable, since the connection to the explanans, no matter [its] complexity, is still deduction. (2021, 841)

In the context of a DNN, this argument requires that we identify law-like statements of any degree of complexity to set up the explanans of a D-N explanation. Here we will focus on two points, the relata of the laws and the factivity condition. Regardless of the form of a lawlike statement in science, laws create (mostly mathematical) relationships between predicates. Such predicates might refer to natural classes such as the electric charge of a particle or the volume of a gas, and they are applied to the state of a system:

> The state of such a system at any given time is characterized by the values assumed at that time by certain quantitative characteristics of the system, the so-called variables of state; and the laws specified by such a theory for the changes of state are deterministic in the sense that, given the state of the system at any one time, they determine its state at any other, earlier or later, time. (Hempel, 1965, 351)

The first question that arises, if one wants to adapt Hempel's D-N model to DNNs, regards the relata of the lawlike statements that will be deployed. Are they relationships between classes of inputs and outputs? Between features of the input space and outputs? Between the weights and biases in the neural net and an output? If we think of a *functional* explanation based solely on the inputs and outputs of the system, it is impossible to think of deterministic laws in DNNs. No class of inputs will inevitably produce the correct decision every time, and sometimes even repeating the sampling might produce a different decision (Kindermans et al., 2017; Molnar, 2020).[6] Something similar occurs if we think of a lawlike relation between types of features and decisions. Simply choosing some set of features from the input space to train the system will not guarantee a deterministic relation between the chosen features and the decision. For example, if a bank uses features such as income, age, credit history, number of family members, and so on to train a fairly accurate ML system, there is still no guarantee that those particular features will generate the same predictions when provided with the same inputs. The only viable alternative is to look at the model itself as composed of deterministic lawlike relationships. This seems to be what Erasmus et al. (2021) have in mind. The example they use is of a DNN trained for image classification. They explain how to apply a D-N explanation of why the DNN classified an input image in one of the output classes in the following way:

> A DN explanation of how the [DNN] assesses an input image involves listing the weights attached to each and every node and the informational routes indicated by each and every edge at every convolution stage, and the weights of the fully connected network along with the assigned numerical values being fed into the input layer and the network architecture. Once we have that, we can list the values for the classifications the [DNN] learned in the training and testing phases of development, and see that its classification of the image is based on comparing the ranges of these classifications with the output value of the image. In doing so, we are explaining the explanandum—here, the [DNN] classifying of image *I* as classification *c*—using an explanans consisting of a law-like premises—in this case,

---

[6] To be sure, there are *functionally* deterministic models in ML, but they have many limitations that make them unsuitable in most real-world applications.

> how the weights of all relevant nodes and edges produced the output value, along with the law that an output is assigned to the most probable class—and additional information about I—which includes the set of input values assigned to I, and the output value c. (2021, 844)

In more simple terms, the D-N explanation of the explanandum value c consists of an exact description of the model, which would take the place of the lawlike sentences $L_1, \ldots, L_n$ in Hempel's model, and a list of the features of the input image I, which would correspond to the sentences $C_1, \ldots, C_n$ that describe the particular facts involved. Together, they explain why the model generated the output value c.

There are two problems with this idea. Firstly, the connection between the features of the input space and the parameters learned by the model cannot be made sense of in an opaque model. Secondly, since there is no direct access to the hidden layers of the model, the values of the weights and biases cannot be verified; therefore, there is no way to determine whether any given explanatory text is true. Let us examine each problem in turn.

Neural networks are designed to identify patterns and make predictions based on these patterns. Inputs pass through multiple layers of interconnected nodes, or "neurons," each of which transforms them in some way before passing them along to the next layer. These transformations often involve nonlinear operations, such as ReLU (Rectified Linear Unit), sigmoid, or tanh functions. These nonlinear transformations, combined with the interactions between the neurons across different layers, allow the neural network to model complex, high-dimensional decision boundaries. Even if we could see all the weights and biases in the network (the parameters that the model learned during training), it is not clear how to interpret them in the context of the original input features. Each input of the network goes through a series of complex, intertwined transformations that make it difficult to understand how the inputs relate to the output. The nonlinear nature of the activation functions compounds this complexity. In a linear system, the effect of each input can be considered independently of the others: if you double an input, the output will also double. But in a nonlinear system, the effect of changing an input depends on the values of all the other inputs. This makes it much more difficult to understand how each input influences the output, and in any case the relationship cannot be expressed using a single lawlike statement. In a neural network, there is no clear sequence of "decisions" to follow –the output is the result of a complex, interconnected web of influences that do not lend themselves to an explanation.

The second problem is that there is no way to verify which parameters are being used in the hidden layers of the neural net. Deep models often have an extremely large number of optima of similar predictive accuracy. This is known as the *model identifiability problem*: "A model is said to be identifiable if a sufficiently large training set can rule out all but one setting of the model's parameters. Models with latent variables are often not identifiable because we can obtain equivalent models by exchanging latent variables with each other" (Goodfellow et al., 2016, 284). It is therefore impossible to verify which of many possible equivalent models is the one that generated the output in this case. Without identifying which model is being used, there is no way to create a true explanatory text of the model's prediction. Therefore, the factivity condition on explanation cannot be fulfilled. There *is*, of course, a true description of the model, but it lives in Popper's World 3, forever inaccessible to human subjects.

Independently of these two unsolvable problems, there is something amiss in saying that a ML model is a set of lawlike statements. It is true that the model is a function and most laws in science are mathematical functions, but that is as far as the resemblance goes. As mentioned above, Hempel insisted once and again that laws must have an essentially general form. They cannot be a conjunction of descriptions of particular facts:

> A statement which is logically equivalent to a finite conjunction of singular sentences, and which in this sense makes a claim concerning only a finite class of cases, does not qualify as a law and lacks the explanatory force of a nomological statement. Lawlike sentences, whether true or false, are not just conveniently telescoped summaries of finite sets of data concerning particular instances. (1965, 377)

This statement alone is a rebuttal of Erasmus and colleagues. There is nothing general in a trained ML model. A description of the model is exactly what Hempel describes: a conjunction of singular sentences about the weights and biases of the model. Its parameters obey the nature of the class of things in the training dataset and cannot be used for other ends. It is a custom-made predictive artifact that serves one and only one purpose. Furthermore, the features used to train the model do not necessarily correspond to natural kinds or to properties that belong to anything outside of the input space. Its limited scope in many cases is another reason why it lacks the generality of laws. It all depends on the data used. In the case of a model like AlphaFold, the features most probably correspond to natural kinds, but in most cases, the features can only be defined within a very specific context of use. ML just seems the wrong context to speak of laws, at least in the Hempelian sense, and a fortiori of covering law models of scientific explanation.

7.3.2 Inductive-Statistical Explanations

The second type of explanation discussed by Erasmus et al. (2021) is Hempel's inductive-statistical model (I-S). Unlike the D-N model, the I-S model uses statistical laws instead of deterministic ones. It preserves, however, the argument form of D-N explanations, but naturally in inductive form. The general schema is:

$$L_1, L_2, \ldots, L_n$$

$$C_1, C_2, \ldots, C_n$$

$$\overline{\overline{\qquad\qquad\qquad}}\; [r]$$

$$E$$

$L_1, L_2, \ldots, L_n$ are general laws, but of statistical nature. They have the form $p(P|Q) = r$ and they must also lack any reference to particular instances. As before, $C_1, C_2, \ldots, C_n$ are sentences describing the particular facts involved, and E is a sentence describing the explanandum phenomenon. The explanans provides inductive support to the truth of the explanandum-statement. The strength of the support is indicated by the number in square brackets. According to Hempel, r must be very high for the explanandum to be expected in light of the premises. As with D-N explanations, both the statistical laws and the empirical conditions must be true.

An I-S explanation of a prediction of a DNN model cannot be based on the model's parameters, as in the case of D-N explanations, because the trained model does not behave stochastically. The only options are to establish a statistical correlation between the input and output classes, or between the features of the input space and the output classes. Describing the first correlation in terms of a statistical law with a very high r is the same as saying that the model is reliable, a claim that has no explanatory value regarding how a particular prediction was reached. For example, saying that the probability that a reliable cat-identifying model will identify a cat when presented with an image of a cat is very high adds nothing to our previous knowledge of the model's reliability. To be sure, the model's reliability leads to the expectation that the prediction is correct, but if most predictions were not correct, the model would not have been deployed in the first place. Therefore, the expectability of the output is the result of a correct validation procedure using the test set, and not of a *post hoc* inductive argument that offers no *independent* reason to expect the output.

Essentially, the same argument applies in the second case, viz. when the statistical lawlike correlation is established between a set of features and the output classes. The particular features selected in the training stage of a reliable DNN model are strongly correlated with its true predictions, but only because the model has been correctly validated. This option runs into an additional problem, that of identifying the features that are strongly correlated with the correct output. In a computer vision task, the features are largely detected by the learning algorithm and not preselected, as in the previous example of the bank loan model. Since the statistical law must be true, one must be able to identify those features. Local *post hoc* interpretability methods like LIME pinpoint the features responsible for an output for any given image, but the results cannot be generalized to other similar images, and they even differ when the sampling process is repeated. Without identifying such features, no true lawlike description of the correlation can be established. Simply saying that there is an epistemically inaccessible set of features responsible for raising the probability of a correct outcome is trivially saying that the process is not random (or magical). Finally, Hempel's point about laws not being a conjunction of singular sentences about specific facts, explained in the previous section, is just as valid in both of these cases.

So, how do Erasmus et al. (2021) apply I-S explanations to DNNs? First, they argue that the explanation will not be of individual predictions, but rather of the accuracy rate of the model. This runs counter to Hempel's original intention of explaining the high predictability of individual facts; their version of I-S explanations have an entirely different explanandum. An I-S explanation of the output class, in their terms, amounts to "details about the training process as statistical laws and the nature of the training data used as empirical information. If these, taken together, inductively entail a probability that the ANN's outputs are accurate, then we have successfully explained the accuracy of the ANN's outputs" (844–845). It is not clear at all how details about the training process can become statistical laws. No further details are provided. If what they have in mind are correlations between features and true predictions, their strategy has already been examined and refuted in the previous paragraphs. In any case, the authors have given up the game of using Hempel's original I-S model, so it is difficult to see what logical structure such an explanation would have. Presumably, it would still be a covering-law model because the use of statistical laws is retained.

7.3.3 Mechanical Explanations and the New Mechanistic View

The last two models of scientific explanation I will discuss are causal in nature. Talk about causation in ML can be seen from two different perspectives. On the one hand, one might want to understand the cause of a specific prediction in a pre-trained model, regardless of whether there is any real causal connection in the world between the identified features and the prediction. On the other hand, one might want to use ML to look for some of the causes of a real-world phenomenon, in which case the features picked up by the ML model must reflect the causal structure of the world. In the present discussion, we will only be concerned with the first use of causal connections.

In *Scientific explanation and the causal structure of the world*, Salmon (1984) proposes a causal mechanical model of explanation based on the idea of a causal *process*. A causal process is characterized by its ability to transmit a mark or the ability to transmit its own structure, in a spatio-temporally continuous way. The local transfer of energy and momentum, for example, constitutes a causal process, while the successive positions occupied by a shadow cast by a moving object do not. Causal *interactions* occur when one causal process spatio-temporally intersects another and produces a modification in its structure following conservation laws. A typical example is a collision of two particles. The sum of all causal processes and interactions forms the causal structure of the world mentioned in the title of the book. To explain a particular occurrence of a phenomenon is to show how it fits into the causal structure of the world. Salmon's original idea was later refined by Dowe (2000) in the conserved quantity model of causation.

How would Salmon's causal mechanical (CM) model be used to explain a prediction of an artificial neural network (ANN)? Erasmus et al. explain:

> For CM explanations of ANNs, we would cite the causal processes and causal interactions involved. This would entail describing the ANN using oft-used terms of art, drawing on biological analogies. ANNs are constituted by connected nodes, sometimes referred to as

> artificial neurons that, like biological neurons, receive and send information, or signals, via connections resembling biological synapses, termed edges. (2021, 845)

Causal talk in this case is purely analogical. The model is not a causal structure because the entire process is mathematical in nature. The "marks" that are transmitted are not physical quantities but approximations of real numbers that are processed at each node using some non-linear function of their weighted sum.[7] These are not CM explanations at all because no genuine causal interaction is involved in the exchange of information. This is not to say that causal analogies are useless in understanding a natural or artificial process. As several authors have argued, analogies are a pathway towards understanding complex phenomena (Bartha, 2010; Hesse, 1966). But analogies are not truthful representations of their target phenomena, and therefore fail the factivity condition for scientific explanation.

Now, DNNs are grounded on genuine causal processes, namely, the electronic interactions in the hardware circuitry involved in computation. But even if there were a feasible way to access and comprehend these physical interactions, that would only add a new layer of uncertainty and complexity to the explanation since there would be no way to establish correlations between the genuine causal processes that take place in the hardware and the mathematical operations that take place in the hidden lawyers of the DNN. Furthermore, Durán (2018) has argued that computational results are underdetermined by physical interactions in the computer because each run of the algorithm instantiates a different set of physical elements in the computer. Erasmus et al. (2021, 846) seem to acknowledge that hardware processes must be treated as elementary, focusing instead on the DNN's architecture.

A similar analysis applies to the New Mechanistic view defended by Machamer et al. (2000). According to the New Mechanistic view, providing an explanation involves showing how some phenomena regularly arise from a collection of entities and activities. Both entities and activities are individuated through their spatiotemporally located physical properties, and causal processes are an essential component of any mechanism. Their approach is not far from Salmon's mechanical model: "Our emphasis on mechanisms is compatible, in some ways, with Salmon's mechanical philosophy, since mechanisms lie at the heart of the mechanical philosophy" (2000, 7). Functional descriptions of a phenomenon, on the other hand, require an understanding of the underlying mechanism, since "functions are the roles played by entities and activities in a mechanism ... [they] should be understood in terms of the activities by virtue of which entities contribute to the workings of a mechanism" (2000, 6). Therefore, functional analyses have to be causally grounded.

Applying the New Mechanistic view to ML models yields descriptions at three different levels of abstraction: as a description of the functional correlation between inputs and outputs, as a "mechanical" description of the model's network

[7] See Chap. 8 by Buijsman 2026 for further discussion of the mathematical nature of ML models and the need to think of ML explanations in non-causal terms.

architecture, and as a description of the hardware circuitry. The last two options were already discussed and dismissed in previous paragraphs. A functional analysis of the ML system would require appealing to the underlying mechanism, but that is impossible because there is no genuine mechanism to speak of. Therefore, the New Mechanistic view seems inapplicable in the context of DNNs.

Given the obvious objections to the application of a physical or mechanistic theory of causation to DNNs, one wonders why Erasmus et al. (2021) did not adopt a different causal approach, one more amenable to the context at hand. In particular, it would seem more natural to use a counterfactual theory of causal dependence such as Lewis's (1973), and a theory of counterfactual explanation, such as Woodward's (2003). Some recent approaches to XAI have used precisely this approach (Buijsman, 2022; Wachter et al., 2018; Watson & Floridi, 2021; see Durán, 2025, for an overview). The reason they offer is that counterfactual approaches to explanation are highly pragmatic and therefore incompatible with the indefeasibility thesis about explanation (Erasmus et al., 2021, 839). But if the indefeasibility thesis is false, at least in the context of DNNs, then it ceases to be a reason to reject this approach. The counterfactual approach to explanation in ML has its own problems, which I will discuss in the final section of the chapter, but at least it does not depend on the existence of physical laws, mechanisms, or processes, which seem completely foreign in the context of machine learning.

7.4 A Plea for Understanding

If the attempt to adapt an extant account of scientific explanation to ML is a hopeless endeavor, there are three remaining options: (i) either to adopt a consensual or stipulative definition of "explanation" in ML; (ii) to abandon the factivity condition for explanation; or (iii) to abandon the idea that there is a unique way of understanding what an explanation is in the context of ML. The first option seems entirely unworkable and arbitrary. The second one is mostly associated with pragmatic theories of explanation (Achinstein, 1983; van Fraassen, 1980). These theories have been fruitfully used to clarify the pragmatic context in which explanations are sought in AI (Miller, 2019, 2021). However, there is a tendency to analyze the concept in terms of its empirical usage, without much normative concern.[8] In previous work (Páez, 2019), I have defended the third option. I have argued that XAI ought to take a turn towards a more pragmatic approach in which the focus of attention shifts from the explanation to the understanding of ML systems. If we focus on the cognitive and practical needs of the different stakeholders involved in designing, implementing, and using an ML model, there will be a wide variety of options available to make the model and its outputs understandable. Whether one calls these paths to understanding "explanations"

[8] I offer a critique of purely pragmatic theories of scientific explanation in Páez (2006).

becomes largely irrelevant. In a well-known survey of XAI methods, Guidotti et al. conclude that one of the most important open problems in XAI is that there is no agreement on what an explanation is:

> Indeed, some works provide as explanation a set of rules, others a decision tree, others a prototype (especially in the context of images). It is evident that the research activity in this field is not providing yet a sufficient level of importance in the study of a general and common formalism for defining an explanation, identifying which are the *properties* that an explanation should guarantee, e.g., soundness, completeness, compactness and comprehensibility. (2018, p. 36)

I am highly skeptical that a formal definition of explanation in terms of necessary and sufficient conditions or properties can be found. Instead of trying to define "explanation" in ML, it would be more fruitful to think of the provision of understanding as the common element that "defines," in a sense, what all XAI methods have in common. This allows a plurality of explanatory methods to flourish as long as they provide understanding to the system's users. A robust account of understanding in ML should provide an adequate grounding for all such methods. In this final section, I want to further develop the account of understanding in ML that I offered in my 2019 paper by adding the idea that understanding is a *success concept* in the sense explained below.

7.4.1 Understanding as a Success Concept

In epistemology, understanding is often distinguished from knowledge.[9] There are two main differences between these concepts. First, understanding is seen as a higher epistemic achievement than knowledge (Kvanvig, 2003; Pritchard, 2010). I can come to know that my coffee is cold just by sipping from the cup, while understanding why my coffee is cold involves relating the coffee's temperature to the laws of thermodynamics. Secondly, the objects of understanding are generally more complex and structured than the objects of knowledge (Zagzebski, 2019). We want to understand the stock market, the theory of relativity, or the New York subway system. Even when we try to understand something simple like my coffee getting cold, the fact must be inserted into a broader, more complex theoretical context.

Some philosophers have argued that understanding is simply knowledge of causes: if I know the cause of *p*, I understand why *p* (Lipton, 2004). Similarly, Khalifa (2017) argues that understanding is knowledge of explanations. De Regt

[9] Needless to say, this opinion is not unanimous among philosophers. I cannot discuss all the arguments for and against this view in this chapter, but if my claim that there are non-factive paths to understanding in ML is correct, the absence of truth will prevent these cases of understanding from being reduced to some species of knowledge.

offers the following example to show that *knowing* why *p* is not equivalent to *understanding* why *p*:

> Merely knowing that global warming is caused by the increase of CO_2 in the atmosphere does not yet amount to understanding it. A student may be able to answer the question "Why does global warming happen"? correctly by answering "Because of the increase of CO_2 in the atmosphere". But this does not imply that she understands why global warming occurs —she merely knows what its cause is. ... The student understands why global warming happens if she not only knows that it is caused by the increase of CO_2, but also "grasps" the causal, explanatory relation between cause and effect. In this case, she needs a theory or model of the climate system that includes the greenhouse effect. (2023, 19–20)

Understanding thus requires connecting different pieces of knowledge. Kvanvig explains: "Understanding requires the grasping of explanatory and other coherence-making relationships in a large and comprehensive body of information. One can know many unrelated pieces of information, but understanding is achieved only when informational items are pieced together by the subject in question" (2003, 192). In a similar vein, Zagzebski argues that understanding "involves grasping relations of parts to other parts and perhaps the relation of parts to a whole" (2009, 144). The kinds of relations she has in mind can be spatial, temporal, or causal. "It seems to me that one's mental representation of the relations one grasps can be mediated by maps, graphs, diagrams, and three-dimensional models in addition to, or even in place of, the acceptance of a series of propositions" (2009, 145). For Grimm, understanding a complex object such as the New York subway system implies an apprehension of how a thing works, "how the various elements of the thing relate to, and depend upon, one another" (2011, 86).

Now, what are the conditions of satisfaction of the "grasping" often mentioned in these characterizations of understanding? Traditional opponents of understanding as an interesting philosophical notion, from Hempel (1965) to Trout (2002), often dismissed understanding as a psychological, subjective state that cannot be rigorously analyzed. This attitude is less common today, but there is still no agreement on what the grasping entails. A useful way of thinking about the conditions of satisfaction for grasping the interconnectedness of different facts is in terms of an agent's success in *using* the information. This idea has been defended in different guises by philosophers of science like Ylikoski (2009), De Regt (2017), and Kuorikoski (2011, 2023). In the inferential conception of understanding defended by Kuorikoski and Ylikoski, for example, understanding "is not only about learning and memorizing true propositions, but about the capability to put one's knowledge to use. To understand is to be able to tell what would have happened if things had been different, what would happen if certain things were changed, and what ways there are to bring about a desired change in the outcome" (2015, 3819). More specifically, understanding can be equated "with the ability to draw correct counterfactual what-if inferences about the object of understanding. ... To understand a phenomenon is to be able to correctly situate it within a space of possibilities" (Kuorikoski, 2023, 218). De Regt also argues that genuine understanding manifests itself as a skill:

an agent must have the ability to use his knowledge.[10] Among the most important uses of knowledge is the construction of simple, idealized models of some complex phenomenon, which serve as mediators between abstract theories and empirical data. The construction of a successful model with the right idealizing assumptions requires an understanding of the interconnectedness of the data and of the way in which the theory can be applied to the model. Successfully putting one's knowledge to use is not limited to reasoning counterfactually and building models. Being able to fix or improve a system, to profit from it, or to game it, are examples of the many possible ways in which usage is a sign of understanding.[11]

An additional pragmatic aspect of this approach is that there is no unique way of measuring success in using a body of knowledge. There is no unique benchmark for understanding. In De Regt's view, understanding is contextual. Criteria for understanding and intelligibility depend on the historical and disciplinary context:

> In the seventeenth century, for example, the generally accepted view was that only a mechanics based on contact action is intelligible, whereas in the eighteenth century (as a result of the success of Newton's theory of gravitation) action-at-a-distance was regarded as the paradigm of intelligibility ... Of course, one might claim that only one of these positions is the correct one, but that would not do justice to the history of science. (2023, 19)

Kuorikoski defends a similar idea: "Historical ruptures in foundational metaphysics, such as the shift from Aristotelian species-essences to Darwinian population thinking or from the classical mechanical conception of the physical reality to the quantum field theory have also constituted fundamental changes in the very criteria of understandability" (2023, 217–218). Furthermore, new sub-fields of science, such as machine learning, "are often built on specific conceptions of scientific understanding" (2023, 217). As we will see in the next section, the contextual aspects of understanding in ML cut even deeper.

Despite their acceptance of different conceptions of understandability, Kuorikoski and Ylikoski are fully committed to a factive conception of understanding: drawing incorrect counterfactual inferences would presumably be a sign of a lack of understanding. However, it is difficult to see how a modal approach based on counterfactuals can preserve the factivity condition. Many counterfactual statements are empirically unverifiable and can only be accepted as possible consequences of a scientific theory, or at least of a causal model, which itself can never be absolutely confirmed. De Regt also uses the ability to infer counterfactuals as a criterion for understanding, but correctly rejects the factivity condition. In his approach, theories used to reason counterfactually have to be intelligible and the inferences drawn from them must conform to the basic epistemic values of empirical adequacy and internal consistency. But both intelligibility and empirical

[10] But see Sullivan (2018) for an argument against this claim.

[11] It is worth noting that Wachter et al., the proponents of one of the most influential counterfactual methods in XAI, drive a wedge between understanding and use. They argue that explanations in ML should be "a means to help a data-subject *act* rather than merely understand" (2018, 843).

adequacy fall short of truth. Naturally, the discussion of whether understanding is factive is not limited to the nature of counterfactual inferences, and the issue has been extensively discussed in epistemology and the philosophy of science. I cannot do justice to the problem here, but this much is clear: an approach to understanding based on pragmatic success need not be subject to the strict alethic standards of knowledge.

Finally, this approach also has the advantage that a person's understanding of a phenomenon or of a subject matter is empirically verifiable. Ylikoski puts the matter thus:

> When we evaluate somebody's understanding, we are not making guesses about his or her internal representations, but about the person's ability to perform according to set standards. The concept of understanding allows that the ability can be grounded in various alternative ways, as long as the performance is correct. Furthermore, the correctness of the internal model is judged by the external displays of understanding, not the other way around. This makes understanding a behavioral concept. (2009, 102)

Having intersubjective criteria for a person's understanding has the advantage of providing ways to test the effectiveness of different models, methods, and devices that aid with understanding a phenomenon. Since these criteria are contextual, this approach also allows for the design of tools that aid comprehension with different users and populations in mind (Lage et al., 2019). This will prove to be essential when we think of the different stakeholders involved in the use of machine learning systems.

7.4.2 Pragmatic Understanding in ML

According to the account provided in the previous section, understanding a complex object requires being able to identify its various parts and their interdependence to a degree that enables the agent to use that knowledge in different ways. This is precisely what surrogate models in ML help bring about. Algorithms such as rule lists, sparse decision trees or linear models provide a simplified version of the original model by identifying its main features, the possible interactions between them, and their combined effect on the output. Knowledge of these individual elements without grasping their interconnectedness will not provide understanding of the original model. Furthermore, decision trees are structures that allow counterfactual reasoning. A user can think counterfactually simply by following paths in the tree that do not lead to the original output. Rule lists can also cover counterfactual cases, although this requires a much less intuitive combination of rules.[12] Both designing a surrogate model and using it correctly are evidence that the person understands the target model. A person's understanding of the

[12] Lakkaraju et al. (2016) have shown that decision trees in general are easier to understand than rule lists.

target model admits of degrees, depending on the person's grasp of the connections displayed in the surrogate model.

Consider an example. Letham et al. (2015) introduced interpretable prediction models that take the form of sparse decision lists. Each list consists of a series of association rules in the form of if-then statements. The rules are pre-mined from the input space using the FP-Growth algorithm (Borgelt, 2005). The resulting models have the same level of complexity as medical scoring systems, thus making them easy to use for clinical practitioners. In terms of performance in a stroke risk classification task, the decision lists were comparable to that of support vector machines (SVMs), and not substantially worse than L^1 logistic regression and random forests trained on the same data. Other examples of simple surrogate models are the decision trees for diabetes risk prediction introduced by Bastani et al. (2017) and the checklists proposed by Jung et al. (2017).

Like most models in the social and natural sciences, surrogate models in ML are not factive. They are "true enough" of their target, to use Elgin's (2017) suggestive phrase. Potochnik (2017) argues that in many cases the commonality between representations and what they represent can be understood in terms of functional similarities. Following Potochnik, I think that the relation between surrogate models and their ML targets should be seen in functional terms as well. Functional representation is a pragmatic concept that depends on the specific functions of interest to the modeler or user. Understanding function means being able to build or use the model and manipulate its features to obtain the desired result. We can think of surrogate models as epistemic tools (Currie, 2017, Knuuttila, 2011) designed to capture functional similarities of interest. A correct use of the tool will reveal that the user understands the function of the parts and their overall fit with the target.

In the previous section, we saw that drawing counterfactual inferences is often seen as a sign of understanding. So, why not use local counterfactual XAI methods directly instead of global surrogate models to achieve understanding of the target model? There are several reasons. The first one is that counterfactual methods suffer from a lack of robustness. Like perturbation-based methods, local counterfactual methods can be easily manipulated and may converge toward drastically different explanations under small perturbations (Slack et al., 2021). Counterfactual probes also critically depend on closeness metrics but there is no principled way to decide which metric to use in any given case, especially since most variables are not causally independent. And like saliency-based methods, the lack of causal grounding can deliver sub-optimal or even erroneous explanations to decision-makers (Chou et al., 2022; Karimi et al., 2021).

To complicate things, there seems to be a lack of interest in the AI community to test whether the counterfactual methods they employ are actually understood by non-experts. A recent survey shows that only 36 out of 117 (31%) research papers evaluating counterfactual explanations included user studies (Keane et al., 2021). As the authors state in the beginning of the paper, "the XAI community is busily developing technical solutions that may have no practical benefits to people in real-life" (2021, 1).

The third reason is epistemologically deeper. Genuine counterfactuals are theory or model dependent. Counterfactuals need a possibility space that can only be defined using lawlike statements and background conditions, or by means of a fully specified causal model. Otherwise, there will be no basis to answer what-if-things-had-been-different questions. Furthermore, as Beckers (2022) points out, understanding the space of possibilities also requires understanding which factors cannot be changed or manipulated in a counterfactual situation. Current counterfactual methods in XAI make simplifying assumptions that land them far from the required conditions to qualify as based on bona fide counterfactuals. In particular, many methods assume that the input variables are causally independent. The strategy is to keep all variables at a fixed value except one and see how a change in that variable produces a different output.[13] Wachter et al. for example, define an explanation of the output of an algorithm as having the following format: "Score p was returned because variables V had values (v_1, v_2, ...) associated with them. If V instead had values ($v_1{}'$, $v_2{}'$, ...), and all other variables had remained constant, score p would have been returned" (2018, 848). The method tells you how to change the output, but not why the output changes. It does not provide an understanding of the process because there is no grasp of the logical and causal connections between the features, and between the features and the prediction. Again, *knowing* why p changed is not the same as *understanding* why p changed.

A similar argument can be used against local saliency and perturbation methods: they tell users which features are relevant to an individual prediction but not why they are relevant, that is, users *know* what caused the output, but they do not *understand* why. The lack of understanding is demonstrated by the difficulties users have in predicting the "explanation" for new, similar input-output pairs. In an empirical study of saliency maps generated by a layer-wise relevance propagation (LRP) algorithm (Bach et al., 2015) on an image classification task using a convolutional neural network, Alqaraawi et al. found that "the maps seem to provide very limited help for participants to anticipate the network's output for new images" (2020, 275). Post hoc local interpretability methods are useful for developers trying to adjust the parameters of a model, but they seem of limited use for other stakeholders.

What is missing from all of these local methods is a grasp of the causal and logical relationships in the input space. Users must build a *mental model* of the relevant features and the possible outcomes before they can make inferences about the target model. In an experiment designed to measure the effectiveness of different global XAI methods, van der Waa et al. (2021) used two behavioral measures of a participant's understanding of the model:

> The first behavioral measurement assessed the participant's capacity to correctly identify the decisive factor of the situations in the system's advice. This measured to what extent the participant recalled what factor the system believed to be important for a specific advice and

[13] Karimi et al. (2021) suggest that "it is perhaps more appropriate to refer to these approaches as contrastive, rather than counterfactual explanations" (2021, 360).

> situation. Second, we measured the participant's ability to accurately predict the advice in novel situations. This tested whether the participant obtained a mental model of the system that was sufficiently accurate enough to predict its behavior in novel situations. (2021, 16–17)

The authors found that presenting participants only with counterfactuals about the system's advice did not improve their identification or predictive abilities when compared to no explanations at all. When presented with rule-based explanations without counterfactuals, the first behavioral measure of understanding improved, but not the second. Lim et al. (2009) also present empirical evidence that suggests that offering users of ML systems decision rules encoded in decision trees yields superior results in terms of both understanding and trust when compared to counterfactual explanations. Buijsman uses this evidence to conclude that "a counterfactual is only helpful when it suggests a reasonable generalization. . . . A pure case-by-case approach, where counterfactuals are presented but without overarching generalizations, doesn't truly explain the functioning of an algorithm" (2022, 569).

In the context of algorithmic recourse, Karimi et al. (2021) also argue that the possibility of changing an unfavorable outcome requires causal knowledge. The authors attribute the shortcomings of existing XAI counterfactual approaches "to their lack of consideration for real-world properties, specifically the causal relationships governing the world in which actions will be performed" (2021, 354). The question is, in what form should causal relations be displayed to be understandable to lay users, and how precise and comprehensive should those causal relationships be? It is well-known that is extremely difficult to specify a full causal model. Despite requiring real-world causal relationships, the authors admit that "in practice, the underlying causal model is rarely known" (360). Since it is extremely difficult to build a factual causal model, algorithmic recourse also needs the understanding provided by non-factual, simpler surrogate models.[14]

As I see it, a surrogate model needs to be "causal enough"–to paraphrase Elgin's phrase– for a user to understand how the parts are interconnected and interdependent, but context will determine the depth of the causal knowledge that needs to be displayed in the model. In scientific contexts, one should expect a highly detailed map of the causal relationships, based on solid background scientific knowledge (Sullivan, 2022), but other contexts might only require sparse decision trees with very few variables and a higher tolerance for spurious correlations. The complexity of the model will be determined by the explanatory requirements of the stakeholders involved (Zednik, 2021). Even the normative standards for transparency that have been proposed recently, such as IEEE P7001 (Winfield et al., 2021), recognize that transparency depends entirely on the background knowledge of the stakeholders trying to understand the model. Specifying the exact shape that

[14] Sullivan and Kasirzadeh (see Chap. 10 2026) go even further and argue that providing users with explanations that aim at understanding AI decisions is ethically preferable than providing them with recourse explanations.

these causal surrogate models should take is a very difficult challenge that falls beyond the scope of this chapter, but causal machine learning is a very active field of research that might provide better paths toward understandable AI.

7.5 Conclusion

The use of the term "explanation" in AI is a conceptual jumble. I believe no amount of conceptual analysis will result in a clear definition or in a more widely accepted set of usage rules. In this chapter, I have shown that the attempt to ground the term in the way it is used in the philosophy of science is unsuccessful and wrongheaded. Explanations in AI are neither logical nor physical nor mechanical nor nomological. They are not even factual. A much better option is to abandon the definitional enterprise and focus instead on the result of what "explanations" are supposed to achieve: understanding of opaque machine learning systems. I have argued for a pragmatic notion of understanding that is empirically testable and broad enough to allow many different approaches and methods.

Acknowledgements I would like to thank Emily Sullivan and Juan M. Durán for their careful reading of the manuscript and for their detailed comments, which have been immensely useful. I also profited from useful discussions with Victoria Martín del Campo and Stefan Buijsman during the session organized by Juan as part of the XVIIth Congress on Logic, Methodology, and Philosophy of Science and Technology in Buenos Aires. As always, I am indebted to the late Arnold Koslow, from whom I learned everything I know about scientific explanation.

References

Achinstein, P. (1983). *The nature of explanation*. Oxford University Press.

Alqaraawi, A., Schuessler, M., Weiß, P., Costanza, E., & Berthouze, N. (2020). Evaluating saliency map explanations for convolutional neural networks: A user study. In *IUI '20: Proceedings of the 25th international conference on intelligent user interfaces* (pp. 275–285). ACM.

Ancona, M., Ceolini, E., Öztireli, C., & Gross, M. (2019). Gradient-based attribution methods. In W. Samek, G. Montavon, A. Vedaldi, L. K. Hansen, & K.-R. Müller (Eds.), *Explainable AI: Interpreting, explaining and visualizing deep learning* (pp. 169–191). Springer.

Bach, S., Binder, A., Montavon, G., Klauschen, F., Müller, K.-R., & Samek, W. (2015). On pixel-wise explanations for non-linear classifier decisions by layer-wise relevance propagation. *PLoS One, 10*(7), e0130140.

Bartha, P. F. (2010). *By parallel reasoning: The construction and evaluation of analogical arguments*. Oxford University Press.

Bastani, O., Kim, C., & Bastani, H. (2017). Interpreting blackbox models via model extraction. *arXiv preprint:1705.08504*.

Beckers, S. (2022). Causal explanations and XAI. In *Proceedings of the first conference on causal learning and reasoning, PMLR* (Vol. 177, pp. 90–109).

Borgelt, C. (2005). An implementation of the FP-growth algorithm. In *OSDM'05 proceedings of the 1st international workshop on open source data mining: Frequent pattern mining implementations* (pp. 1–5). ACM.

Buijsman, S. (2022). Defining explanation and explanatory depth in XAI. *Minds and Machines, 32*(3), 563–584.

Buijsman, S. (2026). Machine learning models as mathematics. In: J. M. Durán, & G. Pozzi (Eds.), *Philosophy of science for machine learning: Core issues and new perspectives*. Synthese Library, Springer.

Chou, Y.-L., Moreira, C., Bruza, P., Ouyang, C., & Jorge, J. (2022). Counterfactuals and causability in explainable artificial intelligence: Theory, algorithms, and applications. *Information Fusion, 81*, 59–83.

Currie, A. (2017). From models-as-fictions to models-as-tools. *Ergo, 4*(27), 759–781.

De Regt, H. W. (2017). *Understanding scientific understanding*. Oxford University Press.

De Regt, H. W. (2023). Can scientific understanding be reduced to knowledge? In I. Lawler, K. Khalifa, & E. Shech (Eds.), *Scientific understanding and representation: Modeling in the physical sciences* (pp. 17–32). Routledge.

Dowe, P. (2000). *Physical causation*. Cambridge University Press.

Durán, J. M. (2018). *Computer simulations in science and engineering: Concepts-practices-perspectives*. Springer.

Durán, J. M. (2025). Causal-manipulationist approaches to explanation in machine learning. In P. Illary & F. Russo (Eds.), *The Routledge handbook of causality and causal methods* (pp. 364–376). Routledge.

Elgin, C. Z. (2017). *True enough*. MIT Press.

Erasmus, A., & Brunet, T. D. P. (2022). Interpretability and unification. *Philosophy & Technology, 35*(2), 1–6.

Erasmus, A., Brunet, T. D. P., & Fisher, E. (2021). What is interpretability? *Philosophy & Technology, 34*(4), 833–862.

Frosst, N., & Hinton, G. (2017). Distilling a neural network into a soft decision tree. *arXiv preprint:1711.09784*.

Goodfellow, I., Bengio, Y., & Courville, A. (2016). *Deep learning*. MIT Press.

Grimm, S. R. (2011). Understanding. In S. Bernecker & D. Pritchard (Eds.), *The Routledge companion to epistemology* (pp. 84–94). Routledge.

Guidotti, R., Monreale, A., Ruggieri, S., Turini, F., Giannotti, F., & Pedreschi, D. (2018). A survey of methods for explaining black box models. *ACM Computing Surveys (CSUR), 51*(5), 1–42.

Hempel, C. G. (1965). *Aspects of scientific explanation*. The Free Press.

Hempel, C. G., & Oppenheim, P. (1948). Studies in the logic of explanation. *Philosophy of Science, 15*(2), 135–175.

Hesse, M. B. (1966). *Models and analogies in science*. University of Notre Dame Press.

Ivanovs, M., Kadikis, R., & Ozols, K. (2021). Perturbation-based methods for explaining deep neural networks: A survey. *Pattern Recognition Letters, 150*(2021), 228–234.

Jung, J., Concannon, C., Shroff, R., Goel, S., & Goldstein, D. G. (2017). Simple rules for complex decisions. *arXiv preprint:1702.04690*.

Karimi, A.-H., Schölkopf, B., & Valera, I. (2021). Algorithmic recourse: from counterfactual explanations to interventions. In *FACCT '21: Proceedings of the 2021 ACM conference on fairness, accountability, and transparency* (pp. 353–362). ACM.

Keane, M. T., Kenny, E. M., Delaney, E., & Smyth, B. (2021). If only we had better counterfactual explanations: Five key deficits to rectify in the evaluation of counterfactual XAI techniques. *arXiv preprint arXiv:2103.01035*.

Khalifa, K. (2017). *Understanding, explanation, and scientific knowledge*. Cambridge University Press.

Kim, S., Jeong, M., & Ko, B. C. (2022). Lightweight surrogate random forest support for model simplification and feature relevance. *Applied Intelligence, 52*(1), 471–481.

Kim, B., Khanna, R., & Koyejo, O. O. (2016). Examples are not enough, learn to criticize! Criticism for interpretability. *Advances in Neural Information Processing Systems, 29*.

Kim, B., Wattenberg, M., Gilmer, J., Cai, C., Wexler, J., & Viegas, F. (2018). Interpretability beyond feature attribution: Quantitative testing with concept activation vectors (TCAV). In *International conference on machine learning, PMLR* (Vol. 80, pp. 2668–2677).

Kindermans, P.-J., Hooker, S., Adebayo, J., Alber, M., Schütt, K. T., Dähne, S., Erhan, D., & Kim, B. (2017). The (un)reliability of saliency methods. In W. Samek, G. Montavon, A. Vedaldi, L. K. Hansen, & K. R. Müller (Eds.), *Explainable AI: Interpreting, explaining and visualizing deep learning* (pp. 267–280). Springer.

Kitcher, P. (1989). Explanatory unification and the causal structure of the world. In P. Kitcher & W. Salmon (Eds.), *Scientific explanation* (pp. 410–505). University of Minnesota Press.

Knuuttila, T. (2011). Modelling and representing: An artefactual approach to model-based representation. *Studies in History and Philosophy of Science, Part A, 42*(2), 262–221.

Koh, P. W., Ang, K.-S., Teo, H., & Liang, P. S. (2019). On the accuracy of influence functions for measuring group effects. *Advances in Neural Information Processing Systems, 32*.

Koh, P. W., & Liang, P. (2017). Understanding black-box predictions via influence functions. In *Proceedings of the 34th international conference on machine learning, PMLR* (Vol. 70, pp. 1885–1894).

Kuorikoski, J. (2011). Simulation and the sense of understanding. In P. Humphreys & C. Imbert (Eds.), *Models, simulations, and representations* (pp. 168–187). Routledge.

Kuorikoski, J. (2023). Factivity, pluralism, and the inferential account of scientific understanding. In I. Lawler, K. Khalifa, & E. Shech (Eds.), *Scientific understanding and representation* (pp. 217–233). Routledge.

Kuorikoski, J., & Ylikoski, P. (2015). External representations and scientific understanding. *Synthese, 192*, 3817–3837.

Kvanvig, J. L. (2003). *The value of knowledge and the pursuit of understanding*. Cambridge University Press.

Lage, I., Chen, E., He, J., Narayanan, M., Kim, B., Gershman, S., & Doshi-Velez, F. (2019). An evaluation of the human-interpretability of explanation. *arXiv preprint arXiv:1902.00006*.

Lakkaraju, H., Bach, S. H., & Leskovec, J. (2016). Interpretable decision sets: A joint framework for description and prediction. In *Proceedings of the 22nd ACM SIGKDD international conference on knowledge discovery and data mining* (pp. 1675–1684). ACM.

Letham, B., Rudin, C., McCormick, T. H., & Madigan, D. (2015). Interpretable classifiers using rules and Bayesian analysis: Building a better stroke prediction model. *The Annals of Applied Statistics, 9*(3), 1350–1371.

Lewis, D. (1973). Causation. *Journal of Philosophy, 70*, 556–567.

Lim, B. Y., Dey, A. K., & Avrahami, D. (2009). Why and why not explanations improve the intelligibility of context-aware intelligent systems. In *CHI '09: Proceedings of the SIGCHI conference on human factors in computing systems* (pp. 2119–2128). ACM.

Lipton, P. (2004). *Inference to the best explanation* (2nd ed.). Routledge.

Lundberg, S. M., & Lee, S.-I. (2017). A unified approach to interpreting model predictions. In *Advances in neural information processing systems, 30*.

Machamer, P., Darden, L., & Craver, C. F. (2000). Thinking about mechanisms. *Philosophy of Science, 67*(1), 1–25.

Miller, T. (2019). Explanation in artificial intelligence: Insights from the social sciences. *Artificial Intelligence, 267*, 1–38.

Miller, T. (2021). Contrastive explanation: A structural-model approach. *The Knowledge Engineering Review, 36*, e14.

Molnar, C. (2020). *Interpretable machine learning. A guide for making black box models explainable*. Lulu.com.

Mothilal, R. K., Sharma, A., & Tan, C. (2020). Explaining machine learning classifiers through diverse counterfactual explanations. In *Proceedings of the 2020 conference on fairness, accountability, and transparency* (pp. 607–617). ACM.

Páez, A. (2006). *Explanations in K. An analysis of explanation as a belief revision operation*. Athena Verlag.

Páez, A. (2009). Artificial explanations: The epistemological interpretation of explanation in AI. *Synthese, 170*(1), 131–146.

Páez, A. (2019). The pragmatic turn in explainable artificial intelligence (XAI). *Minds and Machines, 29*(3), 441–459.

Potochnik, A. (2017). *Idealization and the aims of science*. University of Chicago Press.
Pritchard, D. (2010). Understanding. In A. Haddock, A. Millar, & D. Pritchard (Eds.), *The nature and value of knowledge: Three investigations* (pp. 66–88). Oxford University Press.
Ribeiro, M. T., Singh, S., & Guestrin, C. (2016). "Why should I trust you?" Explaining the predictions of any classifier. In *Proceedings of the 22nd ACM SIGKDD international conference on knowledge discovery and data mining* (pp. 1135–1144). ACM.
Ross, L., & Woodward, J. (2023). Causal approaches to scientific explanation. In E. Zalta & U. Nodelman (Eds.), *The Stanford encyclopedia of philosophy (Spring 2023 ed.)*. https://plato.stanford.edu/archives/spr2023/entries/causal-explanation-science/
Rudin, C. (2019). Stop explaining black box machine learning models for high stakes decisions and use interpretable models instead. *Nature Machine Intelligence, 1*, 206–215.
Salmon, W. C. (1984). *Scientific explanation and the causal structure of the world*. Princeton University Press.
Salmon, W. C. (1989). *Four decades of scientific explanation*. University of Minnesota Press.
Schoenborn, J. M., Weber, R. O., Aha, D. W., Cassens, J., & Althoff, K.-D. (2021). Explainable case-based reasoning: A survey. In *AAAI-21 workshop proceedings*.
Selvaraju, R. R., Cogswell, M., Das, A., Vedantam, R., Parikh, D., & Batra, D. (2017). Grad-cam: Visual explanations from deep networks via gradient-based localization. In *Proceedings of the IEEE international conference on computer vision* (pp. 618–626).
Slack, D., Hilgard, A., Lakkaraju, H., & Singh, S. (2021). Counterfactual explanations can be manipulated. *Advances in Neural Information Processing Systems, 34*, 62–75.
Sullivan, E. (2018). Understanding: Not know-how. *Philosophical Studies, 175*(1), 221–240.
Sullivan, E. (2022). Understanding from machine learning models. *The British Journal for the Philosophy of Science, 73*(1), 109–133.
Sullivan, E., & Kasirzadeh, A. (2026). Explanation hacking: The perils of algorithmic recourse. In J. M. Durán & G. Pozzi (Eds.), *Philosophy of science for machine learning: Core issues and new perspectives*. Synthese Library, Springer.
Tan, S., Caruana, R., Hooker, G., & Lou, Y. (2018). Distill-and-compare: Auditing black-box models using transparent model distillation. In *Proceedings of the 2018 AAAI/ACM conference on AI, ethics, and society* (pp. 303–310).
Trout, J. D. (2002). Scientific explanation and the sense of understanding. *Philosophy of Science, 69*(2), 212–233.
van der Waa, J., Nieuwburg, E., Cremers, A., & Neerincx, M. (2021). Evaluating XAI: A comparison of rule-based and example-based explanations. *Artificial Intelligence, 291*, 103404.
van Fraassen, B. (1980). *The scientific image*. Clarendon Press.
Wachter, S., Mittelstadt, B., & Russell, C. (2018). Counterfactual explanations without opening the black box: Automated decisions and the GDPR. *Harvard Journal of Law & Technology, 31*(2), 841–887.
Watson, D., & Floridi, L. (2021). The explanation game: A formal framework for interpretable machine learning. In L. Floridi (Ed.), *Ethics, governance, and policies in artificial intelligence* (pp. 185–219). Springer.
Winfield, A. F., Booth, S., Dennis, L. A., Egawa, T., Hastie, H., Jacobs, N., Muttram, R. I., et al. (2021). IEEE P7001: A proposed standard on transparency. *Frontiers in Robotics and AI, 8*, 665729.
Woodward, J. (2003). *Making things happen*. Oxford University Press.
Woodward, J., & Ross, L. (2021). Scientific explanation. In E. N. Zalta (Ed.), *The Stanford encyclopedia of philosophy (Summer 2021 ed.)*. https://plato.stanford.edu/archives/sum2021/entries/scientific-explanation/
Wu, M., Hughes, M., Parbhoo, S., Zazzi, M., Roth, V., & Doshi-Velez, F. (2018). Beyond sparsity: Tree regularization of deep models for interpretability. In *Proceedings of the AAAI conference on artificial intelligence* (Vol. 32, No. 1, pp. 1670–1678).
Ylikoski, P. (2009). The illusion of depth of understanding in science. In H. De Regt, S. Leonelli, & K. Eigner (Eds.), *Scientific understanding: Philosophical perspectives* (pp. 100–119). Pittsburgh University Press.

Zagzebski, L. (2009). *On epistemology*. Wadsworth.
Zagzebski, L. (2019). Towards a theory of understanding. In P. Grimm (Ed.), *Varieties of understanding: New perspectives from philosophy, psychology, and theology* (pp. 123–135). Oxford University Press.
Zednik, C. (2021). Solving the black box problem: A normative framework for explainable artificial intelligence. *Philosophy & Technology, 34*(2), 265–288.

Andrés Páez is Professor of Philosophy and Researcher in the Center for Research and Formation in Artificial Intelligence (CinfonIA) at the Universidad de los Andes in Bogotá, Colombia. His current areas of research are the philosophy of artificial intelligence, legal and social epistemology, and the philosophy of science. He is the former President of the Latin American Association for Analytic Philosophy (ALFAn) and member of the editorial board of *Minds & Machines*.

Chapter 8
Machine Learning Models as Mathematics: Interpreting Explainable AI in Non-causal Terms

Stefan Buijsman

Abstract We would like to have a wide range of explanations for the behaviour of machine learning systems. However, how should we understand these explanations? Typically, attempts to clarify what an explanations for questions such as 'why am I getting this output for these inputs?' have been approached from the philosophy of science, through an analogy with scientific (and often causal) explanations. I show that ML systems are best thought of as noncausal, specifically mathematical objects. We should therefore interpret these explanations differently, through analogy with mathematical explanations. I show that this still allows us to use much of the same theoretical apparatus, and argue that the asymmetry of many of the standard ML explanations can be accounted for in virtue of the link these systems have with concrete implementations.

8.1 Introduction

Machine Learning (ML) systems are notorious for requiring explanations. We do not know why they give the outputs that they do, and we even do not know why machine learning systems are as accurate as they are. In response, a wide variety of tools has been developed to try to elucidate the outputs of ML systems (Das & Rad, 2020), and theories have been developed to explain the success of ML systems (Shwartz-Ziv & Tishby, 2017). There are feature importance methods that calculate which input features were the most important for a particular output (e.g. Ribeiro et al., 2016; Lundberg & Lee, 2017), counterfactual methods that calculate the nearest alternative input for which the ML system returns a different (desired) output (Karimi et al., 2020) and much more. In turn, this has put forward the more philosophical issue of what it would mean to explain the behaviour (in a broad sense) of ML systems. This chapter focuses on that question, in particular in the light of

S. Buijsman (✉)
Technology, Policy and Management, Delft University of Technology, Delft, The Netherlands
e-mail: S.N.R.Buijsman@tudelft.nl

J. M. Durán, G. Pozzi (eds.), *Philosophy of Science for Machine Learning*,
Synthese Library 527, https://doi.org/10.1007/978-3-032-03083-2_8

the mathematical aspects of machine learning systems. As discussed in Sect. 8.2, we can understand ML systems as implementing mathematical functions. We use optimization procedures (typically stochastic gradient descent) to fit this function to the data set that is available, with the aim to then classify, predict or produce new instances. All of this suggests that ML systems are abstract objects. Moreover, this idea of an ML system being linked to a mathematical function is also relied on by some of the explainability methods just mentioned. For example, Simonyan et al. (2013) highlight pixels for which the gradient of the implemented function is particularly steep, thus directly using the mathematical function approximated by ML systems in the XAI method.

I will dive into this mathematical aspect of ML systems throughout the chapter. It is noteworthy, however, because most philosophical accounts at the moment approach the interpretation of explanations of ML systems using theories of scientific explanation, leaving the abstract, mathematical, nature of machine learning aside. For example, the overview in Beisbart and Räz (2022, p. 2) mentions for a sub-type of ML systems: "DNNs are mathematical models" but includes no accounts of XAI in terms of mathematical explanations. More concretely, Buijsman (2022) and Watson and Floridi (2021) both use the causal interventionist framework from Halpern and Pearl (2005) and Woodward (2003) to present an account of explanations of ML systems. Erasmus et al. (2021) apply four different accounts of scientific explanation (deductive-nomological, Inductive Statistical, Causal-Mechanical and New Mechanist) to the machine learning context. There is a strong tendency to appeal to accounts of explanation from the philosophy of science, even though these have been developed primarily for describing concrete phenomena.

At the same time, technical explainability tools suggest that an analogy of this kind is suitable. Biswas et al. (2022) and Geiger et al. (2023), based respectively on theoretical work by Buijsman (2022) and Beckers and Halpern (2019), are examples of concrete explainability tools that use causal inference techniques in order to provide answers to why-questions about the outputs of ML models. In short, applying accounts developed for scientific explanations to the case of ML systems is standard practice, and also seems to make practical sense. Yet, there is a tension with the abstract, mathematical, nature of ML systems.

In Sect. 8.2 I discuss this tension in more detail, and argue that it is better not to think of ML systems in strict causal terms. However, I do think that we can make sense of interventions to ML systems, despite their being non-causal. Here, I am sympathetic to the intervention liberalism of Emmerson (2023) as well as to manipulationist accounts of mathematical explanations (Baron et al., 2020). However, with the lack of causation there is a question of explaining the asymmetry of explanations of ML systems. I tackle this question in Sect. 8.3. Finally, if we view ML systems as mathematical, then it also makes sense to consider the wider literature on mathematical explanations. In Sect. 8.4 I argue that this fruitfully applies to machine learning, as we can see for example explanations by constraints (Lange, 2018) for the behaviour of ML systems.

8.2 Abstraction and Explanations of ML Outputs

8.2.1 The Status of ML Systems

The starting point for this chapter is that ML systems are a kind of abstract object, that can be understood as a mathematical model (see Chap. 12 by Kieval in this volume). Machine learning systems are often introduced as such. As an example, Zhou (2021) introduces ML systems in the following setup. First, there is a data set $D = \{\mathbf{x_1}, \mathbf{x_2}, \ldots \mathbf{x_n}\} \in \mathcal{X}$ consisting of d-dimensional vectors $\mathbf{x_i} = (x_{i1}; x_{i2}; \ldots x_{id}$. This data set is often labelled with outcomes $y \in \mathcal{Y}$, formally structured in a training set consisting of pairs $(\mathbf{x_i}, y_i)$.[1] This allows us to tackle, among other learning problems, prediction problems. This is the problem to establish a mapping $f : \mathcal{X} \rightarrow \mathcal{Y}$ by learning from the training set. To do so, we need a learning algorithm $\mathcal{L}$ (also known as an optimization procedure) which aims to minimize the error between the learned f and the input-output pairs of the data set. While we can dive into further mathematical details here, the point is that presentations of ML systems proceed typically in this fashion: we start with a set of ordered pairs, and through a mathematical procedure a function is fit to this set as best we can. And so, we can interpret ML systems as approximating functions. Indeed, there is a host of theoretical results showing that different classes of neural networks (a type of ML system) can act as universal function approximators (Hornik et al., 1989; Castro et al., 2000; Schäfer & Zimmermann, 2006), solidifying the idea that ML systems in general approximate mathematical functions. Looking at ML systems from this perspective, it seems that they are first and foremost mathematical in nature.

However, the tendency to think otherwise is not a surprising one. When reading Erasmus et al. (2021, p. 845), they describe the situation as follows for a medical imaging algorithm (MAIS): "Feeding the mammogram image into the input layer of the ResNet18 [A type of ML system, SB] and the convolutional operations of each layer are causal processes, while the signals sent between the different nodes and layers are causal interactions. These causal processes and interactions lead to the output of the MAIS." And indeed, we enter the input on a specific computer, that will then execute step by step the algorithm until it displays the resulting outcome of the calculations. There are specific causal processes on the computer chip that instantiate these calculations. So, there certainly is a causal process going on here in the concrete application of the ML system, as operations are executed in order in a physical system. If we want to explain an ML system, can't we focus our explanations on those causal processes?

One answer is to point again to the idea that ML systems are approximating mathematical functions and thus that we want to gain insight into the behaviour of those approximated functions. That is different from wanting to understand the

[1] Unsupervised learning, which can for example aim to cluster the data set into groups that are relevantly similar, does not use labelling of this kind, but instead only uses the data set D.

causal processes used in the function approximation. A second argument for why the causal processes are not what we should consider in XAI comes from the debate on the ontology of software more generally. Machine learning systems are a type of software, and we want explanations for the outputs/behaviour of these software systems. In other words, the phenomenon that we want to explain is a type of software, and so if we want to know the ontological status of what XAI aims to explain then it makes sense to look at the ontology of software. The debates here also highlight the same tension between abstraction and concreteness, as for example Colburn (1999) defines software as a "concrete abstraction", because it always requires reference to a medium of description (a programming language) and a medium of execution (the hardware on which the software will run). Colburn reasons that this reference to concrete implementations where the software runs on physical hardware is essential, but that software is nevertheless an abstraction as programmers prefer to abstract away from implementation to consider software as something that will run on any machine.

Still, it is hard to make sense of this position that software is both abstract and concrete at the same time. Instead, more recent accounts such as those by Irmak (2012), Turner (2011) and Duncan (2017) have focused on software as an abstract object, but more specifically an abstract artifact. The idea here is that software has a function, and is created with that intended function in mind, thus making it a kind of artifact. However, software cannot be identified with any concrete implementations on particular computers, and comes with a standard of correctness (bugs in the code and faults in execution). Specifically, Duncan (2017, p. 27) defines a software program as: "a specification that consists of one or more programming language instructions and whose concretization is embodied by an artifact that is designed so that a physical machine may read the concretized instructions." This entails that software is an abstract specification rather than a concrete implementation, but it has to be such that it links to concrete versions (e.g. physical memory states of a computer) that can be read and executed.

This idea of software as a specification is also found in the work of Primiero (2016), based on Floridi's Levels of Abstraction. There, the highest level of abstraction is the intention of the programmer, which we hope to satisfy through the design of an algorithm. "An algorithm is the abstract representation of a mathematical function required to fulfil a task" (Primiero, 2016, p. 98). This algorithm is then implemented in a programming language, which denotes the instructions in machine code and finally controls electrical charges in the hardware. The software is viewed as a type of abstract object, and specifically a mathematical object, but guided by an intention and linked to concrete implementations. We should, then, consider that ML systems are (as a kind of software/algorithm) likewise abstract objects that are linked to concrete implementations but not identified with them. That means that we cannot, strictly speaking, talk about causes in the ML system even if there are causal links between the execution steps of the concrete implementations. As a result, direct applications of philosophical accounts of explanation such as the causal-mechanical account of explanation to ML systems fail.

In short, I see two complementary reasons why we should resist a causal/scientific interpretation of explainable AI. First, ML systems are plausibly interpreted as mathematical structures that approximate functions. They are mathematically defined and analysed, and so we should interpret explainable AI as the endeavour to further illuminate the behaviour of these mathematical objects. Second, AI systems are implemented as pieces of software and explainable AI aims to offer insight into the behaviour of AI qua software that produces outputs. As software is plausibly interpreted as abstract (and mathematical) in nature, we should thus also for this reason consider explainable AI as an endeavour to explain the behaviour of abstract (mathematical) objects. On both counts, then, an interpretation of AI in strictly causal/scientific terms is inappropriate.

8.2.2 Interpreting Explanations of Machine Learning Systems

How, then, can we make sense of explanations of ML outputs? I hold that there is still a suitable dependence relation to be found in the case of ML systems, such that ML outputs depend on the inputs, and the function approximated by the ML system depends on the training data and optimization procedure. We can cash out this dependence in terms of counterfactuals or interventions (depending on how liberal one wants to be with talk of interventions, which on many accounts are restricted to causation; cf. Emmerson, 2023). We can then get an account of ML explanations closely aligned with the philosophy of science-inspired accounts, specifically those related to the manipulationist account of Woodward (2003). Furthermore, it can help us make sense of the formal machinery used in explainability methods such as Biswas et al. (2022) and Geiger et al. (2023), which is currently presented as based on causation.

The basic intuition behind this is as follows: although the ML system is independent from the physical implementation, we still select the data set ourselves, we pick an optimization procedure and learning parameters (e.g. in the case of a neural network we pick the network architecture, hyperparameters and initialization of the network weights). All of these choices impact the resulting ML system: if we had chosen something else here, we could have gotten a different ML system (approximating a different function). Furthermore, once we have an ML system that is optimized on the data set, there is an intuitive sense in which the output depends on the input. If the input had been different, then the output could be different as well. Similarly, if the internal parameters of the ML system (e.g. an artificial neuron in the case of a neural network, or a decision tree in the case of a random forest) had been different, then the output could have been different even when the input is kept fixed. As a range of philosophers has suggested (Baron et al., 2020; Jansson and Saatsi, 2019; Reutlinger, 2018; Woodward, 2018), we can then extend the manipulationist account of explanation from a purely causal one to one about counterfactual dependence. Explanations display patterns of counterfactual dependence between the explanandum and the explanans and we can find these kinds of patterns in ML systems.

Counterfactual dependence is, of course, still difficult to make sense of in mathematics. $2 + 2$ could not be 5, no matter what we do, and similarly theorems and results of calculations are necessary truths. Still, Baron et al. (2020) suggest one way in which we can make sense of such counterfactuals even in a purely mathematical setting. They define the truth-conditions of these counterfactuals using a closeness-based semantics: "$A \mathrel{\Box\!\!\rightarrow} B$ is true at a world ω iff some possible or impossible world in which both A and B are true is closer to ω than any possible or impossible world in which A is true and B is false, if there are any possible or impossible worlds in which A is true." (Baron et al., 2020, p. 4). Impossible worlds are needed due to the impossibility that actual mathematical truths are false and the closeness of worlds is then determined based on similarity: if B is more similar to A than C is, then B is closer to A than C. This gives an account of counterfactuals that is applicable to mathematics (Baron et al. (2020) show as much by applying it to a number of mathematical proofs) and can be used to further specify the idea of counterfactual dependence in this setting.

Specifically in the case of ML systems there is then a range of different types of counterfactual dependence that we can be interested in when demanding explanations. Most technical explainability methods focus on the pattern of dependence of the output on the input. What would happen to the output were we to have a different input to the ML system? Counterfactuals on the closeness-based account would evaluate this in the expected manner: the outcome is simply that which would result from a different input to the actual ML system. This helps us to figure out what function the ML system approximates, and consequently to anticipate what it will output in new situations. In that sense, there is a clear analogy to causal manipulationist accounts: by figuring out how the outputs (counterfactually) depend on the inputs, we know what we need to do in order to obtain a different output.

Alternatively, we can also be interested in why the ML system approximates one function rather than another. Here, the dependence of the ML system on the learning parameters, optimization method and training data are the focus. While there are few technical tools that focus on these dependencies, computer scientists are well-aware of a number of these dependencies. For example, it is common knowledge that biases in the training data often get picked up by the ML systems that are trained on this data. In the case of facial recognition systems this was clear in terms of the accuracy of these systems. First, a big discrepancy in accuracy was observed between faces of people with light skin colours (for which the system worked well) and faces of people with darker skin colours (for which the system was far less accurate) (Lohr, 2022). Then, this discrepancy was diagnosed as the result of a bias in the data set: the data consisted primarily of faces of people with light skin colours. By fixing the representativeness of the data set, the discrepancy in accuracy has been significantly reduced. If we interpret this in terms of an explanation in response to a why-question we can think of it as follows: the question was 'why is there such a large difference in accuracy between skin tones, rather than a much smaller one?' and the answer can be formulated in terms of the presence/absence of dark skin tones in the data set. Adjust that (i.e. had there been more data points of this kind in the training data) and the discrepancy becomes smaller.

In line with the somewhat unusual status of software systems in general, we can see these counterfactuals as closely linked to concrete actions we can take. To evaluate what a change to the inputs of an ML system would do, the easiest route is to simply execute an implementation of the ML system on the alternative input. Likewise, if we want to know whether a change to the composition of the data set leads to a suitably different function approximation, often the simplest way of finding this out is to simply gather more data and run the training procedure on the new data set. That is quite different from the steps needed to evaluate the counterfactuals in pure mathematics that Baron et al. (2020) discuss. We still shouldn't interpret them as causal in nature, but the causal steps that we can take for specific implementations can help us in finding out the counterfactual dependencies for ML systems. As long as the specific implementation has no malfunctions (the code contains no bugs, the hardware operates properly), the answers will be the same for the implementation as for the abstract ML system. However, this qualification of 'no bugs, no malfunctions' is precisely the qualifier that leads us to assign a different status to the ML system, and thus to the conclusion that we shouldn't speak directly of causes in explanations of these systems.

I see a number of benefits to this particular treatment of explanations of ML systems. First, it sticks closely to the steps we actually take to try to explain various aspects of these systems (though for more global/theoretical explanations of ML systems, see Sect. 8.4). We adjust inputs, parameters, training data and more in an effort to understand what works and what doesn't. A wide range of current explainability tools, from counterfactual methods to feature importance (which is based either on a calculated gradient indicating what will cause the biggest change in outputs or on a sample of the local neighbourhood of similar inputs), looks precisely at what the ML system would output were the inputs to be different. Second, it helps us make sense of the recent causal inference-inspired methods. These are based on the use of structural equation models, stating how different variables depend on one another. Usually they are reserved for causal dependence, but Baron et al. (2020) claim that they can also be applied to mathematics and the kind of counterfactuals considered here.

They don't give much of an explanation as to how we can make sense of structural equation models in a non-causal context, so it is worthwhile to explore that here. SEMs do not themselves require that the relation explored is that of causation, but of course we do need to have a relation with the right set of features. Starting with the list of conditions that Woodward (2003) gives for interventions we can see how many might apply to this counterfactual dependence within ML systems.

1. I causes X
2. I acts as a switch for all other variables that cause X. That is, certain values of I are such that when I attains those values, X ceases to depend on the values of other variables that cause X and instead depends only on the value taken by I
3. Any directed path from I to Y goes through X. That is, I does not directly cause Y and is not a cause of any causes of Y that are distinct from X except, of course, for those causes of Y, if any, that are built into the I $\rightarrow$ X $\rightarrow$ Y connection itself; that is, except for (a)

any causes of Y that are effects of X (i.e., variables that are causally between X and Y) and (b) any causes of Y that are between I and X and have no effect on Y independently of X;

4. I is (statistically) independent of any variable Z that causes Y and that is on a direct path that does not go through X.
(Woodward, 2003, p. 98)

If we remove the notion of 'cause' here and consider what happens in the proposed ML setting, how much remains? The first condition can be met in terms of the alternative situation being considered being precisely one where X (e.g. input or data set) is changed. Criterion number 2 can easily be satisfied even in the abstract setting: changing the input value, data set or even some internal part of the ML system's algorithm is done in such a way that this changed variable ceases to depend on exogenous variables. Condition 3 can, at least in the context of ML systems, also be guaranteed: we know that e.g. any change to the input will only affect the output by going through the function implemented by the ML system. Likewise, any change to the data set only changes the learned function via the optimization procedure. Finally, condition 4 is satisfied as long as we make sure that the change of input, data set etc. does not relate to other parts of the ML system. The remaining question is whether counterfactual dependence can indeed stand in for causation. The main worry here (and on any account that understands mathematical explanations using counterfactual dependence) is whether it is appropriately asymmetric. Do these dependencies go only in one direction? If so, then it seems that we have a situation where we can use structural equation modelling based on changes to parts of an ML system. And indeed, various accounts of mathematical explanation (I already mentioned Baron et al. (2020), but Gijsbers (2017) also presents a 'quasi-interventionist' account of mathematical explanation that looks at this kind of counterfactual dependence for the explanatory power of some proofs). However, that asymmetry in the counterfactual dependence is far from obvious and so has to be examined in the next section.

First, though, one may wonder whether this kind of counterfactual dependence is the only possible interpretation of ML explanations, now that we need to look at theories of mathematical rather than of scientific explanation. Causal-mechanistic (Salmon, 1984) and unificationist theories (Kitcher, 1989) are the prominent alternatives for scientific explanation, so can we find analogues for these accounts when it comes to mathematical explanations such as those discussed here? They are certainly harder to find in the literature, but Frans and Weber (2014) defend a mechanistic account of mathematical explanations based on the mechanistic account of Bechtel and Abrahamsen (2005). It should be noted, however, that while the inspiration is mechanistic it gets very close to a Woodwardian account that accounts for explanations in terms of difference-makers. They specifically argue that a particular proof is explanatory by: "(a) identifying a dependency to be explained, (b) identifying entities, (c) substituting the notion of activities with the notion of difference-makers, and (d) show that these difference-makers are organized such that the truth of the theorem is established." (Frans & Weber, 2014, p. 240). In the specific case of ML systems proofs are less applicable, as we do not explain a theorem but rather a (set of) output(s). Still, there are mechanistic accounts applied to ML systems (which we could interpret as a mechanism between inputs and outputs) and so the translation to the mathematical setting is worth exploring, with less emphasis on difference-making. One might argue that when explaining why a particular output is obtained from a particular input we are in fact after a mechanism sketch of the operations performed on the input in order to produce the output. Explainability methods that aim to capture the black box behaviour in decision trees (Nauta et al., 2021) or with structural equation models (Geiger et al., 2023; Wu et al., 2023) would be good candidates for such a view, as they aim to find easier to comprehend operations that yield similar (and of course ideally the same) results. Of course, some adjustments will be needed: the mechanisms are not actually causal and talk of activities as in Bechtel and Abrahamsen (2005) is also not quite fitting for the mathematical setting. Substituting something like 'operations' in its place, as these are performed on the inputs and then on successive hidden layers, suffices for the specific case of ML systems, but of course will not do for mathematical explanations in general. So, if we do need a substitution in terms of difference-making here, as Frans and Weber (2014) suggest, then ultimately there will be little to distinguish a mechanistic approach to the explanations of ML systems from a Woodwardian/interventionist approach.

Unificationism does seem to offer a different story. On such an account, what does the explaining is the unification of a range of phenomena under a single argument schema. Perhaps this can be done through identifying difference-makers (Bangu (2017) for example argues for a somewhat revised version of unificationism in which case causes can be seen as one type of unifier), but the pinpointing of difference-makers is not central to the account. So, we might for example accept explanations in this case that group together a large number of similar cases without telling us what happens should relevant conditions change. This seems to make the application to mathematics even simpler, as there is no need to figure out what interventions and difference-making mean when it comes to mathematical objects. As such, we can propose unificationism as another candidate theory for explanations of ML systems: these explanations should aim to group input-output pairs together under a unifying argument schema. The success of such an account of course depends on how plausible unificationism then is in this domain; Knowles (2021) has recently argued against unificationist accounts of mathematical explanations. With the space available here I won't go into that debate, the point is rather that the shift from interpreting ML as scientific to interpreting it as mathematical need not hamper our efforts to apply the familiar accounts of explanation. They do need some adjustment though, especially when it comes to accounting for the asymmetry of ML explanations. I turn to that issue next.

8.3 The Asymmetry of ML Explanations

Mathematical equations are symmetrical, as opposed to causal relations, and so there is a question whether the counterfactual dependence in the mathematical case is appropriately asymmetrical. For example, it is (often, as techniques such as differential privacy aim to counter precisely this inference) possible to work out the data set on which the ML system was optimized even without having full access to the ML system (Shokri et al., 2017; Olatunji et al., 2021; Hu et al., 2022). Were we to know exactly what the initialization, optimization procedure and outcome were, then we can expect that the training data can be retrieved as well. In short, it seems that there is a problem (as with other mathematical explanations; cf. Lange, 2021) with the asymmetry of explanations. For while the counterfactual dependence may be symmetrical, there is an intuitive idea (reflected in XAI practices) that we should explain the output of an ML system in terms of the inputs, and the behaviour of the ML system in terms of the training data and procedure. What ensures that the counterfactual dependence discussed in Sect. 8.2 will only ever go in this intuitive direction?

Baron et al. (2020), in a similar vein to Jansson and Saatsi (2019), argue that asymmetry is guaranteed on the counterfactual account in virtue of the indeterminacy of counterfactuals going in the 'unintuitive' direction. So, while a change in input to an ML system has a clear effect on the output, the idea is that a specific change in output can be realized in many different ways by changing the input. As a result, there is not one unique change in input that corresponds to this change of output. Hence, we do not have a unique (true) counterfactual from output to input, guaranteeing the asymmetry of this kind of counterfactual dependence. Jansson and Saatsi (2019, p. 17) state the same though as: "Fixing the explanans variable to its actual value should fix the explanandum variable to its actual value".

Lange (2021) has recently criticized this approach to ensuring the asymmetry of mathematical/abstract explanations. He argues that a good number of causal explanations do not match this kind of asymmetry. For example, when considering that the window broke because the rock hit it, then there is no particular counterfactual that holds in the situation that the rock did not hit the window (other than that the window doesn't break). This particular indeterminacy need not be a problem, as both Baron et al. (2020) and Jansson and Saatsi (2019) focus only on whether the explanandum variable is fixed, and that is indeed the case here (as Lange admits, not throwing the rock will fix that the window does not break). However, it does seem to go the other way: if we're clear enough about the situation, it seems likely that the counterfactual 'had the window not been broken, then the stone would not have been thrown' is true in the closest possible world. After all, this requires by far the fewest changes; far fewer than, e.g. ensuring that the glass was bullet-proof and therefore does not break when the stone hits. In addition, it brings us back to a symmetrical setup in terms of counterfactuals: had the stone not been thrown, the window would not have broken and had the window not broken, the stone would not have been thrown. Indeterminacy in one direction does not seem to offer a way out. Lange, in a second argument, echoes this sentiment: in the example of mother distributing 23 strawberries over 3 children we can reason just as well from her having given 7 strawberries to each child with 2 left over to the number of children and strawberries, and link adjustments to the result (her attempt at dividing) to what would then change to the initial situation (the number of children and strawberries). In the end, the indeterminacy seems to go away if we fill in sufficiently many details.

The case of ML systems further reinforces that issue. Here, we have a range of tools precisely designed to calculate the minimal change to the input in order to achieve a desired output known as counterfactual methods/algorithmic recourse (see Karimi et al., 2020 for a review). These methods do, for a given distance function over the inputs, return a unique alternative input that leads to the specified alternative output. In practice it is very difficult to create realistic distance functions, but this doesn't matter for the principled argument here. For these technical tools further solidify the claim that if we specify the situation and the notion of 'closer' enough then there will be true counterfactuals in both directions. So, the asymmetry of mathematical/abstract explanations is an issue when we understand such explanations on the basis of counterfactuals (and see Lange (2021) for further discussion of other attempts to ground the asymmetry).

The specific case of software has a serious benefit here in getting out of these difficulties. Inherent to software is an asymmetry in terms of execution: an earlier line of code/set of instructions is (in any implementation) to be executed before the next. We can apply this more broadly to ML systems: we know that the data set has to be there first, to then feed (step by step) into the optimization procedure and finally for the ML system to process an input line by line to reach an output. The specification of these steps may be an abstract object, but the asymmetry in terms of what happens earlier and what happens later (and thus which parts of the ML system can influence which other parts) is inherent. The values used in the first line of code cannot depend on the values used in the last line of code (at least, if we fully work out recursive functions for clarity's sake). That difference in dependence: we determine one value using another *in the execution of an ML system*, and not the other way around, is a candidate for the explanatory asymmetry.

This also aligns with the idea that an explanation displays a pattern of counterfactual dependence. In these implementations, a change to one of the earlier parts of the process (the data, the optimization procedure, the inputs) will affect the later parts of the ML system, but not vice versa. Changing the input will not affect the training data, and manually changing the output value (during the execution of the ML system) will not change the input of the system.

However, how can we best understand this dependence? We could leave it as is, and opt along with Lange (2021) for a kind of pluralism in where the asymmetry of non-causal explanations comes from. In the case of software, we might say, the asymmetry is grounded in what has to be done first in any implementation of the ML system. We can rely on the asymmetry of causation in these concrete implementations, and be done with it. This account of asymmetry is not available in other cases of mathematical explanations, as these do not link to concrete implementations in the same way. Algorithms, as step-by-step procedures, are in that sense different from most other mathematical objects. Even staying with the case of ML systems we can see this difference, as the next section shows in a brief exploration of explanations by constraint of ML systems.

8.4 Explanations by Constraint in ML Systems

There are further explanations of the behaviour of ML systems that we should not forget. These are more closely related to the typical examples of mathematical explanations of mathematical facts, for example to the explanations by constraint presented in Lange (2018). One example, discussed in detail by Räz (2022), is the attempt to explain why deep neural networks (as a type of ML model) do not overfit despite being overparameterized. GPT-4, for example, is reported to have over a trillion parameters, which we would normally expect to promote literal memorization of the training data and thus very poor generalizability over new inputs. This does not happen, so why are DNN models so good at handling new data? Here, the Information Bottleneck (IB) theory developed by Shwartz-Ziv and Tishby (2017) offers one possible explanation. According to this theory, the success is the result of a trade-off between compression and prediction. This is a general trade-off that has to be made by prediction methods, but the argument goes that deep neural network lead to particularly good results on the trade-off; at or close to the theoretical (Pareto) optimum. Shwartz-Ziv and Tishby (2017) observe that in the initial training phases the DNN optimizes on predictive accuracy, storing more and more information about the input space in order to give the best possible results. After this, there is a second phase in the training where they observed that the stored information is then reduced (compressed), but in such a way as to minimize the loss of predictive accuracy. The optimization procedure thus searches for the information about the input space that is necessary to accurately predict the output, but the end goal is to keep only the minimally sufficient information to do so. Because of this second compression step DNNs do not overfit as much as the number of parameters would lead us to expect, as compression ensures that the irrelevant details of the training data are not used to determine the output. Furthermore, we can interpret this as a high-level explanation by constraint (the IB theory specifies a trade-off for any learning mechanism and DNNs make this trade-off in a particular way) as well as a counterfactual explanation (had the second, compression, phase not happened, DNNs would have overfit much more).

A second example is the impossibility theorems regarding fairness metrics for machine learning systems (Kleinberg et al., 2016). This theorem focuses on three widely used statistical definitions of fairness between groups A and B: (1) equal probability of being assigned to the positive predicted class, (2) equal true positive rates, and (3) equal true negative rates. The impossibility theorem shows that except for some very special cases (e.g. equal base rates in groups A and B) it is impossible to satisfy all three of these definitions of fairness at the same time. Hence, there will be a trade-off between them. We can then use this theorem to explain why, for ML systems, we see that when an ML system is (perfectly) fair according to one definition, it is not (perfectly) fair according to the others. To cite on of their examples: "suppose we want to determine the risk that a person is a carrier for a disease X, and suppose that a higher fraction of women than men are carriers. Then our results imply that in any test designed to estimate the probability that someone is a carrier of X, at least one of the following undesirable properties must hold: (a) the test's probability estimates are systematically skewed upward or downward for at least one gender; or (b) the test assigns a higher average risk estimate to healthy people (non-carriers) in one gender than the other; or (c) the test assigns a higher average risk estima. So, when one wonders why we observe this trade-off in practice, the impossibility theorems can explain this fact in a principled manner. There is a constraint (specified by the theorem) on the properties that the approximated functions can have. Had this constraint not been there, presumably the situation would have been different.

We can make good sense of these mathematical explanations, as would be expected given their similarity to cases already discussed in the literature on mathematical explanation. However, the accompanying downside of this similarity is that it is far less plausible to appeal to the asymmetry inherent in the implementations of ML systems. These constraints, such as that it is impossible to meet two fairness criteria at the same time, do not relate to specific ML systems that can be implemented nor do they link to a procedure in which one part is executed before another. As a result, we cannot use the causal links in the implementation of an ML system to ground the asymmetry of these non-causal explanations. A pluralist interpretation of these (noncausal) explanations seems the most likely, at least in terms of our account of the asymmetry of the explanations.

It may be tempting to avoid this kind of pluralism through an appeal to a more general notion of 'grounding' here to capture both types of mathematical explanations. After all, there is a sense in which the outputs of an ML system are grounded (as in, determined by) the inputs as well as the training data and optimization procedure. Schaffer (2016) has proposed an understanding of metaphysical grounding using structural equation models, which provides a nice link to the explanations of ML outputs as discussed in Sect. 8.2. Additionally, there have been suggestions to capture more standard mathematical (and conceptual) explanations in terms of grounding (Poggiolesi & Genco, 2023). On a general gloss of the situation we could then say that the asymmetry of explanations is accounted for in both cases through the idea that the explanandum is grounded in the explanans.

Such a move towards uniformity would be misleading, however. The conceptual grounding pushed by Poggiolesi and Genco (2023) relies on the use of conceptual complexity. The idea is that more complex concepts may be introduced (in the formal, proof-theoretic sense of having an introduction rule in a deductive system) on the basis of less complex concepts. In this sense, more complex concepts are grounded in less complex concepts and complexity is understood in terms of being further down the proof-theoretic chain. The fact that only introduction rules provide grounding helps to account for the asymmetry of the grounding relation and, similarly, the asymmetry of conceptual explanations. We cannot have explanations of simpler concepts in terms of more complex concepts (e.g. explaining A in terms of $A \wedge B$) because those inference require us to use elimination rules, and thus $A \wedge B$ is more complex than A. Here the difference with the input-output type explanations becomes clear again: such explanations involve calculations which, even if we could cash them out in terms of a full deductive system, will not generally conform to the idea that outputs are arrived at from inputs using introduction rules (if we strictly follow the Peano axioms then one will typically use $\forall$-elimination on $\forall x \forall y(x + S(y) = S(x+y))$ and similar axioms for other mathematical operations). But on a more basic level there is no clear reason why the conceptual complexity of an ML output (in this proof-theoretic sense) should be higher than the conceptual complexity of the ML inputs. Since ML systems often aim to predict, it may very well happen that values of a simpler (but harder to measure) concept are predicted using values of more complex concepts. We still want explanations in the XAI sense to explain how the output values depend on the input values, but these do not track conceptual complexity. ML systems may well be an exceptional case of noncausal explanations here, but the point remains that a unification (using at least these conceptions of grounding) only succeeds by obscuring the underlying differences.

8.5 Conclusion

Explanations of ML systems have typically been considered through the lens of scientific explanations. I have argued that ML systems are best seen as abstract (and plausibly mathematical) objects, to which many of the accounts of scientific explanation do not apply. However, similar accounts of mathematical explanation can be used to interpret explanations of ML systems and to make sense of the use of causal inference techniques in the case of ML systems. Counterfactuals of the form "had the input been x' instead of x, the output would have been y' instead of y" can still be evaluated even without causal relations. Furthermore, the changes that are made to values in an ML system meet all the criteria for interventions aside from the presence of causal relations, thus allowing us to interpret structural equation models in the ML case.

This does raise a worry, however, as without the presence of causality there is no guarantee that the account will be appropriately asymmetrical. Counterfactuals can go both ways, so how is the directionality of explanations ensured? I argued that in the case of ML systems we can typically appeal to the direction inherent to the instructions of which an ML system is made up. To further strengthen this idea, we can also link counterfactuals to concrete implementations of the abstract ML system, where these instructions are executed with causal links in between them.

While that covers a wide range of explanations of ML systems, it fails to include a set of mathematical explanations that is not directly linked to concrete implementations. Explanations answering questions such as why deep neural networks do not overfit, or why we observe a trade-off between fairness metrics in ML systems, are better understood as explanations by constraint. These are still compatible with current counterfactual accounts of non-causal explanations, but there seems to be a big difference in how we will account for their asymmetry. Pluralism on this front, as also suggested by Lange (2021) based on a different set of mathematical explanations, seems the most plausible option.

References

Bangu, S. (2017). Scientific explanation and understanding: Unificationism reconsidered. *European Journal for Philosophy of Science, 7*, 103–126.

Baron, S., Colyvan, M., & Ripley, D. (2020). A counterfactual approach to explanation in mathematics. *Philosophia Mathematica, 28*(1), 1–34.

Bechtel, W., & Abrahamsen, A. (2005). Explanation: A mechanist alternative. *Studies in History and Philosophy of Science Part C: Studies in History and Philosophy of Biological and Biomedical Sciences, 36*(2), 421–441.

Beckers, S., & Halpern, J. Y. (2019). Abstracting causal models. In *Proceedings of the AAAI Conference on Artificial Intelligence* (Vol. 33, pp. 2678–2685).

Beisbart, C., & Räz, T. (2022). Philosophy of science at sea: Clarifying the interpretability of machine learning. *Philosophy Compass, 17*(6), e12830.

Biswas, S., Corti, L., Buijsman, S., & Yang, J. (2022). Chime: Causal human-in-the-loop model explanations. In *Proceedings of the AAAI Conference on Human Computation and Crowdsourcing* (Vol. 10, pp. 27–39).

Buijsman, S. (2022). Defining explanation and explanatory depth in XAI. *Minds and Machines, 32*(3), 563–584.

Castro, J. L., Mantas, C. J., & Benıtez, J. (2000). Neural networks with a continuous squashing function in the output are universal approximators. *Neural Networks, 13*(6), 561–563.

Colburn, T. R. (1999). Software, abstraction, and ontology. *The Monist, 82*(1), 3–19.

Das, A., & Rad, P. (2020). *Opportunities and challenges in explainable artificial intelligence (XAI): A survey*. Preprint. arXiv:2006.11371.

Duncan, W. D. (2017). Ontological distinctions between hardware and software. *Applied Ontology, 12*(1), 5–32.

Emmerson, N. (2023). A defence of manipulationist noncausal explanation: The case for intervention liberalism. *Erkenntnis, 88*(8), 3179–3201

Erasmus, A., Brunet, T. D., & Fisher, E. (2021). What is interpretability? *Philosophy & Technology, 34*(4), 833–862.

Frans, J., & Weber, E. (2014). Mechanistic explanation and explanatory proofs in mathematics. *Philosophia Mathematica, 22*(2), 231–248.

Geiger, A., Potts, C., & Icard, T. (2023). Causal abstraction for faithful model interpretation. Preprint. arXiv:2301.04709.

Gijsbers, V. (2017). A quasi-interventionist theory of mathematical explanation. *Logique et Analyse, 237*, 47–66.

Halpern, J. Y., & Pearl, J. (2005). Causes and explanations: A structural-model approach. Part I: Causes. *The British Journal for the Philosophy of Science, 56*(4), 843–887.

Hornik, K., Stinchcombe, M., & White, H. (1989). Multilayer feedforward networks are universal approximators. *Neural Networks, 2*(5), 359–366.

Hu, H., Salcic, Z., Sun, L., Dobbie, G., Yu, P. S., & Zhang, X. (2022). Membership inference attacks on machine learning: A survey. *ACM Computing Surveys (CSUR), 54*(11s), 1–37.

Irmak, N. (2012). Software is an abstract artifact. *Grazer Philosophische Studien, 86*(1), 55–72.

Jansson, L., & Saatsi, J. (2019). Explanatory abstractions. *The British Journal for the Philosophy of Science, 70*(3), 817–844.

Karimi, A.-H., Barthe, G., Schölkopf, B., & Valera, I. (2020). *A survey of algorithmic recourse: Definitions, formulations, solutions, and prospects*. Preprint. arXiv:2010.04050.

Kieval, P. H. (2024). Representation learning without representationalism. A non-representationalist account of deep learning models in scientific practice. In J. M. Duran & G. Pozzi (Eds.), *Philosophy of science for machine learning: Core issues and new perspectives*. Synthese Library. Springer.

Kitcher, P. (1989). Explanatory unification and the causal structure of the world. In P. Kitcher & W. E. Salmon (Eds.), *Scientific explanation* (pp. 410–505). University of Minnesota Press.

Kleinberg, J., Mullainathan, S., & Raghavan, M. (2016). *Inherent trade-offs in the fair determination of risk scores*. Preprint arXiv:1609.05807.

Knowles, R. (2021). Unification and mathematical explanation. *Philosophical Studies*, *178*(12), 3923–3943.

Lange, M. (2018). Because without cause: Scientific explanations by constraint. In *Explanation beyond causation* (pp. 15–38). Oxford University Press.

Lange, M. (2021). Asymmetry as a challenge to counterfactual accounts of non-causal explanation. *Synthese*, *198*(4), 3893–3918.

Lohr, S. (2022). Facial recognition is accurate, if you're a white guy. In *Ethics of data and analytics* (pp. 143–147). Auerbach Publications.

Lundberg, S. M., & Lee, S.-I. (2017). A unified approach to interpreting model predictions. In *Advances in Neural Information Processing Systems* (Vol. 30).

Nauta, M., Van Bree, R., & Seifert, C. (2021). Neural prototype trees for interpretable fine-grained image recognition. In *Proceedings of the IEEE/CVF Conference on Computer Vision and Pattern Recognition* (pp. 14933–14943).

Olatunji, I. E., Nejdl, W., & Khosla, M. (2021). Membership inference attack on graph neural networks. In *2021 Third IEEE International Conference on Trust, Privacy and Security in Intelligent Systems and Applications (TPS-ISA)* (pp. 11–20). IEEE.

Poggiolesi, F., & Genco, F. (2023). Conceptual (and hence mathematical) explanation, conceptual grounding and proof. *Erkenntnis*, *88*(4), 1481–1507.

Primiero, G. (2016). Information in the philosophy of computer science. In L. Floridi (Ed.), *The Routledge handbook of philosophy of information* (pp. 90–106). Routledge.

Räz, T. (2022). Understanding deep learning with statistical relevance. *Philosophy of Science*, *89*(1), 20–41.

Reutlinger, A. (2018). Extending the counterfactual theory of explanation. In *Explanation beyond causation: Philosophical perspectives on non-causal explanations* (pp. 74–95). Oxford University Press.

Ribeiro, M. T., Singh, S., & Guestrin, C. (2016). "Why should I trust you?" Explaining the predictions of any classifier. In *Proceedings of the 22nd ACM SIGKDD International Conference on Knowledge Discovery and Data Mining* (pp. 1135–1144).

Salmon, W. C. (1984). *Scientific explanation and the causal structure of the world*. Princeton University Press.

Schäfer, A. M., & Zimmermann, H. G. (2006). Recurrent neural networks are universal approximators. In *Artificial Neural Networks–ICANN 2006: 16th International Conference, Athens, Greece, September 10–14, 2006. Proceedings, Part I 16* (pp. 632–640). Springer.

Schaffer, J. (2016). Grounding in the image of causation. *Philosophical Studies*, *173*(1), 49–100.

Shokri, R., Stronati, M., Song, C., & Shmatikov, V. (2017). Membership inference attacks against machine learning models. In *2017 IEEE Symposium on Security and Privacy (SP)* (pp. 3–18). IEEE.

Shwartz-Ziv, R., & Tishby, N. (2017). *Opening the black box of deep neural networks via information*. Preprint. arXiv:1703.00810.

Simonyan, K., Vedaldi, A., & Zisserman, A. (2013). *Deep inside convolutional networks: Visualising image classification models and saliency maps*. Preprint arXiv:1312.6034.

Turner, R. (2011). Specification. *Minds and Machines*, *21*, 135–152.

Watson, D. S., & Floridi, L. (2021). The explanation game: A formal framework for interpretable machine learning. In *Ethics, governance, and policies in artificial intelligence* (pp. 185–219). Springer.

Woodward, J. (2003). *Making things happen: A theory of causal explanation*. Oxford University Press.

Woodward, J. (2018). Some varieties of non-causal. *Explanation beyond causation: Philosophical perspectives on non-causal explanations* (p. 117). Oxford University Press.

Wu, Z., D'Oosterlinck, K., Geiger, A., Zur, A., & Potts, C. (2023). Causal proxy models for concept-based model explanations. In *International Conference on Machine Learning* (pp. 37313–37334). PMLR.

Zhou, Z.-H. (2021). *Machine learning*. Springer Nature.

Stefan Buijsman (1995) is an associate professor at Delft University of Technology working on the philosophy of artificial intelligence. He co-leads the Delft Digital Ethics Centre which focuses on ethics by design by translating ethical principles into requirements for technical design and governance. His own research is centered around the intersection of epistemology (transparency, explainability) and ethics of AI, both in theory and in applied projects in high-risk application areas.

Chapter 9
From Explanations to Interpretability and Back

Tim Räz

Abstract This chapter discusses connections between the interpretability of machine learning and (scientific and mathematical) explanations, provides novel perspectives on interpretability, and highlights under-explored issues. Interpretability types are proposed: kinds of interpretability should be distinguished using both the parts of ML we want to explain and the parts of ML we use to explain. It is argued that not all explanations are contrastive, and that we should also consider contrasts with respect to models and data, not only with respect to inputs. Theoretical explanations are highlighted; they include issues like generalization, optimization, and expressivity. It is proposed that there are two threats to the objectivity of explanations: One comes from radical subject-dependence, the other from a lack of factivity. Finally, pluralism is advocated: There are different notions of interpretability and different notions of (scientific and mathematical) explanations. However, the heterogeneity of one area does not transfer to the other in a straightforward manner.

Keywords Interpretability · Scientific explanation · Mathematical explanation · Understanding · XAI · Pluralism · Machine learning

9.1 Introduction

This chapter discusses connections between the interpretability of machine learning models from computer science on the one hand, and scientific and mathematical explanation from philosophy of science on the other. Both notions have been discussed for decades in their respective fields, but the relation between them has only been explored in the last couple of years. *Prima facie*, it makes sense to look for connections: Interpretability is concerned with explaining phenomena that arise

T. Räz (✉)
Institute of Philosophy, University of Bern, Bern, Switzerland
e-mail: tim.raez@posteo.de

J. M. Durán, G. Pozzi (eds.), *Philosophy of Science for Machine Learning*,
Synthese Library 527, https://doi.org/10.1007/978-3-032-03083-2_9

in machine learning, and it would be useful to know what such explanations should look like. Also, if explanations of ML phenomena exist, it should be of interest to philosophers of science to determine whether these explanations conform to a mode of explanation they are familiar with, or else to characterize the novel explanation type. Thus, looking for connections can be fruitful for both parties.

This chapter provides novel perspectives on interpretability (see the next section for a working definition) and highlights under-explored issues. One general theme is to steer a middle course between extreme positions on interpretability and explanation. For example, it is argued that while we should not strive for one, homogeneous notion of interpretability, there is hope that we can prevent an excessive proliferation of different notions. Then, interpretability should allow for a certain context- and audience dependence; interpretability has a psychological dimension. But we should not view interpretability as purely subjective and untethered by facts. Finally, the focus on explanations does not mean that the twin notion of understanding is not important (see Chap. 10 by Sullivan and Kasirzadeh and Chap. 11 by Schmidt on understanding in this volume). Understanding is highly relevant to the notion of explanation, and where it is not, there is room for a complementary role of the two notions.

Here is a section-by-section overview of the chapter. *2. Preliminaries:* I introduce working definitions of the key notions of explanation and interpretability and provide a very brief overview of supervised learning. *3. Types:* I propose to distinguish kinds of interpretability using the parts of ML we want to explain, and also the parts of ML we use to generate the explanation; this leads to interpretability types. *4. Contrast:* Contrastive explanations are useful. However, not all explanations are contrastive. Also, not only contrasts with respect to input values matter, but also contrasts with respect to models, data, and other parts of ML. *5. Context:* A certain extent of context dependence of interpretability is useful or even necessary, but radical context dependence goes too far. *6. Theoretical explanations:* Explanations of general or theoretical phenomena are under-explored. They include issues like generalization, optimization, and expressivity *7. Levels:* If there are explanations on different levels of generality, it is useful to explore how these are related. *8. Objectivity and Idealization:* There are two threats to objectivity: One from radical subject-dependence, the other from a lack of factivity. Both cannot be fully avoided. *9. Pluralism:* There are different notions of interpretability and there are different notions of (scientific and mathematical) explanations. The heterogeneity of one area does not transfer to the other in a straightforward manner.

9.2 Preliminaries

In this section, I provide working definitions of the notion of explanation as used in philosophy of science (Sect. 9.2.1), of interpretability as used in computer science (Sect. 9.2.2), and of the most important aspects of machine learning, supervised learning in particular (Sect. 9.2.3). These definitions constitute stepping stones for the subsequent discussion.

9.2.1 *Explanation*

Scientific explanations have been an important topic in philosophy of science since Carl G. Hempel's (1948) work on the Deductive-Nomological (DN) theory of explanation (see Woodward and Ross, 2021 for an overview of scientific explanations). It soon turned out that the DN model (or theory) is not an adequate theory of explanations (Ibid.). Despite efforts during the ensuing decades, no universally agreed-upon theory of scientific explanation has emerged; more on this in Sect. 9.9. I will use the following working definition. A *scientific explanation* is an answer to a why-question of the form "Why ϕ?", or "Why is it the case that ϕ?" The idea to characterize explanations in this way goes back to van Fraassen (1980), Bromberger (1966). Note that requests for explanations can take other forms, e.g., that of a request of an explanation-how. The entity ϕ to be explained is called *explanandum*, while the answer to the question, the entity doing the explanatory work, is called *explanans*. Take the following, classic example: "Why did the window shatter?" The *explanandum* is the event that the window shatters. A possible answer is: "Because a rock was thrown at it." In this case, the explanation is causal, because the *explanans* consists in citing a cause of the *explanandum*. Not all explanations are causal; there are also structural, statistical, mathematical explanations, which are based on non-causal explanatory relations; examples are given below. For an in-depth discussion of mathematical explanations of ML phenomena see Chap. 8 by Buijsman in this volume. Also, explanations may not come in the form of an answer to a why-question; the working definition is a first approximation.

9.2.2 *Interpretability*

Interpretability is the problem of understanding properties of ML models, or of classes of ML models, possibly relative to some particular dataset, or to some type of data (see Biran and Cotton (2017), Adadi and Berrada (2018) for computer science surveys and Beisbart and Räz (2022) for a philosophical survey). Research on interpretability encompasses a characterization of what we mean when we say that we want to "understand some aspect of an ML model", a characterization of the properties we want to understand, and the formulation of methods that provide understanding of these properties. Research on interpretability thus encompasses both conceptual problems, such as saying what "understanding a phenomenon" means in general terms, and technical problems, such as formulating methods that provide understanding.

This admittedly vague definition of interpretability does not necessarily capture how the notion is used in computer science. One of the goals of the chapter is to introduce useful distinctions that help to clarify the landscape of interpretability, to distinguish different kinds of interpretability, and to point out connections between different areas of research in computer science that are traditionally not taken to be concerned with interpretability. Note that I will not discuss the notion of explainable AI (XAI) separately from interpretability; rather, questions and methods of XAI are subsumed under interpretability (see Beisbart and Räz (2022) for a discussion of the relation between interpretability and explainability).

9.2.3 ML

The focus of the chapter is on supervised learning, one important kind of machine learning (Hastie et al., 2009). The goal of supervised learning is to construct a function F that predicts values $\hat{y}$ (outputs) of variable Y based on values x (inputs) of variable X. Predictions are written with a hat, $\hat{y} \in Y$, and ground truths, which are part of the dataset, without a hat, $y \in Y$. In order to construct such a model, a dataset $D = \{(x_i, y_i), i \in 1, \ldots, n\}$ is used. D is sampled from a system in the world S, such that the $x_i \in X$ are instances of the variable for which we want predictions, and $y_i \in Y$ are instances of the variable to be predicted. For example, the x_i could be images of different animals, y_i would be correct labels of the animals depicted ($y_i \in \{\text{cat, dog}, \ldots\}$) and the problem would be to predict which animal is depicted in a given image.

The idea is to let a model M learn the function $F : X \to Y$, thus approximating the relation between X and Y encoded in D. More specifically, M constructs an F that minimizes the error on instances in D according to some loss function. The function F can be seen as a property (the input-output profile) of the model M. The model M "learns" F through some optimization procedure O. Learning means that the parameters of M (model parameters) are adapted such that the distance between F and D is minimized as much as possible. In order to check whether M has succeeded in approximating the data, the dataset D is usually split into a training and a test set. The training set is used to learn the function F; the test set is used to check whether the learning was successful, that is, whether F actually approximates D on samples not used in constructing F. An ML model M is successful relative to a dataset D if it has a high accuracy (small loss) on the test set. To return to an example, an ML model has successfully learned to predict (or classify) animals if it is able to predict, with high accuracy, animals depicted in images *the model has not seen during training*. A model that succeeds in doing this is said to generalize well. The most important aspects of supervised learning are summarized in Fig. 9.1.

This overview is very general and incomplete. More can be said about which model is appropriate for which task. For example, while deep neural networks (DNNs) may be appropriate for unstructured, high-dimensional data, other models (random forest, boosted trees) may be more appropriate for tabular data, where

$$S \overset{sample}{\rightsquigarrow} D \overset{approx}{\longleftarrow} F \overset{compute}{\longleftarrow} M \overset{learn}{\longleftarrow} O$$

Fig. 9.1 Components of supervised learning. Legend: S—System to be modeled; D—dataset with structure $\{(x_i, y_i), i \in I\}$ sampled from S; $F : X \to Y$—function approximating the relation D; M—ML model computing F; O—optimization procedure for M to approximate F

individual variables have an "intuitive meaning" (see LeCun et al., 2015; Goodfellow et al., 2016; Hastie et al., 2009 for details). The same is true for finer points concerning architecture, hyperparameter tuning, optimization procedures, and so on. Then, supervised learning is not the only learning paradigm; there are also unsupervised learning, reinforcement learning, and generative modeling. Also, classical supervised learning may no longer be the most successful paradigm in view of self-supervised large language models. Finally, the situation as depicted in the figure lacks context. ML, and interpretability, are affected by issues such as: What kind or class of ML model is used in what empirical context? What is the goal of the use of an ML model (use in natural science, public administration, ...)? Who is affected by model output (directly affected decision subjects, indirectly affected by consequences of ML models, ...)? The discussion below will fill in some of these aspects.

9.3 Types

How are interpretability and explanations related? Simply put, there are different ML phenomena we want to understand, and understanding may be provided by explanations of these phenomena (see Lipton, 2018; Zednik, 2021; Creel, 2020). We have just seen (Fig. 9.1) that supervised learning has different parts. A simple consequence of this is that different explanations may be appropriate for phenomena from different parts of ML, and that different explanations correspond to different kinds of interpretability. How should different kinds of interpretability be classified or distinguished? The simplest way is to distinguish them by the part of ML they belong to. For example, a well-known distinction (Lipton, 2018) is between post-hoc interpretability, which concerns properties of the predictor function F, while transparency concerns properties of the model M, e.g., its parameters.

In this section, I propose a simple refinement of this well-known idea. The refinement is based on a systematic use of the *explanandum-explanans* distinction. The idea is to distinguish kinds of interpretability by specifying not only their *explanandum* type, but also their *explanans* type. Recall that the *explanandum* is the property or phenomenon we want to explain, while the *explanans* is what is doing the explanatory work. To specify the *explanandum* and *explanans* type, we state which ML part they belong to; see Fig. 9.1 for the parts. If we write the explanatory relation between *explanandum* and *explanans* as $\rightarrowtail$, we can specify an *interpretability type* in the following form:

$$\text{explanans type} \rightarrowtail \text{explanandum type}. \tag{9.1}$$

The idea of specifying an interpretability type in this way is borrowed from mathematics. To define a mathematical function, one first specifies a domain and a co-domain (target). For example, a metric on X is a function $d : X \times X \rightarrow \mathbb{R}_{\geq 0}$, such that .... The domain and target do not (in general) determine the function, but they tell us where it lives. This, in turn, tells us about certain properties of the function. For example, the domain of a metric makes sense because we want to examine how close pairs of elements of X are, and we want closeness to be measured on a scale of non-negative real numbers. Functions not defined on this kind of domain and target cannot be metrics.

Here is how this translates to interpretability. Assume that we want to explain a particular output $\hat{y} \in Y$ on the basis of a given, trained model M, but we do not want to take the optimization procedure into account. Then the explanation we are looking for has interpretability type $M \rightarrowtail \hat{y}$, where $\hat{y} \in Y$. Note that this is the usual setting of many XAI methods, such as saliency maps. If we want to explain properties of the predictor function in terms of the model and the optimization procedure, then we are looking for an explanation of the type $\{M, O\} \rightarrowtail F$. This is the setting of the so-called Information Bottleneck (IB) method, which examines learning phases of models to illuminate the generalization properties of DNNs (see Shwartz-Ziv & Tishby, 2017, and Räz, 2022 for a philosophical discussion).

Why is it useful to distinguish interpretability types with relation (9.1)? First, we can capture commonly discussed kinds of interpretability. For example, the distinction between understanding *of* an ML model and understanding *with* an ML model can be rephrased as: if we are after (explanatory) understanding *of* the model, the *explanandum type* will be in the set $\{F, M, O\}$. Understanding *with* the model, on the other hand, would be provided by an *explanandum type* in the set {D, S}, while the *explanans type* should include one of $\{F, M, O\}$; otherwise, it would be an explanation without ML. Second, and more importantly, interpretability types can help us draw finer distinctions and clarify misunderstandings. As mentioned above, the distinction between *transparency* and *post-hoc explanations* was introduced by Lipton (2018). On the basis of interpretability types, we can refine this idea as follows. We may be in a situation in which we have access to the structure of a model (it is transparent), but we do not know how it was trained. In this situation, all we can hope to provide is a *post-optimization* explanation. This is an interpretability type such that O is neither in the *explanandum* nor the *explanans*. Usual XAI methods, which try to explain outputs, are of this type, but also methods that try to align inner parts of a model with human-interpretable concepts, such as TCAV (see Kim et al., 2018 on TCAV and Räz, 2023 for a philosophical discussion). In a different situation, we may only have black-box access to a model, that is, we can feed it inputs and get outputs, but we do not know how information is processed. In this situation, one could try to feed F with novel data D' and try to understand its behavior on D'. Such a *black-box* explanation would be of type $\{D', F\} \rightarrowtail F$. In this case, the interpretability type should not contain M or O. Finally, we may be in a situation where we are interested in model behavior and we have full access to both

M and O. Explanations of type $\{M, O\} \rightarrowtail F$ are not concerned with transparency insofar as we do not want to understand or explain the inner working of a model. It may still be useful, or even necessary, to use M and O as part of the *explanans*, and transparency is required for this interpretability type.

This last scenario is relevant for practical reasons. It is sometimes claimed that certain interest groups (stakeholders) such as decision subjects and policy makers are not interested in the inner workings of ML models. Using interpretability types, one recognizes that this statement is ambiguous and potentially misleading. It may be true that decision subjects primarily want to understand how predictions came about, which means that they are not interested in M and O as *explananda*. However, they may still be interested in explanations of model outputs that draw on M and O, that is, in the interpretability type $\{M, O\} \rightarrowtail F$. Such explanations may provide insights that cannot be gained by a black-box method. Thus, it is a misconception that information about the inner workings of ML models are irrelevant for decision subjects just because their requests for explanations may not concern these aspects.

In the above, I have been rather loose in describing interpretability in terms of both explanations and understanding. While there are differences between these concepts, at least as discussed in philosophy of science, the concept of interpretability types may be fruitful when discussing different kinds of understanding as well: we simply have to distinguish the object we want to understand (in parallel to the *explanandum*), and the means, or information, that we want to draw on to gain understanding (in parallel to the *explanans*). The importance of understanding ML is also emphasized in Chap. 7 by Paez in this volume.

9.4 Contrast

Distinguishing kinds of interpretability by type is a useful classificatory device, but it does not uniquely determine all kinds of interpretability. More fine-grained distinctions can be drawn by using contrast (see Miller, 2018, Sec. 2.3.; Guidotti, 2022). In philosophy, the idea that explanations are contrastive was probably first emphasized in van Fraassen's (1980) pragmatic theory of explanation. Discussions of contrast are part of a wider discussion about the extent to which explanations are context dependent; more on this below. Let us start with an example (adapted from Skow, 2016):

(Q1) Why did Adam eat the apple?

(Q1) is a reasonable request for an explanation, but its scope, the kind of answer one might be after by asking it, is very wide, and depending on the situation, different answers may be appropriate. For example, the question can be read with the following emphasis:

(Q2) Why did *Adam* eat the apple?

(Q2) focuses on Adam's relation to the event in question, and an appropriate explanation focuses on how Adam, as opposed to someone else, ended up eating the apple. An appropriate explanation could be: "There was no one else there." The question can also be read with a different emphasis:

(Q3) Why did Adam eat the *apple*?

(Q3) focuses on the apple's relation to the event, and an appropriate explanation focuses on the choice of apple as opposed to some other food. An appropriate explanation could be: "There was pie there, but Adam is watching his diet." These two examples suggest that the role of contrast is to constrain the kind of explanatory information we are looking for: (Q2) and (Q3) restrict the appropriate answer to certain parts of the causal past of the *explanandum* phenomenon, whereas the original question (Q1) does not provide such a restriction.

Now consider some analogous examples in the context of ML. An important case is the explanation of a particular outcome $\hat{y}$ of an ML model M:

(Q4) Why did decision subject i get prediction $\hat{y}_i$?

(Q4) is a request for explanation with no (explicit) contrast. Like (Q1), this question has a wide scope—in principle, it concerns the entire history leading to the prediction $\hat{y}_i$. It is possible to formulate more specific versions of (Q4), which highlight what kind of explanatory information we seek. For example, assume that F is a binary predictor, and that decision subject i has received a decision $F(x_i) = \hat{y}_i = 0$ based on input x_i. Decision subject i may now wonder what it would take to get a different decision $\hat{y}_i = 1$ from the same predictor F. This kind of explanation has been discussed as "counterfactual explanations" of algorithmic output, starting with Wachter et al. (2018) (see Buijsman, 2022; Zerilli, 2022 and Chap. 8 by Buijsman in this volume for discussions of counterfactual and manipulationist notions of explanations in AI). We can thus reformulate (Q4) as:

(Q5) Why did i get prediction $F(x_i) = \hat{y}_i = 0$, as opposed to getting prediction 1 from F?

(Q5) can be read as a question about the general properties of F that yield predictions 0 and 1. Now, (Q5) may still be too general if decision subject i is only interested in particular contrasts, for example, in inputs that are close to their input x_i according to some metric. This interest may be motivated by the desire to determine which changes to x_i would flip the decision and require the least effort. In this situation, (Q5) specializes to:

(Q6) Which values x are close to x_i (according to some metric), but such that $F(x) = 1$?

(Q6) is an even more focused request for information about the predictor F, in that information about F "far away" from the value x_i is considered to be irrelevant. Thus, (Q6) excludes a lot of information that would be relevant in response to (Q4). And, despite not being a why-question, it can be interpreted as explanatory because, while it is more focused than (Q5), it requests information about properties relevant to classification, in particular, information about the ball around x_i in input space that gets the same prediction as x_i.

Are all explanations contrastive, as suggested by Miller (2018)? Not necessarily. A first example is (Q4), the open-ended request for an explanation of a particular output. This is not a contrastive request for an explanation. It may be objected that while (Q4) is not contrastive, it is not possible to fully answer (Q4), because it is too open ended. However, if the request only concerns information about the predictor F, and if F is sufficiently simple (e.g. a linear model of few variables), one can provide a full answer by providing a description of F.

A second example of a non-contrastive request requires very specific information. A decision subject asking (Q4) may be interested in the input x_i that led to the prediction $F(x_i) = \hat{y}_i$:

(Q7) Which input x_i yielded the prediction $F(x_i) = \hat{y}_i$ for decision subject i?

This is not a contrastive request, and it is also not a why-question, which means that, on the surface level, (Q7) is not a request for an explanation, as opposed to (Q4). The reason why I would nevertheless suggest that (Q7) is a request for a (partial) explanation is that (Q7) asks for a specific fact that partially determines, or is responsible for, the output, which is the phenomenon we want to understand. It therefore constitutes a request for explanatorily relevant information. Also, (Q7) is a special case of (Q4), and the answer, the value x_i, does provide a (partial) explanation of the prediction. Note that accepting (Q7) as a request for an explanation means that we do not strictly adhere to the working definition of what constitutes an explanation.

Insisting that all explanations are contrastive may also be problematic because contrasts narrow down the scope of acceptable explanations. Presupposing a particular contrast leads to the exclusion of certain answers. Let us consider a different version of (Q4) to see why this is problematic. One legitimate concern one may have when asking (Q4) is with properties of the decision process, such as the predictor function F or the model M used to obtain the decision $\hat{y}_i$. One may ask:

(Q8) What is the predictor function F (or model M) that yielded the prediction $\hat{y}_i$?

An answer to this question may give rise to a contrastive follow-up: Once one knows F, it may be asked why F (or a certain class of predictors) was used as opposed to a predictor F' (or a different class of predictors), which may have yielded a different prediction. To give an example, in certain contexts, black-box models may have no advantage in performance over interpretable models, but their decision process is inscrutable; this is proposed by Rudin (2019). This alone may constitute sufficient grounds to challenge decisions reached with black-box models. Compare this to the contrast in (Q6), which focuses on inputs with respect to the *same* predictor F. By narrowing down the explanatory request, this question necessarily excludes other aspects of ML, such as the choice of model. However, the aspects that are excluded by a contrast may be exactly those that need to be explained. Instead of asking how the decision subject should change in view of a fixed predictor, as suggested by the contrast in (Q6), it may be more appropriate to ask whether a different model should be used, as suggested by (Q8). In principle, there is no limit on the aspects of ML that can and should be scrutinized by asking explanatory questions about them, from the predictor F, to the model M, to the optimization, data, or even the very fact that decision are made at all. Thus, an open-ended question, while unfocused, encompasses more possible contrasts with respect to an *explanandum* and may thus be preferable for some purposes.

9.5 Context

To what extent are explanations context dependent? Context dependence means that both the interpretation of an explanatory request, and what constitutes an appropriate answer to this request, depend on the situation in which the request for explanation is posed. In particular, what the explanatory request is and what constitutes an appropriate answer, may depend on (a) the kind of ML model in question, (b) the empirical domain or problem to which the model is applied, (c) the audience asking the question or receiving the explanation, and (d) the purpose behind the request for explanation, and possibly further factors. The context dependence of scientific explanations was discussed extensively in the wake of van Fraassen (1980) pragmatic theory; see also Woodward and Ross (2021, Sec. 6).

Context dependence is an important feature of explanations, but it also creates challenges. For example, it may lead to a proliferation of explanations, it can make it hard to specify what counts as a good explanation, or to provide a "rigorous definition" (Doshi-Velez & Kim, 2017) of explanations. Context dependence also threatens to trivialize explanations, say, if the person requesting the explanation can decide freely whether a given answer is adequate or not. The challenge is to allow for some context dependence, while not pushing contextualization too far by letting what constitutes an appropriate explanation depend on, say, the mood of the person asking for it. The idea that there may be two kinds of context dependence, one that is mostly harmless, and one that is more radical, is articulated in Woodward and Ross (2021). Two reasonable kinds of context dependence have been discussed above. First, if we specify the source and the target domain of an explanation (Sect. 9.3), this may yield several, distinct classes of explanations. Interpretability types place restrictions on both the model in question, the empirical domain (if any) about which the question is asked, and the resources that can be used to answer the question. However, the type does not determine the explanation. The second kind of context dependence is contrast (Sect. 9.4). A precise articulation of the explanatory question narrows down what kind of explanatory information is considered relevant. This may yield different kinds of explanations. Also, by narrowing down what kind of information is considered relevant, contrasts may help with audience dependence, because different kinds of information are relevant for different audiences.

Some kinds of context dependence are not captured by contrast and interpretability types. For one, there are more fine grained kinds of audience dependence; see Langer et al. (2021) for a stakeholder-centric perspective on XAI, and Zednik (2021) for a proposal of which parts of ML different stakeholders are interested in. For example, experts and laypeople may have the very same explanatory request, with respect to the same explanation type, and still have different explanatory needs: An explanation that is comprehensible to an expert may be inaccessible to a layperson. This sort of context dependence is important, but it has traditionally not been central to the discussion of scientific explanations in philosophy of science, because the subject matter of scientific explanations are the best explanations that science has to offer to experts.

Audience dependence creates challenges. For example, what is the relation between an expert explanation, which provides the best available understanding of a certain phenomenon, and an explanation of the same phenomenon for laypeople? A tempting answer is that a layperson explanation is an explanation sketch, or an approximation, of a full explanation, where the sketch provides the gist of the full story (the idea of sketches of historical explanations goes back to Hempel (1942)). The problem with this solution is that if it is not communicated how a sketch deviates from the full story, the layperson explanation can be misleading, because it leaves out certain aspects of the full story. If, however, it is communicated how a sketch deviates from the full story, it can lead to the kind of information overload the approximation was designed to avoid. A different approach would require that a layperson is given an explanation sketch, or an approximative explanation, together with expert guidance on the difference between the sketch and the full story, depending on the goals of the layperson. This approach has the advantage that the gap between sketch and full story is bridged in a customized manner. A drawback is that it is much more costly than a "one size fits all" sketch.

Finally, radical context dependence means that the context dictates not only how the request for an explanation should be interpreted, but also what constitutes an admissible answer. A version of this problem was raised by Kitcher and Salmon (1987) as an objection against van Fraassen's pragmatic theory of explanation; the account given here draws on Woodward and Ross (2021). Even the kinds of facts that form the basis of an explanation can be chosen more or less freely. Radical context dependence does not seem adequate. Take the example of a request for information about close-by inputs that flip a prediction, (Q6) above. Not any kind of information is explanatorily relevant to answer this question, even if the notion of distance is not completely rigorous. Relevant information concerns the decision boundary of the predictor, plus, possibly, further information about the predictor F. This means that an adequate answer to (Q6) will have to draw on these facts, which means that what constitutes an adequate explanation is not completely context dependent. We will return to the discussion of the subjectivity and objectivity of explanations in Sect. 9.8.

9.6 Theoretical Explanations

So far, the examples of ML phenomena and explanation types were mostly situated at a concrete level of particular predictions. However, explanations also occur at a more general and theoretical level. Higher generality means that the *explanandum* type is different. For example, we may ask about the behavior of an entire class of models, without reference to applications or single predictions. The tools to answer such questions are also different. If, say, our question is statistical and not tied to a particular dataset, the answer may be statistical or mathematical as well. Questions and answers of this type result in theoretical explanations.

A first example concerns the generalization properties of DNNs (see Zhang et al., 2021 on the concept of generalization, Kawaguchi et al., 2023 for a recent survey, and Buckner, 2019; Räz, 2022 for philosophical discussions). Many DNNs show a very good test set performance, that is, they generalize well. This phenomenon is general in that DNNs generalize well on many different datasets (e.g. image classification benchmarks), but also for different data modalities, such as images, text, and so on. This raises the question: Why do DNNs generalize well? This question is all the more relevant because *prima facie*, many DNNs are overparametrized, that is, they have many free parameters in comparison to the size of training sets, such that one would expect the models to overfit. This, however, is not the case. So far, there are several proposals and promising avenues, but no satisfactory answer to this why-question (Zhang et al., 2021). The generalization phenomenon is statistical in a double sense: The test set performance of single DNNs itself is a statistical property, and the generalization phenomenon concerns the frequency of this statistical property in applications of DNNs. This suggests that the eventual *explanans* will also be statistical in nature. In fact, many attempts to explain generalization behavior come from statistical learning theory (see the references given above). However, it is also possible that the *explanans* will include domain-specific knowledge, that is, a characterization of the kinds of features or data for which DNNs work well. There are different, interesting contrasts for the generalization phenomenon. One contrast is the question why the models perform well as opposed to overfitting the data. A second contrast, which may also be worth exploring, would be to better understand why DNNs do *not* perform well in some cases. The latter contrast may lead to a better understanding of the kind of features that contribute to good generalization.

A second example of a phenomenon requiring a theoretical explanation concerns the optimization properties of DNNs. The optimization (or learning) process used to train DNNs is usually a version of stochastic gradient descent (SGD). SGD is used to modify the weights of DNNs through backpropagation. There are many open questions regarding the optimization of DNNs. Usually, the optimization problem to be solved by DNNs is non-convex: if the error landscape is optimized locally using SGD, there is no guarantee that one ends up at a global minimum. Thus, the question arises: Why are DNNs optimized with SGD able to find global or close to global minima, as opposed to getting stuck in non-optimal local minima?

This is a request for a general explanation. There has been interesting and relevant work on this problem, see e.g. Vidal et al. (2023), but it is still open, like the generalization puzzle. For example, the optimization properties of special kinds of simple networks (linear networks with one hidden layer) have been explored, and it has been shown that for some kinds of data and architectures, optimization will not get stuck in local minima. This approach to explaining the optimization behavior of DNNs with "idealized" models can be interpreted as providing how-possibly explanations (Scholl & Räz, 2013; Verreault-Julien, 2019, 2025): it is established that the idealized model (a linear, shallow network) reproduces the behavior of the target system (deep, non-linear neural networks), for reasons that are mathematically well understood, but it is not clear to what extent the same explanations also apply to non-idealized models. More recent work has been geared towards de-idealizing these results, e.g., by extending the results to deep and non-linear networks.

A third example where an explanatory perspective may be useful is expressivity; see Gühring et al. (2023) for a recent review, on which the following draws. Expressivity explores the kinds of functions that DNNs can express in principle. Given a certain class of functions, such as continuous functions on a certain domain, it is investigated whether a class of models can approximate functions in that class. For example, one early universal approximation theorem states that a continuous function on a compact real domain can be approximated arbitrarily well by a neural network with sigmoidal activations (see Gühring et al. (2023) for details and Nielsen (2015) for a heuristic argument why DNNs are universal function approximators). Results of this kind can, again, be interpreted as providing how-possibly explanations, because they explain why these models are in principle capable of approximating a certain class of functions. However, expressivity results do not provide how-actually explanations. First, the process by which models are matched with functions does not correspond to the actual optimization procedures used in practice. Thus, there is no guarantee that one can approximate all functions in a class with the optimization procedures used in practice. Second, the model architectures used to prove approximation results can be highly idealized. For example, early universal approximation theorems with shallow networks needed the number of parameters to grow exponentially in the size of the input to achieve good approximations. These results are nevertheless useful, because they may suggest a path from highly idealized results to less idealized ones, and progress has been made since the early days. Note that other theoretical works, e.g., no-free-lunch theorems, may also have explanatory importance (see Sterkenburg and Grünwald, 2021 for discussion of NFL theorems).

9.7 Levels

The examples of general explanations provided in the previous section suggest that one can explain ML phenomena at different levels of generality. For example, we can ask for explanations of single outputs, but also for explanations of statistical properties of generalization. If we accept that explanations come at different levels, the question arises as to how the levels are related. Below are some examples of relations between explanatory levels, highlighting their importance (see Dazeley et al., 2021 for a recent proposal regarding levels of explanations in AI).

A first example can be gleaned from the discussion of contrastive explanations (Sect. 9.4). If we use contrast, we can formulate more focused requests for explanation and exclude certain possible answers as irrelevant. This also means that an explanation request without contrast is more general than the same request with contrast. For example, contrastive explanations that only consider particular values of the predictor function are special cases of an explanation of the global behavior of a predictor function. The relation between these explanations is deductive: global information about all input-output pairs entails information about specific counterfactual questions. Thus, if we are able to provide a global explanation of what a predictor function is doing, then this explanation encompasses everything we may want to know about counterfactuals, which means that the latter become superfluous. However, global explanations may not always be available. If a predictor function is too complex, we may not be able to grasp it, and we have to resort to approximations, or particular counterfactual scenarios, which are more tractable. Also, it is very hard, if not impossible, to specify formal criteria of functional interpretability, which corresponds to a global understanding of predictor functions (Räz, 2024). In a nutshell, the problem is that such understanding is possible if the corresponding predictors are simple, but the notion of simplicity itself is heterogeneous and has no formal characterization.

A second example is the relation between safe application of DNNs in particular contexts and our understanding of the generalization properties of DNNs. An explanation of the generalization properties of DNNs would be very useful because such an explanation would presumably reveal to what extent the application of these models allows us to understand or explain empirical phenomena. An explanation of the generalization properties might provide information about the contexts in which DNNs rely on spurious features, i.e., have high predictive accuracy "for the wrong reasons", and contexts in which DNNs do not rely on spurious features. Only in the latter case should we consider empirical information revealed by DNNs to be genuinely explanatory. The fact that we do not (yet) have a consolidated theory of deep learning means that the different explanatory levels are not firmly integrated (see Räz & Beisbart, 2022, Sec. 5.).

9.8 Objectivity and Idealization

Are explanations of ML phenomena objective? The answer hinges on what we mean by "objective". It is helpful to tackle a different question first: In what sense could explanations be subjective? One possible source of subjectivity is the audience dependence of explanations (Sect. 9.5). Surely, an explanation is only successful to the extent that people are able to grasp it. If so, this commits us to a view of explanations according to which at least a certain degree of grasping (or understanding) is necessary for a successful explanation; more on this below. However, if this is taken to the point where grasping depends on the individuals involved, then explanations become truly subjective. Compare this with scientific explanations: whether or not a (scientific or mathematical) explanation can be grasped is not a matter of individuals, but a matter of consensus in the scientific community. This does not make explanations completely objective: If what constitutes an explanation is determined by a community, the community can disagree about what can be grasped, and the community can be wrong. However, it also shows that explanations are not entirely subjective or up to individuals either. The problem of how to bridge the gap between explanations for experts and for laypeople was briefly discussed in Sect. 9.5.

The problem of graspability leads to a general feature of explanations. On the one hand, it should be granted that explanations do, in fact, serve a psychological purpose: they should be graspable in principle, after adequate scientific training, or with guidance. On the other hand, explanations do not reduce to what has been called a sense of understanding (Trout, 2002). They are grounded in facts, at least to some extent. Thus, explanations have two main ingredients, their psychological role, and factivity; see Wilkenfeld (2017) for a discussion of these two components in the debate on understanding, and Räz (2024) for an articulation of this idea in the context of understanding ML. If an explanation does not have one of these ingredients to a sufficient degree, it loses its explanatory status. To a large extent, the challenge of articulating good explanations boils down to the fact that these two ingredients are necessary, but also in tension.

To see how the tension arises, take the example of explaining single outputs of ML models (Sect. 9.4). In principle, we can simply cite the entire history of a model leading up to the output; this includes the entire model, training process, training data, and how each of these components were generated. This, however, is not an explanation because the entire history of an output is not graspable, it is too large, which leads to cognitive overload. Thus, certain facts have to be left out or averaged over. Also, depending on the exact explanatory question, not all facts are relevant. However, even only relevant facts may lead to cognitive overload. Thus, usually, idealizations have to be introduced. Idealizations are deliberate distortions of facts, which allow us to grasp explanations. But every idealization compromises factivity. Thus, idealization is the second threat to objectivity.

Some critics have rejected approximative explanations as promoted by XAI for the very reason that they are necessarily wrong. In particular, Cynthia Rudin (2019) has argued that explainability methods are inadequate because they are non-factive. As a remedy, she has recommended to use interpretable models, such as rules lists, which are inherently and globally interpretable, while not compromising too much on accuracy. If it is really possible to build globally interpretable models that show predictive performance similar to black-box models, they should be preferred. The worry with this approach is that a loss of predictive performance will be incurred. Such a loss can be interpreted as a loss of a different kind of factivity, viz. the ability of the model to capture an empirical phenomenon. Note that recent results suggest that we can expect similar performance of black-box models and interpretable models at least for tabular data (Chang et al., 2021). If we accept these results, then we should prefer interpretable models.

A second kind of example of explanations relying on idealizations are theoretical explanations (Sect. 9.6). The expressivity results of DNNs are mathematical results about the possibility of DNNs to represent certain function classes. At face value, such results are not idealized, they are mathematical statements, and contain no falsehoods. Considerations of idealization come in when we apply these results to DNNs with architectures as used in practice. Very often, such DNNs will not share some of the properties of the models that are investigated theoretically. This raises the question as to how the theoretical results bear on cases in which the assumptions of a theorem are violated. Above, it was suggested that we can interpret such results as how-possibly explanations. However, it is not clear whether these can be turned into how-actually explanations, e.g., by generalizing the mathematical results to more realistic architectures. At least some recent efforts (cf. Gühring et al., 2023) can be interpreted as working towards such de-idealized results, including work on the role of depth of DNNs in expressivity.

All in all, the use of idealizations is standard practice in science, and need not be necessarily problematic. A minimal requirement for the use of idealizations is that they should be labeled as such. One danger of idealized explanations is that even if such labels are provided in the original proposal of the idealized explanation, such labels are often left out in subsequent work, such that idealized explanations are interpreted as how-actually explanations.

9.9 Pluralism

Interpretability is pluralistic if there is not one kind of explanation of ML phenomena, but many. Pluralism with respect to interpretability has been advocated in the computer science literature by Lipton (2018), and also in philosophy by Krishnan (2020). Sometimes, the advocacy has been accompanied by calls for a formal definition (Doshi-Velez & Kim, 2017), sometimes with scepticism as to the possibility of a unified notion of interpretability (Räz, 2024). In the above discussion, explanation types, contrastive explanations, audience dependence, and different levels of explanation all point towards a certain degree of heterogeneity. Independently of the situation in computer science, scientific explanations presumably are also heterogeneous. There are many different proposals of how to define what a scientific explanation is, both in general and in particular scientific disciplines (see Woodward & Ross, 2021 for scientific explanations and Mancosu, 2018 for mathematical explanations). Some proposals may aspire to be generally applicable, but there is no generally accepted theory of what a scientific or mathematical explanation is, be it in the form of necessary and sufficient conditions, or in the form of a more relaxed "theory of explanation". Some very general properties appear to be shared by most explanations, such as those used as a working definition above (Sect. 9.2.1), but these are too unspecific to merit being called a theory.

What does the state of the philosophical discussion mean for the discussion about explanations of ML phenomena? A critical stance towards the relevance of scientific and mathematical explanation for XAI and interpretability is defended by Páez (2019) and Krishnan (2020); see also Chap. 7 by Paez in this volume. Even if scientific explanations are heterogeneous, this does not imply that explanations of ML phenomena are heterogeneous as well, simply because ML does not coincide with science (or math). However, the heterogeneity of scientific and mathematical explanations does have methodological consequences for the use of philosophical theories of explanation in application to ML. For example, if we can reconstruct a candidate explanation from ML in terms of a theory of explanation from philosophy of science, this does not imply that this candidate is therefore a good explanation, simply because there is no consensus in philosophy of science about what constitutes a good explanation. Such a reconstruction may be fruitful in some cases, but not in others. For example, showing that some mode of reasoning conforms with the DN theory of explanation is not useful per se, because it is known that some candidate explanations that have the form of a DN explanation are actually not very good explanations—put bluntly, the DN theory does not provide a useful picture of scientific explanations. For criticism of applying the DN theory in the context of ML, as proposed by Erasmus et al. (2020) see Chap. 7 by Paez in this volume.

Now, even if there is no unified notion of explaining ML phenomena, this does not need to lead to a trivialization of the notion. One could hope that for each kind of explanation, we can give a reasonably clear account of why it is an explanation. Even if there are several kinds of explanations, it is still possible that we end up with a limited set of reasonable kinds of requests for explanation, and with a limited set of acceptable answers to each request (see Räz (2024) for elaboration, and Durán (2021) for the need of a unified picture of scientific XAI). Of course, right now, such sets have not been identified. If this could be done, it would amount to what could be called a heterogeneous theory of explanations. Such a theory, because it is heterogeneous, would presumably involve distinctions between explanations along some of the dimensions mentioned above: types, context, contrast, levels, and so on.

Finally, note that ML also has the potential to play a unifying role in providing explanations of empirical phenomena, by generating explanations *with* ML. If progress with our theoretical understanding of ML (Sect. 9.6) were made, e.g., with an explanation of the generalization properties of ML, this explanation may tell us something about the features shared by domains in which ML is successfully applied. This, in turn, might lead to a non-trivial, theoretically-grounded account of the circumstances in which inductive reasoning, including scientific explanation, is successful, and also of the circumstances in which it is not.

9.10 Conclusion

To conclude, I would like to stress the advantages of viewing the problem of interpretability from the perspective of scientific and mathematical explanations. We can view these as the best explanations that science (including computer science, mathematics, etc.) has to offer. For some ML phenomena, there is no consensus as to what the best explanation would look like, and for other phenomena, there are no explanations altogether. In this situation, it can be misleading to provide explanations based on simplifications and visualizations that are merely graspable, but do not pay sufficient heed to factivity. It can be misleading because the simplified explanations may paint an inadequate picture of the best available explanation of a phenomenon, but also because the best available explanation may still be very incomplete. Instead of communicating in terms of such simplified explanations, it is preferable to try to convey a picture of the complexity of the best available explanation, and also of the limits of our current knowledge.

Acknowledgments I thank Claus Beisbart, Stefan Buijsman and Juan M. Durán for very valuable comments on earlier drafts of this chapter.

Funding This work is funded by the Swiss National Science Foundation through grant number 197504.

References

Adadi, A., & Berrada, M. (2018). Peeking inside the black-box: A survey on explainable artificial intelligence (XAI). *IEEE Access*, *6*, 52138.

Beisbart, C., & Räz, T. (2022). Philosophy of science at sea: Clarifying the interpretability of machine learning. *Philosophy Compass*, *17*(6), e12830.

Biran, O., & Cotton, C. (2017). Explanation and justification in machine learning: A survey. In *IJCAI-17 Workshop on Explainable AI (XAI)* (Vol. 8, pp. 8–13).

Bromberger, S. (1966). Why questions. In R. Colodny (Ed.), *Mind and cosmos: Essays in contemporary science and philosophy*. University of Pittsburgh Press.

Buckner, C. (2019). Deep learning: A philosophical introduction. *Philosophy Compass*, *14*, e12625.

Buijsman, S. (2022). Defining explanation and explanatory depth in XAI. *Minds and Machines*, *32*(3), 563–584.

Buijsman, S. (2024). Machine learning models as mathematics. In J. M. Duran & G. Pozzi (Eds.), *Philosophy of science for machine learning: Core issues and new perspectives*. Synthese Library. Springer.

Chang, C.-H., Tan, S., Lengerich, B., Goldenberg, A., Caruana, R. (2021). How interpretable and trustworthy are gams? In *Proceedings of the 27th ACM SIGKDD Conference on Knowledge Discovery & Data Mining* (pp. 95–105).

Creel, K. A. (2020). Transparency in complex computational systems. *Philosophy of Science*, *87*(4), 568–589.

Dazeley, R., Vamplew, P., Foale, C., Young, C., Aryal, S., Cruz, F. (2021). Levels of explainable artificial intelligence for human-aligned conversational explanations. *Artificial Intelligence*, *299*, 103525.

Doshi-Velez, F., & Kim, B. (2017). Towards a rigorous science of interpretable machine learning. arXiv:1702.08608v2.

Durán, J. M. (2021). Dissecting scientific explanation in AI (SXAI): A case for medicine and healthcare. *Artificial Intelligence*, *297*, 103498.

Erasmus, A., Brunet, T. D., & Fisher, E. (2021). What is interpretability? *Philosophy & Technology, 34*(4), 833–862.

Goodfellow, I., Bengio, Y., & Courville, A. (2016). *Deep learning*. MIT Press.

Gühring, I., Raslan, M., & Kutyniok, G. (2023). Expressivity of deep neural networks. In *Theory of deep learning* (pp. 149–199). Cambridge University Press.

Guidotti, R. (2022). Counterfactual explanations and how to find them: Literature review and benchmarking. *Data Mining and Knowledge Discovery, 38*(5), 2770–2824.

Hastie, T., Tibshirani, R., & Friedman, J. (2009). *The elements of statistical learning. Springer series in statistics* (2nd ed.). Springer.

Hempel, C. G. (1942). The function of general laws in history. *The Journal of Philosophy*, *39*(2), 35–48.

Hempel, C. G., & Oppenheim, P. (1948). Studies in the logic of explanation. *Philosophy of Science*, *15*, 167–179.

Kawaguchi, K., Bengio, Y., & Kaelbling, L. (2023). Generalization in deep learning. In *Theory of deep learning*. Cambridge University Press.

Kim, B., Wattenberg, M., Gilmer, J., Cai, C., Wexler, J., Viegas, F., et al. (2018). Interpretability beyond feature attribution: Quantitative testing with concept activation vectors (TCAV). In *International Conference on Machine Learning* (pp. 2668–2677). PMLR.

Kitcher, P., & Salmon, W. C. (1987). Van Fraassen on explanation. *Journal of Philosophy*, *84*, 315–330.

Krishnan, M. (2020). Against interpretability: A critical examination of the interpretability problem in machine learning. *Philosophy & Technology*, *33*, 487–502.

Langer, M., Oster, D., Speith, T., Hermanns, H., Kästner, L., Schmidt, E., Sesing, A., & Baum, K. (2021). What do we want from explainable artificial intelligence (XAI)? – A stakeholder perspective on XAI and a conceptual model guiding interdisciplinary XAI research. *Artificial Intelligence*, *296*, 103473.

LeCun, Y., Bengio, Y., & Hinton, G. (2015). Deep learning. *Nature*, *521*, 436–444.

Lipton, Z. C. (2018). The mythos of model interpretability. *Queue*, *16*(3), 31–57. arXiv:1606.03490.

Mancosu, P. (2018). Explanation in mathematics. In E.N. Zalta (Ed.), *The Stanford encyclopedia of philosophy*. Metaphysics Research Lab, Stanford University.

Miller, T. (2018). Explanation in artificial intelligence: Insights from the social sciences. *Artificial Intelligence*, *267*, 1–38.

Nielsen, M. A. (2015). *Neural networks and deep learning*. Determination Press.

Páez, A. (2019). The pragmatic turn in explainable artificial intelligence (XAI). *Minds and Machines*, *29*(3), 441–459.

Páez, A. (2024). Axe the X in XAI: A plea for understandable AI. In J. M. Duran & G. Pozzi (Eds.), *Philosophy of science for machine learning: Core issues and new perspectives*. Synthese Library. Springer.

Räz, T. (2022). Understanding deep learning with statistical relevance. *Philosophy of Science*, *89*(1), 20–41.

Räz, T. (2023). Methods for identifying emergent concepts in deep neural networks. *Patterns*, 4(6).

Räz, T. (2024). ML interpretability: Simple isn't easy. *Studies in History and Philosophy of Science*, *103*, 159–167. arXiv:2211.13617.

Räz, T., & Beisbart, C. (2022). The importance of understanding deep learning. *Erkenntnis*. https://doi.org/10.1007/s10670-022-00605-y

Rudin, C. (2019). Stop explaining black box machine learning models for high stakes decisions and use interpretable models instead. *Nature Machine Intelligence*, *1*, 206–215.

Schmidt, E. (2024). Stakes and understanding the decisions of artificial intelligent systems. In J. M. Duran & G. Pozzi (Eds.), *Philosophy of science for machine learning: Core issues and new perspectives*. Synthese Library. Springer.

Scholl, R., & Räz, T. (2013). Modeling causal structures. *European Journal for Philosophy of Science*, *3*(1), 115–132.

Shwartz-Ziv, R., & Tishby, N. (2017). Opening the black box of deep neural networks via information. arXiv:1703.00810.

Skow, B. (2016). Scientific explanation. In P. Humphreys (Ed.), *The Oxford handbook of the philosophy of science*. Oxford University Press.

Sterkenburg, T. F., & Grünwald, P. D. (2021). The no-free-lunch theorems of supervised learning. *Synthese*, *199*(3), 9979–10015.

Sullivan, E., & Kasirzadeh, A. (2024). Explanation hacking: The perils of algorithmic recourse. In J. M. Duran & G. Pozzi (Eds.), *Philosophy of science for machine learning: Core issues and new perspectives*. Synthese Library. Springer.

Trout, J. (2002). Scientific explanation and the sense of understanding. *Philosophy of Science*, *69*, 212–233.

van Fraassen, B. (1980). The pragmatics of explanation. In *The scientific image*. Clarendon Press.

Verreault-Julien, P. (2019). How could models possibly provide how-possibly explanations? *Studies in History and Philosophy of Science Part A*, *73*, 22–33.

Verreault-Julien, P. (2025). Three strategies for salvaging explanatory value in deep neural network modeling. In T. Knuuttila, T. Grüne-Yanoff, R. Koskinen, & Y. Sjölin Wirling (Eds.), *Modeling the possible*. Routledge.

Vidal, R., Zhu, Z., & Haeffele, B. D. (2023). Optimization landscape of neural networks. In *Theory of deep learning* (pp. 200–228). Cambridge University Press.

Wachter, S., Mittelstadt, B., & Russell, C. (2018). Counterfactual explanations without opening the black box: Automated decisions and the GDPR. *Harvard Journal of Law & Technology*, *31*(2), 841–887.

Wilkenfeld, D. A. (2017). Muddy understanding. *Synthese*, *194*(4), 1273–1293.

Woodward, J., & Ross, L. (2021). Scientific explanation. In E. N. Zalta (Ed.), *The Stanford encyclopedia of philosophy*. Metaphysics Research Lab, Stanford University.

Zednik, C. (2021). Solving the black box problem: A normative framework for explainable artificial intelligence. *Philosophy & Technology*, *34*, 265–288.

Zerilli, J. (2022). Explaining machine learning decisions. *Philosophy of Science*, *89*(1), 1–19.

Zhang, C., Bengio, S., Hardt, M., Recht, B., & Vinyals, O. (2021). Understanding deep learning (still) requires rethinking generalization. *Communications of the ACM*, *64*(3), 107–115.

Tim Räz is a postdoc at the Institute of Philosophy, University of Bern, Switzerland. A philosopher (PhD 2013, University of Lausanne) and mathematician (MSc. 2019, University of Bern) by training, he works on conceptual issues in machine learning and artificial intelligence, including interpretability and fairness.

Part IV
Scientific Understanding and Interpretability

Chapter 10
Explanation Hacking: The Perils of Algorithmic Recourse

Emily Sullivan and Atoosa Kasirzadeh

Abstract We argue that the trend toward providing users with feasible and actionable explanations of AI decisions—known as recourse explanations—comes with ethical downsides. Specifically, we argue that recourse explanations face several conceptual pitfalls and can lead to problematic explanation hacking, which undermines their ethical status. As an alternative, we advocate that explanations of AI decisions should aim at understanding.

Keywords Explainable AI · Algorithmic recourse · Understanding · Counterfactual explanation

10.1 Introduction

When an AI system recommends against you getting a job interview, a bank loan, or categorizes you as 'high risk,' you probably want to know *why*. At first glance, it might seem simple to give reasons why, but in reality there are several different kinds of reasons, including a myriad of justifications, that could fulfill this explanatory purpose. These *reasons-why* serve as explanations. In the context of AI, particularly opaque algorithms, explainable AI techniques are employed to articulate such reasons behind the AI's outputs. One primary research directive on the explainability of AI systems focuses on questions concerning the (non-) epistemic norms required to satisfy our desire–and right–to know why an AI system made its decision. One normative approach that is gaining more and more traction is *algorithmic recourse* (Ustun et al., 2019; Venkatasubramanian & Alfano, 2020). Recourse explanations give reasons that are actionable and feasible for the end user

E. Sullivan (✉)
University of Edinburgh, Edinburgh, UK
e-mail: esulliv2@ed.ac.uk

A. Kasirzadeh
Carnegie Mellon University, Pittsburgh, PA, USA

J. M. Durán, G. Pozzi (eds.), *Philosophy of Science for Machine Learning*, Synthese Library 527, https://doi.org/10.1007/978-3-032-03083-2_10

or data subject to change. If the end user takes on these actionable reasons by implementing changes in their life, then, as the story goes, they would get a decision reversal in the future from the same model, *ceteris paribus*. Take for example being rejected for a bank loan. A recourse explanation of this decision could be something like the following: If you had worked full-time, instead of part-time, then you would have been approved for the loan. This explanation suggests that the user could transition to full-time work and secure a loan approval upon reapplication.

Algorithmic recourse has been held up as an ethical gold standard for AI explanations. For example, Venkatasubramanian and Alfano (2020) say the following:

> Recourse systematically delivers the benefit of reversing harmful decisions by algorithms and bureaucracies across a range of counterfactual scenarios. If someone enjoys recourse, then not only are they able to get a single decision reversed, but they also enjoy the power to reverse decisions across a range of counterfactual scenarios. As such, someone who enjoys recourse need not passively suffer the slings and arrows of outrageous fortune, but is positioned to take up arms against a sea of troubles.

Let's call this sentiment the recourse-first view for explainable AI.

> **Recourse-first:**
> Providing users with recourse explanations of AI decisions—i.e. feasible and actionable counterfactual explanations—is (ethically) preferable to providing users with non-recourse explanations.

In this chapter, we argue that the recourse-first view should be rejected. Instead, we advocate for the following understanding-first view:

> **Understanding-first:**
> Providing users with explanations that aim at understanding AI decisions is (ethically) preferable to providing users with recourse explanations.

While providing users with recourse explanations seems like the ethically responsible strategy, we suggest that, in reality, recourse explanations face several pitfalls and hidden assumptions that undermine their ethical status. An understanding-first approach, we argue, has a better chance of fulfilling our need, and right, to know why an AI system made the decision it did.

A few points of clarification are in order. First, in this chapter, we are concerned with explaining AI systems behavior (what we can call *model-model* questions). We are not discussing how we might use an AI system to explain some real-world phenomena (what we can call *model-world* questions).[1] Second, in this chapter, we focus on counterfactual explanation methods in xAI. We do this because the work on recourse emerged out of counterfactual explanation methods. That said, we think our arguments also generalize to other methods (see Sect. 10.5.2).

The chapter proceeds as follows. First, we introduce counterfactual explanation methods and why they have become a popular explanation method for AI decisions. We discuss that counterfactual explanations can have several aims that come apart

[1] For a discussion on how xAI methods may fail to answer model-world questions, see Durán (2021).

(Sect. 10.2). Second, we discuss the recourse-first view, why it seems attractive, but also argue that it faces many problems (Sect. 10.3). Third, we turn to the understanding-first view. We describe the view and how this view relieves some of the problems of recourse (Sect. 10.4). Lastly, we discuss possible objections, open questions, and directions for future work (Sect. 10.5).

10.2 Counterfactual Explanations

Counterfactual explanation (CE) methods are a type of post-hoc explainability method that seeks to give insight into the input features influencing AI systems' decisions (Biran & Cotton, 2017; Karimi et al., 2020; Wachter et al., 2017). Counterfactual explanations seek to answer "what if" questions by pointing to minimal changes in the input features that flip the prediction. As a result, CE methods do not provide the direct reason why AI systems make decisions but provide modal information to understand the decision boundary. For example, consider a bank loan AI decision support tool. For any decision, CE methods will perturb various input features and combinations of features to see what minimal changes will turn a loan rejection into a positive decision. The resulting CE might look something like: "If you carried $10,000 less debt and had a salary of over $50,000, you would have gotten the loan." The idea is that having this modal information can help users understand why the system made the decisions it did.

CE methods boast that they don't have the same issues of model fidelity that other post-hoc explanation methods have since they are perturbing on the input variables of the learned ML model directly (Mothilal et al., 2020, Mittelstadt et al., 2019). However, CE methods can still engage in distortions. They must select which counterfactual scenarios are the most salient from a larger set of possible counterfactuals. Many CE methods assume causal independence of input features, thus can provide explanations that are impossible in the real-world (Hooker et al., 2021, Karimi et al., 2020); for example, a system suggesting, in a medical context, to increase one's weight but lower BMI. However, weight and BMI are entangled, making it impossible to do both. CE methods suffer from the *Rashomon effect* where different counterfactuals explaining the same decision give conflicting and mutually exclusive reasons (Molnar, 2020). Some recent work even suggests that CEs can be manipulated (Slack et al., 2021). However, in this chapter, we want to set aside worries about adversarial manipulation and model fidelity. Even if those issues are resolved there are several conceptual issues and pitfalls that face CE methods, especially recourse CEs.

Since CE methods probe the AI system for possible decision flips, CE methods can generate *many* different counterfactuals of the same decision. Depending on how complex the ML model is, there could be thousands or more different counterfactuals for a single decision. Thus, one of the central challenges of CE methods is filtering these counterfactuals so that they meet particular high-level goals. In the landmark paper on CE methods, Wachter et al. (2017) highlight three

Table 10.1 Aims of counterfactual explanations

Top level aim	Underlying value
Why a decision was reached	Understanding
Providing grounds to contest the decision	Informed self-advocacy
Providing actionable changes to reverse the decision	Algorithmic recourse, Behavioral adaptation guidance, feasibility

distinct aims for CEs. (i) elucidating why a decision was reached; (ii) offering grounds to contest the decision; (iii) suggesting actionable changes to reverse the decision. While Wachter et al. (2017) suggest that any given CE could satisfy all three of these aims, we think it is best to think of these aims separately. Table 10.1 breaks down these three top level aims and identifies their underlying value.[2] In this chapter, we focus on aims (i) and (iii), showing that they can come apart in interesting ways (König et al., 2023, Freiesleben & König, 2023).

Consider again the example of an AI system that makes loan decisions. Imagine that a 65-year-old wishes to take out a loan from a bank and their application gets rejected based on an AI system's output. There are several CEs that would result in decision flips, but to simplify the case, we can just focus on the following two CEs.

Retirement:
If you were further away from retirement age (i.e. 5–10 years younger), you would have received the loan.

Debt:
If you had $50,000 less in debt and $20,000 per year higher income, you would have received the loan.

Let's further imagine that in this case, age was the feature that weighed the most for the decision, and that the retirement counterfactual was the closest CE to the ML decision boundary.[3] In this case, the retirement counterfactual, achieves the aim of *understanding* why the decision was reached, but fails to achieve *recourse*. First, the retirement counterfactual fails to achieve recourse because it is not physically possible–despite what some tech billionaires believe–for someone to become 5–10 years younger. However, the retirement counterfactual still provides the loan applicant with understanding the decision because it provides the loan applicant with the most influential feature that determined the loan outcome, namely that they were simply too close to retirement age. Barocas et al. (2020) call explanations that

[2] There could be even more aims for CE, we leave this discussion for future work (see e.g. Buchholz, 2023).

[3] Closeness to the decision boundary is a highly contextual concept that relies on specific similarity metrics, which are themselves contested. There are various ways to define a similarity order (see, for example, Lewis, 1979 and Günther, 2022). However, Morreau (2010) demonstrates that there is no single, objective way to compare overall similarity, as similarities cannot be weighed against each other in a universally agreed-upon manner. We leave this issue aside for now and discuss cases even if we had an objective way to measure this closeness we run into problems (see Kasirzadeh & Smart, 2021).

provide users and data-subjects with the crucial or key difference-makers against them a *principal-reasons* based explanation, which is essential for understanding an AI decision (see also Verma et al., 2020).

But what about the debt explanation? First, the debt explanation achieves the goal of recourse, since it could be feasible for the loan applicant to pay off some debt and increase their salary and apply again. However, notice that with the debt CE there is something epistemically suspicious. It fails to provide the loan applicant with the most influential difference-maker for the decision. The principal reasons explanation is absent. Instead, the recourse CE provides (helpful) advice regarding possible ways for reversing the decision. Because of this, Sullivan and Verreault-Julien (2022) have argued that recourse CEs are not actually explanations but should be considered recommendations. Similarly, Karimi et al. (2020) make a distinction between contrastive explanations and consequential recommendations. Contrastive explanations are those that provide information about the relationship between the model and its inputs–thus, aiming to understand the AI decision. And consequential recommendations are recourse CE that requires additional causal information linking up model inputs to the world. It is our contention that the debt CE does not aim at understanding the AI decision because it hides the most important feature further separating the aims of recourse and understanding. Given that CE aims can come apart, one of the central questions for CE methods concerns which aims should CEs methods optimize for. The growing trend in CE research is that the central aim should be recourse.

10.3 Recourse Explanations

Algorithmic recourse embodies three different concepts. The first, it identifies which actionable input variables need alteration to modify a prediction (Ustun et al., 2019). The second, it shows the specific actions that the data-subject must undertake to change the prediction (Karimi et al., 2020). The third, it picks on the actions required by the data-subject to change the target variable itself (König et al., 2023). In all these senses, recourse takes on a very specific meaning to provide *feasible and actionable advice that could reverse AI decisions*. This is different from recourse in the sense of *contestability*, where data-subjects might dispute a decision, lodge an objection, or pursue certain legal protections (aim ii) in Table 10.1). The fact that the CE literature has co-opted the most ordinary connotation of the concept 'recourse' can cause confusion and may even give the appearance of ethical credibility to feasible and actionable CEs more than is merited.

10.3.1 Recourse-First

The recourse-first view for CEs is as follows:

Recourse-first:

Providing users with recourse explanations of AI decisions—i.e. feasible and actionable counterfactual explanations—is (ethically) preferable to providing users with non-recourse explanations.

There are several reasons why a recourse-first view is attractive. First, recourse can provide data subjects with temporally extended agency. Venkatasubramanian and Alfano (2020) argue via Pettit (2015) that temporally extended agency is a morally robust good. We are creatures that make plans for the future that matter to us. In order to achieve our goals, we often need to make a series of sub-plans that are themselves intertwined. Having the right level of foreknowledge to make reliable judgments regarding which plans can in fact achieve our goals, is a necessary condition for having temporally extended agency. As agents living in complex societies we have a right to be able to adequately know how to plan for things that greatly impact our lives. Recourse, they argue, is ethically valuable, and preferable, because it provides data subjects with knowledge that supports temporally extended agency. Recourse CEs provide information regarding how to change loan decisions, job interview decisions, and more, by giving actionable and feasible advice for decision reversals.

Second, recourse explanations can be easily aligned with data subject's and user preferences. Users could, in theory, tell a CE explainer model which actions they are willing to do to reverse a decision, and the CE explainer can be tailored to these preferences. Aligning with stakeholder preferences is a leading approach for thinking about explainability of AI systems (Langer et al., 2021; Zednik, 2021). Recourse CEs seem especially suited toward customizable stakeholder interests. This can in turn increase data subjects' and users' sense of trust and control over the system. If they are given actionable and feasible ways to reverse decisions, then users can find the decisions intuitive, fair, and give them a sense of control.

Third, recourse explanations can be adapted to other ethical theories, such as the capability approach, which can increase the ethical standing of CEs. For example, Sullivan and Verreault-Julien (2022) argue that in order for recourse CE to live up to their ethical promise, aligning feasibility with the capability approach provides useful ethical standards for recourse.

However, despite the initial appeal of a recourse-first view for explaining AI decisions, recourse CEs face a series of problems that, we argue, should lead us to reject the recourse-first view. In this chapter, we will focus on three central problems: deflecting responsibility, the problem of hidden assumptions, and explanation hacking.[4]

[4] There are other problems that can face CE methods, such as fairwashing (Aïvodji et al., 2019) and worries of paternalism.

10.3.2 Problem of Deflecting Responsibility

The first problem with recourse CEs is that they can deflect responsibility away from the AI system or institution that the AI system is a part of, placing responsibility on the end-user or data subject. Recall that the distinctive characteristic of recourse CEs is that they provide data subjects with behavioral adaptation guidance for their life for the purpose of reversing model decisions in the future. In a bank loan case, providing a recourse explanation signals that it is not the bank's responsibility that someone got rejected for a loan, it is the applicant's responsibility to make changes to their financial life to rectify their financial situation, if they want a different decision. While there are indeed cases where the responsibility ought to be placed on end-users or data subjects, there are notable cases where this deflection in responsibility is problematic. To see why, let's first consider a case outside of an AI decision context.

> **Apartment fire:**
> There was a fire in an apartment building, leaving several units uninhabitable. The tenants want an explanation why the fire spread so quickly throughout the building. The building developer and landlord was told by the fire chief that the main reason the fire spread was due to the flammable building materials that went against current fire safety regulations.

It seems straightforward that the counterfactual explanation that provides tenants with understanding why the apartment fire spread as fast as it did would be something like: *If the building had been built according to the fire safety regulations, then the fire would not have spread.* This CE exposes that the responsibility for the fire damage is on the developer and landlord for violating fire safety standards. If the tenants were given such a CE, then this could trigger certain legal protections and property damage claims. However, notice that this CE does not provide the tenants with any feasible and actionable changes that *they themselves* could have done to prevent the fire. The fire safety regulation CE does not give tenants 'recourse' in the sense discussed in explainable AI. If the landlord was inspired by recourse explanations in AI and wanted to provide a recourse CE to their tenants, it would have to look something like: *If the apartment had not had any furniture or books along the walls, and instead had several fountain walls, then the fire would not have spread.* While strictly speaking true, and feasible for tenants to implement in future apartments (and would have prevented the fire in this apartment), it seems quite absurd that this explanation could ever be appropriate.

The apartment fire case is, in a lot of ways, silly. However, the absurdity of the example highlights what is lurking behind recourse CE for AI decisions. Recourse CE has an implicit assumption that the underlying AI system is fair, robust, and trustworthy. However, given the well documented problems that AI systems have with bais and robustness, we should not opt for a CE method that assumes otherwise. Looking back to the bank loan case where we are faced with choosing between the retirement CE and the debt CE, the debt CE deflects responsibility away from the bank onto the data subject. However, if the data subject was given the retirement counterfactual, then it exposes values that are embedded within the AI system that might lead to legal or other contestability avenues for the data subject. In the face of uncertainty regarding the fairness and robustness of AI systems, recourse CEs unduly deflect responsibility away from the AI system.

However, even if a model achieves high standards of fairness, robustness, and trustworthiness, there still may be other reasons that a recourse explanation unjustly deflects responsibility. Recourse CE are focused primarily on actions that can be taken by *individuals*; however, sometimes a socio-structural explanation is more appropriate for explaining AI decisions (Smart & Kasirzadeh, 2024). For example, Haslanger (2016) argues that there are cases where a social level explanation is required in order to capture social injustices.[5] AI decisions occur in several domains where there is a history or wide social injustice, such as the COMPAS model that gives risk scores to those who have been arrested for crimes that they will be arrested again in the future (Biddle, 2022). Or consider this question: Why are the outputs of widely used AI systems for automated healthcare allocation racially biased? Smart and Kasirzadeh (2024) argue that a socio-structural explanation can reveal the main reasons: These algorithms are embedded in a racially biased economic and social structure that perpetually disadvantages Black individuals and other minorities, leading to poorer health outcomes. This bias is mirrored in the training data of these algorithms through patterns of exclusion, limited access, and inferior care, elements of which are stratified along ethnic and racial lines. In cases where an AI system is deployed in a context rife with social injustice, giving any individual level explanation can be deflecting responsibility away from the true responsibility holder, which in the COMPAS case will likely include neighborhood policing practices.

10.3.3 Problem of Hidden Assumptions

In the previous section, we looked at how recourse CEs, even if they are designed properly, may deflect responsibility away from the true responsibility holder. In this section, we change gears to discuss the problems facing developing recourse CEs in the first place. Specifically, recourse CEs face the serious problem of hidden assumptions. While recourse CEs offer insights into potential decision-altering actions, they often carry implicit assumptions about the feasibility and actionability of these suggested changes. These assumptions can be overly simplistic and may not account for the complex, multifaceted realities of individuals' lives.

The hidden (normative) assumptions underlying counterfactual explanations in explainable AI have not gone unnoticed. Kasirzadeh and Smart (2021) articulate the range of ontological, semantic, and ethical considerations that shape and influence the evaluation of AI CEs in general. For instance, take the manipulation of social categories like gender or sex in a CE. The choice of which alternative worlds to

[5] See Ross (2023) for a discussion of social explanations as a type of causal constraint explanation.

consider for evaluating what would counterfactually happen hinges on a background semantic theory. This theory guides us in determining relevant worlds that align with the social categories in question. However, there might be scenarios where the semantic theory does not support an apt counterfactual intervention involving these categories. Consequently, this disconnect can lead to an inability to properly satisfy the assumptions necessary for evaluating the truth or falsity of counterfactuals.

For example, imagine an AI system analyzing wage disparities. If we counterfactually change a person's gender to assess differences in wages, the validity of this counterfactual depends on whether the chosen alternative world — where the person's gender is different — is deemed relevant and plausible by the semantic theory. If the theory does not recognize gender change as a legitimate variable for this context, the counterfactual intervention fails to provide meaningful insight into the wage disparity. The analysis by Kasirzadeh and Smart's (2021) on the hidden assumptions of CEs is a broader problem beyond recourse, and we will return to this problem in Sect. 10.4.3. However, their framework for examining the underlying assumptions inherent in CEs is also helpful for isolating the unique problems with recourse CEs.

The first hidden assumption of recourse CE is the potential disregard for systemic and structural constraints that individuals face. Recourse CEs, by design, suggest changes at the individual level. As a result, their content fails to acknowledge or reference societal or economic elements that play a core role in shaping an individual's opportunities. For instance, suggesting a career change via a CE explanation of "If you had a career in engineering instead of teaching, you would have been approved for the loan" ignores systemic issues like employment discrimination, job market saturation, or educational barriers that might impede such a transition. This potential disregard for systemic and structural constraints can place an unfair burden on the individual to overcome systemic barriers that are often beyond their control, but still places the responsibility of the AI decision on that same data subject.

The second hidden assumption concerns the oversimplification of user experiences, often leading to generalized solutions that do not fit all. For instance, recourse CE methods typically use probability distributions to assess the cost of certain actions, but these costs vary greatly among individuals.[6] This variance is due to differences in cultural, socio-economic, and familial backgrounds, making some CE suggestions impractical or irrelevant for specific users. For example, obtaining an additional educational degree might be feasible for someone with ample time and resources, but it is a far more challenging proposition for individuals with limited leisure time or financial constraints. Similarly, when CEs propose lifestyle alterations like financial behavior changes or dietary shifts, they must consider the individual's unique socio-economic circumstances and cultural

[6] Individual characteristics can be integrated into the selection of explanations and metrics, as demonstrated, for instance by (Shang et al., 2022). However, the critical challenge lies in the effective and robust incorporation of these characteristics. While there is ongoing work in this area, conceptual challenges (such as measuring and formalizing all the relevant aspects of user experiences) persist.

practices. Suggesting financial savings without understanding a person's income and expenses, or recommending dietary changes without considering cultural eating habits, overlooks the diverse realities in which people operate.

The third hidden assumption pertains to immediate actionability. Often recourse CEs suggest alterations that are assumed to be immediately actionable, disregarding the necessary time, resources, and effort. For example, the suggestion to reduce a significant amount of debt, say $50,000, presupposes immediate feasibility. However, this overlooks the nuanced individual circumstances and the complexity of factors that contributed to the accumulation of such debt. The assumption that substantial debt reduction can be swiftly actioned fails to recognize the real-world challenges and constraints faced by many users, rendering the suggestion unfeasible and detached from their practical realities.[7]

The fourth hidden assumption concerns normative injection made by recourse CEs. When CEs propose changes that an individual should make, they may (unwittingly) endorse specific norms about financial status, lifestyle choices, or personal decisions. This normative injection can lead to ethical clashes, especially when the AI system's recommendations conflict with an individual's personal values or ethical beliefs.[8] For instance, consider a scenario where a CE suggests that an individual should reduce their debt by cutting down on non-essential expenses. While this advice may seem financially prudent, it inherently carries a judgment about what constitutes non-essential spending. For some, what is deemed non-essential by the system might be culturally or personally significant, such as spending on church donations or family gatherings. In another example, a CE might recommend someone invest more in savings or retirement plans to improve their financial security. This suggestion, while sound in economic terms, overlooks the individual's immediate financial responsibilities or challenges, such as supporting a family or paying off student loans. It also implies a normative judgment about the importance of long-term savings over present needs.

10.3.4 Explanation Hacking

The last problem we consider that faces the recourse-first view is the potential for *explanation hacking*. The problem is as follows. There is an implicit view in work on CE methods and xAI that we need to make sure that AI systems align with stakeholder values. One legitimate way to do this is by tailoring *explanations* to certain values and stakeholder interests. Aligning explanations to stakeholder interests is the received view in xAI research (Langer et al., 2021; Zednik, 2021).

---

[7] Beretta and Cinquini (2023) explore the incorporation of the temporal dimension into causal algorithmic discourse. Despite this, the temporal assumption remains largely implicit.

[8] See Jesse and Jannach (2021) for a review on the way recommendation systems in general can create nudges.

However, CE methods, as explained above, face a unique challenge here: there are too many CEs that can be generated requiring a method for filtering these counterfactuals. Having to search through thousands of counterfactuals runs the risk of searching the possibility space in a motivated way. Adopting the stakeholder model makes this even more likely. In the stakeholder case, CE methods search for counterfactuals that stakeholders want to find, while discarding those that do not align with their values. One tempting way to address some of the problems with the hidden assumptions of recourse CEs is to make even more tailor-made explanations. While this method might seem like a virtue, it can be easily abused or manipulated.

We want to suggest that CE methods run the risk of engaging in practices that resemble p-hacking in the social sciences (Head et al., 2015). The social sciences in the past 10 years have suffered from false positives and replication failures. In 2014 the term 'p-hacking' was coined by Simonsohn et al. (2014). P-hacking describes the decisions that researchers make that end up intentionally, or not, fishing for significance in a sea of data points. Researchers choosing to stop or continue data collection once they get a result they like, or by changing which variables are being compared, is a way of ensuring a hit on a desired p-value, which is often necessary for publication. But engaging in these practices runs the risk of publishing a false positive, and is seen as a parasitic practice that undermines the overall integrity and credibility of the scientific field as a whole.

Similarly, generating thousands of counterfactuals for a single decision creates an environment where CE methods are calibrated post-hoc in a motivated fashion. Explainers can fit the method to find the counterfactuals that they are motivated to find that align the interests of modelers, industries, or data-subjects. Companies can look for recourse explanations that hide bias and engage in fairwashing (Aïvodji et al., 2019). Moreover, they can provide different counterfactuals to each stakeholder, and these could be mutually exclusive. While it might seem advantageous that the method is so flexible to align with various values, the problem with explanation hacking lies in its excessive flexibility. This over-flexibility starts to diminish the credibility and significance of the explanation itself. Modelers can simply reach into their bag of explanations to find one that suits your needs.

This problem of CE hacking goes beyond recourse. It is a general CE pitfall that faces all counterfactual explanation methods. And it is a pitfall that seems unique to counterfactual methods.[9] That said, recourse CEs have a *greater risk* of problematic explanation hacking. Given the number of hidden assumptions regarding feasibility and actionability (Sect. 10.3.3.), the way recourse CE can deflect responsibility (Sect. 10.3.2.), and the emphasis on aligning each new CE with new stakeholder values, is the perfect recipe for doing a motivated search through the counterfactual explanation space. Explainers can ignore other counterfactuals that might be a more

[9] Other types of post-hoc explanation methods have been found to be misleading and subject to manipulation (Slack et al., 2020, Dombrowski et al., 2019, Krishna et al., 2022). However, the manipulation of these methods isn't explanation hacking in the sense of actively searching among several possible explanations. Instead, they break with the introduction of noise or by creating adversarial examples. This is a different problem than the one identified here.

principal reason for the model decision, such as the feature that holds the most weight for a decision.

10.4 Understanding-First

As an alternative to the recourse-first view for CE, we propose the following understanding-first view.

> **Understanding-first:**
> Providing users with explanations that aim at understanding AI decisions is (ethically) preferable than recourse explanations.

Understanding is an epistemic state that involves grasping or knowing why something is the case. Understanding is thought to have a high degree of epistemic value (Grimm, 2012), can increase our epistemic agency (Kelp, 2015), and has social value (Hannon, 2018). Moreover, since a principal way of gaining understanding is through knowing an explanation (Khalifa, 2017), the social value of *explanation* underscores the social value of understanding. In the context of AI, having an explanation is a central way for end-users to have informed self-advocacy against AI decisions (Vredenburgh, 2022). Moreover, the understanding-first view can still contribute to temporally extended agency, like recourse. Understanding the reasons for decisions can help us with our long-term planning even if the information we receive is not by default actionable. Uncovering which possibilities are not open to us can still facilitate temporarily extended agency.[10]

In the remainder of this chapter, we introduce important principles that underlie the understanding-first CE methodology, and argue that the understanding-first view does not face the same problems as recourse CE. Thus, we argue that the primary aim and value CE methods should aim towards is understanding-first.

10.4.1 What Is an Understanding-First CE?

An understanding-first CE places the emphasis on getting a better understanding of the principal-reasons for the model decision. In this chapter, we do not provide concrete guidance for developing understanding-first CE methods. Our aim is to direct future research into constraining CE methods based on a set of understanding-first principles, and to push back on the momentum of recourse CE methods.

The epistemology of understanding is a rich field within epistemology.[11] For the purposes of this chapter, we will not adopt any particular view of understanding and

[10] The interplay between epistemology and ethics in AI is complex, see Grote (2024).

[11] See Chap. 7 by Páez in this volume for a nice overview on the epistemology of understanding.

instead focus on two specific features of understanding that are largely shared across the understanding literature: (i) someone understands if they are able to answer important why or what-if questions; (ii) the content of one's understanding consists of grasping a core set of dependencies, and also more peripheral dependencies about the thing being understood.

First, if someone understands something, they are able to answer important why- or what-if questions about it. For example, for someone to understand why it rains, they are able to answer various questions about the importance that various cloud formations have on rain patterns. They may also be able to answer what-if questions concerning unactualized possibilities, such as how various stages of climate change might impact rain cycles. In the case of AI decisions, counterfactual explanation methods seem especially well-suited for contributing to understanding because CEs provide users with what-if-things-had-been-different scenarios enabling them to answer various what-if questions about the model's behavior.

Second, we can separate between the *core* dependencies of why something is the case and more *peripheral* dependencies. For example, if we want to understand why apples fall off trees at a certain stage of ripeness, part of the *full* account of why will depend on the gravitational pull of the moon. However, the pull of the moon is not *central* to understanding why, it is just a small influence that lies outside the most impactful factors. The core dependency for understanding why apples fall off trees at a certain stage of ripeness is primarily related to the ripening process of the apple and Earth's gravity. The ripening process causes changes in the apple, such as the weakening of the stem, making it more likely to detach from the tree. Simultaneously, Earth's gravitational force acts upon the now-loosened apple, causing it to fall.

Some even suggest that if someone's beliefs about peripheral dependencies are all false, they can still have understanding, as long as the core is true (Mizrahi, 2012). The specific features that separate the core from the periphery are not straightforward. It is still an active problem in the epistemology of understanding literature. Mizrahi associates the core with grasping a scientific law, but Lawler (2021) takes issue with this restrictive conception. Alternatively, notions of indispensability run large (e.g. Rice, 2019). Lawler and Sullivan (2021) provide several intuitive features that capture the core of an *explanation,* such as a central causal mechanism, that there is a non-triviality between certain explanation propositions and the main explanatory power of the explanation, and that the explanatory propositions that uniquely discriminate a particular explanation from some other explanation are part of the core.[12]

An understanding-first CE seeks to uncover the *core* features that give rise to an AI decision instead of the features that lie on the periphery. How might we separate the *core* set of counterfactuals that explain an AI decision from more peripheral counterfactuals? Baron (2023) argues that CE methods that are embedded

[12] Miller (2019) highlights that good explanations for understanding hinge not on the most powerful causal factors but rather on those most surprising to the recipient.

within a Pearl-Woodwardean framework of causal interventions are necessary for CE methods to give us true understanding of the system and to establish true causal dependence of decisions (see also Buijsman, 2022). On this approach, developing CE methods built on a Pearl-Woodwardean framework could, in principle, give us the core counterfactuals.

For other scientific explanations, Lawler and Sullivan (2021) suggest that the core could be illuminated with robustness analysis, where stable features are the core features (see also Weisberg, 2006). In the case of CE methods, aggregating across CEs for different decision subjects could be a way of locating the features that are globally salient for determining the model's decision. Restricting CEs to only include sets of stable global features might be a way of capturing this type of robustness.[13] Closeness to the decision boundary is an existing desiderata in current CE methods that seems like a well-grounded epistemic assumption that gets to the core of an AI decision, but faces its own challenges (see footnote 3). Other desiderata like sparsity might also have epistemic importance for understanding since fewer features might generalize across a large set of instances.[14] Páez (Chap. 7 in this volume) suggests that another desiderata important for measuring understanding in the context of CEs is the specific epistemic context. In some contexts, the core causal map might need to be sufficiently detailed, whereas in other contexts a more rough and sparse core is enough for understanding. The understanding-first approach urges CE researchers to develop methods that seek to uncover the core counterfactuals, and that inspiration from philosophy of science and epistemology could be useful.

One counter-intuitive implication that results from taking this understanding-first perspective, is that it is not necessarily problematic CE methods to allow for impossible real-world counterfactual changes. If the AI model is trained and developed in such a way that it allows for correlating input features in ways that are physically impossible, then it is fair game, if not necessary, to allow for those same impossible correlations to be reflected in the CE of the AI system.[15] For example, if the training process of an AI model allows for decreases in age to develop model correlations and affect model decisions, then we should still allow that assumption into the CE method–since on the understanding-first view–our main goal is understanding the AI system. If, on the other hand, modelers used methods that prevented unrealistic manipulations of data, then these same assumptions must also be built into the CE method. Notice that this is a large departure from recourse CE methods, since for recourse-first the underlying aim is feasible behavior adaptation on the part of the data subject, not understanding the AI decision.

[13] For a discussion on robustness analysis in the case of ML see Freiesleben and Grote (2023).

[14] Here work on the unification view of explanation and understanding would be helpful (e.g. Gijsbers, 2013). On the unification view, covering more instances increases the amount of understanding.

[15] See Chen et al. (2020) for a similar discussion for SHAP.

One final note about the understanding-first approach developed here is the distinction between user-understanding and the epistemic norms that can capture core model dependencies. There is a difference between having an explanation that is accurate in the sense of providing all the input features necessary that captures the core of why something is the case, and having an explanation that is legible or intelligible to various people. The latter is something that greatly depends on people's previous familiarity with the subject matter and other psychological aspects. These research streams are separate. Finding the core understanding-first CE is one research stream, where the other research stream is presenting this information to users.[16] It is possible that CE methods are best suited for finding the core dependencies that govern AI decisions, but are not best suited for conveying this information to users (van der Waa et al., 2021).[17] Or vice versa. It might be that other xAI techniques are better suited for finding the core dependencies, but users benefit most if the information is given to them in a counterfactual structure. Often this distinction is collapsed in xAI research, where xAI methods are validated in part by user-studies (Byrne, 2019). User-studies on xAI methods are necessary to make sure users actually understand explanations, but are not a way of locating core model dependencies that give rise to AI decisions.

10.4.2 Responsibility

Understanding-first CEs can alleviate some of the worries related to deflecting responsibility associated with recourse-first CEs. Recall that recourse explanations necessarily, by providing behavioral adaptation guidance to individuals, implicitly place responsibility for the decision on those individuals. However, since current AI systems face issues of model bias and robustness problems, the responsibility for poor AI decisions likely lies elsewhere. Understanding-first CEs need not face this problem. First, CEs that aim at understanding, in principle, would uncover model bias or model unfairness if it was part of the core reasons for a model decision. If an AI system relies on core reasons that underscore bias or unfairness (even through proxies), then data subjects and end-users will have access to this information. For example, in the bank loan case, the retirement counterfactual would be provided as an understanding-first CE because it was the core reason for the decision. Notice that even though the retirement counterfactual is still an individual-based explanation, it shifts the burden of responsibility back on the bank (and the local and state government) for making decisions based on age.

Second, understanding-first CEs also have the flexibility to include socio-structural explanations as a component into the AI decisions. In cases where societal factors are a core contributor to the model decision, then it is even necessary to

[16] See Chap. 1 by Beisbart in this volume.

[17] See Chap. 7 by Páez in this volume for a discussion on user perceptions of CEs.

include these factors in an explanation. While perhaps we can introduce the concept of group recourse, where recourse explanations can go beyond individual actions to collective actions, this can further exacerbate worries of deflecting responsibility onto groups or collectives that may themselves still be marginalized. On the other hand, understanding-first CEs could still point to social structures, but how these structures should be revised would be left open.

10.4.3 Hidden Assumptions

In this section, we discuss how the understanding-first view can address some of the ethical pitfalls linked to the hidden assumptions inherent in the recourse-first approach, as previously explored in Sect. 10.3.3.

First, the understanding-first perspective offers a considerable advantage over recourse-first explanations, notably in its potential to recognize and integrate systemic and structural factors within counterfactual explanations. Unlike recourse CE, which tends to focus on individual-level changes while often overlooking broader societal contexts, the understanding-first approach is poised to acknowledge and address wider systemic and structural constraints. For instance, a typical recourse-first CE might suggest: "If you had a career in engineering instead of teaching, you would have been approved for the loan." This advice, while direct, disregards systemic issues like employment discrimination, job market saturation, or educational barriers, thereby placing an unfair burden on the individual. In contrast, understanding-first explanations could incorporate socio-structural factors into their framework, highlighting how AI systems make their decisions. For example, that banks have the freedom to make loan decisions based on job-type instead of a collection of other more equitable factors. Of course, this assumes that the AI system possesses extensive background knowledge from socio-structural data analysis. While this is a significant requirement, the potential for such comprehensive explanations certainly exists.

Second, the understanding-first view can circumvent the pitfalls of oversimplified explanations that often arise in recourse-first approaches. By eschewing a one-size-fits-all approach to CE, it treats users as knowledgeable participants. This view assumes respect for the users' ability to understand multiple layers of complex information, aligning more closely with principles of dignity and autonomy in legitimate decision-making. Users who are well-informed about the nuances of AI decision-making are better equipped to adjust their actions and expectations accordingly and engaged in informed self-advocacy.

Third, the understanding-first view facilitates the legitimacy of engaging with contestability. Instead of prioritizing immediate actionability, users are more likely to trust proper CEs when they have a clearer epistemic sense of how AI decisions are made, ensuring that their interaction with AI is based on a solid foundation of knowledge, rather than any systemic or organizational incentives for manipulating the users by prioritizing actionability. The understanding-first view enables users to grasp in a richer way the underlying principles and logic of AI decisions. As compared to the recourse-first view, understanding-first is analogous to teaching someone the rules of a game rather than just what to do for a specific move; the former equips them to strategize better.

Fourth, the concern about normative injections in CE pertains to both recourse-first and understanding-first views. For example, an AI system in a hiring scenario might generate a CE like, "If you had not been Black, your application would have been shortlisted." As Kasirzadeh and Smart (2021) argue, this statement requires deep-seated ontological and epistemological assumptions for its validity. The AI system must have a preconceived definition of race, particularly the construct of being Black. The manner in which race is conceptualized and integrated into the model's space of analytical knowledge significantly influences the outcomes and interpretations of these CEs. However, we believe that the issue of normative injection, particularly regarding actionability and specific individual-level recourse explanations, is less pronounced in understanding-first CEs. By allowing for systemic considerations and the possibility of downplaying the individual attributes such as race or gender, understanding-first CEs can overcome many of the ethical pitfalls associated with imposing normative judgments based on these categories. This approach lends itself to a more holistic and contextually aware form of counterfactual reasoning that acknowledges the complex interplay of societal factors and individual circumstances.

10.4.4 Explanation Hacking

Every CE method will have some vulnerability to explanation hacking since the method necessarily involves exploring a larger counterfactual space. However, taking an understanding-first approach reduces this vulnerability. First, the understanding-first approach chiefly concerns finding the core features that give rise to an AI decision. This prevents motivated searching from explanations that already confirm one's beliefs or values. This is important for thinking about the difference between illusory understanding and genuine understanding (e.g. Trout, 2002). Illusory understanding comes with the sense of understanding, but without actually grasping the true dependencies underlying a phenomena, or in this case, the reasons for an AI decision. Understanding-first CE methods would be built surrounding what the actual core dependencies are that can make for genuine understanding.

As an extra safeguard, we recommend that in order to prevent explanation hacking, researchers should 'pre-register' the assumptions and values that are embedded in their CE methods, so that the methods can be audited for fairwashing and other biases. This means embracing a checklist like what Kasirzadeh and Smart (2021) suggest, but with awareness of the aims of explanation: recourse or understanding or both. Even in designing understanding-first methods, it is important to first lay a groundwork and motivation for the specific epistemic norms that are central in the method, and what choices are made in the face of mutually conflicting CEs. Taking an understanding-first approach pushes research in a direction that aims to find well-grounded metrics based in epistemically sound principles and developing methods based on those principles. Moreover, we should stop engaging in real time, post-hoc, model exploration that allows users, industry, or developers to justify their model based on flipping through several candidate CEs. While this practice may still be useful for users to find actionable advice or recommendations, it is important to separate the difference between explaining an AI decision versus what action guiding possibilities are open to individuals.

10.5 Discussion

In this chapter, we argued that the trend in xAI research on developing feasible recourse explanations is not the preferable solution for explaining AI decisions. When an AI system makes a decision that affects our lives, we want to know *why* this decision was made. We argued that taking an understanding-first approach to answering this why-question better conforms to other social values concerning responsibility, trust, and informed self-advocacy. Before concluding, we consider a few objections to the understanding-first view and point to future research directions.

10.5.1 Objections and Replies

Objection 1: Since recourse explanations still give you counterfactual inferences, don't they by default also give you understanding? What is the real tension here?

Reply 1: According to minimal accounts of understanding in epistemology *any* answer to a why-question or a what-if question is tantamount to having some small degree of understanding.[18] On this view, recourse explanations provide end-users with understanding, simply in virtue of providing what-if information. Even the recourse explanation in the APARTMENT FIRE case urging tenants to install water accent walls to prevent fires, would still count as providing understanding on these minimal accounts of understanding, since the tenants grasp dependencies that dry walls have on fire strength. However, minimal views of understanding fail to capture what other epistemologists have noted about understanding that there are core dependencies and more peripheral dependencies. If we want attributions of understanding to signal real epistemic success and the holders of *good* explanations (Hannon, 2018), what we are really after is that the understander grasps the core

[18] See Hubert and Malfatti (2022) for a nice discussion on the degrees and threshold for understanding.

dependencies. Similarly, Baron (2023) makes a distinction between basic causal knowledge of an AI system—that could be given by *any* CE—and a complete causal picture that is necessary for understanding why the AI made its decision. Given the problem of explanation-hacking with CE methods, basic causal knowledge of knowing a single remote counterfactual renders the understanding we gain to be borderline trivial. So, if the minimal view of understanding is correct, a simple revision to the understanding-first view—to a *deep* understanding-first view—is necessary. It is not just any CE that gives us deep understanding, only the core CEs give us deep understanding. Recourse explanations that are on the periphery fail to provide deep understanding.

Objection 2: How can you really say that one counterfactual explanation is 'the reason' for a decision, when other counterfactual explanations are strictly speaking also true reasons?

Reply 2: In addressing this objection, it is crucial to distinguish between "a reason" and "core or central reasons" in the context of CEs. A CE, by its nature, posits an alternate scenario that could have led to the same outcome, suggesting that multiple explanations can coexist as valid reasons. However, identifying "the core or central reasons" typically involves prioritizing one explanation over others based on its relative significance or causal weight in the specific context. In any decision-making scenario, numerous factors contribute to the final outcome, each representing a potential CE. The challenge lies in evaluating these factors to determine which holds the most weight. This evaluation often involves an analysis of the directness of the cause, its historical precedence, or its explanatory power in simplifying or understanding of the event. So, when assessing that one CE is "the core or central reason" for a decision, it implies that this reason is identified as the most significant or influential among several true reasons. It does not negate the validity of other explanations but recognizes a hierarchy of influences, where the primary reason has a most direct or substantial impact on the outcome than others. So, while multiple CEs can be true, assessing one as "the core or central reason" involves a reasoned judgment about its relative importance in the specific context of the decision. Understanding-first principles urges researchers to develop methods that are able to make such judgments on sound epistemic principles. Importantly, it might very well be that CE methods are ill suited for the task of determining causal weight and isolating core reasons from peripheral reasons. If so, understanding-first recommends abandoning CE methods for those that are more suited for providing understanding.

Objection 3: If people want actionable explanations, give the people what they want!

Reply 3: We are not advocating against giving feasible recourse CEs full stop. Nor are we advocating against doing research on developing methods for extracting feasible recourse CEs. We are arguing for three central claims. First, feasible recourse CEs are not the gold ethical standard of CEs. Second, we have an obligation to always include an understanding-first CE with AI decisions. In some cases, the understanding-first CE and the recourse CE will overlap, in cases where there isn't overlap it is possible to provide both a recourse CE and an understanding CE. However, in this latter case we suggest, following Sullivan and Verreault-Julien (2022) in using the language of recourse *recommendations*, instead of recourse *explanations*.[19] Third, we are highlighting the urgency to move away from real time model exploration of the counterfactual space that some researchers are purposely building into explanation interfaces (e.g. Bove et al., 2023, Cheng et al., 2020, and Gomez et al., 2020). Instead, we urge CE researchers to 'pre-register' their assumptions and calibrated methods for generating CE, and then stick to these pre-registered methods, even if it means giving an end-user a CE that doesn't align with their values. Doing so would signal to the end-user that maybe the system itself does not align with their values, providing them with genuine grounds to trust or distrust the system.

10.5.2 Future Research Directions

In this chapter, we focused exclusively on counterfactual explanation methods. However, recourse explanations do not need to be counterfactual explanations. While counterfactual explanations are a common method used in providing recourse in AI systems, they represent just one approach among several possible methods. Other methods may include, for example, non-counterfactual causal reasoning, where the explanation focuses on the causal relationships within the data, or rule-based explanations, where certain rules or conditions are provided for achieving a different outcome. Our arguments may also apply to these other recourse explanations, as long as they are not part of the core reasons for an AI decision.

Lastly, in this chapter, we do not solve the hard problem of how we could go about developing understanding-first CEs. If anything, some of the main problems with CE methods still loom large even on an understanding-first view. CE methods still make several hidden assumptions regarding what kind of features lend themselves to counterfactual reasoning, which are likely misguided, like in the case of sensitive attributes like race and gender. CE methods are still susceptible to explanation-hacking even if we focus on developing understanding explanations, since there can still be motivated searching of the counterfactual space even considering epistemic norms. Lastly, there are hard conceptual and technical challenges to developing CE methods that actually achieve the aim of understanding. It may very well be that given all these considerations that we should opt for a different type of explanation method altogether. We suspect the right solution will involve utilizing a variety of explanation methods (Freiesleben & König, 2023), where CE methods are just one aspect to explain why a certain AI decision was reached.

[19] See also Baron (2023), who argues that we should provide both advice and complete causal information for why the decision was reached.

Acknowledgments We presented this work at various stages in development at Oxford University, Utrecht University, and the 2023 ESDiT away days. We thank the attendees for their helpful comments and suggestions. We would also like to thank Juan Durán, Timo Freiesleben, Thomas Grote, Andrés Páez, and Yeji Streppel for their comments and suggestions. This work is supported by the Netherlands Organization for Scientific Research (NWO grant number VI.Veni.201F.051) and the research programme Ethics of Socially Disruptive Technologies funded by the Gravitation programme of the Dutch Ministry of Education, Culture, and Science and the Netherlands Organization for Scientific Research (NWO grant number 024.004.031).

References

Aïvodji, U., Arai, H., Fortineau, O., Gambs, S., Hara, S., & Tapp, A. (2019, May). Fairwashing: The risk of rationalization. In *International conference on machine learning* (pp. 161–170). PMLR.

Barocas, S., Selbst, A. D., & Raghavan, M. (2020, January). The hidden assumptions behind counterfactual explanations and principal reasons. In *Proceedings of the 2020 conference on fairness, accountability, and transparency* (pp. 80–89).

Baron, S. (2023). Explainable AI and causal understanding: Counterfactual approaches considered. *Minds and Machines, 33*(2), 1–31.

Beisbart, C. (in this volume). In which ways is machine learning opaque? In J. M. Durán & G. Pozzi (Eds.), *Philosophy of science for machine learning: Core issues and new perspectives*. Synthese Library, Springer.

Beretta, I., & Cinquini, M. (2023, July). The importance of time in causal algorithmic recourse. In *World conference on explainable artificial intelligence* (pp. 283–298). Springer Nature Switzerland.

Biddle, J. B. (2022). On predicting recidivism: Epistemic risk, tradeoffs, and values in machine learning. *Canadian Journal of Philosophy, 52*(3), 321–341.

Biran, O., & Cotton, C. (2017). Explanation and justification in machine learning: A survey. In *Proceedings of the twenty-eighth international joint conference on artificial intelligence, IJCAI-17 workshop on explainable AI (XAI)* (pp. 8–13).

Bove, C., Lesot, M. J., Tijus, C. A., & Detyniecki, M. (2023, March). Investigating the intelligibility of plural counterfactual examples for non-expert users: An explanation user Interface proposition and user study. In *Proceedings of the 28th international conference on intelligent user interfaces* (pp. 188–203).

Buchholz, O. (2023). A means-end account of explainable artificial intelligence. *Synthese, 202*(2), 33.

Buijsman, S. (2022). Defining explanation and explanatory depth in XAI. *Minds and Machines, 32*(3), 563–584.

Byrne, R. M. (2019). Counterfactuals in explainable artificial intelligence (XAI): Evidence from human reasoning. In *IJCAI* (pp. 6276–6282).

Chen, H., Janizek, J. D., Lundberg, S., & Lee, S. I. (2020). True to the model or true to the data?. *arXiv preprint arXiv:2006.16234*.

Cheng, F., Ming, Y., & Qu, H. (2020). Dece: Decision explorer with counterfactual explanations for machine learning models. *IEEE Transactions on Visualization and Computer Graphics, 27*(2), 1438–1447.

Dombrowski, A. K., Alber, M., Anders, C., Ackermann, M., Müller, K. R., & Kessel, P. (2019). Explanations can be manipulated and geometry is to blame. *Advances in Neural Information Processing Systems, 32*.

Durán, J. M. (2021). Dissecting scientific explanation in AI (sXAI): A case for medicine and healthcare. *Artificial Intelligence, 297*, 103498.

Freiesleben, T., & Grote, T. (2023). Beyond generalization: A theory of robustness in machine learning. *Synthese, 202*(4), 109.
Freiesleben, T., & König, G. (2023, July). Dear XAI community, we need to talk! Fundamental misconceptions in current XAI research. In *World conference on explainable artificial intelligence* (pp. 48–65). Springer Nature Switzerland.
Gijsbers, V. (2013). Understanding, explanation, and unification. *Studies in History and Philosophy of Science Part A, 44*(3), 516–522.
Gomez, O., Holter, S., Yuan, J., & Bertini, E. (2020, March). Vice: Visual counterfactual explanations for machine learning models. In *Proceedings of the 25th international conference on intelligent user interfaces* (pp. 531–535).
Grimm, S. (2012). The value of understanding. *Philosophy Compass, 7*(2), 103–117.
Grote, T. (2024). Machine learning in healthcare and the methodological priority of epistemology over ethics. *Inquiry, 68*(4), 1218–1247.
Günther, M. (2022). Causal and evidential conditionals. *Minds and Machines, 32*(4), 613–626.
Hannon, M. (2018). *What's the point of knowledge?: A function-first epistemology*. Oxford University Press.
Haslanger, S. (2016). What is a (social) structural explanation? *Philosophical Studies, 173*, 113–130.
Head, M. L., Holman, L., Lanfear, R., Kahn, A. T., & Jennions, M. D. (2015). The extent and consequences of p-hacking in science. *PLoS Biology, 13*(3), e1002106.
Hooker, G., Mentch, L., & Zhou, S. (2021). Unrestricted permutation forces extrapolation: Variable importance requires at least one more model, or there is no free variable importance. *Statistics and Computing, 31*, 1–16.
Hubert, M., & Malfatti, F. (2022). Towards ideal understanding. *Ergo, 10*, 21.
Jesse, M., & Jannach, D. (2021). Digital nudging with recommender systems: Survey and future directions. *Computers in Human Behavior Reports, 3*, 100052.
Karimi, A.H., Barthe, G., Schölkopf, B., & Valera, I. (2020). A survey of algorithmic recourse: Definitions, formulations, solutions, and prospects. *arXiv preprint arXiv:2010.04050*.
Kasirzadeh, A., & Smart, A. (2021). The use and misuse of counterfactuals in ethical machine learning. In *Proceedings of the 2021 ACM conference on fairness, accountability, and transparency* (pp. 228–236).
Kelp, C. (2015). Understanding phenomena. *Synthese, 192*(12), 3799–3816.
Khalifa, K. (2017). *Understanding, explanation, and scientific knowledge*. Cambridge University Press.
König, G., Freiesleben, T., & Grosse-Wentrup, M. (2023). Improvement-focused causal recourse (ICR). In *Proceedings of the AAAI conference on artificial intelligence* (Vol. 37, No. 10, pp. 11847–11855).
Krishna, S., Han, T., Gu, A., Pombra, J., Jabbari, S., Wu, S., & Lakkaraju, H. (2022). The disagreement problem in explainable machine learning: A practitioner's perspective. *arXiv preprint arXiv:2202.01602*.
Langer, M., Oster, D., Speith, T., Hermanns, H., Kästner, L., Schmidt, E., Sesing, A., & Baum, K. (2021). What do we want from explainable artificial intelligence (XAI)?–A stakeholder perspective on XAI and a conceptual model guiding interdisciplinary XAI research. *Artificial Intelligence, 296*, 103473.
Lawler, I. (2021). Scientific understanding and felicitous legitimate falsehoods. *Synthese, 198*(7), 6859–6887.
Lawler, I., & Sullivan, E. (2021). Model explanation versus model-induced explanation. *Foundations of Science, 26*, 1049–1074.
Lewis, D. (1979). Counterfactual dependence and time's arrow. *Noûs, 13*(4), 455–476.
Miller, T. (2019). Explanation in artificial intelligence: Insights from the social sciences. *Artificial Intelligence, 267*, 1–38.
Mittelstadt, B., Russell, C., & Wachter, S. (2019). Explaining explanations in AI. In *Proceedings of the conference on fairness, accountability, and transparency* (pp. 279–288).

Mizrahi, M. (2012). Idealizations and scientific understanding. *Philosophical Studies, 160*, 237–252.

Molnar, C. (2020). *Interpretable machine learning*. Lulu. com.

Morreau, M. (2010). It simply does not add up: Trouble with overall similarity. *The Journal of Philosophy, 107*(9), 469–490.

Mothilal, R. K., Sharma, A., & Tan, C. (2020, January). Explaining machine learning classifiers through diverse counterfactual explanations. In *Proceedings of the 2020 conference on fairness, accountability, and transparency* (pp. 607–617).

Páez, A. (in this volume). Axe the X in XAI: A plea for understandable AI. In J. M. Durán & G. Pozzi (Eds.), *Philosophy of science for machine learning: Core issues and new perspectives*. Synthese Library. Springer.

Pettit, P. (2015). *The robust demands of the good: Ethics with attachment, virtue, and respect*. OUP.

Rice, C. (2019). Models don't decompose that way: A holistic view of idealized models. *The British Journal for the Philosophy of Science, 70*(1), 179–208.

Ross, L. N. (2023). What is social structural explanation? A causal account. *Noûs, 58*(1), 163–179.

Shang, R., Feng, K. K., & Shah, C. (2022). Why am I not seeing it? Understanding users' needs for counterfactual explanations in everyday recommendations. In *Proceedings of the 2022 ACM conference on fairness, accountability, and transparency* (pp. 1330–1340).

Simonsohn, U., Nelson, L. D., & Simmons, J. P. (2014). P-curve: A key to the file-drawer. *Journal of Experimental Psychology: General, 143*(2), 534.

Slack, D., Hilgard, S., Jia, E., Singh, S., & Lakkaraju, H. (2020, February). Fooling lime and shap: Adversarial attacks on post hoc explanation methods. In *Proceedings of the AAAI/ACM conference on AI, ethics, and society* (pp. 180–186).

Slack, D., Hilgard, A., Lakkaraju, H., & Singh, S. (2021). Counterfactual explanations can be manipulated. *Advances in Neural Information Processing Systems, 34*, 62–75.

Smart, A., & Kasirzadeh, A. (2024). Beyond model interpretability: Socio-structural explanations in machine learning. *AI & Society, 40*, 2045–2053.

Sullivan, E., & Verreault-Julien, P. (2022, July). From explanation to recommendation: Ethical standards for algorithmic recourse. In *Proceedings of the 2022 AAAI/ACM conference on AI, ethics, and society* (pp. 712–722).

Trout, J. D. (2002). Scientific explanation and the sense of understanding. *Philosophy of Science, 69*(2), 212–233.

Ustun, B., Spangher, A., & Liu, Y. (2019, January). Actionable recourse in linear classification. In *Proceedings of the conference on fairness, accountability, and transparency* (pp. 10–19).

van der Waa, J., Nieuwburg, E., Cremers, A., & Neerincx, M. (2021). Evaluating XAI: A comparison of rule-based and example-based explanations. *Artificial Intelligence, 291*, 103404.

Venkatasubramanian, S., & Alfano, M. (2020, January). The philosophical basis of algorithmic recourse. In *Proceedings of the 2020 conference on fairness, accountability, and transparency* (pp. 284–293).

Verma, S., Boonsanong, V., Hoang, M., Hines, K. E., Dickerson, J. P., & Shah, C. (2020). Counterfactual explanations and algorithmic recourses for machine learning: A review. *arXiv preprint arXiv:2010.10596*.

Vredenburgh, K. (2022). The right to explanation. *Journal of Political Philosophy, 30*(2), 209–229.

Wachter, S., Mittelstadt, B., & Russell, C. (2017). Counterfactual explanations without opening the black box: Automated decisions and the GDPR. *The Harvard Journal of Law & Technology, 31*, 841.

Weisberg, M. (2006). Robustness analysis. *Philosophy of science, 73*(5), 730–742.

Zednik, C. (2021). Solving the black box problem: A normative framework for explainable artificial intelligence. *Philosophy & Technology, 34*(2), 265–288.

Emily Sullivan is an Assistant Professor at Utrecht University. Her work focuses on the way that technology mediates our knowledge forming practices, especially how machine learning methods can provide scientific understanding. She is an associate editor for the *British Journal for the Philosophy of Science* and a research fellow in the *Ethics of Socially Disruptive Technologies* research consortium.

Atoosa Kasirzadeh is an Assistant Professor at Carnegie Mellon University. She is interested in the ethical and philosophical foundations of AI systems with guarantee for responsible development and deployment. Previously, she was a Chancellor's Fellow at the University of Edinburgh and a Governance of AI Fellow at the University of Oxford. Her research has received recognition through the 2024 Schmidt Futures AI2050 early career fellowship and Google visiting faculty program.

Chapter 11
Stakes and Understanding the Decisions of Artificial Intelligent Systems

Eva Schmidt

Abstract Explainable artificial intelligence (XAI) aims to overcome the opacity of black box systems, i.e., to make them understandable to suitable stakeholders. In this chapter, I investigate how understanding depends on how much is at stake in a context. I support the intuition that understanding is sensitive to the stakes with a pair of cases. I further use this pair of cases to spell out how exactly the stakes affect understanding, particularly, outright understanding why. To do so, I connect discussions of the concept of understanding with debates on pragmatic encroachment and on inductive risk. I zoom in on one necessary condition on understanding specifically, viz. the epistemic justification condition, according to which the beliefs involved in understanding have to be supported by sufficient evidence and have to cohere sufficiently well. I argue that, where the stakes are high, there are more stringent standards for sufficient evidence and coherence, and that this explains the stakes-sensitivity of understanding.

Keywords Understanding · Explainable artificial intelligence · Pragmatic encroachment · Inductive risk · Epistemic justification · Evidence · Coherence

11.1 Introduction

As artificial intelligent (AI) systems become more and more widespread, their decisions and recommendations increasingly impact people's lives. The predictions of modern AI systems are often more accurate than those of human decision-makers; but this tends to come at the price that they are epistemically opaque—not even their developers can understand how the systems work or produce particular outputs. This is the so-called black box problem (e.g. Bathaee, 2018, Zednik, 2021): Where AI systems substantially contribute to high-stakes decisions, it seems especially

E. Schmidt (✉)
TU Dortmund, Dortmund, Germany
e-mail: eva.schmidt@tu-dortmund.de

J. M. Durán, G. Pozzi (eds.), *Philosophy of Science for Machine Learning*, Synthese Library 527, https://doi.org/10.1007/978-3-032-03083-2_11

important that we understand why they provide certain outputs or how they function; but this information tends to be unavailable especially for modern, powerful AI systems.

Against this backdrop, the last decade has seen an explosion of interdisciplinary work on explainable artificial intelligence (XAI). The point of much of this research has been to alleviate the opacity of AI systems by providing explanations either locally of individual outputs, or globally of a system's overall functioning (e.g. Adadi & Berrada, 2018, Linardatos et al., 2021, Angelov et al., 2021). An important philosophical contribution to the field has been to highlight that explanations are useful to the extent that they make AI systems or their output *understandable* to relevant *stakeholders* in the *context* in question (Langer et al., 2021, Nyrup & Robinson, 2022, Páez, 2019, Fleisher, 2022). We shouldn't expect that a 'one size fits all' explanation works for all uses of AI systems. Rather, the context in which a system is employed has an impact on what kinds of explanations and explainability methods will be useful. One crucial element of the context is the intended recipient of the explanation—an explanation succeeds in making an AI system explainable given that the recipient understands the system, or understands why it gave a certain output (Beisbart & Räz, 2022).[1] The person's understanding will then contribute to the fulfillment of certain desiderata that are relevant in the context. For instance, it may enable an end user to (reasonably) trust the system, to bear responsibility for her AI-supported decision, or to detect whether the system's outputs are due to algorithmic bias (e.g. Lazar, 2024, Purves & Davis, 2022, Baum et al., 2022, Schmidt, 2022).

This line of thought puts into focus the question under what conditions an agent understands something that's explained to her –which is also my focus here. In particular, I will investigate how understanding depends on what is at stake in a particular context. Let me motivate this thought. It is often presumed that whether we need explainability at all, or how detailed or accurate explanations have to be, depends on whether we are facing a high-stakes or a low-stakes situation (House of Lords, 2018, Adadi & Berrada, 2018, Yuan et al., 2023, Munch et al., 2024, or the AI Act[2]). For instance, where an AI system determines a listener's music playlist, very limited and superficial explanations of the music picks seem good enough, if any are needed at all; however, in case an AI system calculates the recidivism risk score for a convict, which then influences the severity of the convict's sentence, a good explanation will have to be quite accurate and more detailed. Since the point of XAI is to provide understanding, the following claim is plausible: Whether an explanation suffices to ensure understanding depends on how much is at stake in a concrete situation. In other words, understanding depends on the stakes.

---

[1] Chapter 7 by Páez and Chap. 10 by Sullivan and Kasirzadeh in this volume discuss the importance of understanding of AI systems.

[2] See https://www.europarl.europa.eu/news/en/headlines/society/20230601STO93804/eu-ai-act-first-regulation-on-artificial-intelligence.

This claim makes good pragmatic sense: Where AI systems support or take over potentially life-changing decisions, it is important that we can be sure they work flawlessly and that we grasp how they work or how their outputs come about. However, it would be a waste of resources to technically ensure understanding of these matters in situations where AI systems make or support clearly harmless decisions. In a more philosophical vein, the above claim is highly intuitive, and I will support the intuition with a pair of cases in the following section. I will further use the pair of cases to spell out how exactly the stakes affect understanding, specifically understanding why. To do so, I will connect discussions of the concept of understanding (Baumberger et al., 2017) with debates on pragmatic encroachment (Gao, forthcoming) and inductive risk (Hempel, 1965, Douglas, 2000). My discussion will thereby fill two scientific lacunae: First, little research so far has been done on whether pragmatic factors or non-epistemic values affect understanding (though see Phillips, 2020, Kelp, 2015, and Wilkenfeld, 2013), in addition to knowledge or the rationality of accepting or rejecting scientific hypotheses. Second, while the notion of inductive risk has been applied in several papers to the issue of design of AI algorithms and models (e.g. Johnson, 2023, Biddle, 2022, Karaca, 2021), little attention has been paid to whether the concepts of inductive risk or pragmatic encroachment can be employed to illuminate how practical stakes affect the understanding that is supposed to be generated by explainable AI (though see Sullivan, 2022a, 2022b).

Much will be said about the nature of understanding below. Let me immediately introduce one central conceptual distinction regarding understanding that will be crucial to my argument. Sometimes, ‘understanding’ picks out an on-off phenomenon, a mental state one either has or doesn’t have. Other times, we treat understanding as graded—a subject can have more or less of it (see Pedersen Phillips, 2020, 142; Bachmann, 2020). Kelp (2016) develops a graded notion of understanding that takes maximal understanding as its base point and defines degrees of understanding by their distance from the maximum. One aspect underlying this gradation is the depth of understanding; following Kelp (2015, 3812), depth concerns how many connections the subject can trace within the understood subject matter. Simultaneously, Kelp (2015, 3813) defends a context-relative notion of outright understanding. A subject has outright understanding when her graded understanding meets a certain, contextually determined threshold. Which degree of graded understanding suffices for outright understanding depends on the practical task at hand. For instance, a ten year old has a lower *degree* of understanding of evolutionary theory than an adult with a PhD in biology. Nonetheless, when the ten year old gets an A in his fifth grade biology test about evolutionary theory, we can truly say of him that he understands (*period*) evolutionary theory. By contrast, when a search committee discusses whom to hire for a university professorship in biology, the child’s degree of understanding is not good enough to ascribe to him outright understanding of evolutionary theory (Kelp, 2016, 254). These cases indicate that a graded concept and an outright concept of understanding should both be acknowledged.

In my view, when it comes to the stakes bearing on whether an XAI method suffices to provide understanding in a context, it is *outright* understanding that is at issue. Here is how we should conceive the matter: A level of graded understanding that suffices for outright understanding given low stakes may not be good enough to ensure that a subject outright understands why an output was given when the stakes are high.

The aim of this chapter is to provide a pragmatic encroachment/inductive risk based account of how the stakes affect the outright understanding of the recipients of XAI explanations.[3] I will apply these notions to the understanding of individuals in XAI contexts in order to explain how an individual's understanding, when the individual is confronted with an explanation of an AI system's output, can be affected by how much is at stake. In one direction, that notions of pragmatic encroachment and inductive risk can be fruitfully applied in these contexts lends further support to the claim that pragmatic encroachment and inductive risk are genuine phenomena; in the other direction, that conceptual tools like pragmatic encroachment and inductive risk, which have been developed for other contexts, are a match here as well, lends additional plausibility to the claim that what's at stake in XAI contexts indeed affects (outright) understanding.[4]

The plan for this chapter is the following: In Sect. 11.2, I introduce a pair of low stakes/high stakes cases and tease out the intuition that understanding why an AI system provided a certain output depends not just on the given explanation, but also on how much is at stake. To lay the foundation for my account of how the stakes affect understanding, I next provide a brief overview over necessary conditions of understanding why, taken from the recent philosophical debate (Sect. 11.3). I then turn to two ways in which the stakes make a difference to whether an agent understands why an output was provided: The stakes affect how good the subject's *evidence* has to be for understanding (Sect. 11.4) and they affect how well the subject's beliefs about why the system provided its output have to *cohere* internally or with her background beliefs for understanding (Sect. 11.5). Section 11.6 concludes the chapter.

11.2 Two Cases of XAI

Here is a pair of cases. In each of them, the subject wants to understand why an AI system provides a certain output. Imagine that the system in question has a reliable, built-in function for generating an explanation of a particular output upon demand,

---

[3] In the following, all unspecified uses of 'understand' will be about outright understanding.

[4] Beyond this, the purpose of this chapter is not to give a principled defense of pragmatic encroachment or inductive risk on the one hand, or of effects of the stakes on whether an explanation suffices for understanding in XAI contexts. It is to propose an account of how exactly the stakes may affect understanding, granted that they do so.

that the subject in both cases requests and receives such an explanation, and that the explanation is correct. Further imagine that the subject in both cases has little to no prior background knowledge of how AI systems work, or of how her particular system works.

Anger Management (low stakes): Luana is a member of a prison parole board who has to decide whether a Black prisoner, Connor, will be offered state-funded anger management training with his parole. She is supported by COMPAS in her decision-making (Northpointe, 2015, Angwin et al., 2016). COMPAS assigns a maximum recidivism risk score of 10 to Connor. Upon reviewing the case files, Luana estimated Connor's recidivism risk as rather low, and she wants to understand why COMPAS assigns such a negative prognosis for him. She receives the explanation: "COMPAS assigned a risk score of 10 to Connor because convicts with music interests similar to Connor's often committed further crimes within two years."

Sentencing (high stakes): Juana is a judge who has to decide on the sentence length for a Black convict, Connor, and is supported by COMPAS in her decision-making. COMPAS assigns a maximum recidivism risk score of 10 to Connor. Upon reviewing the case files, Juana estimated Connor's recidivism risk as rather low, and she wants to understand why COMPAS assigns such a negative prognosis. She receives the explanation: "COMPAS assigned a risk score of 10 to Connor because convicts with music interests similar to Connor's often committed further crimes within two years."[5]

In *Anger Management*, it seems intuitive that Luana understands why COMPAS provides the negative assessment of Connor's recidivism risk, whereas in *Sentencing*, Juana does not understand. (Recall that my claims are about outright understanding.) For instance, it would be absolutely reasonable for Luana to think: "Weird, but ok – apparently many others who share Connor's music taste recidivated, and that's why COMPAS gives him the high risk score, so I get it." If a friend were to challenge Luana, saying, "You don't understand why COMPAS outputted the high risk score", she would rightly feel that she was slighted *qua* epistemic subject. By contrast, it would be perfectly reasonable for Juana to think: "Why would COMPAS give Connor such a high risk score because of his *music interests*? I still don't understand why it gave this prognosis!" At the same time, it would be unfitting for her to think: "Weird, but ok – apparently many convicts with music interests like Connor's have recidivated, and that's why COMPAS outputted the high risk score, so I get it." If a friend were to challenge Juana's understanding of why COMPAS provided this

[5] These are moral cases. If readers prefer, they can, in the following, replace my cases by cases where the subject's self-interest is at stake: Bachelor student Philip starts to write a philosophy essay and uses ChatGPT to find a topic. The system recommends to discuss why Hume was not a skeptic about induction. Philip was taught that Hume is an uncompromising skeptic, and so wonders why this suggestion was provided. He receives the following explanation: "ChatGPT recommends to discuss why Hume was not a skeptic about induction because Hume abandoned the Title Principle after the *Treatise*." Philip only has beginner's level knowledge about Hume and the *Enquiry*, and knows little else about him. Low stakes: The essay is only for practice and it wouldn't matter if Philip were to start from a misguided presupposition (viz. that Hume was not a skeptic). High stakes: The essay will be Philip's final paper for a course for which he needs a very good grade. He has only 2 days left to write it and would be unable to adjust his topic if he were to rely on a mistaken presupposition.

output, it would be appropriate for her to concede that she is unsure of what's going on here, and doesn't understand why COMPAS provided the high recidivism risk score for Connor. My suggestion is that subjects in high-stakes cases need deeper (graded) understanding to achieve outright understanding—not just understanding why the output was given, but also understanding why the system would give this output for this reason. This follows Kelp's aforementioned suggestion that depth of understanding correlates with how well-connected the beliefs pertinent to one's understanding are.

Let me elaborate on these cases before placing them in the context of the pragmatic encroachment debate. They depart from reality in several respects: First, COMPAS doesn't actually make available automated explanations of its outputs. However, there are other digital tools that automatically provide explanations—for instance, Google search allows users to get an explanation of why specific search results are presented, so such a service could plausibly be made available for COMPAS as well. There is also so far no consensus on how to evaluate the reliability of an explainability method, though proposals are discussed in the literature (e.g. Nauta et al., 2023, Nielsen et al., 2022).

Second, the explanation provided brings up a feature that is not actually available to the system—the questionnaire on which COMPAS relies as input has no entry for music interests.[6] I use this concrete example explanation because it doesn't make any sense (to the subjects or, presumably, to my readers): It is not immediately clear why music interests should predict recidivism.[7] At the same time, the explanation appeals to a real correlation that an AI system might use for its predictions. Music interests predict race. For instance, rap music has a greater fan base among Black listeners. Assume in the following that the system's output is due to a racial bias in both cases, and that COMPAS uses music interest as a proxy attribute for race (Marshall & Naumann, 2018, Tschantz, 2022; regarding algorithmic bias, see Fazelpour & Danks, 2021).

The described cases parallel the so-called 'bank cases' from the debate over pragmatic encroachment on knowledge (see e.g. Stanley, 2005, Fantl & McGrath, 2002, Schroeder, 2012). A subject is considering whether to deposit her check in the bank immediately despite the long lines in front of the bank. She has solid, but not perfect, evidence that the bank will be open tomorrow, on Saturday. In the low stakes version of the case, it is of no importance whether the money is deposited quickly or not; intuitively, the subject knows that the bank will be open tomorrow. In the high stakes version of the case, it is of great importance that the money is deposited before Sunday, since a large mortgage payment will be taken out of the subject's bank

[6] See https://www.documentcloud.org/documents/2702103-Sample-Risk-Assessment-COMPAS-CORE.

[7] That said, if readers prefer a more realistic explanation, they can imagine that COMPAS gives the explanation: "COMPAS assigned a risk score of 10 because convicts with the same ZIP code as Connor often committed further crimes within two years." Keep in mind that the subjects are unable to make sense of the explanation, as they lack the background knowledge to connect it to the given output.

account then. In this version of the case, the subject intuitively doesn't know that the bank will be open tomorrow, exactly because the stakes are high—if the subject falsely believes that the bank will be open tomorrow and acts on this belief, this will have extremely negative practical consequences for her. Crucial characteristics of these cases are that, first, they are epistemically identical—the subject has the same evidence, the same intellectual capacities and background knowledge—and, second, the only pertinent difference is how much is at stake in each case.

Anger Management and *Sentencing* share these characteristics. The explanation-providing function of COMPAS is equally reliable in both cases—say that, in 90% of cases, the generated explanations highlight central factors bearing on COMPAS's outputs. So the explanations provided are equally good evidence that COMPAS ascribed Connor a high risk score because of his musical interests. The explanation is true in both cases, and makes no sense from the point of view of its recipient—music interest seems completely irrelevant to recidivism probabilities. Finally, Luana and Juana lack relevant background knowledge and have equal intellectual capacities.

The cases differ with respect to the stakes, with respect to how much practically hangs on the subject getting the explanation right. *Anger Management* is a low-stakes scenario. If Luana were to accept the explanation and take herself to understand why COMPAS provided such a high recidivism score, but the explanation were incorrect or misleading, for instance because it leaves out relevant explanatory factors, this would have no negative practical consequences. The worst that could happen is that Luana offers Connor an unnecessary anger management course. By contrast, *Sentencing* is a high-stakes scenario, and a lot hangs on Juana getting the explanation right. If Juana were to accept the provided explanation and take herself to understand why COMPAS assesses Connor as high risk, even though the explanation was misleading—in the given scenario, it misleads by omitting the fact that algorithmic bias was a crucial factor—this would have very bad consequences. For if she were thus to get wrong core factors that really stood behind the system's output, she would very likely discriminate against Connor by giving him a severe sentence because of algorithmic bias.[8,9]

Notice the difference between how pragmatic factors affect outright understanding in these cases and pragmatic encroachment on knowledge: There, the pragmatic encroaches on knowledge just in case the subject's having a *false belief* and acting

[8] Following Altman (2020), I take discrimination to "impose a relative disadvantage on persons based on their membership in a salient social group".

[9] Does this mean that my point is really whether subjects in high stakes situations are justified to *act* on the outputs of AI systems, given certain explanations (not whether they understand)? This is not so. My interest is in an epistemic status, understanding, and how it is affected by the stakes. Nonetheless, there is an effect on action, but this can be spelled out by appeal to understanding-action principles that parallel the knowledge-action principles discussed in the debate on pragmatic encroachment. For instance, Fantl and McGrath (2002) argue that one knows or is justified to believe that p only if it is rational for one to act as if p. I submit that, similarly, if the subject doesn't understand why p, it may not be rational for her to act on the basis of p.

on it would lead to very bad consequences (see Schmidt, 2025). For understanding, the issue is: If the subject *got* the actual explanation of the system's output *wrong* and acted on this basis, would that lead to very bad consequences? 'Getting the explanation wrong' covers scenarios in which the explanation accepted by the subject is false, but also scenarios in which it merely leaves out central contributing factors, so that the subject is *misled* about what really explains why the output was given. These points will become relevant below.[10]

To sum up, in *Anger Management* and *Sentencing*, the subjects are epistemic equals as far as the quality of their evidence, their intellectual capacities, or their relevant background knowledge go. But they intuitively differ with respect to their understanding of why the relevant output was given: While Luana understands why COMPAS assigned a high recidivism score to Connor, Juana does not. Since the striking difference between the cases is how much is at stake, it seems that the stakes affect understanding as much as they affect knowledge. To make this claim plausible and to see how exactly the pragmatic encroaches on understanding, we first need to take a closer look at understanding.[11]

11.3 Understanding: The Concept

The mental state of understanding—whether graded or outright—is currently much discussed in the philosophy of science and epistemology (see Baumberger et al., 2017, Grimm, 2021, Hannon, 2021 for recent overviews). One topic of debate is the nature of understanding (in both senses). Similarly to the long-running discussion on the correct analysis of knowledge, there are efforts to provide the conditions under which a subject understands. My aim in this section is to bring out the conditions of understanding that, on my proposal, are affected by the stakes. I will present some core elements of the debate over the nature of understanding, and particularly focus on four widely accepted necessary conditions on understanding. As before, my presentation focuses on outright understanding.

Let me start with a further important distinction, that between objectual and explanatory understanding (e.g. Baumberger et al., 2017, 5). A person has *objectual* understanding when she understands an overall subject matter or a broader domain, as when she understands machine learning or genetics. We say the person has *explanatory* understanding, or understanding why, when she understands why

[10] Similarly, see Biddle and Kukla's (2017) push for broadening our conception of inductive risk to one of epistemic risk, which goes beyond risks from acting on false belief.

[11] I have introduced my proposal via a pair of cases that supposedly elicit a certain intuition, but intuitions tend not to be shared by everyone. In acknowledgement of this, readers can think of this chapter as starting from the observation that many contributors to the field of XAI hold that the stakes make a difference to whether explainability is needed; and of its aim as providing a theory of how the stakes can affect the need for explanation by way of a stakes-sensitive account of understanding. The cases then only serve an illustrative purpose.

something happened or is the case, for instance, when she understands why inflation went up in 2022 or why the dinosaurs became extinct. Objectual understanding is plausibly relevant to XAI in situations where we want understanding of how an AI system works overall. My interest here is in understanding why a system gave a certain output, which I take to be an instance of explanatory understanding.[12] This is what I will focus on here.

So what are the necessary conditions on understanding? As with conditions on knowledge, there is disagreement about the details, but many views converge on four conditions that a subject has to meet to understand: (1) belief/acceptance, (2) grasp, (3) truth/factivity, and (4) justification (see e.g. Baumberger et al., 2017, Pritchard, 2010, Chap. 4, Hills, 2015, Malfatti, 2023). Note the parallels with the traditional analysis of knowledge as justified true belief; a difference is that for knowledge that *p*, the necessary conditions apply only to a single proposition *p*. By contrast, for understanding why *p* is the case, propositions beyond *p* are relevant, viz. those propositions *q*, *r*, *s* that constitute the *explanans* of why *p*, as well as the proposition that *q*, *r*, *s* explain why *p*. With this in mind, let's turn to the four necessary conditions on understanding.

S understands why *p* is the case only if

(1) *S believes* or accepts that *p* and that *q*, *r*, *s* (and maybe also believes or accepts that *q*, *r*, *s* explain why *p*),

For instance, I understand why the dinosaurs became extinct 66 million years ago only if I believe that the dinosaurs became extinct then, and I additionally believe propositions such as: that an asteroid hit Earth, that this caused huge amounts of dust to enter the atmosphere, leading to temperature drops, which led to mass extinction, and that in this way the asteroid impact explains why the dinosaurs went extinct.

Some philosophers relax the belief condition and instead endorse an acceptance condition on understanding, so that *S* has to merely accept the relevant propositions as premises from which to draw further inferences, or on which to act. Unlike belief, acceptance is an attitude that a person can correctly have towards propositions that approximate the truth, but are not literally true (Elgin, 2017, Chap. 2; Malfatti, 2023, Chap. 8). To illustrate, the dinosaurs didn't go extinct *exactly* 66 million years ago, and not all of them did—the birds survived. And the explanation sketched above oversimplifies, since many species went extinct due to tsunamis, earthquakes, and wildfires caused by the asteroid impact. I wish to remain neutral between belief and acceptance, here; for ease of exposition, I will speak only of belief in the following.

(2) *S grasps* the interconnections between *p* and the propositions *q*, *r*, *s*, which explain it,

[12] Though see Páez (2019, 454) for the view that the relevant understanding is objectual, though localized. In the following, unqualified instances of 'understanding' will refer to outright *explanatory* understanding.

Understanding is more than just knowing or believing individual pieces of information, it involves seeing how these connect or fit together. The notion of grasp is supposed to capture this. In the example, I understand why the dinosaurs became extinct only if I grasp how their extinction relates, e.g., to the asteroid's impact on Earth, to the dust in the atmosphere, or to the temperature dropping, and how these explanatory factors in turn hang together. More specifically, I have to see what explanatory or counterfactual relationships exist between the dinosaurs' extinction, the asteroid impact, and so on (Riggs, 2003). Grasp is often spelled out by way of relevant abilities the subject has to have, such as being able to explain how the extinction happened, or to say what would have happened in certain counterfactual situations, e.g. if the asteroid had bypassed Earth (Hills, 2015).

(3) At least central propositions among *p*, *q*, *r*, *s* (and the proposition that *q*, *r*, *s* explain why *p*) are *true*, or approximately true,

I cannot understand why the dinosaurs became extinct if this isn't even approximately true, or if the explanation I believe isn't even in the neighborhood of the truth. Whether the exact truth is needed or something more approximate is sufficient is contentious. Positions range from the view that all the mentioned propositions need to be true (Kelp, 2015), to the view that central propositions have to be true, while peripheral propositions may be false (Kvanvig, 2003), to the view that the relevant propositions need be merely approximately true (Elgin, 2007). I remain neutral between these positions.

(4) *S*'s beliefs that *p, q, r, s* (and that *q, r, s* explain why *p*) are epistemically *justified*: they are (a) supported by *sufficient evidence* and they (b) *cohere* among each other and with *S*'s background beliefs.

If a subject randomly accepts certain propositions (whether on the *explanandum* or on the *explanans* side), although it's utterly nebulous to her why they should be true, she doesn't thereby gain understanding (see Pritchard, 2010, 76, referring to Zagzebski, 2001). Instead, her beliefs must be epistemically justified to constitute understanding. In my view, this has two aspects. On the one hand, the relevant propositions have to be sufficiently supported by the subject's epistemic reasons or evidence; on the other hand, they have to cohere sufficiently well, by being consistent or standing in explanatory or probabilistic relations with each other (Kvanvig, 2003, Elgin, 2007, 35) and fitting in well with the other beliefs of the subject (Malfatti, 2023, 38, Elgin, 2017).[13] For instance, my understanding of why the dinosaurs became extinct presupposes that I don't just randomly guess that an asteroid struck Earth 66 million years ago, but that I have evidence that this is so, e.g. that I read this in a reliable science book. Furthermore, there have to be interconnections between the relevant propositions (which I grasp, see (2)). If I justifiably believe several true propositions in addition to my belief that the

[13] As this indicates, coherence is of central importance for understanding; to my mind, understanding differs from knowledge in this regard, which requires no coherence.

dinosaurs became extinct, but these are completely unrelated to each other or to the dinosaurs' extinction, this doesn't amount to understanding. Imagine that, beyond these beliefs about dinosaurs, the asteroid, and so on, I am a dyed-in-the-wool creationist who believes that God created the world, including ready-made dinosaur fossils, 6000 years ago. Given the lack of fit (or rather, the contradictions) between my beliefs about the dinosaur extinction and my background beliefs, I don't *really* understand how the dinosaurs became extinct.

Some philosophers object to a justification condition on understanding and defend the claim that we can understand why something is the case even where the beliefs involved in our understanding are supported only by *defeated* evidence (Hills, 2015). In Malfatti's (2023, 189) example, if a medical doctor hears a patient's report of his condition and infers that he suffers from anemia, and then receives the misleading evidence that the patient is a compulsive liar, her evidence (the patient's report) is defeated. But, this line of thought has it, her understanding of why the patient exhibits certain symptoms is not thereby undermined.

Discussing this general point here would lead me too far afield. However, I find the claim implausible at least in XAI contexts, even for low stakes scenarios. Imagine that the advertisements in your Facebook stream seem to pick up on the fact that you just booked a vacation, and that you do not want this. In the Facebook settings, you look up which of your personal information Facebook advertisements use and see that they do *not* rely on your recent purchases. You thereby take yourself to understand something about how your Facebook advertisements are generated, and *trust* Facebook advertisements not to use information about recent purchases. Now imagine that from a reliable source, you receive the (albeit misleading!) evidence that Facebook's statements about which information it uses to personalize advertisements are highly unreliable. This defeats your evidence that Facebook doesn't use information about your vacation for its advertisements. So, your belief that the ads don't respond to your booked vacation is no longer epistemically justified. I submit that simultaneously (and contrary to Malfatti), you lose your understanding of how Facebook came to display the ads in question to you. It would be perfectly reasonable for you to think: "I thought I understood which factors Facebook uses to determine which advertisements to display – but now I'm confused. I don't understand why it shows me these specific ads!"

My claim is supported by the fact that now you rightly no longer trust Facebook's algorithm not to rely on your recent purchases. Put more generally, understanding can be instrumental to reasonable trust in AI contexts (e.g. Ribeiro et al., 2016, EU High-Level Expert Group on Artificial Intelligence, 2019, 13, Schmidt, 2022). Given this, the fact that your trust is lost indicates that your understanding is lost also. More generally, the call for XAI is not a call for what one might call *idle* understanding, but for understanding that enables stakeholders to *do* things relevant in the context (Langer et al., 2021); but purported understanding based on defeated evidence is insulated from (reasonable) action. At least in XAI contexts, defeaters thus undermine the subject's understanding together with her evidence, and (4) stands as a necessary condition for understanding.

How does all this connect to the claim, introduced in Sect. 11.2, that the stakes affect whether a subject understands? I will next propose that the stakes affect understanding by making a difference to whether the subject's beliefs are epistemically *justified* in XAI contexts. The idea to be developed is this: The more is at stake, the stricter the justification condition—both in terms of whether the evidence is strong enough to suffice for justification, and in terms of whether the believed propositions cohere well enough to render the beliefs justified.

11.4 How the Stakes Affect Understanding (I): Strength of Evidence

Understanding (in XAI contexts) requires epistemic justification of the relevant beliefs, and justification requires sufficient evidence or reasons, just as in the case of knowledge. Correspondingly, one way for the stakes to make a difference to a subject's understanding is by making a difference to whether her evidence or epistemic reasons suffice for justification, i.e., via 'standard' pragmatic encroachment.[14]

To illustrate, imagine (contrary to the original scenarios) that the explanation that the subjects are given for why COMPAS assigned Connor the high risk score is that individuals with a similar criminal history have often reoffended. Say that Luana/Juana knows that algorithmic bias is a real problem of some AI systems, possibly including COMPAS (Angwin et al., 2016, Chouldechova, 2017). Finally, imagine that the explanation is correct, and that COMPAS really is sensitive to features of Connor's criminal history shared by convicts who frequently recidivated, and that this explains the system's prognosis.

Luana is in a low-stakes scenario; if she simply adopted the explanation and acted on it, even if the explanation were false, nothing bad would happen in consequence. She might offer Connor an optional anger management course, but that's all. Accordingly, her belief—that COMPAS assigned Connor the maximum recidivism risk score because people with a similar criminal history often recidivated—is justified by the explanation she receives from COMPAS (indeed, it amounts to knowledge). Luana understands why the high recidivism risk score was outputted.

By contrast, Juana is in a high-stakes situation: If—counterfactually—an algorithmic bias against Black people really explained the system's prognosis (so the automatically generated explanation of the prognosis was false), this might lead to very bad moral consequences. For if Juana took the provided explanation at face value, formed the false belief that COMPAS provided the high risk score because of Connor's criminal history, and gave him a severe sentence because of COMPAS's high risk score, she would discriminate against him. Being imprisoned

[14] A related issue is whether the justificatory power of an AI system's outputs may be undermined by its opacity. See Chap. 6 by Zednik and Verreault-Julien in this volume.

unnecessarily would be terrible for Connor, and the important moral and legal value of fair treatment in criminal proceedings would be violated.

In light of this, Juana's belief that COMPAS provided the high risk score because of Connor's criminal history is *not* justified (and it is not knowledge). True, the fact that a reliable explanatory mechanism gave the explanation—that Connor was assigned a maximum risk score because individuals with a similar criminal history have often reoffended—indicates that Connor's criminal history explains COMPAS's high risk score; it is good evidence that COMPAS provided the high risk score because of Connor's criminal history. This evidence justifies belief in this explanation to a degree. However, because so much is at stake, this evidence does not suffice to justify Juana's corresponding belief, all things considered.[15] Think about it this way: Given that Juana is aware of the real-world problem of algorithmic bias, and given that (as in the original *Sentencing*) her own perusal of Connor's files led her to asses him as low risk, is it epistemically rational for her to simply accept the explanation provided by COMPAS? I think not; rather, it would be rational for her to remain in doubt as to the real explanation of the high risk score. As I have phrased it elsewhere (Schmidt, 2025), where a lot hinges on an agent's getting a belief right, she needs to be especially *diligent* in weighing her reasons in forming the belief, which is to say that it takes more for her epistemic reasons to believe to suffice for all things considered justified belief.[16] In the current context, this means that Juana needs to be especially sensitive to possible limitations of her evidence, such as the possibility that the explanation she received might be incorrect or misleading. She should therefore accord the testimony provided by COMPAS's explanatory mechanism less weight than she would if nothing much was at stake. Given that her evidence is thus weakened, it is insufficient to justify belief; *and so her understanding of why COMPAS provided its output is undermined.*[17]

[15] Two elucidations: (1) I here conceive of evidence as considerations that probabilify the target proposition. So, the fact that a reliable explanatory mechanism states that the risk score was given because of criminal history is evidence that, indeed, the risk score was given because of criminal history. The fact that an explanation is given here serves as evidence that a certain causal relation holds. (2) I rely on the distinction between a graded notion and an all things considered notion of justification that parallels the graded/outright distinction regarding understanding: A proposition can be more or less justified for a believer, where this does not fix whether this justification suffices to make belief the correct attitude, all things considered. Think of all things considered justification as knowledge-level justification—the kind of justification that is a necessary condition for knowledge (Schroeder, 2012).

[16] My full account of pragmatic encroachment in Schmidt (2025) spells this thought out by appeal to attenuating conditions that weaken the pertinent reasons to believe. It further appeals to how much weight a perfectly virtuous reasoner would accord her reasons, and uses this to explain why the subject's reasons are weakened. There is no need to take these claims on board to accept my proposal here.

[17] Another respect in which the evidence provided by COMPAS may be insufficient is that it is bare statistical evidence, where it is important to make judicial decisions on the basis of individualized evidence (Schmidt et al., 2023, Sullivan, 2022a, 314).

The same point can be made with the help of the notion of inductive risk from the philosophy of science (Hempel, 1965, Douglas, 2000, Steel, 2010):[18] There is always some probability of error in accepting or rejecting a scientific hypothesis on the basis of empirical evidence. Errors of two kinds are possible: First, false positive error—the hypothesis that a certain phenomenon occurs is accepted, although it does not occur; second, false negative error—the hypothesis that a phenomenon occurs is rejected, but it does occur. The risk that one of these two errors will occur is called inductive risk. According to Hempel (1965), scientists cannot avoid this risk; the only latitude they have concerns whether to err on the false positive side or on the false negative side. In other words, there is a tradeoff—scientists can only minimize the risk of false-negative errors at the cost of an increased risk of false-positive results, and *vice versa*. They must thus independently determine how much evidence is needed before the hypothesis can be accepted. To do so, they need to look to *non-epistemic values*. It is what is of practical or moral value to us that determines which type of error we should rather risk—missing out on true hypotheses or accepting false hypotheses (see Douglas, 2000, 561/562).

Applied to *Sentencing*, this means that, since false positive errors (mistakenly accepting an automatically generated, seemingly unbiased explanation of why COMPAS deemed Connor high risk) raise the risk of violating important moral values, such as fair treatment in criminal procedures, the risk of false positive error should be minimized. Juana shouldn't accept the explanation on the given evidence, but needs better evidence to do so. In other words, the belief that Connor was accorded a maximum risk score by COMPAS because of his criminal history is not justified for her. Since justification is a necessary condition for understanding, she therefore does not understand why COMPAS assessed Connor's recidivism risk so negatively. By contrast, in *Anger Management*, no practical or moral values are violated if Luana believes COMPAS's explanation of why it provided the high risk score, i.e. if she makes a false positive error. There is no need to minimize risking such errors, and so she is justified to believe the explanation, and thereby (outright) understands why COMPAS provided this output.

Inductive risk is used by opponents of the so-called 'value-free ideal of science' to argue that not only what is of epistemic value is directly relevant to science, but that matters of practical, moral, or political value affect which theories scientists correctly accept (e.g. Longino, 1990, Rudner, 1953). Again, the idea is that in many cases, practical or moral concerns set the standards within which scientists justifiably accept hypotheses, choose models, and so on. In recent years, several philosophers have argued that, similarly, the design of AI systems is not, and cannot be, value-free. The correct development of a system—e.g. how the model is designed, which data are chosen, or how they are labeled—is partly determined by what is practically or morally at stake (Sullivan, 2022a, Sullivan, 2022b, Karaca, 2021, Johnson, 2023, Biddle, 2022, Ratti & Graves, 2022). Sullivan (2022a,

[18] See Miller (2014), Fantl and McGrath (2011) for the relation between pragmatic encroachment and inductive risk.

2022b) extends this analysis to how practical or moral values impact individuals' understanding of AI systems. In line with this, my proposal is that there is no suitable value-free account of when explanations suffice for understanding why certain AI outputs were given; any such account must be sensitive to the practical or moral values at stake in the context.

However, standard pragmatic encroachment and inductive risk, affecting the sufficiency of a subject's evidence, don't fully capture how the stakes impact understanding. I now turn to a second way in which the pragmatic encroaches on understanding why. I propose that pragmatic factors also impact epistemic justification (as a necessary condition on understanding) via its *coherence* component.

11.5 How the Stakes Affect Understanding (II): Coherence

My account as developed so far does not concern any aspects specific to understanding, but proposes that practical or moral values affect understanding exactly in the same way as they affect knowledge: by bearing on how much evidence is needed for all things considered justification. So the account so far may be rather unsurprising. For instance, philosophers who take understanding to be a kind of knowledge will have expected exactly this. However, there is an additional way in which the pragmatic encroaches on understanding which goes beyond this standard story. My proposal is that understanding is harder to acquire in high-stakes contexts because of higher demands on the *coherence* of the provided explanation with the explanandum as well as with the subject's background beliefs, which again makes it harder to meet the justification condition.

To illustrate this additional way in which practical or moral matters affect whether a subject understands, let me contrast again (the original versions of) *Anger Management* and *Sentencing*. Recall that the explanation given for COMPAS's negative assessment for Connor—"individuals with similar music interests as Connor often reoffended"—coheres little with Luana's/Juana's background beliefs, or with what each subject is trying to understand. As far as either of them knows, there is no explanatory or probabilistic relationship between a convict's or prisoner's musical interests and risk of recidivism generally speaking, and *a fortiori* none specifically between Connor's music preferences and his (alleged) high recidivism risk. On the positive side, at least there is no inconsistency between *explanans*, *explanandum*, and background beliefs, and the subject may think that technical devices sometimes do things for odd reasons.

First, let's ask whether in *Anger Management*, Luana is justified to believe that COMPAS assigns Connor the maximal risk score because people with similar music interests have often recidivated. As argued above, there is nothing at stake in Luana's believing the *explanans*, and in her believing that the fact that others with similar music interests often recidivated explains why COMPAS assigns the high risk score. If she were to believe it, and it were misleading or false, there would be no bad consequences—as before, the worst that could happen is that Connor would receive an offer for an optional anger management course. In line with the intuitive judgment I outlined above in Sect. 11.2, that the explanation does provide Luana with understanding why COMPAS assigned Connor the high recidivism risk score, my proposal predicts that her belief is justified and so meets this necessary condition for understanding. In a low-stakes scenario little coherence suffices for justification.

In contrast to *Anger Management*, in *Sentencing* a lot is at stake: If Juana were to accept the explanation and take herself to understand why COMPAS assigned Connor a maximum risk score, but the explanation was misleading (by leaving out the important factors that music interest are used by COMPAS as a proxy attribute for race and that the system has a bias against Black people), then Juana's sentencing decision would likely discriminate against Connor. Since a lot is at stake, there are more severe demands on the coherence of the propositions relevant to the subject's understanding, internally and with Juana's background beliefs. The little coherence of the explanation provided by COMPAS with Juana's other beliefs and with what she is trying to understand does not suffice for justification. Since justification is necessary for understanding, and in line with the intuitive judgment teased out in Sect. 11.2, she does not understand why COMPAS assesses Connor so negatively.

The situation for Juana is different if the explanation provided coheres better. Imagine that Juana is not as uninformed as in the original version of the scenario, but that she has the following background knowledge: She is aware that music interests correlate with race in the United States and that some AI systems are affected by algorithmic bias and use proxy attributes for race. Further, imagine that the explanation is correct and that COMPAS gives its output in response to a proxy attribute for race due to algorithmic bias. If so, the explanation is highly coherent for her—she can infer that COMPAS is probably using music interest as a proxy attribute for race and is biased against Connor as a Black convict, and that COMPAS gives Connor a maximum risk score because of algorithmic bias. In this case, the coherence is high enough to ensure justification and to allow Juana to understand despite the high stakes.

My proposal is distinct from, but relates in interesting ways to that of Sullivan (2022a). Emily Sullivan applies the concept of pragmatic encroachment to high stakes XAI contexts, but appeals to my condition (2) on understanding—to grasp—and in particular to the abilities constitutive of grasping. She claims that "full-fledged understanding … depends on the threshold of the number of what-if questions in the set of all possible what-if[-things-had-been-different] questions on a given topic in a particular context that is necessary to attribute understanding" (p. 316). The core of Sullivan's proposal is that, depending on how much is at stake, a subject has to be able to answer fewer or more questions about what would happen in counterfactual situations, to count as having outright understanding. Such abilities show the quality of the subject's grasp of the relations between the relevant propositions.

Although Sullivan appeals to a different condition on understanding, her view picks up on related features of the subject and her situation: To have counterfactual reasoning abilities (i.e. abilities to answer relevant what-if questions) is, at bottom, to see how the different elements of the explanation, the state of affairs to be explained, and surrounding matters cohere. For example, in *Anger Management*, Luana can give only very limited answers to questions like, "What risk score would COMPAS have assigned to Connor if he had different musical interests?" This corresponds to the limited coherence of the explanation provided by COMPAS with her background beliefs and with the fact in need of explanation, that COMPAS assigned Connor the maximum risk score. To provide some contrast, re-focus on the scenario in Sect. 11.4, in which COMPAS explains the high risk score by appeal to Connor's criminal history. This explanation coheres extremely well with Luana's background beliefs about how people's criminal history and their future recidivism risk relate. And correspondingly, Luana can provide more detailed answers to the question, "What risk score would COMPAS have assigned to Connor if he had a different criminal history?"

Still, my proposal and Sullivan's are not identical. In my high-stakes scenario, it's the missing coherence between the given propositions that calls into question the subject's understanding, not that she fails to grasp existing coherence relations or to answer what-if questions on the basis of grasping them. Both necessary conditions are distinct, as there might be cases where subjects have beliefs that cohere in the right way, but fail on the grasp/abilities dimension. I am happy to allow that both the grasp condition and the justification condition are stakes-sensitive. Sullivan's and my proposals can thus be seen as complementary.

The following picture of pragmatic encroachment specifically on understanding in XAI contexts has emerged: Say that a subject's belief that a certain factor explains an AI system's output coheres only weakly with her beliefs about this output or with her background beliefs. Given low stakes, this can nonetheless be good enough for justification on the coherence dimension, so that the justification condition on understanding may be met. By contrast, when much is at stake, this raises the requirements for how well the relevant beliefs have to cohere, so that it is harder for the subject's beliefs to be justified. And accordingly, in cases with weak coherence, the relevant beliefs will typically not be justified, which is to say that the subject does not understand why the AI system gave the output it did.

11.6 Concluding Remarks

I have started from the assumption, widely accepted by theorists working in the field of XAI, that there are higher demands on explainability in contexts in which important decisions are made or supported by AI systems. This raised the question how exactly the stakes in a context might affect whether an explanation is successful, which is to say, how they affect whether a subject receiving the explanation in the context understands what is explained. To answer this question, I focused on explanations of why certain outputs were provided, and looked to philosophical debates on pragmatic encroachment and inductive risk.

According to my proposal, it is the epistemic justification condition on outright understanding that is sensitive to what is at stake in a situation and which thus leads to the stakes-sensitivity of understanding—when the stakes are higher, there are more stringent requirements on all things considered justification and thus on understanding. I developed this idea, first, by appeal to the claim that in high-stakes contexts, the subject must have better evidence in order to have justification, and second, by appeal to the claim that in such contexts, there must be greater coherence between the propositions involved in the believed explanation, and greater coherence of these with the subject's background beliefs, for her beliefs to be justified.[19]

Let me conclude with practical implications of my proposal. In real-world high-stakes situations in which AI systems are used, it is important to ensure not only that relevant stakeholders receive explanations of the systems' outputs, but also that they sufficiently understand these outputs (we are now focusing on a graded concept of understanding). For this, it must *first* be ensured that the technical systems providing such explanations are highly reliable, so that the recipients of the explanations can accept the given explanations on the basis of sufficient evidence.[20] Moreover, it is important that the provided explanations cohere well with the web of the recipients' background beliefs. This can (*second* implication) be facilitated by designing explanatory tools to provide explanations that match their recipients' level of expertise and intellectual capacities. Such coherence can (*third* implication) be facilitated by improving recipients' background knowledge, for instance, by educating subjects who will be faced with high-stakes AI situations with regard to the functioning of such systems or with regard to relevant domain knowledge. In these three ways, stakeholders' understanding of why an AI system came to a certain decision or gave a certain recommendation can be improved in practice.

Acknowledgements I thank audiences at the Workshop *Epistemological Issues of Machine Learning in Science* (TU Dortmund), at the Applied Epistemology Workshop (University of Zurich), at the conference *Trust and Opacity in AI* (TU Dresden), at AISoLA (Crete), and at EXTRA.12 (Ruhr University Bochum) for helpful and challenging comments. I'm especially grateful to Kevin Baum, Lindsay Brainard, and Joachim Horvath for their suggestions. Finally, I'd like to thank Giorgia Pozzi, Milou Jansen and Koray Karaca for their comments on an earlier draft of the chapter.

[19] Another possible account of the stakes-sensitivity of understanding appeals to graded understanding. On this account, the subject needs a greater degree of graded understanding when stakes are high. This idea is certainly worth exploring. Since I cannot discuss this in detail, let me at least briefly give my motivations for focusing on outright understanding instead: (1) This focus allows us to treat pragmatic encroachment on knowledge and understanding in parallel. (2) It transparently offers a very clear treatment of the stakes affecting the *epistemic* states of the subject; by contrast, proponents of the 'graded' proposal needs to tell us in what sense subjects *need* a higher degree of understanding in high-stakes situations. Is this an instrumental, purely practical or an epistemic demand? I don't want to deny that such a proposal could be developed in one or another way; though I am doubtful that it would be able to uphold the—to my mind rather attractive—claim that the stakes impact understanding itself.

[20] Computational reliabilism is discussed in Chap. 4 by Durán and Chap. 5 by Alvarado in this volume.

Funding This paper was supported by the Volkswagen Foundation, as part of the project *Explainable Intelligent Systems* (EIS), project number AZ 98510.

References

Adadi, A., & Berrada, M. (2018). Peeking inside the black-box: A survey on explainable Artificial intelligence (XAI). *IEEE Access, 6*(2018), 52138–52160.

Altman, A. (2020). Discrimination. In E. N. Zalta (Ed.), *The Stanford encyclopedia of philosophy (Winter 2020 Edition)*. https://plato.stanford.edu/archives/win2020/entries/discrimination/

Alvarado, R. (in this volume). Challenges for computational reliabilism: Epistemic warrants, endogeneity and error-based opacity in machine learning. In J. M. Durán & G. Pozzi (Eds.), *Philosophy of science for machine learning: Core issues and new perspectives*. Synthese Library, Springer.

Angelov, P. P., Soares, E. A., Jiang, R., Arnold, N. I., & Atkinson, P. M. (2021). Explainable artificial intelligence: An analytical review. *WIREs Data Mining and Knowledge Discovery, 11*, 1–13. https://doi.org/10.1002/widm.1424

Angwin, J., Larson, J., Mattu, S., & Kirchner, L. (2016). *Machine bias: There's software used across the country to predict future criminals and it's biased against blacks*. ProPublica.

Bachmann, M. (2020). The epistemology of understanding. A contextualist approach. *Kriterion – Journal of Philosophy, 34*(1), 75–98.

Bathaee, Y. (2018). The artificial intelligence black box and the failure of intent and causation. *Harvard Journal of Law & Technology, 31*(2), 889–938.

Baum, K., Mantel, S., Schmidt, E., & Speith, T. (2022). From responsibility to reason-giving explainable artificial intelligence. In N. Conradie, H. Kempt, & P. Königs (Eds.), *Philosophy and technology* (Vol. 35., Topical collection on AI and responsibility, pp. 1–30).

Baumberger, C., Beisbart, C., & Brun, G. (2017). What is understanding? An overview of recent debates in epistemology and philosophy of science. In S. Grimm, C. Baumberger, & S. Ammon (Eds.), *Explaining understanding: New perspectives from epistemology and philosophy of science* (pp. 1–34). Routledge.

Beisbart, C., & Räz, T. (2022). Philosophy of science at sea: Clarifying the interpretability of machine learning. *Philosophy Compass, 17*(6), e12830. https://doi.org/10.1111/phc3.12830

Biddle, J. (2022). On predicting recidivism: Epistemic risk, tradeoffs, and values in machine learning. *Canadian Journal of Philosophy, 52*(3), 321–341. https://doi.org/10.1017/can.2020.27

Biddle, J. B., & Kukla, R. (2017). The geography of epistemic risk. In K. C. Elliott & T. Richards (Eds.), *Exploring inductive risk: Case studies of values in science*. Oxford University Press. https://doi.org/10.1093/acprof:oso/9780190467715.003.0011

Chouldechova, A. (2017). Fair prediction with disparate impact: A study of bias in recidivism prediction instruments. *Big Data, 5*(2), 153–163.

Douglas, H. (2000). Inductive risk and values in science. *Philosophy of Science, 67*(4), 559–579.

Durán, J. M. (in this volume). Beyond transparency: Computational reliabilism as an externalist epistemology for algorithms. In J. M. Durán & G. Pozzi (Eds.), *Philosophy of science for machine learning: Core issues and new perspectives*. Synthese Library, Springer.

Elgin, C. Z. (2007). Understanding and the facts. *Philosophical Studies, 132*(1), 33–42. https://doi.org/10.1007/s11098-006-9054-z

Elgin, C. Z. (2017). *True enough*. MIT Press.

EU High-Level Expert Group on Artificial Intelligence. (2019). Ethics guidelines for trustworthy AI. https://ec.europa.eu/digital-single-market/en/news/ethics-guidelines-trustworthy-ai

Fantl, J., & McGrath, M. (2002). Evidence, pragmatics, and justification. *The Philosophical Review, 111*, 67–94.

Fantl, J., & McGrath, M. (2011). Pragmatic encroachment. In S. Bernecker & D. Pritchard (Eds.), *The Routledge companion to epistemology* (1st ed., pp. 558–568). Routledge. https://doi.org/10.4324/9780203839065

Fazelpour, S., & Danks, D. (2021). Algorithmic bias: Senses, sources, solutions. *Philosophy Compass, 16*(8), e12760. https://doi.org/10.1111/phc3.12760

Fleisher, W. (2022). Understanding, idealization, and explainable AI. *Episteme, 19*(4), 534–560. https://doi.org/10.1017/epi.2022.39

Gao, J. (forthcoming). Varieties of pragmatic encroachment. In K. Sylvan, E. Sosa, J. Dancy, & M. Steup (Eds.), *The Blackwell companion to epistemology* (3rd ed.).

Grimm, S. (2021). Understanding. In N. Edward (Ed.), *The Stanford encyclopedia of philosophy (Summer 2021 Edition)*. https://plato.stanford.edu/archives/sum2021/entries/understanding/

Hannon, M. (2021). Recent work in the epistemology of understanding. *American Philosophical Quarterly, 58*(3), 269–290. https://scholarlypublishingcollective.org/uip/apq/article/58/3/269/283447/RECENT-WORK-IN-THE-EPISTEMOLOGY-OF-UNDERSTANDING

Hempel, C. G. (1965). Science and human values. In *Aspects of scientific explanation and other essays in the philosophy of science* (pp. 81–96). The Free Press.

Hills, A. (2015). Understanding why. *Noûs, 49*(2), 661–688.

House of Lords, Select Committee on Artificial Intelligence. (2018). *AI in the UK: Ready, willing and able?* House of Lords.

Johnson, G. M. (2023). Are algorithms value-free? *Journal Moral Philosophy, 21*(1–2), 1–35.

Karaca, K. (2021). Values and inductive risk in machine learning modelling: the case of binary classification models. *European Journal for Philosophy of Science, 11*(4), 1–27.

Kelp, C. (2015). Understanding phenomena. *Synthese, 192*(12), 3799–3816.

Kelp, C. (2016). Towards a knowledge-based account of understanding. In S. Grimm, C. Baumberger, & S. Ammon (Eds.), *Explaining understanding* (pp. 251–271). Routledge.

Kvanvig, J. L. (2003). *The value of knowledge and the pursuit of understanding*. Cambridge University Press.

Langer, M., Oster, D., Speith, T., Hermanns, H., Kästner, L., Schmidt, E., Sesing, A., & Baum, K. (2021). What do we want from explainable artificial intelligence (XAI)?—A stakeholder perspective on XAI and a conceptual model guiding interdisciplinary XAI research. *Artificial Intelligence, 296*, 103473. https://doi.org/10.1016/j.artint.2021.103473

Lazar, S. (2024). Legitimacy, authority, and the political value of explanations. *Oxford Studies in Political Philosophy, 10*, 28–56.

Linardatos, P., Papastefanopoulos, V., & Kotsiantis, S. (2021). Explainable AI: A review of machine learning interpretability methods. *Entropy, 23*, 18. https://doi.org/10.3390/e23010018

Longino, H. E. (1990). *Science as social knowledge: Values and objectivity in scientific inquiry*. Princeton University Press.

Malfatti, I. (2023). *Verstehen verstehen. Eine erkenntnistheoretische Untersuchung*. Schwabe.

Marshall, S. R., & Naumann, L. P. (2018). What's your favorite music? Music preferences cue racial identity. *Journal of Research in Personality, 76*, 74–91.

Miller, B. (2014). Science, values, and pragmatic encroachment on knowledge. *European Journal for Philosophy of Science, 4*(2), 253–270.

Munch, L., Bjerring, J. C., & Mainz, J. (2024). Algorithmic decision-making: the right to explanation and the significance of stakes. *Big Data and Society, 11*(1), 1–12.

Nauta, M., Trienes, J., Pathak, S., Nguyen, E., Peters, M., Schmitt, Y., Schlötterer, J., van Keulen, M., & Seifert, C. (2023). From anecdotal evidence to quantitative evaluation methods: A systematic review on evaluating explainable AI. *ACM Computing Surveys, 55*(13s), 295. https://doi.org/10.1145/3583558

Nielsen, I., Dera, D., Rasool, G., Ramachandran, R., & Bouaynaya, N. (2022). Robust explainability: A tutorial on gradient-based attribution methods for deep neural networks. *IEEE Signal Processing Magazine, 39*, 73–84. https://doi.org/10.1109/MSP.2022.3142719

Northpointe. (2015). Practitioner's guide to COMPAS core. https://s3.documentcloud.org/documents/2840784/Practitioner-s-Guide-to-COMPAS-Core.pdf

Nyrup, R., & Robinson, D. (2022). Explanatory pragmatism: A context-sensitive framework for explainable medical AI. *Ethics and Information Technology, 24*(1), 13.
Páez, A. (2019). The pragmatic turn in explainable artificial intelligence. *Minds and Machines, 29*(3), 441–459.
Páez, A. (in this volume). Axe the X in XAI: A plea for understandable AI. In J. M. Durán & G. Pozzi (Eds.), *Philosophy of science for machine learning: Core issues and new perspectives*. Synthese Library, Springer.
Pedersen Phillips. (2020).
Phillips, K. P. (2020). Knowing what is said & other essays on knowledge, understanding, and communication. PhD Dissertation.
Pritchard, D. (2010). Understanding. In *The nature and value of knowledge: Three investigations*. Oxford Academic. https://doi.org/10.1093/acprof:oso/9780199586264.003.0004
Purves, D., & Davis, J. (2022). Public trust, institutional legitimacy, and the use of algorithms in criminal justice. *Public Affairs Quarterly, 36*, 136–162.
Ratti, E., & Graves, M. (2022). Explainable machine learning practices: Opening another black box for reliable medical AI. *AI and Ethics, 2*(4), 1–14.
Ribeiro, M. T., Singh, S., & Guestrin, C. (2016). "Why should I trust you?": Explaining the predictions of any classifier. In Association for Computing Machinery (Ed.), *Proceedings of the 22nd ACM SIGKDD international conference on knowledge discovery and data mining* (pp. 1135–1144).
Riggs, W. D. (2003). Balancing our epistemic goals. *Noûs, 37*(2), 342–352.
Rudner, R. (1953). The scientist qua scientist makes value judgments. *Philosophy of Science, 20*, 1–6.
Schmidt, E. (2022). Wie können wir autonomen KI-Systemen vertrauen? Die Rolle von Gründe-Erklärungen. In B. Konz, K.-H. Ostmeyer, & M. Scholz (Eds.), *Gratwanderung Künstliche Intelligenz—Interdisziplinäre Perspektiven auf das Verhältnis von Mensch und KI*. Kohlhammer.
Schmidt, E. (2025). Reasons, attenuators, and virtue: A novel account of pragmatic encroachment. *Analytic Philosophy, 66*(2), 159–180. https://doi.org/10.1111/phib.12314
Schmidt, E., Sesing-Wagenpfeil, A., & Köhl, M. (2023). Bare statistical evidence and the legitimacy of software-based judicial decisions. *Synthese, 201., Topical collection on evidence in law and ethics* (A. Dyrda & T. Żuradzki, Ed.), 1–27. https://doi.org/10.1007/s11229-023-04141-2
Schroeder, M. (2012). Stakes, withholding, and pragmatic encroachment on knowledge. *Philosophical Studies, 160*, 265–285.
Stanley, J. (2005). *Knowledge and practical interests*. Oxford University Press.
Steel, D. (2010). Epistemic values and the argument from inductive risk. *Philosophy of Science, 77*(1), 14–34.
Sullivan, E. (2022a). How values shape the machine learning opacity problem. In I. Lawler, K. Khalifa, & E. Shech (Eds.), *Scientific understanding and representation: Modeling in the physical sciences* (pp. 306–322). Routledge.
Sullivan, E. (2022b). Inductive risk, understanding, and opaque Machine Learning models. *Philosophy of Science, 89*(5), 1–13. https://doi.org/10.1017/psa.2022.62
Sullivan, E., & Kasirzadeh, A. (in this volume). Explanation hacking: The perils of algorithmic recourse. In J. M. Durán & G. Pozzi (Eds.), *Philosophy of science for machine learning: Core issues and new perspectives*. Synthese Library, Springer.
Tschantz, M. C. (2022). What is proxy discrimination? In Association for Computing Machinery (Ed.), *Proceedings of the 2022 ACM conference on fairness, accountability, and transparency (FAccT '22). New York*. https://doi.org/10.1145/3531146.3533242
Wilkenfeld, D. A. (2013). Understanding as representation manipulability. *Synthese, 190*(6), 997–1016.
Yuan, C. W., Bi, N., Lin, Y.-F., & Tseng, Y.-H. (2023). Contextualizing user perceptions about biases for human-Centered explainable artificial intelligence. In Association for Computing

Machinery (Ed.), *Proceedings of the 2023 CHI conference on human factors in computing systems (CHI '23). New York.* Article 248 (pp. 1–15). https://doi.org/10.1145/3544548.3580945

Zagzebski, L. (2001). Recovering Understanding. In M. Steup (Ed.), Knowledge, truth, and duty: essays on epistemic justification, responsibility, and virtue (pp. 235–252). New York: Oxford University Press.

Zednik, C. (2021). Solving the black box problem: A normative framework for explainable artificial intelligence. *Philosophy and Technology, 34*, 265–288.

Zednik, C., & Verreault-Julien, P. (in this volume). Can explainable AI contribute to justification? In J. M. Durán & G. Pozzi (Eds.), *Philosophy of science for machine learning: Core issues and new perspectives*. Synthese Library, Springer.

Eva Schmidt is Professor of Theoretical Philosophy at TU Dortmund and a PI at the Lamarr Institute for Machine Learning and Artificial Intelligence. She works on explainable artificial intelligence, perception, and the theory of reasons and normativity. Her book *Modest Nonconceptualism* was published in 2015 with Springer. Her work has appeared in *Noûs*, *Philosophical Studies*, *Synthese*, and *Philosophy & Technology*.

Part V
Scientific Models and Representation

Chapter 12
Representation Learning Without Representationalism

A Non-representational Account of Deep Learning Models in Scientific Practice

Phillip Hintikka Kieval

Abstract Both machine learning research and philosophy of model-based science appeal to representations. Given this shared emphasis, it seems natural to conclude that we should understand the epistemic worth of deep learning models in science in terms of their capacity to represent the world. I argue that this is a mistake. I distinguish the prevailing notion of representation used in machine learning research from the concept of scientific representation that figures in the philosophy of model-based science. I argue that the former cannot do the work of the latter in model-based inferences. I then defend an artifactualist approach to machine learning models in science, which aims to understand their epistemic worth in terms of the material practices of constructing and applying them. Machine learning models are instruments that we use to facilitate our epistemic activities in science. They do so without scientific representation.

Keywords Deep learning · Models in science · Scientific representation · Representationalism · Artifactualism

12.1 Introduction

Many philosophers of science agree that the defining characteristic of scientific models lies in their capacity to represent (Frigg & Nguyen, 2017; Frigg & Hartmann, 2020; Giere, 2004, 2010; Morgan & Morrison, 1999; Parker, 2020; Suárez, 2004; van Fraassen, 2008; Weisberg, 2013). Models' epistemic value rests on our ability to use them to reason about aspects of the world, and this is possible because models represent their targets. As Frigg and Nguyen (2017) put it, "models

P. H. Kieval (✉)
Department of History and Philosophy of Science, University of Cambridge, Cambridge, UK
e-mail: pzhk2@cam.ac.uk

J. M. Durán, G. Pozzi (eds.), *Philosophy of Science for Machine Learning*,
Synthese Library 527, https://doi.org/10.1007/978-3-032-03083-2_12

must be representations: they can instruct us about the nature of reality only if they represent the selected parts or aspects of the world we investigate" (Frigg & Nguyen, 2017, p. 49).

Discussions of machine learning place similar emphasis on representations. Machine learning is a set of algorithmic methods for learning representations of data. Deep neural networks, for instance, involve passing signals through a network of simple computational nodes. Each node transforms a representation of input data and passes it along to the next layer in the network. By composing these simple transformations at subsequent layers, the model learns increasingly abstract representations of the data. This allows deep learning methods to learn and exploit complex patterns in large datasets.

For this reason, researchers sometimes refer to deep learning and similar methods as *representation learning*. As LeCun et al. (2015) write,

> Representation learning is a set of methods that allows a machine to be fed with raw data and to automatically discover the representations needed for detection or classification. Deep-learning methods are representation-learning methods with multiple levels of representation, obtained by composing simple but non-linear modules that each transform the representation at one level (starting with the raw input) into a representation at a higher, slightly more abstract level (LeCun et al., 2015, p. 436).

Given this shared emphasis on representation, it seems natural to conclude that we should assess the epistemic worth of deep learning models in science in terms of their capacity to represent the world. This approach seems to reflect the reality of much of the extant philosophy of science for machine learning. Emily Sullivan's (2022) illuminating discussion of "link uncertainty," for example, centers around the degree to which empirical evidence supports a link between a model and its target. More recently, Tamir and Shech (2023) take an explicitly representationalist approach to machine learning models in applying what they call the "C-schema" to understand the connection between models and their targets. They propose that we should understand machine learning models as targeting the relationship between features represented by data. Conversely, Boge (2022) claims that deep learning models do not contain structure that can be mapped to features of their target system. He thus takes up a representationalist perspective to argue that deep neural networks fail to explain their targets.

Though they diverge in their exact diagnosis of deep learning models in science, these authors seem to share a broad commitment to understanding their epistemic value in terms of their capacity to represent the world. This makes sense in light of the history of philosophy of science, which has continually centered representation as a characteristic feature of scientific models. I argue, however, that this is a mistake when it comes to deep learning. We should distinguish the prevailing notion of representation used in machine learning research from the concept of *scientific representation* that figures in the philosophy of model-based science. Scientific representations must be usable by researchers to make inferences (Contessa, 2007; Giere, 2004, 2010; Suárez, 2004, 2010). The intended use of a model specifies how it represents. The shared practices of modelling communities fix standard conventions for reading a model. These conventions clarify how we can use the model to reason about its target (Suárez, 2004; Weisberg, 2013, p. 39).

But no such conventions exist in the case of deep learning. In most cases in science, there just isn't a sensible way of interpreting the parameters of a deep learning model in a way that would enable a competent user to make inferences about a target. Learned representations, in contrast with scientific representations, are used by the machine itself. Human investigators cannot use them to make model inferences. Thinking about deep learning through a representationalist lens is thus bound to generate confusion. Yet, I think there are good reasons to think of these systems as models, given their extensive use in contemporary science. We, therefore, need to explore alternative avenues for making sense of their epistemic utility.

I argue that *artifactualism*—an approach that understands scientific models as tools that mediate our epistemic activities (see especially Knuuttila, 2011; Morgan & Morrison, 1999; Sanches de Oliveira, 2022)—can help to bridge this gap. According to artifactualism, preoccupation with representation doesn't do justice to the diverse roles of models in science. The view, instead, aims to understand models' epistemic utility in terms of the material practices of constructing and applying them. Yet, artifactualists tend to be somewhat circumspect about the exact nature of the epistemic benefits of their view. I thus make the case that adopting an artifactualist approach helps us to recognize the epistemic and practical value of skill in scientific modelling. I examine the use of deep learning in scientific practice through such a lens. The design and construction of deep learning models involves an iterative process that depends on a combination of skill and tacit knowledge about how to get the most out of these systems. Developing such a skill scaffolds genuine understanding about the relationship between our means and our epistemic ends. Deep learning models are instruments that we use to facilitate epistemic activities like prediction, exploration, and manipulation. They do so without scientific representation.

My argument goes like this: I begin by introducing the role of representation in model-based science. Next, I consider the concept of representation used in machine learning research and distinguish it from scientific representation. I then argue that the former cannot do the work of the latter in model-based inferences. Finally, I present some considerations in favor of an artifactual account of deep learning models in scientific practice cashed out in terms of the epistemic benefits of skilled engagement with epistemic tools.

12.2 Scientific Models and Representation

Scientists use models to learn about the world, to predict and explain what's going on around us. We construct models of everything from atomic particles to economic systems using diverse methods and materials. There are many reasons for doing so. Often the targets of our investigations are too complex to study directly. We thus need to simplify them to make inquiry possible. It might be appropriate, for instance, to consider an abstract model that includes only the causal factors we think are essential. At other times, we may introduce false assumptions that help us understand the structure underlying a phenomenon (cf. Elgin, 2004; Potochnik, 2017; Weisberg, 2007).

Despite these distortions, models contribute to the epistemic success of science. Many philosophers agree that models help scientists achieve their aims by *representing* the targets of inquiry (Frigg & Nguyen, 2017; Frigg & Hartmann, 2020; Giere, 2004, 2010; Morgan & Morrison, 1999; Parker, 2020; Suárez, 2004; Weisberg, 2013). This is appealing because it provides a ready answer to how we can reason with models to learn about the world. Modelling involves representing a select aspect of the world. This aspect is the model's *target*. We can thus think of a model as the source of a representation that is directed at a target (Suárez, 2010).

And yet, defining representation is a matter of controversy (for a thorough review, see Frigg & Nguyen, 2017). Representation alone comes cheap, but models need to do more than just denote what they stand for. Callender and Cohen stress that anything can represent anything else by stipulative fiat (Callender & Cohen, 2006, p. 73). I can stipulate that the coffee mug on my desk represents alpha Centauri, but that alone doesn't mean that I can use my mug to infer anything about stellar parallax. Any adequate account of model representation needs to tell us how those representations enable surrogate reasoning. That is, models represent in ways that license inferences about their targets. I will call a representation that satisfies this desideratum a *scientific representation.*[1] Their inferential power sets scientific representations apart from brute signs, but philosopher disagree about how models meet this demand.

We can characterize theories of scientific representation as either substantive or deflationary (Suárez, 2010). Substantive accounts take the relationship between a model and its target to be a matter of fact; they aim to specify a robust feature or relation that holds between models and their targets. Two time honored examples include similarity and isomorphism. Similarity maintains that a model represents its target just in case that model is similar to its target in some relevant respect. Isomorphism, on the other hand, is a well defined mathematical relation between extensional structures. Setting formal sophistication aside, isomorphism holds that a model represents its target when there is a one-to-one mapping between elements in the model's structure and elements in the target's structure. Both of these substantive accounts aim to ground the capacity for surrogate reasoning in an objective relation that holds between models and their targets.

[1] My use thus departs from the most general sense of the term understood as any representation produced by scientific inquiry. This more general notion includes instrument readings, visualizations of data, photographs and digital images, and so on. It is impossible to draw a sharp line between models and these other kinds of representation, given that many of these other forms appear to be "model-laden" (Bokulich, 2020; Chang, 2004, p. 188; Leonelli, 2019; Tal, 2017). A full account of the relation between narrow, model-centric scientific representations and the more general notion is beyond the scope of this chapter.

Deflationary accounts, on the other hand, deny any such robust, factual relationship obtains between models and their targets. These accounts, instead, tie representation to its use in modelling practices. Mauricio Suárez (2004, 2015) inferentialist account, for instance, spells this out in terms of the capacity of a model to direct competent and informed users to consider a target and draw out specific inferences. Inferentialism nevertheless maintains an objective component. This is because only some kinds of structures will let informed agents draw good inferences about their targets. Suárez writes that it's a model's "internal structure that allows informed agents to correctly draw inferences" (Suárez, 2004, p. 774). For the Lotka-Volterra model to count as a scientific representation of predation ecologists need to be able to use it. Ecologists can, in fact, use oscillations in the model to represent the relationship between changes in predator and prey populations. This helps them understand, for example, how an environment stays below its carrying capacity. Without this inferential capacity, the model would be a vacuous mathematical structure. Our ability to make sound inferences constrains what structures will be admissible in a particular modelling context.

Many philosophers now appreciate that scientific modelling is an activity practiced by agents with particular research agendas and goals (Giere, 2004, 2010; Godfrey-Smith, 2006; Knuuttila, 2005, 2011; Parker, 2020; Suárez, 2004; Weisberg, 2013). They thus stress the importance of human judgement and competence in the successful use of models to represent the world. Learning with models comes down to our ability to interpret their structure in ways that enable us to infer things about their targets. More and more, both substantive and deflationary views share this attention to scientists' ability to use models in practice.

Along these lines, recent substantive accounts allow human agents to fill in the details of scientific representation (Giere, 2004, 2010; Godfrey-Smith, 2006; van Fraassen, 2008; Weisberg, 2013). Michael Weisberg, for example, says that "models consist of an *interpreted structure* that can be used to represent a real or imagined phenomenon" (Weisberg, 2013, 15 emphasis original). For Weisberg, this involves agents adopting particular *construals* of a model, which set up relations of denotation between models and their targets (2013, p. 39). A crucial part of a construal is what Weisberg calls the assignment. Assignments determine the relationship between parts of the model's structure and the parts of their target system. Assignments are thus "explicit specifications of how parts of real or imagined target systems are to be mapped onto the parts of the model" (Weisberg, 2013, p. 39). Assignments, as such, are often implicit in the practices and conventions for reading out model descriptions adopted by specific epistemic communities. They serve as a kind of record of the coordinating actions of modelling communities. Assignments thus resemble Suárez (2004) characterization of the practical maintenance of what he calls representational force—the capacity of a model to lead informed users to consider its target.

Of course, there are important differences between all of these accounts. Yet, they agree that the successful use of a model depends on our ability to interpret that model's structure in a way that licenses correct inferences. Scientific representation requires that agents intend to use a model for a particular purpose. Communities of modellers may codify, implicitly or explicitly, conventions for reading out that model's structure. This structure must be such that a competent and informed inquirer could use it to infer something about the target. If we can't use a model, then it's not a scientific representation.

12.3 Learning Representations of Data

Machine learning as a discipline aims to engineer systems that learn from data. We can take "learning" here to mean improving performance on the basis of observed data. Evaluating performance depends, in turn, on specifying a measure of success—maximizing, for example, predictive accuracy on unobserved data. Modellers expose their model to data and then optimize some measure of performance over that data.

Machine learning involves many different methods, including decision tree methods such as random forests, Bayesian non-parametric techniques like Gaussian processes, and artificial neural networks. All of these involve procedures for learning a plausible model that fits observed data. We can characterize this model as a function expressed in terms of a set of parameters whose values are initially unknown. This model is both a description of observed data and a hypothesis about the underlying phenomena that produced said data. That is, the model provides a hypothetical description of the data generating process that aims to cohere with background knowledge and facilitate reliable extrapolations. Learning, then, is a procedure for finding the parameter values that maximize the likelihood of our data, often while balancing the need to generalize to unseen data. In most real-world cases, we can't solve for these parameters analytically. Machine learning algorithms thus use numerical methods to estimate these unknown parameter values.

The last decade bore witness to remarkable advances in machine learning driven by a set of neural network methods called deep learning (LeCun et al., 2015; Buckner, 2019). All neural networks consist in an array of simple computational nodes. So-called deep neural networks feed input values through multiple layers of nodes, which are connected together by weights that determine the strength of signal passed from one layer to the next. Each node takes the signal from the preceding layer and transforms it according to various functions. By composing these simple transformations, a deep neural network can learn complex features of observed data from low-level information. An image classifier, for instance, can learn to identify objects like cars, people, or animals from the pixel values of a digital image.

Deep neural networks (hereafter, DNNs) learn by adjusting their weights in response to data. Depending on the specific network architecture and modelling context, modellers may assign random values to these weights prior to training, or they may opt for a more sophisticated initialization scheme. In either case, the network's initial performance may be no better than random guessing. As we expose the network to data, however, a learning algorithm determines how the network will adjust. Such an algorithm tells the network when it's made an error and by how much. The network then learns from its mistake by altering the strength of connections between nodes that were most responsible for the error.

When a network experiences labeled data, we call this process "supervised learning" because the learning algorithm acts like a tutor watching over the network and pointing out when it has gone wrong. The network thus gets to compare its outputs with the label assigned to its target and corrects accordingly. Many scientific contexts call for supervised learning. It works well in any situation where we want to predict the value of a target variable from a set of known features. When the target takes on a discrete set of possible values, we call the task classification. If we want instead to predict a continuous, real value, then we call the task a regression.

Sometimes, however, we don't have access to good labelled data or our aims diverge from straightforward prediction. In such cases, we might train a model to learn features of the dataset without explicit guidance. This is called "unsupervised learning".[2] Unsupervised models experience an entire dataset and learn useful properties of its structure. This often means learning the entire probability distribution that generated our data. Useful applications include density estimation, learning to generate samples from a learned distribution, de-noising data, clustering and outlier detection, and reducing the dimensionality of complex datasets so that we can visualize them (cf. Goodfellow et al., 2016; Raghu & Schmidt, 2020).

DL promises to accelerate scientific progress by automatically extracting subtle patterns in large datasets and exploiting those patterns to make novel predictions. One recent lesson data scientists have learned from these models is that flexible methods of learning from large quantities of data tend to produce the best predictions when dealing with very complex phenomena (Ghahramani, 2015). Such flexibility means modellers make minimal assumptions about their targets. Rather than relying on domain-specific theoretical assumptions to tightly constrain the space of possible functions a model could learn, we engineer domain-general models that learn interesting features of the data on their own.

Machine learning researchers often couch this in terms of learning a representation of data (Bengio et al., 2013; Goodfellow et al., 2016, ch. 1; LeCun et al., 2015). We can understand what this means by contrasting DL with simpler statistical methods. In classical statistics and simple machine learning, modellers impose significant structure on their model before fitting it to data. They gather and prepare a dataset and then invoke prior hypotheses about the features of the phenomena that generated their data. Modellers use these hypotheses as the basis for a representation of their data. This representation takes the shape of a probability model whose parameters correspond to features of the data. Fitting this model to the data involves maximum likelihood estimation, an analytic method of finding the parameter values that maximize the probability of a dataset.[3] These parameter values are the so-called

[2] The distinction between supervised and unsupervised learning is not a formal one and often the lines become fuzzy, as in the cases of, for example, semi-supervised and self-supervised learning.

[3] Since an exact solution to the likelihood function is often intractable, modellers typically use an approximation like the method of least squares estimation instead. Least squares estimation,

"best estimator" of the dataset given a class of models with a certain structure. After fitting the model, modellers can evaluate it by plotting the residuals, or the difference between our model's predictions and what they actually observe in the data. The aim here is to identify any systematic pattern that modellers can use to improve the model.

Let's consider a simple example of classical statistical modelling with supervised learning. Suppose I want to model how the average monthly temperature in Cambridge has changed over time so that I can predict average temperatures over the next few years. I can take some data of temperature readings over time and design a model to represent features of that data. Fortunately, the UK Met Office makes some historic data from a weather station in Cambridge publicly accessible.[4] I can take this data and design a probability model to represent changes in temperature as a function of time. I can then use supervised learning to find the parameter values of my model that constitute the best estimator of my labeled temperature dataset. Now, I know that annual temperatures are cyclical. So I can describe my model in terms of a simple periodic function like this

$$\text{Temp} \approx \alpha + \beta \sin(2\pi \text{t} + \theta) + N(0, \sigma^2)$$

where t is the date in years, α, β, and θ are unknown parameters that fix the shape of our periodic model, and $N(0, \sigma^2)$ is a Gaussian noise term that represents the idea that our data is a noisy realization of the target phenomenon.[5] When I fit my model, a simple algorithm finds the values of these predetermined parameters that are the best estimator of my data. Figure 12.1 shows the fitted model.

Right off the bat, I have a pretty strong sense that this toy model isn't very good. That's because I'm aware of warming trends that are not captured by my initial representation. To get a grip on how this omission affects my model I can compute and plot the residuals, as shown in Fig. 12.2.

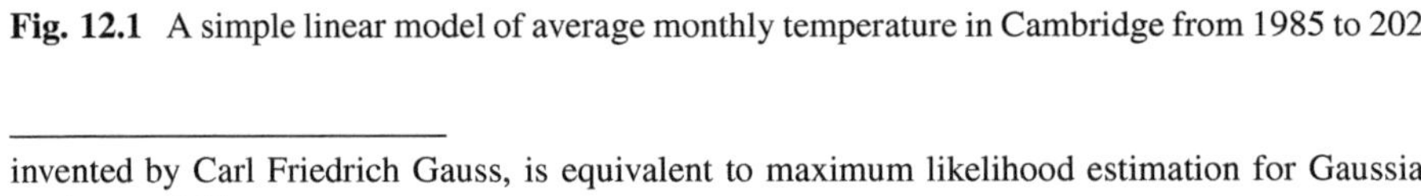

Fig. 12.1 A simple linear model of average monthly temperature in Cambridge from 1985 to 2024

invented by Carl Friedrich Gauss, is equivalent to maximum likelihood estimation for Gaussian random variables.

[4] https://www.metoffice.gov.uk/pub/data/weather/uk/climate/stationdata/cambridgedata.txt.

[5] Technically, we need to rearrange the terms a bit to make all of the parameters linearly independent before we can fit the model. We can do this with some simple trigonometry to get $\text{Temp} \approx \alpha + \beta_1 \sin(2\pi \text{t}) + \beta_2 \cos(2\pi \text{t}) + N(0, \sigma^2)$.

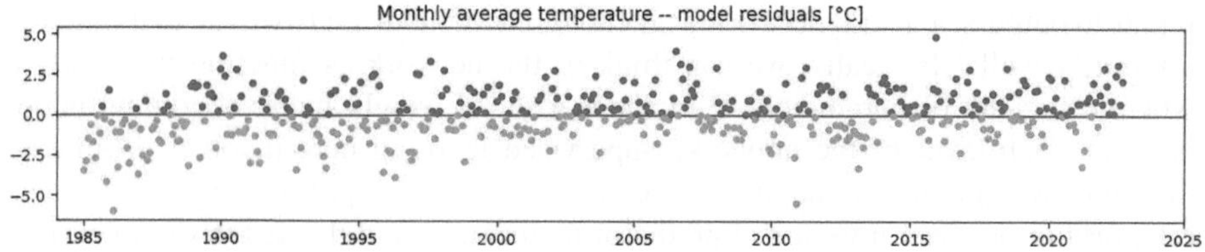

Fig. 12.2 A plot of the difference between what my initial model predicts and what we observe in the data

There's a certain immediacy to this plot. A secular trend that shows temperature steadily increasing jumps right out at me. To confirm, I might fit a linear regression to this residual plot. I would then modify my original model to include an additional parameter representing this linear warming trend.

Real cases in data science will be a good deal messier than this example. We may not know what features to represent in our model and diagnosing weaknesses might turn out to be an arduous process. But the point I've tried to show is that in simple supervised learning, as in classical statistical inference, we begin with a representation that we want to fit to our data. Domain-specific theoretical knowledge about the target systems motivates our choice of representation. Training the model is about learning a good mapping between this representation and our dataset. Such a representation is intelligible in light of the substantive modelling assumptions that we've imposed on our data. Intelligibility enables us to iteratively manipulate our model in response to its performance and use it to make inferences (de Regt, 2017). We can see how well it does and make theoretically informed changes to how our model represents different features of the target system. Moreover, we can use the representation to make inferences because the parameters were hand-chosen to account for these features.

In deep learning, things look much different. We build a deep learning model (henceforth, DL model) that learns not only a good mapping but also the representation itself (Goodfellow et al., 2016, p. 4). This is the notion of representation invoked in talk of "representation learning" (Bengio et al., 2013; Goodfellow et al., 2016, p. 5; LeCun et al., 2015, p. 436). Representation learning is much more flexible than methods that impose considerable parametric modelling assumptions. This is because learning a representation in this sense requires little domain knowledge about the target system. Overparameterized models that exploit learned representations, moreover, tend to outperform models with representations designed by hand without the need for significant hypotheses about what to expect from our data.

This works out because DNNs are so-called "universal function approximators" (Hornik et al., 1989). That is, we have a theoretical guarantee that any feedforward neural network with three or more layers "can approximate virtually any function of interest to any desired degree of accuracy, provided sufficiently many hidden units are available" (Hornik et al., 1989, p. 360). This means that, if we suppose that we can describe our data in terms of a complex function mapping inputs to outputs, a neural network with enough nodes can approximate that function arbitrarily well.[6] Basically, we can think of the network as function $y = f_\theta(x)$, where θ is a set of parameters. θ includes a set of weight matrices joining nodes between each layer of the network. Supervised learning here involves estimating the set of weight values that allows the network to best approximate the functional relationship between a vector of predictor features, x, and the target variable(s), y. A trained model thus learns how to represent functional relationships in the data in terms of a set of parameters. But unlike classical methods, we don't choose those parameters on the basis of features of the target system. We instead choose a network architecture on the basis of the general kind of task we want to solve. Architectural choices amount to domain-general inductive biases that constrain the actual set of functions the network can learn.

More flexible learning works so well because training leads a network to learn representations of data that have properties that make the relevant task easier. Statistical modelling and machine learning often draws on a geometric metaphor here. We can conceive of each data sample as a point in a multidimensional space whose axes are constituted by the qualitative features of the dataset. Representation learning involves finding a set of transformations of this "feature space" that reduces the complexity of the task. In a classification task, for example, the final layer of a network will be a linear classifier. Learning a good representation for this task might mean learning a representation of the inputs that makes them linearly separable in feature space at the penultimate layer. This basically means that we can draw a straight line that separates inputs into mutually exclusive categories. Such separability will then make classification simple.

This stresses, however, an important theoretical detail of representation learning that machine learning researchers often fail to make explicit. Learned representations are provided to later stages of processing within the network itself. In a feedforward classifier, the network learns a representation that makes input features linearly separable. This representation is used by the network's output layer, which typically computes something like a softmax regression on this representation. Learned representations are thus representations *for the network*. The high-dimensional nature of feature space precludes our ability to visualize or even imagine the structural properties of a learned representation. A neural network, and not a human inquirer, is the user of learned representations.

We should take care to attend to this distinction, else we are liable to equivocate. In cases of simple statistical learning like the example above, modellers hand-pick a representation in the form of a probability model. The model then finds a mapping between this representation and a dataset by learning the parameter values that make the model a good fit for the data given the chosen representation. Consider another example of this kind. Chhatwal et al. (2009) built a statistical model to estimate breast cancer risk. The modelers mapped a range of diagnostic

[6] Of course, this does not come without limitations (see e.g. Wolpert, 1996). Building models, which learn from finite data, that we can rely on to generalize well is far from trivial.

criteria to discrete variables and performed a logistic regression on these features to predict the presence of breast cancer. They assigned each diagnostic feature to a predictor variable weighted by an unknown parameter. The model then learns how to weight these features to maximize accuracy. Models like this one are straightforward to evaluate through cross-validation. We can even perform tests to evaluate how significant specific features are as breast cancer predictors. It's clear how clinicians and researchers can use the model to make inferences about what diagnostic features are reliable indicators of cancer risk.

Contrast this with a DL model designed to predict the presence of cancer. Esteva et al. (2017) designed a DL model to detect the presence of melanomas and carcinomas from pictures of skin lesions. Unlike the logistic regression example, this model does not have a built-in representation of clinical features that might predict cancer risk. In a regression, the modellers give the model information about the patient in the form of independent variables. These variables function as explicit representations of those clinically relevant features. The DL model receives only a large set of digital images of skin lesions with corresponding labels. The model learns to represent features of skin lesions that predict the presence of different kinds of skin cancer directly from these images. Such representations turn out to be complex patterns of weighted activity distributed throughout the network that facilitate classification at the output layer. A good representation is thus "one that is useful to a supervised predictor" (Bengio et al., 2013, p. 1798). It's difficult to see, on the other hand, how clinicians or researchers could use these learned representations as scientific representations.

Unless human inquirers can use learned representations to make inferences about a target, they just won't qualify as scientific representations. In the section that follows, I argue that learned representations can't satisfy this use requirement. In fact, there is no good way of construing a DL model or its parameters as a scientific representation.

12.4 Learned Representations Are Not Scientific Representations

DL models cannot function as scientific representations. This section lays out the case for this claim in more detail. I consider first whether we can use learned representations in model-based inferences. I show that, even if we can provide a characterization of learned representations, they won't help us make inferences about their targets. We should, therefore, de-center representation in our epistemology of DL models in scientific practice.

It will help to first have an idea of how DL models learn useful representations. Let's continue on with the example of deep, feedforward classifiers. The most successful kind of modern classifier network is called a deep convolutional neural network, or ConvNet (LeCun et al., 1990, 1998, 2015). ConvNets underlie significant breakthroughs in multiple areas of artificial intelligence research made over the last decade. They set the state-of-the-art in object recognition (Krizhevsky et al., 2012) and facial recognition (Taigman et al., 2014) technologies. ConvNets even played a role in the success of AlphaGo, the first artificial intelligence system to defeat a human world champion in the game of Go (Silver et al., 2016). They also find useful applications in science (Raghu & Schmidt, 2020). Modelers can appropriate the same strategy used in image recognition to learn predictive features from scientific data.

ConvNets take innovations made by Fukushima and Miyake (1982) and add end-to-end supervised learning, which was made possible for DNNs by the application of the backpropagation algorithm for training multi-layer neural networks (Rumelhart et al., 1986). The characteristic depth of these networks allows them to learn representations of increasingly abstract, or conceptually framed, features at successive stages of processing (Bengio et al., 2013; Buckner, 2018). As Bengio et al. (2013) put it, DL models "learn to identify and disentangle the underlying explanatory factors hidden in the observed milieu of low-level sensory data." ConvNets involve a sequence of stacked layers that combine two types of computational operations: linear convolutional filtering and non-linear pooling. I'll elaborate briefly on both.

Convolution is a linear algebraic operation whose properties make ConvNets very effective for processing input data that has a grid-like structure (see e.g. Goodfellow et al., 2016, pp. 321–330 for a more precise overview of the technical details). Image processing, for instance, takes digital images as input. A 2-D grid of pixels comprises each image. Pixels are the smallest unit of usable information in a digital image. Each pixel is a vector of red, green, and blue color intensity values at a given location in an image. Convolutional nodes—often called kernels or filters—operate on a rectangular spatial region of the whole grid. They transform the pixel values at a small chunk of the image in a way that maximizes some values while minimizing others. In image recognition and classification tasks, this has the effect of emphasizing salient features present in a given region of the image. A vertical-edge kernel, for instance, uses convolution to maximize pixel values that fall along vertical edges, while minimizing adjacent ones. Each kernel will then pass its output to a rectified linear unit (ReLU), which activates if the signal from the preceding kernel exceeds a given threshold. This ReLU node acts like a kind of "detector," since it sends signal forward through the network only if the preceding kernel detects a certain feature such as a vertical edge. The combination of convolutional filters with ReLUs results in a set of low-level feature detectors that operate on different spatial regions of the input.

Recognizing and classifying objects, however, requires more than just detecting low-level features like vertical edges. We need to be able to detect edges in different orientations and assemble them into composite features like curves and shapes. Networks achieve this through a kind of downsampling. Feature detectors pass their output to a non-linear pooling function. Pooling replaces the output of a detector at a location with a summary of nearby outputs. Max pooling, for example, selects the greatest output within a given rectangular region. Pooling introduces a degree of invariance. This means that the output of a layer becomes more or less insensitive to small translations in the input. This kind of invariance is useful if we want to detect the presence of a feature but don't care exactly where it appeared or how it was oriented. We can thus move from vertical edge detection to detecting edges in general. Invariance is especially important for tasks like object recognition, since we want to accurately identify objects in the face of irrelevant "nuisance variation" like changes in pose, angle, or lighting conditions (Buckner, 2018).

Each layer in a ConvNet implements this sequence of processing. Their joint operations equip ConvNets with the capacity for a kind of hierarchical abstraction, which Buckner (2018) calls transformational abstraction. From a grid of pixel values, the combined effort of convolutional filtering and max pooling allow the network to learn a representation of low-level, visual features of a digital image like edges. By passing this representation through subsequent layers, the network can assemble these low-level features into a representation of increasingly abstract features. Processing thus involves an iterative transformation of available information into a representation that preserves and emphasizes features that are relevant for a given task. Training a ConvNet, then, means finding the right set of weights between layers such that the network can find a multi-level configuration of feature representations that enables the network to perform its task well while also avoiding overfitting idiosyncratic features of its training set. The network needs to find a set of transformations that strike a good balance between context sensitivity and invariance.

Given this characterization, it's hard to see how scientists could fruitfully use learned representations in the course of model-based inferences. Learned representations amount to a learned set of transformations over the feature space of sampled data that emphasizes task-relevant features. Inspecting learned representations can thus at best reveal how specific transformations of input data enable more efficient estimations of the target variable.

To see this, let's consider a recent scientific application of deep learning. The accelerating loss of Arctic sea ice threatens dire consequences for both local ecosystems and global climate conditions. Accurate deterministic models of sea ice concentration exist, but their performance breaks down at even moderate timescales of a few months due to various stochastic elements. To enhance our predictive capabilities, Andersson et al. (2021) developed a probabilistic DL model for sea ice forecasting they call IceNet. IceNet is a stack of twenty-five ConvNets. Each constituent in this ensemble is a U-Net, a specific type of encoder-decoder ConvNet architecture originally developed for medical imaging segmentation (Ronneberger et al., 2015). A U-Net takes grid-like data structures—like digital images—as input and produces a new image as output. Each U-Net consists in the standard, contracting feature-extraction pathway described above, but subsequent layers augment this architecture with a symmetrical, expanding pathway that replaces pooling operations with upsampling to generate precise, localized outputs. A U-Net thus takes an image and extracts predictive features from contiguous regions of the image, then uses those extracted features to output a reconstructed image that facilitates some pre-specified goal, such as producing a precise segmentation map of the original image (Ronneberger et al., 2015). Andersson et al. (2021) trained IceNet to predict seasonal changes in Arctic sea ice concentration (SIC) from a combination of satellite imaging,[7] fifty monthly averaged climate variables, and synthetic climate model data. IceNet outputs SIC forecast maps for the following six months wherein each grid cell within the map is labelled with one of three SIC classes (less than or equal to 15% SIC, greater than 15% but less than 80% SIC, or greater than or equal to 80% SIC). Moreover, it assigns a discrete probability distribution for each labelled cell.[8]

What would it look like to apply a representational account of model-based inference to IceNet? One way would be to treat them analogously to simpler parametric statistical models. On this approach, we would treat a neural network like any other probability model by investigating how the learned parameter values of an estimator represent the complex relationship between predictor features and target variables. We might thus try to consider how the internal structure of the model relates to its capacity to accurately forecast SIC. After all, there is one sense in which the IceNet's internal structure is fully transparent. All DL models are—in a restricted sense—mathematically tractable (Creel, 2020; Leslie, 2019; Lipton, 2018; Zerilli, 2022). No matter how complex, "an algorithmic model is a closed system of effectively computable operations where rules and transformations are mechanically applied to inputs to determine outputs." (Leslie, 2019, p. 41). It's thus, in principle, possible to manually inspect the complete structure of a DL model. Perhaps, then, we can use IceNet's structure to understand how it represents its target phenomenon.

But such naive optimism would soon dissipate once we consider what interpreting the internal structure of IceNet would require. In his recent discussion of ML opacity, Zerilli (2022) distinguishes between several ways that a model can be more or less transparent.[9] Following Lipton (2018), he identifies fathomability with the degree to which someone could immediately grasp the relationship between features in the model and its predictions. Hence, for a model to count as fathomable, "a human should be able to take the input data together with the parameters of the model and in reasonable time step through every calculation required to produce a prediction" (Lipton, 2018, pp. 38–39). The high dimensionality and extreme nonlinearities characteristic of deep learning thus present paradigmatic barriers to fathomability. IceNet—a relatively compact DL model by today's standards—contains roughly forty-four million trainable weights (Andersson et al., 2021, p. 9). Clearly, the sheer number of parameters rules out the possibility of contemplating the structure of the model all at once. In other words, the brute size of DL models makes them unfathomable (Lipton, 2018; Zerilli, 2022).

---

[7] Satellite observations of microwave brightness get converted into a 432×432 grid estimating SIC based on the fractional area covered in sea ice at each grid cell of the image.

[8] The latter two classes are summed to obtain $P(\text{SIC} > 15)$.

[9] See also Creel (2020) for a similar discussion on the varieties of transparency in computational systems.

Yet, it's not the case that intelligibility requires the ability to visualize—mentally or otherwise—a model (cf. de Regt, 2017, Ch. 7). A lack of fathomability thus marks a practical—rather than in principle—obstacle to intelligibility (on essential versus non-essetial epistemic opacity, see also Humphreys, 2004, 2009). For example, linear statistical models can also become so complex as to be practically unfathomable. Yet, such models remain intelligible as scientific representations in the sense that each parameter is assigned a clear, semantically interpretable role within the overall functioning of the model. This is, at least in part, because linear models typically involve plugging predictor variables into a chosen linear link function, which then maps features of the target system to a linear probability model. Linearity here means that every term in the model is either a constant or a feature variable weighted by a tune-able parameter. Preserving linearity is crucial for intelligibility, as it functions as a proxy for conditional monotonicity.[10] That is, conditional on all other explanatory variables, a change in the value of one independent variable begets a predictable, corresponding change in the dependent variable. Without conditional monotonicity, our intuitions about the relationship between features of a target phenomenon tend to fail. Linearity also guarantees that we can apply validation techniques like sensitivity analysis, which allows modellers to determine the degree to which uncertainty in the various parameters of the model accounts for uncertainty in the model's outputs. Such techniques provide a degree of information about patterns of counterfactual dependency between the structure of the model and their target, which is crucial for their explanatory power (cf. Bokulich, 2011; Woodward, 2003). Linearity thus ensures that statistical models remain somewhat intelligible even when they are very complex.

This reveals that DL models are unintelligible in a much deeper sense, one that goes beyond their unfathomable number of parameters. DNNs contain extreme nonlinearities due to the many stacking layers of functions embedded within functions. These many nonlinear, higher-order interactions make it practically impossible to disentangle how a model relates features of the target to produce its predictions. DL models thus lack what Creel (2020) calls "functional transparency," which involves our ability to distinguish the high-level, logical rules involved in the transformation of inputs into outputs. Moreover, a complete mathematical articulation of a trained DNN admits no intelligible semantic interpretation. In other words, there is no way to meaningfully interpret the parameters of the model in terms of the qualitative features of its target system. This is because there is no presumed connection between the model structure and the target system. In scientific practice, the number of neural network parameters often far outstrips the number of quality dimensions found in the data they are modelling. IceNet's parameters, for instance, outnumber its target features by a ratio of close to one

[10] This, however, is not a guarantee. Achen (2005) argues that a serious problem with large regression models—especially in the social sciences—lies in their tendency to sneak in unnoticed nonlinearities, which would undermine the reliability of any conclusions drawn from such analyses.

million-to-one. These parameters correspond to autonomously learned weights that govern the transformation of signal between layers, but those transformations need not have any intrinsic connection to the target phenomenon. Hence, although DL models are, strictly speaking, parametric, they share much in common with nonparametric statistical models in the sense that modellers choose them for the richness of their mathematical structure rather than for any relevance to their target (Desai et al., 2022; Napoletani et al., 2011, 2021; Pietsch, 2015).[11] We, therefore, cannot use the mathematical structure of a trained DL model as an informative representation of the target system.

12.5 Post Hoc Interpretability Methods and Scientific Representations

An alternative possibility remains for someone who wants to preserve a representational account of DL models in science. If we can assess how a model has learned to represent data via a series of transformations between its hidden layers, then we may be able to evaluate what information the network has learned to preserve. If we can identify the informational content within substructures associated with these transformations, then we might be able to use the model's learned representations to understand how the relative importance of different features present in our data leads to better estimations of the target (cf. Boge, 2022; Tamir & Shech, 2023).

In principle, inspecting the activity of hidden layer units might thus shed light on the informative relationship between features in the sample data and the target phenomenon. But in practice, this runs into deep methodological and philosophical difficulties with identifying the content of learned representations (see Chap. 13 by Freiesleben in this volume). As I argued above, characteristic features of deep learning—domain-generality, extreme nonlinearity, and overparameterization—mean that it is impossible to associate particular parameters with qualitative features of a model's target. This applies also to learned representations, since many qualitative features the model learns to represent are not explicitly provided to the network. Instead, the model learns informative patterns directly from the "low-level" sensory data. IceNet, for instance, learns predictive features directly from satellite imaging without having those features explicitly encoded as an input variable. We hope that the model will learn transformations of the data that preserve and exploit genuinely explanatory features of the target phenomenon, but there is no guarantee that this will be the case.

What's more, learned representations are often inscrutable because information is distributed throughout the nodes in a network such that it cannot be broken down

[11] A wealth of recent, theoretical work aims to draw explicit connections between neural networks and nonparametric models (for example, see Dutordoir et al., 2021; Hu et al., 2021; Jacot et al., 2018).

into semantically interpretable units. There is a wealth of research on interpretable machine learning and so-called "explainable AI" (XAI) that aims to ameliorate this problem by understanding how information travels through hidden layer representations (see parts 3 and 4 of this volume for valuable discussion). Some of this work aims to "disentangle" DNNs in ways that make information flow interpretable (Chen et al., 2018, 2021; Olah et al., 2020). Disentanglement is a constraint on nodes in a network such that they acquire "disentangled representations" (Chen et al., 2021). Disentangled representations carry interpretable semantic information about explanatory features of sample data. In a fully disentangled network we would be able to inspect the activity of individual nodes and understand how they represent interpretable features of data. But disentanglement is a difficult, open problem in machine learning research. Given the current state-of-the-art, demanding fully disentangled networks is impossible without loss of performance, which undermines the motivations for using DNNs in the first place.

Additive explanations offer a more promising alternative for the representationalist. Additive explanations refer to a set of XAI tools developed to explain how opaque models produce their outputs. Additive explanations work by deploying a simpler model that explains an opaque model's outputs as a linear function of a set of feature variables (Lundberg & Lee, 2017; Ribeiro et al., 2016). The idea here is that if we can reliably explain the predictions of DL models in terms of these additive feature attributions, then we can understand deep learning systems as representing the relative salience of features in a dataset. To return to the IceNet example, an additive explanation would produce a linear model of IceNet's predictions that would tell us which features in the input data were the most salient factors in its making those particular predictions.

But this view remains untenable for two reasons. On the one hand, additive explanations are not reliable. Prominent methods, such as SHAP (Lundberg & Lee, 2017), LIME (Ribeiro et al., 2016), and breakDown (Staniak & Biecek, 2019) produce explanations of DL models that are not robust, are highly uncertain, have low-fidelity with respect to the original model, and often produce mutually inconsistent explanations (Alvarez-Melis & Jaakkola, 2018; Gosiewska & Biecek, 2019). These results have led some authors to caution against relying on additive explanation methods (Rudin, 2019; Lakkaraju & Bastani, 2020). On the other hand, philosophers of science broadly recognize that non-factive explanations can contribute to the epistemic success of science. While this may be so, it cannot rescue the representationalist view of DL models. This is because additive explanations simply target the wrong phenomenon. Additive explanations are model-centric: their target of understanding is the original model, not the world. They are models of models. Such explanations may thus help us gain modest understanding of how a DL model works, but they will not thereby furnish explanations of phenomena in the world. Additive explanations, as such, fall on the wrong side of Lawler and Sullivan's (2021) distinction between model explanations and model-induced explanations. For techniques like SHAP or LIME, the content of the model is itself an explanation (of the original DL model). But what we want is to use the original model to induce understanding of some target phenomenon.[12]

A final way of considering DL models representationally involves treating them as abstract models of biological neural circuits or the algorithms they compute. We can associate this perspective with the arrival of so-called "neuroconnectionism" on the scene of cognitive computational neuroscience (cf. Cichy & Kaiser, 2019; Doerig et al., 2023; Kieval, 2022; Lindsay, 2021; Stinson, 2020; Yamins & DiCarlo, 2016). This view has recently come under fire from critics who argue that methodological flaws in how neuroscientists assess hierarchical correspondence between DNNs and brains undermines their status as successful models (Bowers et al., 2023; Schaeffer et al., 2022; Sexton & Love, 2022). But these arguments trade on a substantive notion of model correspondence (e.g. in terms of isomorphism) that many representationalist would contest. Neuroconnectionism may thus be a genuine exception to our inability to make representational use of DL models. This leaves the domain of application of DL models severely restricted if we maintain a representationalist perspective. In general, this approach to DL modelling diverges substantially along multiple axes from their broader use within data science. I think it is, therefore, productive to open up non-representational avenues for understanding DL models in science. As I show in the next section, such a view can also further illuminate the use of DL models in neuroscience and even assuage some of the worries raised by critics of neuroconnectionism. I think we thus have good theoretical reasons to extend my account to all uses of DL models in science.

12.6 Understanding DL Modelling as a Skill

I have argued that representationalism is not the way to go when it comes to DL models in science. There just doesn't seem to be any plausible way to use them as scientific representations in most cases. It's thus incumbent upon us to reach for other conceptual tools. I take the first steps towards doing so here.

First, however, I need to face up to a potential objection. Namely, why think DL models are models at all? Perhaps the way that computer scientists use the term 'model' is just a bit of misleading, idiosyncratic lingo. If scientific models are models in virtue of their capacity to represent the world in a particular way, and DL models fail to represent in this way, then maybe they just don't play the role of scientific models. This position is not unheard of in the philosophical literature (Napoletani et al., 2011), for instance, take pains to avoid calling neural networks models, instead referring to them as "techniques." Picking up on this subtlety, Paul Humphreys calls them "non-linear prediction and classification techniques" (Humphreys, 2013, p. 580).

[12] Sullivan (2023) herself makes a similar point in a recent exchange with Tamir and Shech (2023).

This worry is hard to shake. Philosophers of science are not obliged to always take scientists' descriptions of their practice literally. We needn't insist that DNNs are models just because practitioners use the word 'model' to describe them. I do think, however, that philosophers of science should take scientific practice seriously. We should look to modelling practices to understand scientific models. Likewise, we should appreciate how scientists use DL systems to adjudicate their status as scientific models. Attending to model use in scientific practice reveals a great diversity of objects and methods suited for different epistemic aims. It's not clear that we should always consider representing the primary goal of these activities. This calls for broadening the scope of our philosophical analyses of models in science.

Understanding the epistemic significance of DL models in science requires attending to what scientists do with them. Proponents of representationalism, as I see it, are too preoccupied with a conception of scientific knowledge in terms of the accumulation of facts. They locate the epistemic value of models exclusively in their ability to generate new propositional knowledge about aspects of the world. As Frigg and Nguyen (2017, p. 49) put it, "a model can instruct us about the nature of its subject matter if it represents the selected part or aspect of the world that we investigate." The word 'instruct' suggests a picture of the scientist as a passive pupil in whom the model deposits information about its subject. But models are objects embedded in a complex web of activities. When it comes to philosophical analyses of DL models, I want to stress the importance of *modelling* rather than that of models. Recent accounts of models as epistemic artifacts can help to reorient the conversation around modelling activities. I emphasize that DL modelling demands a peculiar set of skills. When combined with specific domain knowledge, these skills help us achieve our epistemic goals. I will argue that a distinctive sort of scientific knowledge consists in the manifestation of DL modelling skills in scientific practice.

Artifactualism provides an account of models that de-centers representation. The basic proposal laid out by Tarja Knuuttila (2005, 2011, 2021a, 2021b) treats models as "epistemic artifacts" that we use in a plethora of ways to gain knowledge. The artifactual approach emphasizes two key elements of modelling: multifunctionality and materiality. By stressing models' multifunctionality, artifactualism "suggests that the links between models and reality are more complex than appears when we focus only on representation" (Knuuttila, 2005, p. 1267). By emphasizing materiality, artifactualism also brings much needed attention to how a medium enforces constraints and enables affordances. Knuuttila draws on insights from Morgan and Morrison (1999), who stress that models can take on "a life of their own" as material mediators between theory and observation (Morgan & Morrison, 1999, p. 18). As material artifacts whose construction proceeds quasi-independently, models have "distinctive ways of functioning. They are not open to all possible interpretations and uses, which simplifies or modifies the cognitive task scientists face in their work" (Knuuttila, 2005, p. 1267).

Artifactualism stresses the usability of models as goal-oriented, epistemic tools. We construct models via diverse material means and use them for some intended purpose. One such purpose can be to represent a target system. In short, some models are tools built for representing the world, but some are made for other purposes such as extending our predictive powers. Hence, artifactualism is consistent with a representational perspective on some models but is broader in scope since it does not require representation as the purpose of all model construction. Artifactualism thus denies representation pride of place in our theory of scientific models. So understood, representation is neither necessary nor sufficient for a tool to count as a model (Sanches de Oliveira, 2022). Following Sanches de Oliveira (2022), I suggest that we can get at the core of this dispute over representationalism in the following way. First, the ontology of models is function dependent. In a slogan, models are what they do. Representationalism insists that what models do is represent the world. Artifactualism suggests instead that models facilitate our epistemic activities or ends through their diverse, material affordances.

Artifactualism better accounts for DL models in scientific practice, especially their emphasis on useful predictions. This comes out most clearly through Knuuttila's brief description of the results-orientedness of modelling activities (Knuuttila, 2011, p. 268). On this point, she echoes Paul Humphreys (2004), who stresses the "enormous importance of a relatively small number of computational templates" (Humphreys, 2004, p. 68). Such templates get uptake throughout the sciences due to their ease of use and tendency to produce dependable results. As Knuuttila explains,

> ...the starting point is often the output and effects that models are supposed to produce. Instead of directly trying to represent some selected aspects of a given target system—as is conventionally assumed—modellers often proceed in a roundabout way, seeking to build hypothetical systems in the light of the anticipated results or of certain general features of the phenomena they are supposed to exhibit (Knuuttila, 2011, p. 268).

The practices of DL modelling reflect just this kind of opportunistic appropriation of computational templates. IceNet, for instance, seizes on a successful model architecture originally designed for medical imaging to handle climate data. We have dynamical models that adequately represent seasonal patterns of Arctic sea ice concentration, but these turn out to be inadequate for our non-representational purposes. Hence, Andersson et al. (2021) adapted existing U-Net architectures to meet the needs that a representational model could not. IceNet is just one of many readily available cases of such opportunism. These kinds of model transfers and extensions to new domains are only possible due to the generality of deep learning architectures; DL models are not crafted to represent a particular target, but we can appropriate them to learn useful properties of data in any domain. As Knuuttila puts it "the overall usability of computational templates is based on their generality and the observed similarity between different phenomena" (Knuuttila, 2011, p. 268). But doing this well requires a degree of joint understanding of how different DL model architectures behave, the kinds of data structures on which they are effective, and how to recognize that a particular problem can be operationalized in a way such that a DL template can be suitably applied. I argue that, for this reason, we should resist thinking of DL modelling activities in terms of what Alisa Bokulich (2011) calls "phenomenological" models, which involve an ad hoc fitting of a model to data that merely "saves the phenomena" (Bogen & Woodward, 1988). In what follows, I develop a sketch of how DL models—construed as domain-general epistemic tools—can scaffold understanding through the manifestation of the skills required to construct and apply them.

Skill provides a natural way to conceptualize this epistemic scaffolding. Ellen Fridland (2020, 2021) has developed an account of skilled action in terms of strategic control structures that proves illuminating. On her account, skills are "functions from intentions to controlled, successful actions, where the functional transformations from intention to action are implemented by control structures that have been developed through practice" (Fridland, 2020, p. 247). For Fridland, strategic control involves the successful execution of intentions, which are themselves control structures acquired through practice that both automatize the motor components of skilled action and organize our practical plans. These structures sit at the intermediary between agent-level, abstract representations and low-level sensorimotor processes, and they are instrumentally rational states (Fridland, 2020, p. 250). Moreover, attention and motor control guide the successful instantiation of these control structures in skilled actions.

Acquiring skills often involves apprenticeships and begins with the imitation of more skilled practitioners. Fridland (2013) argues persuasively that through imitation, we take up an intrinsic interest in the means of an observed intentional action. Through this "means-centric orientation" (Fridland, 2013, p. 205) we acquire a stance that's required for us to refine our skills in such a way that facilitates genuine understanding of what we are doing. New knowledge manifests through the understanding of how our means relate with their consequences. As Michael Polanyi put it, learning to use a tool or a probe involves "the transformation of the tool or probe into a sentient extension of our body" (Polanyi, 1966, p. 17). In doing so, we gain the capacity to fully integrate its affordances into our own cognition. Truly understanding a model thus comes about only by treating it as an extension of ourselves. Only then do we cease to merely copy the activities of others and instead integrate the model as a subsidiary means to our own epistemic ends.

We can apply this picture of skill to DL modelling activities. Constructing a working DL model requires skillful manipulation of multiple interconnected elements. Having identified an appropriate problem, we can conceive of the model construction process in terms of three interrelated sets of decision procedures. First, we have choices related to data, like sampling, labeling and curating observations into datasets, pre-processing, and visualization. Next, we must make various choices about the structure of our model itself. This involves decisions about the architecture of the model, whether to appropriate and modify existing computational templates, how to tune hyperparameters, what kind of learning paradigm to use, how to structure the task goals, how to regularize and optimize for performance, and how to diagnose and correct systematic errors or biases. Finally, we implement various kinds of validation strategies like testing the model's performance on held out data. What's more, this entire process is highly iterative in the sense that we may use the outcome of decisions at any one stage to further modify our choices at other stages. Recognizing and diagnosing failures and iterating on the model is thus essential to effective model construction, but also an exceptionally complex matter. Doing so well requires understanding how interactions at each choice point affect the overall performance of the model. The skills required for DL modelling must be learned through practice via engagement with various model architectures across a range of contexts. This engagement may start out as a brute imitation of more skilled modelers, but with an intrinsic interest in understanding how skilled modelers achieve their ends. Over time this develops into a genuine, intelligent skill through which a modeler can recognize subtle ways that model structure relates to performance on real world phenomena.

Using a model to learn about the world depends on our ability to construct a model that is adequate for our purposes. We need to understand how to relate the model to our epistemic aims. Using DL models in science requires knowing how to train a DNN, which means possessing modeling skills. Unlike many classic examples of theoretical models, however, DL modelling skills do not depend on significant, domain-specific knowledge about the target system. DL modelling requires understanding how to flexibly modify and apply domain-general computational architectures to suitable data structures. Skilled modelers can build a DL model that will work well on a given data structure whilst knowing very little about the content of that data. Take, for instance, the toy example from Sect. 12.4, predicting future weather patterns in Cambridge from historic weather data. Suppose that I had access to the .csv file containing the very same historic weather data, but I had no knowledge of what that data was about except that one column is a target label that I want to predict. If I was tasked with predicting future values of that target, I could build a simple, feedforward neural network that would perform better than the toy model described in Sect. 12.4. And I could do this without even knowing what I am predicting. This is the sense of domain-generality that sets DL modelling skills apart from other kinds of modelling skills in science.

The domain-generality of deep learning means that the process of learning DL modelling skills makes the models themselves into the object of inquiry. The means of inquiry become an end in themselves. But we really want to understand how to use DL models to learn about the world. This requires moving away from a means-centric orientation, which requires fully integrating modelling skills into the broader aims of our inquiry. Once integrated, the object of inquiry shifts. In this context, DL models become tools that researchers build and evaluate in terms of their capacity to generate understanding of an aspect of the world. Modelling skills thus become an important tacit dimension of scientific knowledge production, facilitating more effective inquiry by directing our attention towards aspects of a problem. DL modelling skill thus reflects a deep and genuine form of understanding that underpins successful scientific inquiry.

Generating new scientific knowledge with DL models, however, requires integrating domain-general modelling skills with domain-specific, theoretical knowledge. Successful DL modelling in science requires more than just knowing how to get a model to perform well on validation data. It also requires knowing how to use a model to pursue a particular epistemic aim. This requires the ability to apply one's modelling skills in pursuit of a well-posed scientific problem. Doing so involves what Henk de Regt (2017) calls pragmatic understanding. De Regt clarifies this sense of understanding in terms of intelligibility, which he describes as the cluster of qualities that facilitate the use of a theory (cf. de Regt, 2017, p. 40). De Regt's notion of intelligibility emphasizes that pragmatic understanding promotes the fruitfulness of our epistemic activities. Scientific progress requires fruitfulness, because the ability to apply a concept or theory is necessary for generating predictions, crafting experiments, and producing new explanations (cf. Stuart, 2016, p. 31). Pragmatic understanding is what makes our epistemic activities go well in scientific practice. We need intelligible theories to make good use of DL models in science; drawing out inferences about the world using a DL model requires a confluence of modelling skills and domain knowledge. This requires pragmatic understanding, the ability to use theoretical knowledge to construct a model that is adequate for purpose. Pragmatic understanding enables the transition from learning a good estimator of patterns in data to model-based inferences about the world.

12.7 Conclusion

This chapter began by distinguishing representations in machine learning from the notion of representation that figures prominently in the philosophy of model-based science. I argued that the former cannot do the epistemic work required of the latter. Hence, I suggest that the represenationalist idiom is liable to produce a stagnant philosophy of machine learning in science. I suggest instead that artifactualism, couched in account of skilled practice, is a more promising starting point.

It takes a plethora of skills to build models that work well. These skills are not mysterious or illusory. They are implementations of strategic control structures acquired through practice. Skills involve flexible scripts of action. Modelling skills scaffold genuine understanding because we need to be able to grasp the complex relations between our means and methods and our desired epistemic ends. When we marry DL modelling skills with pragmatic understanding, DL models become fruitful tools of inquiry. This initial sketch makes some progress towards a richer account of what artifactualists mean when they talk of modelling as a material engagement with epistemic tools. We can apply this picture to better grasp the contribution of deep learning to genuine, world-directed understanding.

Acknowledgments For their helpful comments and constructive feedback, I'm grateful to Florian Boge, Hasok Chang, Eamon Duede, Juan Durán, Timo Freiesleben, Marta Halina, Damon Kutzin, Scott Partington, Mona Wandrey, and Oscar Westerblad.

References

Achen, C. H. (2005). Let's put garbage-can regressions and garbage-can probits where they belong. *Conflict Management and Peace Science*, *22*(4), 327–339.

Alvarez-Melis, D., & Jaakkola, T. S. (2018). On the robustness of interpretability methods. *ArXiv*.
Andersson, T. R., Hosking, J. S., Pérez-Ortiz, M., Paige, B., Elliott, A., Russell, C., Law, S., Jones, D. C., Wilkinson, J., Phillips, T., Byrne, J., Tietsche, S., Sarojini, B. B., Blanchard-Wrigglesworth, E., Aksenov, Y., Downie, R., & Shuckburgh, E. (2021). Seasonal Arctic sea ice forecasting with probabilistic deep learning. *Nature Communications*, *12*(1), 1–12.
Bengio, Y., Courville, A., & Vincent, P. (2013). Representation learning: A review and new perspectives. *IEEE Transactions on Pattern Analysis and Machine Intelligence*, *35*(8), 1798–1828.
Boge, F. J. (2022). Two dimensions of opacity and the deep learning predicament. *Minds and Machines*, *32*(1), 43–75.
Bogen, J., & Woodward, J. (1988). Saving the phenomena. *The Philosophical Review*, *XCVII*(3), 303–352.
Bokulich, A. (2011). How scientific models can explain. *Synthese*, *180*(1), 33–45.
Bokulich, A. (2020). Towards a taxonomy of the model-ladenness of data. *Philosophy of Science*, *87*(5), 793–806.
Bowers, J. S., Malhotra, G., Dujmović, M., Llera Montero, M., Tsvetkov, C., Biscione, V., ... Blything, R. (2023). Deep problems with neural network models of human vision. *Behavioral and Brain Sciences, 46*, e385. https://doi.org/10.1017/S0140525X22002813
Buckner, C. (2018). Empiricism without magic: Transformational abstraction in deep convolutional neural networks. *Synthese*, *195*(12), 5339–5372.
Buckner, C. (2019). Deep learning: A philosophical introduction. *Philosophy Compass*, *14*(10), 1–19.
Callender, C., & Cohen, J. (2006). There is no special problem about scientific representation. *Theoria-Revista De Teoria Historia Y Fundamentos De La Ciencia*, *21*(55), 67–85.
Chang, H. (2004). *Inventing temperature: Measurement and scientific progress*. Oxford University Press.
Chen, C., Chen, Z., Huang, H., Rudin, C., Semenova, L., & Zhong, C. (2021). Interpretable machine learning: Fundamental principles and 10 grand challenges. *arXiv*, pp. 1–80.
Chen, T. Q., Li, X., Grosse, R., & Duvenaud, D. (2018). Isolating sources of disentanglement in variational autoencoders. In *6th International Conference on Learning Representations, ICLR 2018 - Workshop Track Proceedings*, (NeurIPS).
Chhatwal, J., Alagoz, O., Lindstrom, M. J., Kahn, C. E., Shaffer, K. A., & Burnside, E. S. (2009). A logistic regression model based on the national mammography database format to aid breast cancer diagnosis. *American Journal of Roentgenology*, *192*(4), 1117–1127.
Cichy, R. M., & Kaiser, D. (2019). Deep neural networks as scientific models. *Trends in Cognitive Sciences*, *23*(4), 305–317.
Contessa, G. (2007). Scientific representation, interpretation, and surrogative reasoning. *Philosophy of Science*, *74*(1), 48–68.
Creel, K. A. (2020). Transparency in complex computational systems. *Philosophy of Science*, *87*(4), 568–589.
de Regt, H. W. (2017). *Understanding scientific understanding* (Vol. 1). Oxford University Press.
Desai, J., Watson, D., Wang, V., Taddeo, M., & Floridi, L. (2022). The epistemological foundations of data science: A critical analysis. *SSRN Electronic Journal*, *(iv)*, 1–26.
Doerig, A., Sommers, R. P., Seeliger, K. et al. (2023). The neuroconnectionist research programme. *Nature Reviews Neurosciences, 24*, 431–450. https://doi.org/10.1038/s41583-023-00705-w
Dutordoir, V., Hensman, J., van der Wilk, M., Ek, C. H., Ghahramani, Z., & Durrande, N. (2021). Deep neural networks as point estimates for deep gaussian processes. *Advances in Neural Information Processing Systems*, *12*(NeurIPS), 9443–9455.
Elgin, C. Z. (2004). True enough. *Philosophical Issues*, *14*(1), 113–131.
Esteva, A., Kuprel, B., Novoa, R. A., Ko, J., Swetter, S. M., Blau, H. M., & Thrun, S. (2017). Dermatologist-level classification of skin cancer with deep neural networks. *Nature*, *542*(7639), 115–118.

Freiesleben, T. (2024). Artificial neural nets and the representation of human concepts. In J. M. Duran & G. Pozzi (Eds.), *Philosophy of science for machine learning: Core issues and new perspectives*. Synthese Library. Springer.
Fridland, E. (2013). Imitation, skill learning, and conceptual thought: An embodied, developmental approach. In L. Swan (Ed.), *Origins of mind: Spring book series in biosemantics* (pp. 203–224). Springer.
Fridland, E. (2020). The nature of skill: Functions and control structures. In E. Fridland & C. Pavese (Eds.), *The Routledge handbook of philosophy of skill and expertise. Routledge handbooks in philosophy* (pp. 245–257). Routledge; New York, NY: Routledge, 2021.
Fridland, E. (2021). Skill and strategic control. *Synthese, 199*(3–4), 5937–5964.
Frigg, R., & Hartmann, S. (2020). Models in science. In E. N. Zalta (Ed.), *The Stanford encyclopedia of philosophy*. Metaphysics Research Lab, Stanford University, spring 2020 edition.
Frigg, R., & Nguyen, J. (2017). Models and representation. In *Springer handbook of model-based science* (pp. 49–102). Springer International Publishing.
Fukushima, K., & Miyake, S. (1982). Neocognitron: A new algorithm for pattern recognition tolerant of deformations and shifts in position. *Pattern Recognition, 15*(6), 455–469.
Ghahramani, Z. (2015). Probabilistic machine learning and artificial intelligence. *Nature, 521*(7553), 452–459.
Giere, R. N. (2004). How models are used to represent reality. *Philosophy of Science, 71*(5), 742–752.
Giere, R. N. (2010). An agent-based conception of models and scientific representation. *Synthese, 172*(2), 269–281.
Godfrey-Smith, P. (2006). The strategy of model-based science. *Biology and Philosophy, 21*(5), 725–740.
Goodfellow, I., Bengio, Y., & Courville, A. (2016). *Deep learning*. MIT Press.
Gosiewska, A., & Biecek, P. (2019). Do not trust additive explanations. *arXiv preprint arXiv:1903.11420*.
Hornik, K., Stinchcombe, M., & White, H. (1989). Multilayer feedforward networks are universal approximators. *Neural Networks, 2*(5), 359–366.
Hu, T., Wang, W., Lin, C., & Cheng, G. (2021). Regularization matters: A Nonparametric Perspective on Overparametrized Neural Network. Proceedings of The 24th International Conference on Artificial Intelligence and Statistics. *Proceedings of Machine Learning Research, 130*, 829–837 Available from https://proceedings.mlr.press/v130/hu21a.html
Humphreys, P. (2004). *Extending ourselves: Computational science, empiricism, and scientific methods*. Oxford University Press.
Humphreys, P. (2009). The philosophical novelty of computer simulation methods. *Synthese, 169*(3), 615–626.
Humphreys, P. (2013). Data analysis: Models or techniques? *Foundations of Science, 18*(3), 579–581.
Jacot, A., Gabriel, F., & Hongler, C. (2018). Neural tangent kernel: Convergence and generalization in neural networks. In *Advances in Neural Information Processing Systems* (pp. 8571–8580).
Kieval, P. H. (2022). Mapping representational mechanisms with deep neural networks. *Synthese, 200*(196), 1–25.
Knuuttila, T. (2005). Models, representation, and mediation. *Philosophy of Science, 72*(5), 1260–1271.
Knuuttila, T. (2011). Modelling and representing: An artefactual approach to model-based representation. *Studies in History and Philosophy of Science Part A, 42*(2), 262–271.
Knuuttila, T. (2021a). Imagination extended and embedded: Artifactual versus fictional accounts of models. *Synthese, 198*, 5077–5097.
Knuuttila, T. (2021b). Models, fictions and artifacts. In *Language and scientific research* (pp. 199–220). Springer International Publishing.

Krizhevsky, B. A., Sutskever, I., & Hinton, G. E. (2012). ImageNet classification with deep convolutional neural networks. In F. Pereira, C. J. C. Burges, L. Bottou, & K. Q. Weinberger (Eds.), *Advances in Neural Information Processing Systems* (Vol. 25, pp. 1097–1105). Curran Associates, Inc.

Lakkaraju, H., & Bastani, O. (2020). "How do I fool you?": Manipulating user trust via misleading black box explanations. In *AIES 2020 - Proceedings of the AAAI/ACM Conference on AI, Ethics, and Society* (pp. 79–85).

Lawler, I., & Sullivan, E. (2021). Model explanation versus model-induced explanation. *Foundations of Science, 26*(4), 1049–1074.

LeCun, Y., Bengio, Y., & Hinton, G. (2015). Deep learning. *Nature, 521*(7553), 436–444.

LeCun, Y., Boser, B., Denker, J. S., Howard, R. E., Habbard, W., Jackel, L. D., & Henderson, D. (1990). Handwritten digit recognition with a back-propagation network. In *Advances in neural information processing systems 2* (pp. 396–404). Morgan Kaufmann Publishers Inc.

LeCun, Y., Bottou, L., Bengio, Y., & Haffner, P. (1998). Gradient-based learning applied to document recognition. *Proceedings of the IEEE, 86*(11), 2278–2323.

Leonelli, S. (2019). What distinguishes data from models? *European Journal for Philosophy of Science, 9*(2), 22.

Leslie, D. (2019). Understanding artificial intelligence ethics and safety. *arXiv preprint arXiv:1906.05684.*

Lindsay, G. W. (2021). Convolutional neural networks as a model of the visual system: Past, present, and future. *Journal of Cognitive Neuroscience, 33*(10), 2017–2031.

Lipton, Z. C. (2018). The mythos of model interpretability. *Communications of the ACM, 61*(10), 35–43.

Lundberg, S. M., & Lee, S. I. (2017). A unified approach to interpreting model predictions. In *Advances in Neural Information Processing Systems* (pp. 4766–4775).

Morgan, M. S., & Morrison, M. (1999). *Models as mediators: Perspectives on natural and social science*. Cambridge University Press.

Napoletani, D., Panza, M., & Struppa, D. (2021). The agnostic structure of data science methods. *Lato Sensu: Revue de la Société de philosophie des sciences, 8*(2), 44–57.

Napoletani, D., Panza, M., & Struppa, D. C. (2011). Agnostic science. Towards a philosophy of data analysis. *Foundations of Science, 16*(1), 1–20.

Olah, C., Cammarata, N., Schubert, L., Goh, G., Petrov, M., & Carter, S. (2020). Zoom in: An introduction to circuits. *Distill, 5*, e00024.

Parker, W. S. (2020). Model evaluation: An adequacy-for-purpose view. *Philosophy of Science, 87*, 457–477.

Pietsch, W. (2015). Aspects of theory-ladenness in data-intensive science. *Philosophy of Science, 82*(5), 905–916.

Polanyi, M. (1966). *The tacit dimension*. The University of Chicago Press.

Potochnik, A. (2017). *Idealization and the aims of science*. University of Chicago Press.

Raghu, M., & Schmidt, E. (2020). A survey of deep learning for scientific discovery. *arXiv*, 1–48.

Ribeiro, M. T., Singh, S., & Guestrin, C. (2016). "Why should I trust you?" Explaining the predictions of any classifier. In *NAACL-HLT 2016 - 2016 Conference of the North American Chapter of the Association for Computational Linguistics: Human Language Technologies, Proceedings of the Demonstrations Session* (pp. 97–101).

Ronneberger, O., Fischer, P., & Brox, T. (2015). U-Net: Convolutional networks for biomedical image segmentation. In *Lecture notes in computer science (including subseries Lecture notes in artificial intelligence and lecture notes in bioinformatics)* (Vol. 9351, pp. 234–241).

Rudin, C. (2019). Stop explaining black box machine learning models for high stakes decisions and use interpretable models instead. *Nature Machine Intelligence, 1*(5), 206–215.

Rumelhart, D. E., Hinton, G. E., & Williams, R. J. (1986). Learning representations by back-propagating errors. *Nature, 323*(6088), 533–536.

Sanches de Oliveira, G. (2022). Radical artifactualism. *European Journal for Philosophy of Science, 12*(2), 36.

Schaeffer, R., Khona, M., & Fiete, I. R. (2022). No free lunch from deep learning in neuroscience: A case study through models of the Entorhinal-Hippocampal circuit. *bioarXiv*, (Icml):2022.08.07.503109.
Sexton, N. J., & Love, B. C. (2022). Reassessing hierarchical correspondences between brain and deep networks through direct interface. *Science Advances*, *8*(28), eabm2219.
Silver, D., Huang, A., Maddison, C. J., Guez, A., Sifre, L., Van Den Driessche, G., Schrittwieser, J., Antonoglou, I., Panneershelvam, V., Lanctot, M., Dieleman, S., Grewe, D., Nham, J., Kalchbrenner, N., Sutskever, I., Lillicrap, T., Leach, M., Kavukcuoglu, K., Graepel, T., & Hassabis, D. (2016). Mastering the game of Go with deep neural networks and tree search. *Nature*, *529*(7587), 484–489.
Staniak, M., & Biecek, P. (2019). Explanations of model predictions with live and breakDown packages. *R Journal*, *10*(2), 395–409.
Stinson, C. (2020). From Implausible Artificial Neurons to Idealized Cognitive Models: Rebooting Philosophy of Artificial Intelligence. *Philosophy of Science*, *87*(4), 590–611. https://doi.org/10.1086/709730
Stuart, M. T. (2016). Taming theory with thought experiments: Understanding and scientific progress. *Studies in History and Philosophy of Science Part A*, *58*, 24–33.
Suárez, M. (2004). An inferential conception of scientific representation. *Philosophy of Science*, *71*(5), 767–779.
Suárez, M. (2010). Scientific representation. *Philosophy Compass*, *5*(1), 91–101.
Suárez, M. (2015). Deflationary representation, inference, and practice. *Studies in History and Philosophy of Science Part A*, *49*, 36–47.
Sullivan, E. (2022). Understanding from machine learning models. *The British Journal for the Philosophy of Science*, *73*(1), 109–133.
Sullivan, E. (2023). Link uncertainty, implementation, and ML opacity: A reply to Tamir and Shech. In *Scientific understanding and representation* (pp. 341–345). Routledge.
Taigman, Y., Yang, M., Ranzato, M., & Wolf, L. (2014). DeepFace: Closing the gap to human-level performance in face verification. In *Proceedings of the IEEE Computer Society Conference on Computer Vision and Pattern Recognition* (pp. 1701–1708).
Tal, E. (2017). A model-based epistemology of measurement. In *Reasoning in measurement* (pp. 233–253). Routledge.
Tamir, M., & Shech, E. (2023). Understanding from deep learning models in context. In I. Lawler, K. Khalifa, & E. Shech, (Eds.), *Scientific understanding and representation: Modeling in the physical sciences* (pp. 323–340). Routledge.
van Fraassen, B. C. (2008). *Scientific representation: Paradoxes of perspective*. Oxford University Press.
Weisberg, M. (2007). Three kinds of idealization. *Journal of Philosophy*, *104*(12), 639–659.
Weisberg, M. (2013). *Simulation and similarity: Using models to understand the world*. Oxford University Press.
Wolpert, D. H. (1996). The lack of a priori distinctions between learning algorithms. *Neural Computation*, *8*(7), 1341–1390.
Woodward, J. (2003). *Making things happen: A theory of causal explanation*. Oxford University Press.
Yamins, D. L., & DiCarlo, J. J. (2016). Using goal-driven deep learning models to understand sensory cortex. *Nature Neuroscience*, *19*(3), 356–365.
Zerilli, J. (2022). Explaining machine learning decisions. *Philosophy of Science*, *89*(1), 1–19.

Phillip Hintikka Kieval is a PhD candidate in the Department of History and Philosophy of Science at the University of Cambridge where he is funded by a Gates Cambridge Scholarship. He is also a Student Fellow at the Leverhulme Centre for the Future of Intelligence. Phillip's current research broadly concerns the history and philosophy of machine learning and applied statistics in science and public life.

Chapter 13
Artificial Neural Nets and the Representation of Human Concepts

Timo Freiesleben

Abstract What do artificial neural networks (ANNs) learn? The machine learning (ML) community shares the narrative that ANNs must develop abstract human concepts to perform complex tasks. Some go even further and believe that these concepts are stored in individual units of the network. Based on current research, I systematically investigate the assumptions underlying this narrative. I conclude that ANNs are indeed capable of performing complex prediction tasks, and that they may learn human and non-human concepts to do so. However, evidence indicates that ANNs do not represent these concepts in individual units.

Keywords Concepts · Representation · Artificial neural networks · Deep learning · XAI · Natural kinds

13.1 Introduction

Large parts of the machine learning (ML) community seems to share a remarkable narrative about the representational qualities of Artificial Neural Networks (ANNs) (Olah et al., 2020; LeCun et al., 2015):

1. ANNs can perform tasks that only humans could perform in the past, such as classifying videos or producing human-like text.
2. When humans perform these tasks, they rely on *concepts*, that means higher order representations. Humans perceive images as composed of *objects*. They think of texts as following abstract *grammatical rules*.
3. To do the same tasks as humans, ANNs must have developed the same concepts as humans.

T. Freiesleben (✉)
Cluster of Excellence: "Machine Learning: New Perspectives for Science", University of Tübingen, Tübingen, Germany
e-mail: timo.freiesleben@uni-tuebingen.de

J. M. Durán, G. Pozzi (eds.), *Philosophy of Science for Machine Learning*, Synthese Library 527, https://doi.org/10.1007/978-3-032-03083-2_13

4. These concepts are represented in the weights, units, filters, channels, or layers of the ANNs, depending on the model architecture and who you ask.

And indeed, people started searching for evidence in favor of this narrative. They found that early layers in trained Convolutional Neural Nets (CNNs) closely reassemble or even exactly match edge detectors designed by human engineers (Olah et al., 2017). Later units in CNNs show co-activation with some complex concepts found in images, such as wheels or grass (Bau et al., 2017). Text classifiers have neurons that are connected to the sentiment in texts (Radford et al., 2017) and Large Language Models (LLMs) appear to store factual knowledge largely in mid-layer modules (Meng et al., 2022) and have neurons connected to space and time (Gurnee and Tegmark, 2023).

If the narrative were correct, it would have great practical and philosophical significance: It would elucidate why ANNs work so well—they simply develop the right concepts (LeCun et al., 2015); It would show that human concepts, including our natural kinds (Bird & Tobin, 2023), indeed arise naturally from efficiently solving complex prediction tasks (Watson, 2023); Moreover, it would elucidate how empiricist philosophy can explain human cognition (Buckner, 2018).

Yet, I remained skeptical: Aren't neural nets just estimators of conditional densities? Are our concepts really so natural that ANNs have to learn the exact same concepts? Wasn't the point of using ANNs to distribute meaning rather than store it in individual units?

I therefore started my journey to examine the ML literature on concept representation in ANNs. I found that ML researchers remain vague on what they mean by the representation of learned human concepts, and hence they do not provide testable hypotheses, but rather anecdotal evidence (Leavitt & Morcos, 2020a). I have structured my search, and consequently this chapter, along the three key assumptions I found underlying the narrative above. I already give here my conclusions about these assumptions:

- **Assumption I, *ANNs work well*:** Even though the theoretical grounds are still shaky, there is strong empirical evidence that ANNs are solid approximations of the best possible predictors in static data settings (See Sect. 13.3).
- **Assumption II, *ANNs learn human concepts*:** There is both evidence in favor of this claim (e.g. transfer learning) and evidence against it (e.g. adversarial examples). It is likely that ANNs learn both some human and some non-human concepts (See Sect. 13.4).
- **Assumption III, *ANNs represent these learned human concepts by single units*:** Although some evidence could be interpreted in favor of the claim, it turns out to be very weak. Individual units that are supposed to represent concepts often only weakly coactivate with them and do not share their functional role (See Sect. 13.5).

My Goals with This Work
First, to provide clarity on the representational qualities of current ANNs based on evidence. Second, to provide ML researchers with conceptual guidance on the notion of the *representation of learned human concepts* with the hope of helping them form falsifiable hypotheses and gather further evidence for assumptions I–III.

In this work, I will *not* touch on the question whether ANNs are in any way similar to biological neural nets, nor will I discuss the question of how humans store or process concepts. Readers interested in these questions are pointed to Marblestone et al. (2016), Buckner (2018), Bowers et al. (2022) or Grujičić (2024).

13.2 Background

In this section, I will first introduce what I mean by the representation of learned human concepts. Second, I will discuss other work concerned with concept representation in ANNs. And third, I will provide a short primer on ANNs.

13.2.1 Introducing Terms

What concepts precisely are is widely debated and rarely agreed upon in the literature (Frigg & Nguyen, 2021). To avoid entering this minefield, I will focus in this chapter on the widely accepted roles of concepts rather than on their ontology.

Roles of Concepts
Concepts take an *abstractional* role (Yee, 2019). To some objects in the world a concept applies, to others it does not (Peacocke, 1992). Take the concept of a *metro*. Both the Munich and the London metro are instantiations of the concept metro. Having learned a concept also means being able to decide (relatively) competently whether a concept applies to an object or not.[1]

Concepts take a *functional* role (Lalumera, 2010). Having learned a concept means being able to perform certain tasks and entertain certain properties (Keil, 1992). For example, a kid has learned the concept of addition if she can give the correct answer in a series of moderately difficult addition problems she has never faced before. Clearly, learning a concept comes in degrees, one may have learned a

[1] There are two ways to define the abstractional role of concepts—extensionally and intensionally. *Extensional:* Definition in terms of all the objects or ideas to which the concept applies. I can define *metro* by pointing to all metro systems on earth. *Intensional:* Definition by the properties that an object or idea must have for the concept to apply. One can define what a tree is by specifying a set of conditions for being a tree, e.g. having a stem, growing roots, being made out of wood, etc. This requires further, more basic concepts.

concept more or less well. The kid may be able to perform addition when positive numbers are involved, but not when one of the numbers is negative.

Other roles could be pushed here, like the compositionality of concepts (Del Pinal, 2016; Hampton & Jönsson, 2012). We can, for instance, in some contexts define a metro line via its stops and the stops via their locations. For this chapter, I restrain myself to the abstractional and functional role, which I both consider as essential roles of concepts. It is extremely difficult and perhaps impossible to give necessary and sufficient conditions for the concept *concept*, so I will stick to these agreeable family resemblances (Wittgenstein, 2010; Hampton, 2006).

Human Concepts

A concept is a *human concept*[2] if there is at least one human being who uses the concept and can communicate effectively about it. Metro is a human concept as I and probably other human beings can use the concept and communicate about it.

Indeed, what counts as a human concept is constantly evolving (Keil, 1992). New concepts emerge as new objects and new goals are developed by humans. One hundred years ago *smartphone* was not a human concept, today every kid learns this concept. Similarly, 100,000 years ago, people probably had rich conceptual frameworks related to hunting that we no longer understand.

In the philosophical literature there is the idea of natural kinds—i.e. concepts that exist in the world independently of human goals (Bird & Tobin, 2023). These concepts are considered particularly suitable for drawing scientific inferences, examples of supposedly natural kinds are the elements in the periodic table or elementary particles in physics. Quine (1969) believed that humans have an innate capacity to judge similarities between objects and find natural kinds. If ANN models develop the same kind of concepts as we do, this would suggest that humans also develop concepts to solve complex prediction tasks, rather than having an innate ability to find natural kinds (Williams, 2018).

Representing Learned Concepts in a Model

A prediction model represents[3] a learned concept C in model part U, if

(i) *Coactivation:* the model part U coactivates when exhibited to instances of the concept C.
(ii) *Functional role:* removing or manipulating the model part U takes away or changes the functional role of the concept C in the prediction.

Model parts can be variables, parameters, mathematical relations, or compositions thereof. As should be evident, representation is meant here as in scientific representations rather than mental representations (Frigg & Nguyen, 2021; Duede,

[2] Animals also learn concepts, both concepts similar to humans and some animal specific (Andrews & Beck, 2017).

[3] Perfect representation is clearly the exception, much more often models only approximately represent (See Chap. 14 by Manganini and Primiero in this volume).

2023). Newton's laws of gravity represent the concept of mass in the variable m, since an increase in the mass of an object leads to the prediction of an increased gravitational pull on other bodies. An ANN represents the concept of addition in a model part U when each time it performs an addition, the part is activated and removing the unit results in ANNs no longer being able to perform common addition problems that they were previously capable of solving.

13.2.2 Related Work

There are both, philosophical papers and technical papers on ML that touch upon the question of what ANNs learn and represent.

Proponents that ANNs Represent Concepts in Individual Units
Olah et al. (2020) are maybe the strongest proponents of Assumption III. Based on examining many neurons via activation maximization, they conclude that units in ANNs store individual disentangled concepts and that the algebraic operations of neurons can be seen as performing logical reasoning with these concepts. They moreover push the claim that there is a universal set of concepts that must be developed to perform certain tasks. Similarly, Bau et al. (2017) believe that networks can ultimately be dissected into hierarchies of individual concepts, from shapes and colors to scenes and objects. Räz (2023) discusses both their methods (I will introduce these methods in more detail in Sect. 13.5), namely network dissection (Bau et al., 2017) and activation maximization (Olah et al., 2017), and moreover testing with concept activation vectors (TCAV) by Kim et al. (2018). He believes that the methods provide compelling evidence for the representation of concepts in individual units, but argues for the demand of additional methods and applying a variety of methods for identifying concepts in ANNs.

I will argue why we should be skeptical of the claim that ANNs store learned human concepts in individual units, in particular I will question how conclusive the evidence obtained from the above methods really is.

Skeptics if ANNs Represent Concepts in Individual Units
Freiesleben et al. (2022) argue against the view that ANNs represent human concepts in individual units and instead argue that ANNs only represent conditional probabilities. Both options will be discussed below, together with the third option that ANNs learn human concepts but store them holistically. Duede (2023) argues that the confusion about representation in ANNs arises from mixing functional and relational conceptions of representation. I provide further arguments why it is questionable to see ANNs as representing concepts in individual network units. Leavitt and Morcos (2020a) argue that current research on concepts in ANNs is too driven by intuitions rather than testable hypotheses and advocate for understanding ANNs as distributed representations. Similarly, Kieval emphasizes that while ANNs

can learn useful higher-order representations, these are useful for the ML model to perform the task at hand, and not for the human (See Chap. 12 by Kieval in this volume). He points out that ANNs rely on distributed rather than a unit-wise representation, and argues against representationalist approaches to ANNs overall. I will go into more detail on why at least the representation of concepts in individual units is difficult to defend.

Other Philosophical Positions on Concepts in ANNs

Buckner (2018) argues that a specific form of ANNs, namely convolutional neural networks (CNNs), can help understand the mechanisms in the brain that allow us to learn abstractions from examples and generate examples from abstractions, and in turn solve philosophical problems of empiricist philosophers. However, Buckner lays no claims to whether these abstractions in ANNs are stored in individual units or whether ANNs learn human concepts. Similarly, López-Rubio (2021) argues that CNNs and generative adversarial networks learn emergent natural object categories from images without human supervision. But like Buckner (2018), he does not address the question of whether these categories are proper concepts that fulfill "objective criteria of correctness and intersubjective agreement" (López-Rubio, 2021, p. 10024). López-Rubio (2021) bases his arguments on the empirical work by Bau et al. (2017) and Bau et al. (2018), but notes more clearly the limitation of their approach due to human bias in judging the semantics of a unit and the sometimes quite distributed representation of categories.

Work by Ilyas et al. (2019) indicates that ANNs learn concepts that humans understand as well as concepts humans do not use, like certain predictive texture structures in images. Based on these findings, Buckner (2020) discusses how examining predictive concepts that humans do not yet know can shed new light on the problem of natural kinds. Boge (2023) argues that ANNs do not learn concepts, but rather develop what he calls *functional concept proxies*—representations that take the inferential role of specific concepts in certain contexts, but do not capture their semantics and their relations to other concepts. However, like Buckner (2020), he beliefs, based on Ilyas et al. (2019), that ANNs can develop concept proxies for meaningful scientific concepts that humans have not learned and may not have access to.

13.2.3 A Short Primer on ANNs

Artificial neural networks (ANNs) consist of primitive stacked units called neurons (Goodfellow et al., 2016). Each neuron is composed of two parts: a weighted linear sum and an activation function. The architecture of ANNs is determined by the number of parallel units, the layers in which they are stacked, and the connectivity between layers.

The so-called "universal approximation theorem" shows that these simple stacked units are extremely expressive. In fact, any continuous function can be arbitrarily well approximated with an ANN of depth two (Hornik et al., 1989).[4] However, there is a significant gap between theoretical existence and the practical search for such functions. We need so-called learning algorithms to find the right neural networks that approximate the function of interest. These algorithms use the error that models make in their predictions to update the weights in the linear sums. This is done using methods like *gradient descent*, which move the weights in the direction of their gradient in the hope that fewer prediction errors will occur.

How one initializes the weights and which architecture one chooses has a large impact on how successful the search will be. These choices are usually referred to as *inductive biases*. The two architectures most commonly discussed in the literature are: First, convolutional neural networks (CNNs) (LeCun et al., 2015), where only neighboring nodes are connected. This architecture has been designed for image recognition tasks. Second, the transformer architecture (Vaswani et al., 2017), which relies on an attention mechanism to better store contextual information. This architecture was originally developed for language translation, but is now successful in all kinds of setups and forms the core of current chatbots, like ChatGPT.

The ANN obtained from the learning algorithm, that is after the learning process, is called the *prediction model*. In this chapter, when I talk about prediction models, I refer to these trained models and investigate whether they perform well and ideally develop human concepts. The focus will be on supervised ML models that perform prediction tasks after being fed labeled data.

13.3 What ANNs Do Learn: Conditional Probabilities

Let's check Assumption I: ANNs work well. What does it mean for an ANN to work well? In supervised learning, the goal of ANNs is to predict a given target feature Y (e.g. the correct animal) from a set of predictors $X_1, \ldots, X_p$ (e.g. pixels from image of animal) as accurately as possible. What constitutes a good prediction depends on how we define the so-called loss function L. Categorizing a leopard as a leopard seems uncontroversial, but what about errors? One may argue that mistaking a leopard for a cheetah is less bad than mistaking it for an elephant. But one could also hold that any misclassification is equally bad.

The loss function determines the best possible prediction functions from X to Y, also called the (Bayes) optimal predictor (Hastie et al., 2009; Von Luxburg & Schölkopf, 2011).[5] For example, if we take mean squared error (MSE) as our loss

[4] While in theory one hidden layer with high width is enough to learn arbitrary functions, in practice depth is crucial. Research suggests that this might be connected to the greater expressivity of deeper networks in finite parameter regimes (Lu et al., 2017; Rolnick & Tegmark, 2018).

[5] Note however, that the best predictor does not determine a single best function $f : \mathcal{X} \to \mathcal{Y}$. It only determines the functional values for $\mathbb{P}(X = x) > 0$ or in continuous cases, where the corresponding density is positive. For example, some adversarial attacks rely on probing the model

in a regression task, the optimal predictor is the conditional expectation $\mathbb{E}[Y \mid X]$. For the so-called 0–1 loss that assigns the same error of 1 to every misclassification, the optimal predictor is the $\arg\max_{y \in \mathcal{Y}} \mathbb{P}(Y = y \mid X)$. In general, the best predictor is a quantity that can be derived from $\mathbb{P}(Y \mid X)$.[6]

13.3.1 Theoretical Guarantees Are Lacking

Can we prove that ANNs learn optimal predictors, or at least good approximations of them? In statistical learning theory, originally the main framework for studying such questions, there exists a trade-off between the learning guarantees we get and the expressivity of the function class (Vapnik, 1999). As we already discussed above, ANNs are extremely expressive, they have infinite VC-dimension. Unfortunately, this implies that we cannot obtain learning guarantees from statistical learning theory for ANNs by any of the standard theorems (Sterkenburg, 2023).

Often in the training process of ANNs, we have more model parameters than training data. Hence, there are infinite models that perfectly approximate the training data but do not approximate the optimal predictor—they overfit the data (Hastie et al., 2009). The double-descent phenomenon shows that ANNs are different again, here more parameters can increase the generalization qualities of ANNS (Belkin et al., 2019). Belkin (2021) speculated that this quality may be related to the greater ability to select the model that minimizes structural risk among all interpolating models.

It is not even theoretically clear why ANNs trained with gradient based approaches learn models that perfectly interpolate the training data. Since the surface of the loss is non-convex, there are many local minima where the optimization could get stuck (Kingma & Ba, 2014). Belkin (2021) again provides an idea for why local minima are not a problem: in overparametrized regimes, every local minimum is a global minimum, since there is a global optimum in close proximity around every weight vector.

in regions with zero density, where the model is not necessarily competent (Szegedy et al., 2013) and must extrapolate (Freiesleben & Grote, 2023).

[6] Since people are often confused when it comes to causality: this already shows that current ANNs neither learn nor represent the causal mechanism between X and Y (See Chap. 8 by Buijsman in this volume). They learn a single distribution, rather than a set of distributions. ANNs completely ignore causal dependencies between the variables $X_1, \ldots, X_n$ and exploit effects of the target Y or spurious correlations just as much as causes (König et al., 2023).

13.3.2 *Practical Success Is Evident*

Empirically, we have seen the vast success of ANNs:

- They have been successful for all kinds of data, from tabular to image and geospatial to text and speech data (Erickson et al., 2017; Oikarinen et al., 2019; Ren et al., 2021).
- Moreover, ANNs have been successfully used in all kinds of domains, e.g. optimizing industrial processes, providing more deceiving advertisement, or diagnosing patients (Yang et al., 2020; Zhang et al., 2019; Gao et al., 2020).
- ANNs learn even highly complex interactions efficiently and usually improve with bigger data even of poor quality, they can even perfectly fit synthetically produced random noise (Zhang et al., 2021).

The empirical success of ANNs cannot be denied, suggesting that ANNs have theoretical properties yet to be revealed. However, there are also cases where ANNs fail even in static distributions, namely in the presence of shortcuts (Geirhos et al., 2020; Freiesleben & Grote, 2023); ANNs may learn to distinguish huskies from wolfs based on the snow in the background rather than the appearance of the animals (Ribeiro et al., 2016). Such mistakes can often be avoided by augmenting data or choosing data representational for the task (Kim et al., 2021; Freiesleben & Grote, 2023).

Taking Stock
Theoretical guarantees are indeed lacking. But I believe the empirical success indicates that ANNs approximate optimal predictors well in many settings.

13.4 What ANNs Might Learn: Human Concepts

Let's look into Assumption II: ANNs learn human concepts. It could be argued that an understanding of human concepts is necessary to perform complex classification tasks - hence ANNs learn these too. However, there are many different features on which classification can be based. Imagine a network that distinguishes sweaters from shoes, there can be different strategies:

- *Definitional features:* using features such as the shape of the objects that define the class.
- *Likely features humans use:* using features that humans also use that are likely cues for the target class, such as the pleats of the sweater or shoelaces.
- *Likely features humans do not use:* using features that humans do not use that are likely cures for the target class, such as sweater typical texture structures.
- *Spurious features:* using features that are specific to the data in the dataset. They are neither necessary nor likely cues for the class, think of certain colors, fabrics or background conditions.

Which of these do ANNs learn?

13.4.1 Transfer Learning Indicates that Reliable and General Concepts Are Learned

Transfer learning means that parts of models can be reused to solve related but different tasks (Tan et al., 2018), as shown in Fig. 13.1. To perform a new image classification task, instead of learning a completely new model, it is common to use well-trained models such as ImageNet or ResNet and fine-tune them to the task at hand (Kornblith et al., 2019; Shermin et al., 2019). Similarly, for a new text generation task, one would fine-tune existing language models like ChatGPT rather than starting from scratch (Zhong et al., 2023). This common practice produces excellent results, even with little training data.

Transfer learning shows that ANNs do not only learn spurious features, otherwise the learned features would not generalize to other tasks. Instead, the features must capture robust pattern in the data that are useful across problems and reliable indicators to make accurate predictions.

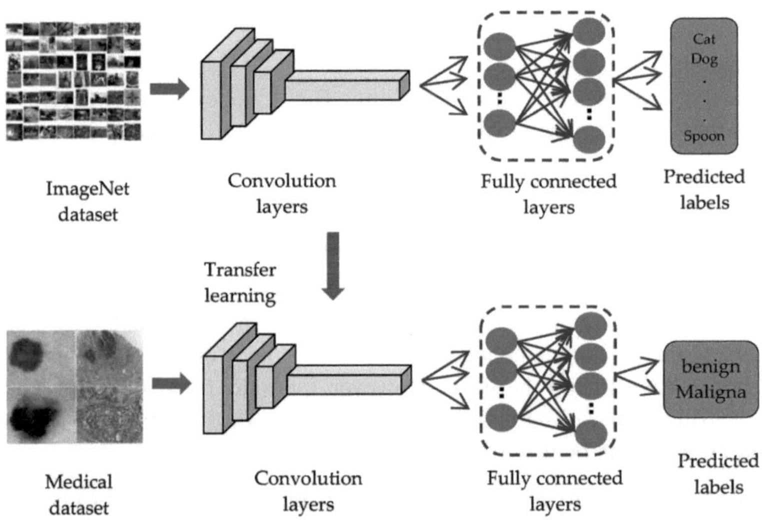

Fig. 13.1 An illustration of transfer learning by Mukhlif et al. (2023). In this case, filters of a CNN are transferred from one to another task. The fact that this strategy works means that models have learned general and reliable concepts. (from Mukhlif et al. (2023), licensed under CC-BY 4.0)

13.4.2 TCAV Indicates that ANNs Learn Human Concepts

Still ANNs might only learn concepts humans do not use. The interpretability technique *testing with concept activation vectors (TCAV)* by Kim et al. (2018) provides evidence that some human concepts are learned.

Intuitively, TCAV measures how (marginally) adding or subtracting concepts from instances changes the local or global behavior of the model. It does so via a few steps:

(i) To use the technique, you need a trained ANN image classification model and a dataset where to some instances a certain concept applies and to others not.
(ii) Run each instance of the above dataset through your trained classifier and store the activations of layers of the ANN in a table. Then, use a simple logistic regression model to separate images that instantiate the concept from those images who don't based on the activations stored in the table.
(iii) The orthogonal of the linear curve that separates the concept class from the non-concept class defines your concept vector. A concept influences the behavior of your model if adding or subtracting the concept vector (multiplied by some scalar) changes the predictions of single or many instances.

Kim et al. (2018) show for instance that ImageNet relies on colors like red, yellow, and blue to classify fire engines; Inception3 shows problematic behavior by relying on gender to classify aprons, specifically associating them with women.

Note that the concepts in TCAV are ultimately defined by the objects that belong to a concept class. This may be problematic as multiple concepts can be present in this class (Räz, 2023). For example, we want to get the concept vector for the concept *basketball*. However, it could be that basketballs are the only orange objects in the concept dataset, in which case our concept vector could represent the concept *orange* and not the concept *basketball*. Also, the TCAV method cannot actually be applied to learn which non-human concepts the model relies on, as this would require giving instances of these concepts.

13.4.3 Adversarial Examples Show that Models May Rely on Other Concepts than Humans

Adversarial examples describe cases where slight manipulations of data lead to incorrect predictions:

- Buying a bird can make you eligible for a loan (Ballet et al., 2019).
- Adding imperceptible noise makes image classifiers believe that a panda is a gibbon (Goodfellow et al., 2014).
- A small printed patch makes models classify every object next to it as a toaster (Brown et al., 2017).

- Deceiving prompts can lead large language models to hallucinate (McKenna et al., 2023).

If ANNs would have learned all and only human concepts, such attacks would not be possible. One may take adversarial examples therefore as evidence that ANNs learn mostly faulty concepts. Recently, however, a more nuanced viewpoint has been put forward by Ilyas et al. (2019): well trained ANNs indeed learn reliable concepts that generalize from the training to the test set and often also to other natural instances. Even though those concepts are reliably predictive of the target class in natural settings, humans do not use these concepts. For example, a sweater typical texture is very predictive of sweaters, but humans do not rely on it, probably because other features are even more predictive and easier to recognize from greater distances.

Ilyas et al. (2019) concluded that ANNs learn both, human and non-human concepts. It is an open question what the nature of these non-human concepts is and whether humans can in principle understand them (Buckner, 2020). Bowers et al. (2022) present interesting ideas on this question: ANNs are more likely to pay attention to complex textures than to shape, and even when they do pay attention to shape, it is more likely to be local than global shape (Geirhos et al., 2018). Another approach to better understand these non-human concepts could be via interpretability techniques: saliency maps can highlight the regions of the adversarial input that most influenced the model's decision (Wang & Gong, 2022) and the distribution of TCAVs of adversarial examples differs from the distribution of TCAVs of normal instances (Kim et al., 2018).

To me, adversarial examples clearly show that models often do neither learn the hierarchical importance of features nor their causal role

- *Hierarchical importance:* Even if a certain kind of texture on a panda appears to a classifier as very gibbonish, this should be overruled by the shape of the animal that is indeed very distinct between a panda and a gibbon.
- *Causal role:* Even if people with more pets are statistically more likely to pay back their loans, this does not mean that increasing the number of pets would make them more creditworthy (Ballet et al., 2019; Freiesleben, 2022).

Taking Stock

We can conclude that ANNs may use all kinds of features to perform their predictions, those that are definitional, predictive features used and not used by humans, and sometimes also spuriously correlated features. There is more than one way to bake a cake. I would speculate that from a purely predictive perspective there are probably many more useful concepts humans do not use than there are concepts humans do use.

13.5 What ANNs Probably Don't Do: Represent Human Concepts in Units

Let us turn to Assumption III: ANNs represent learned human concepts by single units. I have argued so far that ANNs can learn concepts and that they may even learn human concepts, but where do they store them? There is the widespread belief that ANNs store all concepts individually in single units of the network (Olah et al., 2020; Bau et al., 2017). What is the evidence on that?

13.5.1 *Activation Maximization Can Be Misleading*

One technique that got the narrative of representation in single units thriving was *activation maximization* (Olah et al., 2017). It asks the following question: What input would maximally trigger an individual neuron/unit in a neural network? This problem can be solved by gradient descent, however, this time the parameters we optimize over are not the weights of the network but the input features. Applied to image classifiers, activation maximization generates fancy images as shown in Fig. 13.2.

What can be deduced from such images? They require human interpretation. In many of such images, humans see certain concepts present to a greater or lesser degree. If one concept is dominant in an image according to human judgment, the neuron is assumed to represent the concept.

There are many reasons to be skeptical about such conclusions:

- *Representativeness of individual image:* We should not take the image we get as representative of the neuron. The image varies greatly depending on the image initialization, the designed objective, and the random seeds. These choices have

Fig. 13.2 Olah et al. (2017, 2020) applied activation maximization three times to one single unit of an ANN with slightly different initializations and objectives. Humans may find some semantically meaningful structures in the images (e.g. cat heads, car fronts, and bee bodies), but this only works for some units and can be conflicting as in this case. Olah et al. (2017) calls this neuron polysemantic, but how can we be sure that even there is any semantics based on these non-representative examples? (from Olah et al. (2020), licensed under CC-BY 4.0)

a large impact on the generated images and result in very different detected concepts (See Fig. 13.2). Also, minimizing activation rather than maximizing activation again points to different concepts (Olah et al., 2017).

- *Human biases:* Ultimately, humans are interpreting the images generated from activation maximization, which indeed introduces a bias (Räz, 2023; López-Rubio, 2021). For example, there is a strong confirmation bias towards finding human concepts. As we know from adversarial examples, humans are probably incapable of recognizing other (more texture based) concepts that might as well be present in the images. More generally, the images presented do not look natural and invite ambiguity—ideal conditions for leading people to be selective about what they see or fantasize (Leavitt & Morcos, 2020a).
- *Functional role derived from single image:* Most critically, neurons, even when associated with a concept, do not necessarily take the functional role of the concept in the prediction. Activation maximization as a technique sheds no light on the functional role of neurons. We know from other cases that removing neurons that are seemingly meaningful concepts has little to no effect on the models performance (Donnelly & Roegiest, 2019). Is the classifier getting worse in classifying cats if the cat heads unit is missing?

13.5.2 Network Dissection Is Impressive in Its Story but Not in Its Details

A more systematic approach for concept detection than activation maximization was taken by Bau et al. (2017). Their framework, called *network dissection*, aims to decompose image classifiers into disentangled components with semantic meaning. Intuitively, they test whether the region in images where humans see concepts matches the representations of those images in specific channels of the neural network.

What their method requires is a standard labeled dataset for which there are additionally labeled concepts, i.e. the concepts humans see in the images are highlighted in the images. Next, one has to decide on a particular channel in the image classifier that should be matched with a certain concept. The main idea is then to insert images into the image classifier and look at the activations of the channel under consideration. The extent to which the top activations of the channel align with the concept region highlighted by a human annotator is a measure for how well the channel represents the concept. Formally speaking, we take the intersection of the channel activation and the labeled concept region and divide it by their union, which gives us the so-called intersection over union (IoU) score.

Bau et al. (2017) differentiate between color, texture, material, scene, part, and object concepts and find for each type coactivating units in the neural network. They show that the alignment between concepts and units is not random, namely if rotations are performed on the channels, the alignment decreases. Moreover, they find that deeper and wider networks show a greater number of concepts stored in single units.

The work seems indeed impressive. However, looking into the details, I have found great weaknesses:

- The numbers they present are really not impressive. Even the best coactivating example units they show only have a maximum IoU of 30%. This means human concept regions and unit activations match only to 30%. Furthermore, the threshold for a unit representing a concept is set surprisingly low, with an IoU score of 0.04 without further explication.
- Even for ResNet, the most "interpretable network" according to the authors, only a small fraction of units has an IoU above 4% with some concept. Conversely, only about a quarter of the concepts in the concept dataset can be matched to any of the units. At the same time, most concepts can be linked to many different channels rather than just one; In ResNet, up to 37 units can be linked to the concept "airplane".
- As with activation maximization, network dissection does not reveal whether the units actually play the functional role of the concepts, so whether their removal would affect the network functioning with respect to these concepts. To make network dissection useful for interpretability, we would need to understand mechanistically what role these units take in the prediction. Fortunately, this question was explored in their follow-up work (Zhou et al., 2018). But again, the results seem unconvincing to me: removing units has little effect on overall performance, and a relatively small effect on class performance (on average around 8% performance loss, but admittedly of 50% in one extreme case). Thus, removing all 37 airplane units likely results in only a minor reduction in ResNet's ability to detect airplanes. This is evidence against the units representing this concept and, methodologically, it is another call for ablation studies when it comes to concepts in ANNs.

13.5.3 Selectivity Is Not Necessary for Better Performance

Where does this leave us? What Bau et al. (2017) has shown, at least, is that—some units in some networks respond systematically differently for objects of some classes compared to objects of other classes. This behavior is called "selectivity". But is selectivity somehow advantageous for the task at hand? Lead selectivity favoring architectures or training methods to better predictive performance?

Already Bau et al. (2017) noted that selectivity is not a prerequisite for predictive performance. In fact, many widely used techniques such as self-supervision, dropout or batch normalization, have been found to reduce selectivity but not performance (Morcos et al., 2018). Leavitt and Morcos (2020c) approach the question of how selectivity and performance relate to each other systematically by introducing a regularization term that controls selectivity. The term allows selectivity to be both enforced and impeded in training. They show that impeding class selectivity can even improve performance, conversely enforcing class selectivity impedes performance. In a follow-up study they show that impeding selectivity increases robustness to natural changes, while enforcing selectivity can lead to improved adversarial robustness to gradient based attacks (Leavitt & Morcos, 2020b).

Taking Stock

I share Leavitt and Morcos (2020a) opinion: "One's skepticism should be proportional to the feeling of intuitiveness." It needs more than single fancy images to call a unit in an ANN a concept. Representing a concept also goes beyond coactivation or highlighting the same regions in 4% of cases. To make the idea of human concepts in ANNs really worthwhile to the interpretability project, we need to systematically understand when human concepts emerge and check if they represent in the sense of coactivating AND taking the functional role of concepts.

13.6 Discussion

In this chapter, I examined the remarkable narrative that ANNs learn human concepts and store these concepts in individual units. I differentiated between three claims that underlie the narrative: (1) ANNs work well; (2) ANNs learn the concepts humans learn; (3) ANNs store these concepts in individual units. I found solid evidence for the first, mixed evidence for the second and questionable evidence on the third claim. I hope the arguments that I have presented will lead to some skepticism against the narrative. Further evidence must be gathered to give a decisive answer. I hope that my two criteria for concept representation—coactivation and functional role—can guide researchers in forming testable hypotheses on the questions.

Concepts Beyond Supervised Learning

The view that I described in this chapter focused mostly on the representation of concepts emerging in ANNs through supervised learning. The reason is that many techniques that investigate representations focus on ANNs trained with a supervised learning approach. I believe that unsupervised learning techniques and particularly generative models show greater promise in representing concepts, as generating requires greater concept ontologies than classifying. GAN-dissection by Bau et al. (2018) is a strong step in the right direction as it is studying the functional role of units under manipulations. Similarly, causal representation learning aims to enforce the emergence of semantically meaningful concepts by posing additional constraints (Schölkopf et al., 2021).

Enforcing the Emergence of Human Concepts

In some cases, it might be desirable to enforce ANNs to rely on the same features as humans. Think of safety or robustness issues like in autonomous driving. Also, features that humans rely on are features that humans understand—interpretability is and will be a key concern when it comes to interacting with ML systems (Lipton, 2018).

Can the representation of human concepts in ANNs possibly be enforced? Concept bottleneck models provide one example of how this might work if we incorporate human supervision (Koh et al., 2020). However they require a labeled concept dataset that is not normally available. I believe that if representing disentangled concepts in individual units is really what we are after, as a first step (as Leavitt & Morcos, 2020a has done) we should investigate training methods that enforce selectivity without sacrificing performance. However, this might be hard.

Natural Kinds

Even if they cannot store them in single units, ANNs seem to develop reliable concepts. It would be interesting to find out how to make ANNs learn human concepts or even scientific natural kinds (given they exist). This may require additional tasks beyond prediction such as generating data or acting in a real-world environment. Such an approach, where additional tasks are imposed, might ultimately allow us to characterize how human concepts emerge and what constitutes natural kinds. But perhaps such a characterization is impossible. Human concepts may not capture natural kinds in the world or be the result of performing a particular task, but rather they may just be conventions of social entities.

Dangers of the Narrative

I get the impression that the ML community is too convinced of the narrative to really drop it. People want to believe that ANNs are logical concept machines that only need to be scrutinized to become fully transparent. I think it is dangerous for this narrative to go unquestioned. It leads to research that is prone to confirmation bias, and that bias has real consequences for users. Some interpretability methods already build explanations to end-users based on the assumption that units represent human concepts, like for instance (Achtibat et al., 2023). Faulty assumptions will lead to faulty explanations.

The questions that I think researchers should ask are: Is it desirable that ANNs store human concepts in individual neurons? If yes, what learning methods allow us to enforce this behavior? If not, how can we better investigate the distributed representation of concepts?

To advance interpretability as a field, we must start forming proper falsifiable scientific hypotheses in the Popperian sense (Popper, 2005) rather than generating narratives based on fancy images and human biases. My hope is that my notion of human concept representation is seen as a challenge to ML researchers—show me a network where individual units coactivate with human concepts and removing them destroys the functional role of the concepts in prediction and I am sold that this network represents human concepts.

Acknowledgments I thank Giuseppe Primiero and Juan Durán for feedback on earlier versions of this manuscript. Also, I acknowledge support by the Carl Zeiss Foundation (Project: Certification and Foundations of Safe Machine Learning Systems in Healthcare).

References

Achtibat, R., Dreyer, M., Eisenbraun, I., Bosse, S., Wiegand, T., Samek, W., & Lapuschkin, S. (2023). From attribution maps to human-understandable explanations through concept relevance propagation. *Nature Machine Intelligence*, *5*(9), 1006–1019.

Andrews, K., & Beck, J. (2017). *The Routledge handbook of philosophy of animal minds*. Taylor & Francis.

Ballet, V., Renard, X., Aigrain, J., Laugel, T., Frossard, P., & Detyniecki, M. (2019). Imperceptible adversarial attacks on tabular data. Preprint. arXiv:1911.03274.

Bau, D., Zhou, B., Khosla, A., Oliva, A., & Torralba, A. (2017). Network dissection: Quantifying interpretability of deep visual representations. In *Proceedings of the IEEE Conference on Computer Vision and Pattern Recognition* (pp. 6541–6549).

Bau, D., Zhu, J. Y., Strobelt, H., Zhou, B., Tenenbaum, J. B., Freeman, W. T., & Torralba, A. (2018). GAN dissection: Visualizing and understanding generative adversarial networks. In *International Conference on Learning Representations*.

Belkin, M. (2021). Fit without fear: Remarkable mathematical phenomena of deep learning through the prism of interpolation. *Acta Numerica*, *30*, 203–248.

Belkin, M., Hsu, D., Ma, S., & Mandal, S. (2019). Reconciling modern machine-learning practice and the classical bias–variance trade-off. *Proceedings of the National Academy of Sciences*, *116*(32), 15849–15854.

Bird, A., & Tobin, E. (2023). Natural kinds, In E. N. Zalta & U. Nodelman (Eds.), *The Stanford encyclopedia of philosophy* (Spring 2023 ed.). Metaphysics Research Lab, Stanford University.

Boge, F. J. (2023). Functional concept proxies and the actually smart hans problem: What's special about deep neural networks in science. *Synthese*, *203*(1), 16.

Bowers, J. S., Malhotra, G., Dujmović, M., Montero, M. L., Tsvetkov, C., Biscione, V., Puebla, G., Adolfi, F., Hummel, J. E., Heaton, R. F., & Evans, B. D. (2022). Deep problems with neural network models of human vision. *Behavioral and Brain Sciences, 46*, 1–74.

Brown, T. B., Mané, D., Roy, A., Abadi, M., & Gilmer, J. (2017). Adversarial patch. Preprint. arXiv:1712.09665.

Buckner, C. (2018). Empiricism without magic: Transformational abstraction in deep convolutional neural networks. *Synthese*, *195*(12), 5339–5372.

Buckner, C. (2020). Understanding adversarial examples requires a theory of artefacts for deep learning. *Nature Machine Intelligence*, *2*(12), 731–736.

Buijsman, S. (2024). Machine learning models as mathematics. In J. M. Duran & G. Pozzi (Eds.) *Philosophy of science for machine learning: Core issues and new perspectives*. Synthese Library. Springer.

Del Pinal, G. (2016). Prototypes as compositional components of concepts. *Synthese*, 193, 2899–2927.

Donnelly, J., & Roegiest, A. (2019). On interpretability and feature representations: An analysis of the sentiment neuron. In *Advances in Information Retrieval: 41st European Conference on IR Research, ECIR 2019, Cologne, Germany, April 14–18, 2019, Proceedings, Part I 41* (pp. 795–802). Springer.

Duede, E. (2023). The representational status of deep learning models. Preprint. arXiv:2303.12032.

Erickson, B. J., Korfiatis, P., Akkus, Z., & Kline, T. L. (2017). Machine learning for medical imaging. *Radiographics*, *37*(2), 505–515.

Freiesleben, T. (2022). The intriguing relation between counterfactual explanations and adversarial examples. *Minds and Machines*, *32*(1), 77–109.

Freiesleben, T., & Grote, T. (2023). Beyond generalization: A theory of robustness in machine learning. *Synthese*, *202*(4), 109.

Freiesleben, T., König, G., Molnar, C., & Tejero-Cantero, A. (2022). Scientific inference with interpretable machine learning: Analyzing models to learn about real-world phenomena. Preprint. arXiv:2206.05487.

Frigg, R., & Nguyen, J. (2021). Scientific representation. In E. N. Zalta (Ed.), *The Stanford encyclopedia of philosophy* (Winter 2021 ed.). Metaphysics Research Lab, Stanford University.

Gao, Y., Cai, G. Y., Fang, W., Li, H. Y., Wang, S. Y., Chen, L., Yu, Y., Liu, D., Xu, S., Cui, P. F., et al. (2020). Machine learning based early warning system enables accurate mortality risk prediction for covid-19. *Nature Communications*, *11*(1), 5033.

Geirhos, R., Jacobsen, J. H., Michaelis, C., Zemel, R., Brendel, W., Bethge, M., & Wichmann, F. A. (2020). Shortcut learning in deep neural networks. *Nature Machine Intelligence*, *2*(11), 665–673.

Geirhos, R., Rubisch, P., Michaelis, C., Bethge, M., Wichmann, F. A., & Brendel, W. (2018). ImageNet-trained CNNs are biased towards texture; increasing shape bias improves accuracy and robustness. In *International Conference on Learning Representations*.

Goodfellow, I., Bengio, Y., & Courville, A. (2016). *Deep learning*. MIT Press.

Goodfellow, I. J., Shlens, J., & Szegedy, C. (2014). Explaining and harnessing adversarial examples. Preprint. arXiv:1412.6572.

Grujičić, B. (2024). Deep convolutional neural networks are not mechanistic explanations of object recognition. *Synthese*, *203*(1), 1–28.

Gurnee, W., & Tegmark, M. (2023). Language models represent space and time. Preprint. arXiv:2310.02207.

Hampton, J. A. (2006). Concepts as prototypes. *Psychology of Learning and Motivation*, *46*, 79–113.

Hampton, J. A., & Jönsson, M. L. (2012). 385 Typicality and composition a Lity: The logic of combining vague concepts. In *The Oxford handbook of compositionality*. Oxford University Press. https://doi.org/10.1093/oxfordhb/9780199541072.013.0018

Hastie, T., Tibshirani, R., Friedman, J. H., & Friedman, J. H. (2009). *The elements of statistical learning: Data mining, inference, and prediction* (Vol. 2). Springer.

Hornik, K., Stinchcombe, M., & White, H. (1989). Multilayer feedforward networks are universal approximators. *Neural Networks*, 2(5), 359–366.

Ilyas, A., Santurkar, S., Tsipras, D., Engstrom, L., Tran, B., & Madry, A. (2019). Adversarial examples are not bugs, they are features. In *Advances in Neural Information Processing Systems* (Vol. 32).

Keil, F. C. (1992). *Concepts, kinds, and cognitive development*. MIT Press.

Kieval, P. H. (2024). Representation learning without representationalism. a non-representationalist account of deep learning models in scientific practice, In J. M. Duran & G. Pozzi (Eds.), *Philosophy of science for machine learning: Core issues and new perspectives*. Synthese Library. Springer.

Kim, B., Wattenberg, M., Gilmer, J., Cai, C., Wexler, J., Viegas, F., et al. (2018). Interpretability beyond feature attribution: Quantitative testing with concept activation vectors (TCAV). In *International conference on machine learning* (pp. 2668–2677). PMLR.

Kim, E., Lee, J., & Choo, J. (2021). Biaswap: Removing dataset bias with bias-tailored swapping augmentation. In *Proceedings of the IEEE/CVF International Conference on Computer Vision* (pp. 14992–15001).

Kingma, D. P., & Ba, J. (2014). Adam: A method for stochastic optimization. Preprint. arXiv:1412.6980.

Koh, P. W., Nguyen, T., Tang, Y. S., Mussmann, S., Pierson, E., Kim, B., & Liang, P. (2020). Concept bottleneck models. In *International conference on machine learning* (pp. 5338–5348). PMLR.

König, G., Freiesleben, T., & Grosse-Wentrup, M. (2023). Improvement-focused causal recourse (ICR). In *Proceedings of the AAAI Conference on Artificial Intelligence* (Vol. 37, pp. 11847–11855).

Kornblith, S., Shlens, J., & Le, Q. V. (2019). Do better imagenet models transfer better? In *Proceedings of the IEEE/CVF Conference on Computer Vision and Pattern Recognition* (pp. 2661–2671).

Lalumera, E. (2010). Concepts are a functional kind. *Behavioral and Brain Sciences, 33*(2–3), 217.

Leavitt, M. L., & Morcos, A. (2020a). Towards falsifiable interpretability research. Preprint. arXiv:2010.12016.

Leavitt, M. L., & Morcos, A. S. (2020b). On the relationship between class selectivity, dimensionality, and robustness. Preprint. arXiv:2007.04440.

Leavitt, M. L., & Morcos, A. S. (2020c). Selectivity considered harmful: Evaluating the causal impact of class selectivity in DNNs. In *International Conference on Learning Representations*.

LeCun, Y., Bengio, Y., & Hinton, G. (2015). Deep learning. *Nature, 521*(7553), 436–444.

Lipton, Z. C. (2018). The mythos of model interpretability: In machine learning, the concept of interpretability is both important and slippery. *Queue, 16*(3), 31–57.

López-Rubio, E. (2021). Throwing light on black boxes: Emergence of visual categories from deep learning. *Synthese, 198*(10), 10021–10041.

Lu, Z., Pu, H., Wang, F., Hu, Z., & Wang, L. (2017). The expressive power of neural networks: A view from the width. *Advances in Neural Information Processing Systems* 30.

Manganini, C., & Primiero, G. (2024). Defining formal validity criteria for machine learning models. In J. M. Duran & G. Pozzi (Eds.), *Philosophy of science for machine learning: Core issues and new perspectives*. Synthese Library. Springer.

Marblestone, A. H., Wayne, G., & Kording, K. P. (2016). Toward an integration of deep learning and neuroscience. *Frontiers in Computational Neuroscience, 10*, 94.

McKenna, N., Li, T., Cheng, L., Hosseini, M. J., Johnson, M., & Steedman, M. (2023). Sources of hallucination by large language models on inference tasks. Preprint. arXiv:2305.14552.

Meng, K., Bau, D., Andonian, A., & Belinkov, Y. (2022). Locating and editing factual associations in GPT. *Advances in Neural Information Processing Systems* (Vol. 35, pp. 17359–17372).

Morcos, A. S., Barrett, D. G., Rabinowitz, N. C., & Botvinick, M. (2018). On the importance of single directions for generalization. In *International Conference on Learning Representations*.

Mukhlif, A. A., Al-Khateeb, B., & Mohammed, M. A. (2023). Incorporating a novel dual transfer learning approach for medical images. *Sensors, 23*(2), 570.

Oikarinen, T., Srinivasan, K., Meisner, O., Hyman, J. B., Parmar, S., Fanucci-Kiss, A., Desimone, R., Landman, R., & Feng, G. (2019). Deep convolutional network for animal sound classification and source attribution using dual audio recordings. *The Journal of the Acoustical Society of America, 145*(2), 654–662.

Olah, C., Cammarata, N., Schubert, L., Goh, G., Petrov, M., & Carter, S. (2020). Zoom in: An introduction to circuits. *Distill, 5*(3), e00024–001.

Olah, C., Mordvintsev, A., & Schubert, L. (2017). Feature visualization. *Distill, 2*(11), e7.

Peacocke, C. (1992). *A study of concepts*. MIT Press.

Popper, K. (2005). *The logic of scientific discovery*. Routledge.

Quine, W. V. (1969). *Ontological relativity and other essays*. Columbia University Press.

Radford, A., Jozefowicz, R., & Sutskever, I. (2017). Learning to generate reviews and discovering sentiment. Preprint. arXiv:1704.01444.

Räz, T. (2023). Methods for identifying emergent concepts in deep neural networks. *Patterns, 4*(6), 1–7.

Ren, X., Li, X., Ren, K., Song, J., Xu, Z., Deng, K., & Wang, X. (2021). Deep learning-based weather prediction: a survey. *Big Data Research, 23*, 100178.

Ribeiro, M. T., Singh, S., & Guestrin, C. (2016). "Why should I trust you?" Explaining the predictions of any classifier. In *Proceedings of the 22nd ACM SIGKDD international conference on knowledge discovery and data mining* (pp. 1135–1144).

Rolnick, D., & Tegmark, M. (2018). The power of deeper networks for expressing natural functions. In *International Conference on Learning Representations.*

Schölkopf, B., Locatello, F., Bauer, S., Ke, N. R., Kalchbrenner, N., Goyal, A., & Bengio, Y. (2021). Toward causal representation learning. *Proceedings of the IEEE, 109*(5), 612–634.

Shermin, T., Teng, S. W., Murshed, M., Lu, G., Sohel, F., & Paul, M. (2019). Enhanced transfer learning with imagenet trained classification layer. In *Image and Video Technology: 9th Pacific-Rim Symposium, PSIVT 2019, Sydney, NSW, Australia, November 18–22, 2019, Proceedings 9* (pp. 142–155). Springer.

Sterkenburg, T. F. (2023). Statistical learning theory and Occam's razor: The argument from empirical risk minimization. Preprint. arXiv:2312.13842.

Szegedy, C., Zaremba, W., Sutskever, I., Bruna, J., Erhan, D., Goodfellow, I., & Fergus, R. (2013). Intriguing properties of neural networks. Preprint. arXiv:1312.6199.

Tan, C., Sun, F., Kong, T., Zhang, W., Yang, C., & Liu, C. (2018). A survey on deep transfer learning. In *Artificial Neural Networks and Machine Learning–ICANN 2018: 27th International Conference on Artificial Neural Networks, Rhodes, Greece, October 4–7, 2018, Proceedings, Part III 27* (pp. 270–279). Springer.

Vapnik, V. (1999). *The nature of statistical learning theory.* Springer Science & Business Media.

Vaswani, A., Shazeer, N., Parmar, N., Uszkoreit, J., Jones, L., Gomez, A. N., Kaiser, Ł., & Polosukhin, I. (2017). Attention is all you need. *Advances in neural information processing systems* (Vol. 30).

Von Luxburg, U., & Schölkopf, B. (2011). Statistical learning theory: Models, concepts, and results. In *Handbook of the history of logic* (Vol. 10, pp. 651–706). Elsevier.

Wang, S., & Gong, Y. (2022). Adversarial example detection based on saliency map features. *Applied Intelligence, 52*(6), 1–14.

Watson, D. S. (2023). On the philosophy of unsupervised learning. *Philosophy & Technology, 36*(2), 28.

Williams, D. (2018). Predictive processing and the representation wars. *Minds and Machines, 28*(1), 141–172.

Wittgenstein, L. (2010). *Philosophical investigations.* John Wiley & Sons.

Yang, J., Li, S., Wang, Z., Dong, H., Wang, J., & Tang, S. (2020). Using deep learning to detect defects in manufacturing: A comprehensive survey and current challenges. *Materials, 13*(24), 5755.

Yee, E. (2019). Abstraction and concepts: When, how, where, what and why? *Language, Cognition and Neuroscience, 34*(10), 1257–1265.

Zhang, C., Bengio, S., Hardt, M., Recht, B., & Vinyals, O. (2021). Understanding deep learning (still) requires rethinking generalization. *Communications of the ACM, 64*(3), 107–115.

Zhang, S., Yao, L., Sun, A., & Tay, Y. (2019). Deep learning based recommender system: A survey and new perspectives. *ACM Computing Surveys (CSUR), 52*(1), 1–38.

Zhong, Q., Ding, L., Liu, J., Du, B., & Tao, D. (2023). Can chatGPT understand too? A comparative study on chatGPT and fine-tuned bert. Preprint. arXiv:2302.10198.

Zhou, B., Sun, Y., Bau, D., & Torralba, A. (2018). Revisiting the importance of individual units in CNNs via ablation. Preprint. arXiv:1806.02891.

Timo Freiesleben works at the interface between philosophy and machine learning. He is currently a postdoc at the *Cluster of Excellence: Machine Learning for Science* together with Thomas Grote at the University of Tübingen. His research interests focus on conceptual problems in modern machine learning related to interpretability, robustness, and fairness, but he is also interested in classical philosophical topics like causality, scientific methodology, and rational decision making. Outside of academia, Timo enjoys hiking in the alps, cooking for his friends, reading books from people like Dostoyevsky and Kafka, and learning more about our strange universe.

Chapter 14
Defining Formal Validity Criteria for Machine Learning Models

Chiara Manganini and Giuseppe Primiero

Abstract In the context of deterministic scientific simulations, formal validity requirements have typically been defined to help us reason about the relationship between the mathematical model underlying the target system and the computational model used to simulate it. With machine learning simulations entering the picture, we argue that these formal requirements need to be reviewed, as the objects to which they apply have significantly changed. This is due to several reasons: the target system is no longer an available system object of investigation; the probabilistic mathematical model is abstracted from the target system through the mediation of the computational model, which however remains largely opaque to us due to its high complexity. For these reasons, we formulate weaker, probabilistic versions of the traditional relations of Simulation, Bisimulation, and Approximate Simulation. Accordingly, we define three corresponding validity criteria that capture a range of cases, from the strongest to the weakest, depending on the extent to which the machine learning model can be assumed to correctly represent its target system.

Keywords Non-deterministic computation · Probabilistic simulation · Validity · Model · Markov semantics · Process algebra

14.1 Introduction

The introduction of computational systems in the practice of science has had a major significance for the related philosophical reflection. In particular, the use of computational modeling and simulation software has called for the definition of specific methodological principles to establish what is a valid scientific result obtained through algorithmic processes. The problem of validity as a property of a scientific result obtained by computational methods could be formulated as follows:

C. Manganini (✉) · G. Primiero
Logic, Uncertainty, Computation and Information Lab and PhilTech Research Center, Department of Philosophy, University of Milan, Milan, Italy
e-mail: chiara.manganini@unimi.it; giuseppe.primiero@unimi.it

J. M. Durán, G. Pozzi (eds.), *Philosophy of Science for Machine Learning*,
Synthese Library 527, https://doi.org/10.1007/978-3-032-03083-2_14

Problem 1 (Computational Model Validity). What are the conditions for accepting as valid the results of a computational model, and how do we express them formally?

In this broad context, the integration of machine learning (ML) methods in computer simulations has represented another turning point. Widely discussed in the context of "traditional"—i.e., non-ML—simulations (Humphreys, 2004; Durán, 2018; Durán & Formanek, 2018; Alvarado, 2023), two problematic aspects characterize most ML systems, and hence most ML simulations:

1. Opacity, due to their high structural complexity,
2. Non-determinism, resulting from their instantiation of learning methods to spot statistical patterns in the data.

The problem of validity for ML simulations can therefore be reformulated as follows:

Problem 2 (Opaque and Non-Deterministic Computational Model Validity). What are the conditions for accepting the results of an opaque and non-deterministic computational model, and how do we express them formally?

Despite the abundance of technical and philosophical analyses of ML opacity (Guidotti et al., 2018; Boge, 2021; Facchini & Termine, 2022) and ML non-determinism (Summers & Dinneen, 2021; Cooper et al., 2022; Creel & Hellman, 2022), a full-blown theoretical examination of Problem 2 is still at early stages at the moment. With the present contribution, we wish to help to address it.

In order to do so, we start from an overview of the philosophical debate on deterministic simulations in Sect. 14.2. We then turn to non-deterministic ones, first in general terms (Sect. 14.3) and then focusing on four specific aspects constituting relevant departures from the deterministic scenario (Sect. 14.4). We then advance our proposal for a formal approach to Problem 2 that essentially consists in extending one already formulated for the deterministic case in (Primiero, 2019) to non-deterministic simulations. This idea is offered in Sect. 14.5, where we justify the choice of the relevant formal structures, without providing any full-blown formalization yet, which will be addressed in further work. Finally, in Sect. 14.6, we formulate some related questions on the notion of validity emerging from the use of non-deterministic, opaque computational models. Section 14.7 concludes the work with an overview of the future steps of this research.

14.2 The Story So Far: On the Validity Criteria for Deterministic Computations

The conceptual study on what kind of validity conditions can be established for results obtained from implemented computational systems, such as simulation software, is a complex issue. It is strongly intertwined with the question of whether such criteria can be formulated in a formal setting. The origin of this debate is notoriously traced back to Humphreys:

> Inasmuch as the simulation has abstracted from the material content of the system being simulated, has employed various simplifications in the model, and uses only the mathematical form, it obviously and trivially differs from the 'real thing', but in this respect, there is no difference between simulations and any other kind of mathematical model, and it is primarily when computer simulations are used in place of empirical experiments that this element of falsity is important. But if the underlying mathematical model can be realistically construed (i.e. it is not a mere heuristic device) and is well-confirmed, then the simulation will be as 'realistic' as any theoretical representation is. (Humphreys, 1990, pp. 501–502)

From the assumed epistemological identity of the validity of computer simulations and their use in place of empirical experiments, the question arises of how to formulate their validity conditions. In particular, it becomes pressing to establish whether a formal treatment of such criteria is possible. Here the debate has diverged, with different positions being formulated in the literature. On the one hand, it has been argued that such a question cannot be answered in absolute terms, and that the very conditions of knowledge of the target system determine the appropriate methodology to formulate what validity means:

> For the data to hit the target, the experiment, simulation, or "simulating experiment" must mirror the target in a number of relevant aspects. Whether the mirroring should be purely formal or material in character, depends on the kind of question we are asking, and on the amount and quality of background knowledge we have accumulated regarding the target system itself. (Guala, 2002, p. 15)

On the other hand, detractors of formal methods—including Humphreys himself—have argued against the usefulness of formalization, seen as imposing a further departure from the observed system, to be added to the abstractions and idealizations already implied by equation-based models:

> [...] no comprehensive model at the level of the total system exists. Computationally irreducible processes thus occur in systems in which the most efficient procedure for calculating the future states of the system is to let the system itself evolve. [F]ormalization takes one away from what is important about the form of the equations, and hence is at best unnecessary; at worst it distorts the important formal features. [...] logic is not well suited to deal with every goal of science, or even particularly well with any one of them. (Humphreys, 2004, pp. 149–153)

Nonetheless, provided that computational models have proven time and again their extraordinary capacity for accurately predicting the behavior of real systems, their reliability inevitably requires formal criteria to assess their correctness:

> If computer simulations are reliable processes that render correct results, then we need a way to characterize what a 'correct' result would be. [...] Verifications and validations are two fundamental pillars for asserting the reliability of computer simulations [...] [R]esearchers are on good grounds for claiming that their computer simulations are reliable processes, and thus that they are justified in believing that the results of the simulation are correct. [...] [R]eliability of computer simulations support the claim that researchers are able to claim for explanation, prediction, and [...] for understanding of the results. [...] Although neither software nor hardware can be fully verified or validated, researchers are still developing methods that reduce the occurrence of errors and increases the credibility of the simulation. (Durán, 2018, pp. 91–96)

As stressed also by Humphreys (2019) with the notion "construction assumptions" for formal templates, the role of formal methods is crucial to establish what kind of idealizations and approximations are appropriate in knowledge construction. Determining such assumptions for computational methods is essential to distinguish which validity criteria hold. It is, in fact, necessary if we want to know under which conditions the results of such models can be deemed valid.

One attempt to ground such formal validity conditions is the notion of isomorphism, well-known in the philosophical debate on what enables a scientific model to *represent* the target system (van Fraassen, 1980; Brading & Landry, 2006). The isomorphic view anchors scientific representation in the existence of a formally definable one-to-one, bijective correspondence between the set of states representing the dynamic evolution of the target and the one representing the dynamic evolution of the computation. Notoriously, it has been argued that such a conception is somewhat too strong, as it requires perfect matching between the target's and the model's structures. Moreover, it imposes properties like reflexivity and symmetry that seem undesirable for qualifying the representation relation (see, for example, the directionality requirement by Frigg & Nguyen, 2017, p. 55). To respond to these shortcomings, weaker variants of target-model correspondence have been proposed, like partial isomorphism and homomorphism.

In what follows, we are interested in developing an approach inspired by the isomorphic view and its weaker versions. In particular, we borrow the concept of *simulation relations* from the process algebra literature, where they have been used to define different identity conditions for software systems (Angius & Primiero, 2018). Moreover, in (Primiero, 2019) these relations have been applied to the problem of validity of computer simulations, to formalize the correspondence between the deterministic model of a *target system* and the deterministic computational model of a *simulation software* that should predict or describe it.[1] Such an application defines the underlying formal mappings between:

1. An observed target system;
2. An abstract mathematical construct that describes the dynamic transformations of the target system and the relationships between its scientifically relevant variables;
3. A computational model, i.e., an algorithmic structure that translates the mathematical model into discrete and approximate solutions that computers can process;
4. A physical computational system executing the computational model.

[1] This formal strategy relies on an idealization process which assumes the ability to construct models for both the target and the computational systems. Especially for the former, such an idealization is not trivial and it may have to take into account certain limitations, such as partial observability (see Sect. 14.5).

Essentially, the basic idea behind the present contribution is that of translating the above entities and the relations among them into proper formal counterparts known from process algebra. In particular, we will primarily focus on (1–3) and their relations, leaving aside the questions relative to (4) and its relations with the other three entities (in other words, we will assume the physical implementation of the simulation software to be correct, which is of course a crucial requirement for the validity of the simulation results).

Without entering into the formal treatment yet, suffice to say that *simulation relations* can be used to define formal criteria for simulation validity as they allow us to express whether and how the target system and the computational model match in their dynamic evolution (i.e., in terms of their initial states and transitions). The ultimate goal is to verify property satisfaction, thereby checking whether a condition holding for one structure (something the target system satisfies) also holds for the other (i.e., the computational model). In regard to this, three fundamental types of simulation relations can be described as follows:

- In a *Simulation Relation*, the computational model displays only behaviors of the target system;
- In a *Bisimulation Relation*, the computational model displays *all and only* the behaviors of the target system;
- In an *Approximate Simulation Relation*, the computational model displays *some but not only* behaviors of the target system.

Hence, a Simulation Relation holds between a computational model and the target system whenever the former mimics all stepwise behaviors of the latter, but not vice versa. A Bisimulation Relation holds if the above relation is shown to hold in both directions, i.e., when the two systems exhibit identical stepwise behavior. The Approximate Simulation Relation expresses incompleteness and divergence of the computational model with respect to the target, as some computational behaviors have no counterpart in it. By means of these relations, one can for instance guarantee properties of the target system by verifying their presence in some or all behaviors of a sufficiently precise simulative computational structure. Moreover, the relations of Simulation and Bisimulation have been used to define the notions of exact and inexact copies of computational artifacts (Angius & Primiero, 2018), while a notion of Approximate Simulation Relation is introduced to define approximate copies. The corresponding safety and liveness checking algorithm over approximate copies is defined in (Angius & Primiero, 2023).

At this point, one might wonder whether this framework is applicable to the case of ML simulations where, as we said, the deterministic assumption fails because of their inherent stochastic nature. In the next section, we extend the investigation in this direction.

14.3 The New Scenario: Entering Non-Deterministic Computations

Machine learning systems have almost entirely changed any discussion on algorithms. Although deterministic computations still represent a crucial and overwhelming technological component of our daily lives, as well as the object of recent philosophical investigation (Angius et al., 2021), the new generation of AI systems is increasingly employed in scientific and industrial contexts alike. From the point of view of the present investigation, we are interested in assessing which *formal* criteria can be formulated to establish whether the result of an ML simulation (for example, to generate input data in driving scenarios or product lines in simulated factories) can be considered valid. To better understand what kind of validity is adequate for ML simulations, let us first clarify their underlying ontology and how it differs from that of traditional simulations.

We begin by considering a general statement applicable to any simulation, whether traditional or ML-based: the target system is what ultimately grounds the notion of validity for all simulation results. Not only in the sense that the resulting predictions are *a posteriori* evaluated against observations of the target system itself (to statistically validate the model), but also in the sense that the target system plays a pivotal role in the actual process of simulation construction. The dual role played by the target system could be seen as roughly tracing the distinction between, respectively, the context of justification and the context of discovery (Reichenbach, 1938): on the one hand, based on the target system we *justify* whether the simulation is valid; on the other hand, we *find* such a simulation model by looking at the target.

Against this general characteristic of simulations, we notice that ML simulations constitute a departure from traditional ones in regard to the actual process through which the computational model is constructed from observations. In the traditional simulation scenario, the mathematical model given by an already established mathematical description of the target phenomenon—for instance, a set of differential equations—is *top-down* translated into the computational model by a manual process:

> It is well documented, for example, that the mathematical operations that form the basis of computer simulations of dynamic systems are quite different from those found in the theoretical stages of inquiry. [...] The way a computer simulation solves an equation is by providing approximate values to discretised parameters that roughly correspond to the results one gets from a continuous equation. While the results can be similar or approximate to an almost negligible degree, the fact remains that both the results and the methods by which they are arrived at are distinct from a continuous equation. Often the translation of mathematical models into code for a computer to run includes many idiosyncratic engineering practices that are far removed from the sound theoretical principles in virtue of which the initial model was constructed. (Alvarado, 2022, p. 1190)

On the other hand, in the ML case, something different happens. Through the training process, the mathematical function is automatically extracted from a sample of data—the training set—that is considered to be representative enough of the target system. This process is, instead, *bottom-up,* in the sense that the mathematical model is obtained from the computational one, rather than the other way around. As a result of this bottom-up process, ML systems inherently exhibit two fundamental characteristics: non-determinism and opacity.

Regarding the former, in fact, the result of the learning process is a computational model implementing the learned function $\hat{f}$ which constitutes the mathematical model of the ML-based simulation. This describes a probability distribution that associates each input with its outcome *probabilities*. Therefore, the input-output relation is non-deterministic to the extent that it is stochastic, i.e., can be reasoned upon using probabilities.

As for the opacity, we can see the training phase as a process that makes it possible for the designer of the computational model to formulate a specification (i.e., the set of functional requirements) of its functioning without *explicitly* expressing it. The training data set in fact provides a partial specification of the required behavior, thus partly replacing the designer's intentions and requirements.[2] A first, crucial aspect of the ML opacity is that the outcome of the training process is not already the probabilistic function $\hat{f}$, but *the computational model that implements* it—i.e., the list of parameters of a neural network. This means that a work of abstraction is further required to extract the mathematical model out of the computational one, as the former is not immediately accessible to us. However, even when the mathematical model is successfully reconstructed from the algorithm, this may not be a guarantee of epistemic transparency, due to its inherent complexity. A second and significant obstacle to transparency is hence represented by the huge number of variables $X_1, \ldots, X_n$ of $\hat{f}$, that makes its decomposition into simpler joint distributions $P(X_i)$ with $1 \leq i \leq n$ a very complex epistemic task (see Facchini & Termine, 2022).

The computational model is in turn implemented into a physical artifact which materially executes the computational processes fixed by the formal dynamic descriptors. The result returned by such a physical computational execution constitutes the final experimental result of the entire ML simulation. Standard software testing techniques can be used to verify its correctness and validate it. For the *verification* of machine learning models, both the training engine and the trained model can be tested through standard software techniques, while the *validation* of the trained model is usually carried out by means of well-known performance metrics, such as accuracy, precision, recall, F1-score, and AUC-ROC. Note, however, that here we are not interested in asking which among these statistical measures is best suited to evaluate the fitness of a machine learning system for a given prediction task. Rather, what we are seeking to do is to establish a formal counterpart to such statistical criteria of validity. This goal is not entirely new. Indeed, formal approaches to the verification of ML systems are slowly emerging within the literature. Starting from the work by Seshia et al. (2018), computer scientists have

[2] Under this perspective, machine learning could be thought of as a new, automatic process of "requirement engineering", in the sense of the term discussed by Jackson (1995), aimed at producing "learned" specifications (Banks & Ashmore, 2019; Vogelsang & Borg, 2019).

increasingly focused on the formal verification of deep neural networks, in order to check some properties of interest—such as safety, liveness, and robustness. Within the philosophical literature, noteworthy contributions include the work by Angius and Plebe (2023). In a similar spirit to ours, the authors investigate what are the appropriate validity criteria for Deep Learning (DL) systems. Interestingly, they claim that the most appropriate notion of correctness for such systems is that of *experimental validity*, as defined in (Primiero, 2019). An experimentally valid simulation is one in which:

- The mathematical model is *fit for purpose*, i.e. it is able to represent some of the requested variables and inferential properties of the target system;
- The derived computational model is *isomorphic or partially isomorphic* to the mathematical one and *usable*, i.e., it allows to accomplish the desired experimental analysis.[3]

Here, we want to expand the intuition further. We agree with Angius and Plebe in considering the notion of experimental validity as the most appropriate to account for the evaluation of machine learning models. Nevertheless, reasoning transparently and deterministically about the identity of their behaviors—as we saw in the previous section, by means of deterministic simulation relations—seems impracticable for the reasons that will be illustrated in the next section.

14.4 Novel Aspects of Simulation Relations for Non-Deterministic Computations

We now turn to four aspects of ML simulations marking a significant difference with respect to traditional ones. These aspects, we argue, make simulation relations defined in Sect. 14.2 inapplicable to the ML context.

To begin with, it is extremely hard to establish what is the relevant or appropriate target system for an ML simulation. In other words: defining *correctness* is a hard task for computational systems of this sort. In traditional simulations, one usually starts with the object to be modeled (a physical, social, or engineering system), and *only then* designs the relevant computational model to simulate and study it. On the other hand, when a machine learning system is at work, one may only extract intended or desirable behaviors for a correct prediction, while establishing whether false negatives and false positives occur—and their frequency—is a different matter and so is determining whether unintended results are produced. For instance, in the case of a self-driving autonomous system trained on real data, surely there is an informal intention, possibly expressed by the relevant rules of road navigation

[3] Note that these notions are to be considered *weak* as opposed to robustness of a model, where the model is able to include all variables of interest and all the inferential properties of the simulated empirical system.

in urban and non-urban environments, to avoid collisions and in general to drive safely. This certainly can be taken to constitute "the norm" that the trained model should respect. The formal specification of such rules and habits, against which the model trained on the data should be evaluated, is a different matter. Verifying that the algorithm is correct would require such a formal specification and that it does satisfy it for all potential cases and nothing else, which is a matter of testing against numerous, if not infinite, cases of use.

Second, which learned model is working as a computational model is not necessarily known beforehand. While in standard deterministic software systems the definition of the mathematical function is sufficient to establish the behavior of the system for any given input, in the case of machine learning systems the mathematical function is largely determined by the training data. Change the data, change the learned function. Hence, only by knowing in detail both the function and the training data, one can be at least sufficiently aware of what type of computations one can expect to occur once a new input data is provided.

Third, even when the computational model is known, an accurate definition of its behavior with respect to the intended mathematical function—the desired behavior—is not always possible. To this aim, one would need to be able to perform statistical inferences on its behavior on the given test set, to establish whether there are input/output relations that falsify the intended functional behavior represented by the mathematical function of learning. Unfortunately, performing such inferences is often impossible, due to the inaccessibility of the structure of the model and its complexity.

Finally, the computational model itself cannot be directly evaluated as to whether it is appropriate to the task at hand. To do so, one would need to have full access to its structure, namely, parameters which are automatically set, and hyper-parameters which are tuned either manually or automatically. Full understanding of these coefficients is extremely hard in practice, and often totally missing for the end user. Even more crucially, even assuming its availability, it is unclear whether this knowledge would make the validation of the model any easier, since we might be unable to establish cognitively meaningful simulation relations between the model structure and the target system, as discussed in Chap. 12 by Kieval (2025).

All the above makes it very hard to establish what is a valid result for such a system. One would need to know that the test set agrees with the training set and that the distribution stability is satisfied to prevent loss of prediction accuracy or distribution shift. One would also need to have both structural and inferential knowledge of the computational model, while the former is hard and the latter *very* hard, especially without any formal modeling. The result of these considerations is double. On the one hand, it seems that the very essence of the formal model on which simulation relations are defined, i.e., the requirement of representing states of both the target system and of the computational model, the ability to establish valid evolutions of both structures, and the strategy to compare them to identify whether their behaviors are totally or partially identical, is bound to fail because of the opaque and non-deterministic nature of the new computational model at stake.

On the other hand, though, it is evident that without such a formal analysis, the evaluation of such systems based on their accuracy can only rely on non-structural, almost epi-phenomenal reasons, thereby making the claim of validity for their use in scientific contexts void. What is needed, therefore, is a formal model which may accommodate the uncertainty of both the structure and the behavior of machine learning systems.

To take stock, when it comes to ML simulations, it becomes impossible to assume a well-defined and sufficiently fine-grained representation of the computational model to check its structural identity with the target system; this calls for some new formal validity principles, as those that will be introduced in what follows.

14.5 Extending Validity Criteria to ML Simulations

A well-studied and natural way to represent the uncertain behaviors of machine learning systems is provided by probabilistic extensions of state transition systems—such as Markov Chains, Markov Decision Processes, or Probabilistic Process Algebra (Jonsson et al., 2001; López & Núñez, 2004)—and their associated *simulation relations* (Baier et al., 2000; Baier et al., 2006). An equivalent approach consists in devising an appropriate inferential system (see, e.g., D'Asaro et al., 2025; Genco & Primiero, 2023). Either way, a common strategy underlies these formalizations:

- Assign probability distributions to training and test data as frequencies in the population to represent possible data for the trained model and possible test inputs;
- Assign to each feature the probability to hold for the current data point;
- Compute, by an appropriate probabilistic consequence (or inference) relation, the probability that any set of assignments of features leads to an assignment of the target feature (as an equivalence class).

An attribution of features to a data point for classification or prediction represents a state of the model (or a derivable formula); an evaluation of a new feature is an action that allows one to move from state to state (or a rule to proceed in the deduction); and the end state is one where the target feature is reached (or deduced) with a certain probability. Doing so for the target system and for the computational model respectively allows one to ask how much the two resulting abstract structures are *similar* to each other, or else how much *the model of* the ML classifier *represents the model of* the target system. To do so, one may refer to simulation relations for probabilistic systems to evaluate the adherence between the ML computational model and the target system, based on the given probability assignments.

In the literature, it is usual to distinguish between *strong* and *weak simulation relations* to distinguish cases where a simulation relation[4] holds between two systems whose states are completely observable, from those cases where the systems are only partially observable (Baier et al., 2006). ML models clearly refer to the latter case. Hence, we can define three weak counterparts of the simulation relations previously introduced, to express partial identity of the computational ML model and the probabilistic representation of the target system:

- In a *Weak Simulation Relation*, the non-deterministic computational model displays *only* behaviors with *at least the same* probability as in the target model;
- In a *Weak Bisimulation Relation*, the non-deterministic computational model displays *only* behaviors *with the same* probability as in the target model;
- In a *Weak Approximate Simulation Relation*, the non-deterministic computational model displays *not only* behaviors *with the same* probability as in the target model.

Some observations are necessary concerning these variants of the simulation relation. The relation of Weak Simulation admits the case of underfitting: the ML system does not predict beyond the scope of the target system, while it may assign to certain predictions a higher probability than the corresponding value in the target system. In contrast to this, Weak Bisimulation expresses in formal terms a perfect fitting of the computational model with respect to the target system: what the model predicts is as in the system. This relation can only be instantiated by a perfect classifier: despite not knowing all the probabilities for the intermediate evaluations, the observed results do not show any false positives or false negatives. Finally, the Weak Approximate Simulation Relation renders the notion of overfitting: the model may make predictions which are in the scope of the target system, but it may also predict features of variables which the target system does not admit. In this sense, the computational model may create predictions which do not reflect the complexity of the model, and therefore the classification may show high bias.

The formal translation of these relations is left for another occasion. Instead, we now focus on how these relations, and the notion of validity they induce, allow us to answer crucial questions on the correctness of results produced by computational systems based on machine learning algorithms.

14.6 Validity Principles

When using systems based on machine learning algorithms, the question of their validity reduces to understanding how much can be known about the *correctness*

---

[4] This applies to Simulation and Bisimulation. The notion of Approximate Simulation Relation introduced in (Angius & Primiero, 2018) misses such a distinction in the literature. The definition of a Weak Approximate Simulation Relation in this section is meant to precisely fill this gap.

of their results with respect to the intended or actual behavior of the target system. Such functional knowledge would be inferable under the assumption of structural knowledge of the computational system under observation. Unfortunately, as highlighted above, such structural knowledge is almost always not available for ML algorithms, except for trivial cases. A possible strategy to attack such a problem, under the assumption of the ability to determine simulation relations for probabilistic systems as formulated above, is to establish what type of knowledge we can infer from the fact that such a relation holds. To further explain this aspect, we will formulate some questions related to the structural knowledge of the computational system under analysis and show which answers can be provided in terms of simulation relations.

The first pair of questions concern the relation between the characteristics of a seemingly correct prediction/classification and the inferable knowledge on its underlying causal structure and its verifiability:

Question 1 Can we establish why a certain ML classification/prediction is correct?
Question 2 Can we verify that an ML system classifies/predicts correctly for the same reasons we would use to explain the real-world phenomena?

To answer these questions, let us reconsider the criteria associated with ML-based simulations. The assertion that a non-deterministic model of the ML system validly represents the target system would require, at the very least, that the former displays only behaviors expected or assumed to hold for the latter. Failing this, one would have evidence that the structure of the computational model exceeds in some way that of the target system. Such validity of the structure should, in principle, be of a causal nature.[5] Arguably, of the three simulation relations defined in the previous Section, only the Weak Bisimulation captures a necessary (although not sufficient) criterion for establishing causal correctness between the ML model and the target system. Hence, a machine learning model can be assumed to be causally correct for a given target system *only if* a weak bisimulation relation holds between them. Accordingly, a machine learning artifact is causally correct only if it implements without errors a causally correct computational model.

Unfortunately, this is a necessary but not sufficient condition, as a much stronger criterion would need to be formulated. Interventionism or causal manipulation is currently considered the most adequate one, as it requires to qualify which variables are significant by observing structural variations induced by changing their values (Woodward, 2000). This approach is largely used in ML practice through perturbation techniques, consisting in switching some features' values from current to alternative ones (Liu & Lorini, 2023).[6] The following principle summarizes the above:[7]

---

[5] There is a long-standing debate on the causal validity in models and the associated problems. See, for instance (McKim & Turner, 1997; Beckers et al., 2023). For a recent overview of the notion of causality in machine learning see (Termine & Primiero, 2024).

[6] For a non-causal interpretation of ML systems and their explanations see Chap. 8 by Buijsman (2025).

Principle 1 (Justifiably Valid Model). A machine learning based computational model is justifiably valid for a given target system only if at least a weak bisimulation relation holds between them, i.e.:

- The test data satisfy all and only probabilistic assignments of the training data, and
- The assignments of the machine learning model to data points in the test data match only the features in the target system.

Note that the formulation of the first point in the above definition assumes that no distribution shift occurs between training and test data.

The second pair of questions concern the possibility that a correct model for some target system is wrongly extended to be used for a different target, or what in the literature is known as the Out-of-Distribution problem. So, assuming that an implemented system matches the real-world phenomenon's causal structure, how do we know the underlying test set is representative of the real-world phenomena? And moreover: if the classification is not too limited by the training set, is its distribution stable with respect to the currently given test set? In other words:

Question 3 What are the limits of the training set for the validity of the ML classification/prediction?

Question 4 What are the limits of the test set for the validity of the ML classification/prediction?

In this case, the necessary condition for the simulation relation is that the computational model displays behaviors that have at least the same probabilities as in the target system, or that the machine learning system does not underestimate the predictions which are realistically assigned by the target. This brings us to a weak notion of validity:

Principle 2 (Weakly Valid Model). A machine learning based computational model is weakly valid for a given target system only if at most a weak probabilistic simulation relation holds between them, i.e.:

- The test data satisfy probabilistic assignments with at least the same values as in the training data, and
- The assignments of the computational model for features to points in the test data are at least the same as in the target system.

Also in this case, the condition is necessary but not sufficient. This means that a guarantee that the current model is applied with a valid training set and to a valid test set would again require a stronger condition, reducing in the first place to (but not just) the notion of justifiably valid model.

Consider that verifying Weak Bisimulation (which implies perfect fitting) is hard: this is not just 100% accuracy, but a full-blown probabilistic structural identity on data sets and step-by-step probabilistic evaluations. On the other hand, weak simulation (admitting underfitting) implies the ability of the machine learning model to correctly classify/predict multiple data sets, a phenomenon hardly known in

[7] For significant contributions to the role of justification in ML, see Chap. 4 by Durán, (2025), Chap. 5 by Alvarado, (2025), and Chap. 6 by Zednik and Verreault-Julien (2025).

reality. This suggests that justifiable validity and weak validity for machine learning based computational models may be difficult to obtain.

The last and weakest degree of validity is the one expressed by the relation of approximate weak simulation:

Principle 3 (Almost Valid Model). A machine learning based computational model is almost valid for a given target system only if at most an approximate probabilistic simulation relation holds between them, i.e.:

- The test data satisfy probabilistic assignments with at least the same values as the training data, but not only, and
- The assignments of the computational model for features to points in the test data are at least the same as in the target system, but not only.

In other words, we are considering a computational model which might be expressing underfitting evaluation for the variables of interest, while this does not exclude that it assigns values to non-relevant features. The corresponding ontological notion is what has been called a *usable computational artifact* in (Primiero, 2019, p. 266), that is, a computational system which allows for a consistent and fit-for-purpose experimental analysis of the conceptual space determined by the variables of interest, although it does not exclude assignment of probabilities outside of the scope of the target system. Note that *usability* is here intended in a deprecative sense, since an artifact can be usable, yet without ensuring the validity of its result. In fact, the relation of approximate simulation is the most realistic case, as it admits both overfitting and underfitting: this implies asking for which features the computational model is too strong, and for which is too weak, for which features the model might be positively and negatively biased. In this sense, verification of approximate simulation might be a crucial step for improving the fairness of machine learning models.

14.7 Conclusions

In this work, we have approached the problem of formulating validity criteria for ML-based computational models. We argued that such a task requires formal modeling: the validity problem consists in assessing correctness criteria, and hence it assumes the necessity of essentially formal verification techniques. It was then pointed out that a currently available strategy—based on defining simulation relations between the structure of the computational model and that of the target system—is challenged by two characteristics of ML methods: opacity and non-determinism. To address this limitation, we have proposed that weak simulation relations based on the definition of probabilistic models are appropriate to the definition of validity conditions for the computational models of interest. The result of such an approach suggests that formal validity criteria can be formulated for a nuanced range of cases, going from the strongest to the weakest, according to how much a computational model can be assumed to correctly represent a given target system.

We have furthermore formulated a notion of *justifiable validity* expressing necessary (but not sufficient) conditions for a computational model capable of fully expressing the range of results assumed to be correct for the target system. The notion of *weak validity* offers necessary (but not sufficient) conditions for a computational model capable of being at least as correct as the intended results of the target model but possibly admitting overfitting. Finally, the criteria for an *almost valid model* spell out the necessary (but, again, not sufficient) conditions for a computational model capable of at least being as correct as the intended results of the target model but possibly admitting high bias. The formalization of the associated simulation relations and the formulation of appropriate verification algorithms are the next steps in this research.

Acknowledgements This research was supported by the Ministero dell'Università e della Ricerca (MUR) through PRIN Project SMARTEST—Simulation of Probabilistic Systems for the Age of the Digital Twin (20223E8Y4X), and through the Project "Departments of Excellence 2023–2027" awarded to the Department of Philosophy "Piero Martinetti" of the University of Milan.

References

Alvarado, R. (2022). Computer simulations as scientific instruments. *Foundations of Science, 27*(3), 1183–1205. https://doi.org/10.1007/s10699-021-09812-2

Alvarado, R. (2023). Computer simulations in science. In *Synthese library* (pp. 19–28). Springer International Publishing. https://doi.org/10.1007/978-3-031-38647-3_2

Alvarado, R. (2025). Challenges for computational reliabilism: Epistemic warrants, Endogeneity and error-based opacity in machine learning. In J. M. Durán & G. Pozzi (Eds.), *Philosophy of science for machine learning: Core issues and new perspectives*. Synthese Library, Springer.

Angius, N., & Plebe, A. (2023). From coding to curing. Functions, implementations, and correctness in deep learning. *Philosophy & Technology, 36*(3), 47. https://doi.org/10.1007/s13347-023-00642-7

Angius, N., & Primiero, G. (2018). The logic of identity and copy for computational artefacts. *Journal of Logic and Computation, 28*(6), 1293–1322. https://doi.org/10.1093/logcom/exy012

Angius, N., & Primiero, G. (2023). Copying safety and liveness properties of computational artefacts. *Journal of Logic and Computation, 33*(5), 1089–1117. https://doi.org/10.1093/logcom/exac053

Angius, N., Primiero, G., & Turner, R. (2021). The philosophy of computer science. In E. N. Zalta (Ed.), https://plato.stanford.edu/archives/spr2021/entries/computer-science/ ;MetaphysicsResearchlab*The Stanford encyclopedia of philosophy (Spring 2021)*. Stanford University.

Baier, C., Engelen, B., & Majster-Cederbaum, M. (2000). Deciding bisimilarity and similarity for probabilistic processes. *Journal of Computer and System Sciences, 60*(1), 187–231. https://doi.org/10.1006/jcss.1999.1683

Baier, C., Hermanns, H., Katoen, J., & Wolf, V. (2006). Bisimulation and simulation relations for Markov chains. *Electronic Notes in Theoretical Computer Science, 162*, 73–78. https://doi.org/10.1016/j.entcs.2005.12.078

Banks, A., & Ashmore, R. (2019). Requirements assurance in machine learning. In *SafeAI@AAAI*.

Beckers, S., Halpern, J. Y., & Hitchcock, C. (2023). Causal models with constraints. In *Conference on causal learning and reasoning* (pp. 866–879). PMLR.

Boge, F. J. (2021). Two dimensions of opacity and the deep learning predicament. *Minds and Machines, 32*(1), 43–75. https://doi.org/10.1007/s11023-021-09569-4
Brading, K., & Landry, E. (2006). Scientific structuralism: Presentation and representation. *Philosophy of Science, 73*, 571–581.
Buijsman, S. (2025). Machine learning models as mathematics. In J. M. Durán & G. Pozzi (Eds.), *Philosophy of science for machine learning: Core issues and new perspectives*. Synthese Library, Springer.
Cooper, A. F., Frankle, J., & De Sa, C. (2022). Non-determinism and the lawlessness of machine learning code. In *Proceedings of the 2022 symposium on computer science and law, 1–8. CSLAW '22. New York, NY, USA*. Association for Computing Machinery. https://doi.org/10.1145/3511265.3550446
Creel, K., & Hellman, D. (2022). The algorithmic leviathan: Arbitrariness, fairness, and opportunity in algorithmic decision-making systems. *Canadian Journal of Philosophy, 52*(1), 26–43. https://doi.org/10.1017/can.2022.3
D'Asaro, F. A., Genco, F. A., & Primiero, G. (2025). Checking trustworthiness of probabilistic computations in a typed natural deduction system. *Journal of Logic and Computation, 35*(6), exaf003. https://doi.org/10.1093/logcom/exaf003
Durán, J. M. (2018). *Computer simulations in science and engineering – Concepts, practices, perspectives*. Springer.
Durán, J. M. (2025). Beyond transparency: Computational reliabilism as an externalist epistemology for algorithms. In J. M. Durán & G. Pozzi (Eds.), *Philosophy of science for machine learning: Core issues and new perspectives*. Synthese Library, Springer.
Durán, J. M., & Formanek, N. (2018). Grounds for trust: Essential epistemic opacity and computational reliabilism. *Minds and Machines, 28*(4), 645–666.
Facchini, A., & Termine, A. (2022). Towards a taxonomy for the opacity of AI systems. In V. C. Müller (Ed.), *Philosophy and theory of artificial intelligence 2021* (pp. 73–89). Springer International Publishing.
Frigg, R., & Nguyen, J. (2017). Models and representation. In L. Magnani & T. Bertolotti (Eds.), *Springer handbook of model-based science* (pp. 49–102). Springer.
Genco, F. A., & Primiero, G. (2023). A typed lambda-calculus for establishing trust in probabilistic programs. https://arxiv.org/abs/2302.00958
Guala, F. (2002). Models, simulations, and experiments. In L. Magnani & N. J. Nersessian (Eds.), *Model-based reasoning: Science, technology, values* (pp. 59–74). Springer. https://doi.org/10.1007/978-1-4615-0605-8_4
Guidotti, R., Monreale, A., Ruggieri, S., Turini, F., Giannotti, F., & Pedreschi, D. (2018). A survey of methods for explaining black box models. *ACM Computing Surveys, 51*(5), 1–42. https://doi.org/10.1145/3236009
Humphreys, P. (1990). Computer simulations. In *PSA: Proceedings of the biennial meeting of the philosophy of science association (Vol. 1990)* (pp. 497–506). Cambridge University Press.
Humphreys, P. (2004). *Extending ourselves: Computational science, empiricism, and scientific method*. Oxford University Press.
Humphreys, P. (2019). Knowledge transfer across scientific disciplines. *Studies in History and Philosophy of Science, 77*(2018), 112–119.
Jackson, M. (1995). The world and the machine. In *Proceedings of the 17th international conference on software engineering – ICSE '95*. https://doi.org/10.1145/225014.225041
Jonsson, B., Yi, W., & Larsen, K. G. (2001). Probabilistic extensions of process algebras. In J. A. Bergstra, A. Ponse, & S. A. Smolka (Eds.), *Handbook of process algebra* (pp. 685–710). Elsevier Science. https://doi.org/10.1016/B978-044482830-9/50029-1
Kieval, P. H. (2025). Representation learning without representationalism. A non-Representationalist account of deep learning models in scientific practice. In J. M. Durán & G. Pozzi (Eds.), *Philosophy of science for machine learning: Core issues and new perspectives*. Synthese Library, Springer.

Liu, X., & Lorini, E. (2023). A unified logical framework for explanations in classifier systems. *Journal of Logic and Computation, 33*(2), 485–515. https://doi.org/10.1093/logcom/exac102

López, N., & Núñez, M. (2004). An overview of probabilistic process algebras and their equivalences. In C. Baier, B. R. Haverkort, H. Hermanns, J. Katoen, & M. Siegle (Eds.), *Validation of stochastic systems: A guide to current research* (pp. 89–123). Springer. https://doi.org/10.1007/978-3-540-24611-4_3

McKim, V., & Turner, S. P. (1997). *Causality in crisis?: Statistical methods & search for causal knowledge in social sciences*. Notre Dame Press.

Primiero, G. (2019). *On the foundations of computing*. Oxford University Press.

Reichenbach, H. (1938). *Experience and prediction: An analysis of the foundations and the structure of knowledge*. University of Chicago Press.

Seshia, S. A., Desai, A., Dreossi, T., Fremont, D., Ghosh, S., Kim, E., Shivakumar, S., Vazquez-Chanlatte, M., & Yue, X. (2018). Formal specification for deep neural networks. In *Proceedings of the International Symposium on Automated Technology for Verification and Analysis (ATVA)* (pp. 20–34). Springer International Publishing.

Summers, C., & Dinneen, M. J. (2021). Nondeterminism and instability in neural network optimization. In M. Meila & T. Zhang (Eds.), *Proceedings of the 38th international conference on machine learning* (Proceedings of machine learning research) (Vol. 139, pp. 9913–9922). PMLR. https://proceedings.mlr.press/v139/summers21a.html

Termine, A., & Primiero, G. (2024). Causality problems in machine learning systems. In P. Illari & F. Russo (Eds.), *The Routledge handbook of causality and causal methods* (pp. 325–341). Routledge.

van Fraassen, B. C. (1980). *The scientific image*. Oxford University Press.

Vogelsang, A., & Borg, M. (2019). Requirements engineering for machine learning: Perspectives from data scientists. https://arxiv.org/abs/1908.04674

Woodward, J. (2000). Explanation and invariance in the special sciences. *The British Journal for the Philosophy of Science, 51*(2), 197–254.

Zednik, C., & Verreault-Julien, P. (2025). Can explainable AI contribute to justification? In J. M. Durán & G. Pozzi (Eds.), *Philosophy of science for machine learning: Core issues and new perspectives*. Synthese Library, Springer.

Chiara Manganini is a PhD Candidate with the Logic, Uncertainty, Computation and Information Lab, at the University of Milan, Italy. She is part of the PhilTech Research Center for the Philosophy of Technology in the Department of Philosophy of the same university. Her philosophical research pertains to the ontology and epistemology of computational artifacts, especially in relation to the notion of computational error. Relatively to machine learning systems, her research focuses on logical formalizations of bias and fairness.

Giuseppe Primiero is Professor with the Logic, Uncertainty, Computation and Information Lab and Director of the PhilTech Research Center for the Philosophy of Technology in the Department of Philosophy, University of Milan, Italy. Giuseppe works in the formal modeling and verification of multi-agent systems. His preferred tools are proof systems, modal and computational logics. Giuseppe's formal and philosophical research is applied to AI systems and their deployment for resolving issues of misinformation and disinformation online, as well as computable approaches to the evaluation of trustworthiness, bias and fairness. He is author of "On the Foundations of Computing" (OUP, 2019).

Part VI
Scientific Practice and Scientific Values in ML

Chapter 15
Why Are Human Epistemic Agents Not Displaced in Machine Learning Scientific Inquiries? A Practice Perspective on ML in Science

Sahra A. Styger, Marianne de Heer Kloots, Oskar van der Wal, and Federica Russo

Abstract This chapter considers machine learning (ML) practices used in science. Because ML practices enjoy increasing degrees of automation at various stages of the process, the question whether human epistemic agents are displaced arises. We first point out that shifting focus from the ML outputs to the *practice* of designing and using ML models allows one to appreciate the role of different actors in this process, from the human designers and modelers to the algorithms themselves. We illustrate this point with a description of ML-based practices in neuroscience. We then go further with problematizing the role of human epistemic agents in ML and argue that they are not displaced.

Keywords Human epistemic agents · Artificial agents · ML practices · NLP · Cognitive neuroscience · Algorithmic bias · Human displacement · Human-in-the-loop

---

S. A. Styger (✉)
Department of Philosophy, University of Konstanz, Konstanz, Germany
e-mail: sahra.styger@uni-konstanz.de

M. de Heer Kloots · O. van der Wal
Institute for Logic, Language and Computation, University of Amsterdam, Amsterdam, The Netherlands
e-mail: m.l.s.deheerkloots@uva.nl; o.d.vanderwal@uva.nl

F. Russo
Utrecht University, Utrecht, The Netherlands
e-mail: f.russo@uu.nl

J. M. Durán, G. Pozzi (eds.), *Philosophy of Science for Machine Learning*, Synthese Library 527, https://doi.org/10.1007/978-3-032-03083-2_15

15.1 Machine Learning and the Question of the Displacement of Scientists

Recent advancements in machine learning (ML) and in artificial intelligence (AI) have led to an increased societal prominence of artificial systems that display some degree of agency and of intelligent and autonomous behavior. To put it roughly, an agent is capable of acting. In that sense, computer agents or artificial agents are described to do things such as operate autonomously, perceive their environment, persist over a long period of time, adapt to change, and create and pursue goals (Russell & Norvig, 2021, 21–22). Then, ML in particular is an artificial *learning* agent, because ML improves performance after, e.g., observing some data, building a model based on the observed data, and using the model to solve a problem (Russell & Norvig, 2021, 669). According to Ezenkwu and Starkey (2019, 335), there is no unifying definition of machine autonomy in artificial agents to assess the degree of autonomy in, e.g., industrial robots or game playing agents. However, Ezenkwu and Starkey (2019, 336ff) propose two categories of attributes of machine autonomy: low-level attributes (incl. perception, actuation, learning, context-awareness, decision-making) and more advanced high-level attributes (incl. domain-independence, self-motivation, self-recovery, self-identification of goals). The high-level attributes in particular are the subject of numerous research projects in the field of autonomous systems with ML, which points to future progress in the degree of autonomy in artificial systems.

We, human epistemic agents, interact with such advanced artificial systems during everyday activities (think of smart devices for domestic use or chatbots used in customer care), and also in proper scientific settings (think of nested models used in climate science or the use of ML techniques in neuroscience). In this chapter, we focus on the use of ML models in scientific research. In numerous contexts, we are witnessing human scientists assisted by artificial counterparts. 'Intelligent' machines are able to take over many tasks where human expertise was deemed central, and this could be perceived as a threat to the unique role of humans in the process of scientific inquiry. With the rapid development of more advanced ML models, one therefore wonders whether humans may get epistemically displaced or even completely replaced by such intelligent machines. Humphreys (2004, 2009) has already expressed similar concerns about displacement of humans by computational science: computational methods used in science raise new questions for the philosophy of science, because computational science uses methods that "push humans away from the center of the epistemological enterprise" (Humphreys, 2009, 616) and of knowledge production. Or, more recently, the question pertains whether technologies that automatically induce scientific discoveries from data (e.g., AlphaFold discovering protein structures; see Jumper et al. (2021)) may with time become so advanced that scientists will be replaced in doing research.

It is worth noting that this sense of displacement is not new—think for instance of the Marxist critique of technology that followed the first industrial revolution—but with yet another wave of developments in AI and ML techniques in the fourth industrial revolution, it is becoming as timely as ever. Recent philosophical works deal with critical reflections on automated science. For instance, Mieke Boon (2020) has pointed to the question of whether a future is conceivable in which ML algorithms may replace human scientists by taking over more and more epistemic tasks. Within the Philosophy of Information, Luciano Floridi (2016) has already discussed a similar phenomenon using the concept of 'in-betweenness'.

'In-betweenness' refers to the different forms of interactions between agents, qua users, and technologies. The general scheme involves a 'prompter' (i.e., whatever stimulates or suggests using a given technology), a user, and a technology in between them:

$$\text{user} \leftarrow \text{technology} \rightarrow \text{prompter}$$

The simplest type of relation is called *first-order technology*, in which nature acts as a prompter and technology is interposed between humans and nature as a reaction to said prompt:

$$\text{humans} \leftarrow \text{technology} \rightarrow \text{nature}$$

First-order technologies are, for instance, sunglasses to protect our eyes, an axe to split woods, or spectacles to improve our vision. Most of such first-order technologies are analogue technologies that help humans to cope with nature in simple ways. In science, we can think of a simple meter or an analogue telescope to be this kind of technology.

With *second-order technologies*, we alter this configuration, and now technologies are in-between humans and another technology:

$$\text{humans} \leftarrow \text{technology} \rightarrow \text{technology}$$

Thus, for instance, we use a screwdriver to operate on a screw, or a remote controller to turn on the TV. Many household appliances or scientific instruments before the digital revolution are of that type.

Finally, *third-order technologies* are those in which humans are (allegedly) out of this chain, according to the scheme:

$$\text{technology} \leftarrow \text{technology} \rightarrow \text{technology}$$

In this third scheme, technology is not just a prompter (the position on the right) and a mediator (middling position) *but is also an agent that reacts to a prompt* (the position on the left). Numerous digital technologies can interact with other technologies without humans being present. Prima facie, this happens in the Internet of Things, in deep-learning algorithms, or in 'nested' climate models. In science, the practice of using machine learning models to answer increasingly data-driven questions has been suggested as a paradigm-shift away from more traditional experiment-, theory- and simulation-driven approaches to scientific discovery (Hey et al., 2009). Recent deep learning-powered breakthroughs in for example protein structure prediction (Jumper et al., 2021) and weather forecasting (Lam et al., 2022; Nguyen et al., 2023) can be seen as part of this new paradigm. It is this use of machine learning techniques in numerous scientific contexts that may lead to the impression that we are in Floridi's third-order configuration, in which humans in general and, specific to our context, scientists, seemingly do not appear anymore in the process that leads from data to the model to the scientific output.

In this chapter, we problematize the question of the displacement of humans, and specifically of human scientists by artificial agents in ML-based scientific inquiries. Our answer is that while it is undeniable that ML models are driving change in the way scientific inquiry is conducted, scientists are *not* ipso facto completely displaced. On the contrary, by adopting a *practice perspective* on our philosophical analysis of ML-based science, we show that scientists still have a distinct and fundamental role. Our point is partly descriptive, in that we rely on descriptions of ML practices, but is also normative in that it prescribes a certain role of human scientists in said practices. Briefly put, according to the practice perspective that we embrace, we enlarge the focus from the outputs of ML models as to include relevant and informative descriptions of all the actors involved in the generation of these outputs; methodologically, this is what allows us to identify the enduring role of human epistemic agents in the scientific process. Ultimately, in the chapter, we make two interrelated points. First, the question of 'displacement' does not call for a yes or no answer: The possible displacement comes in degrees, depending on which tasks are outsourced to artificial agents and by how much. Second, we argue that human scientists are not in fact completely displaced: The practice-based description we offer below shows that human scientists remain involved even when their tasks are largely outsourced to ML techniques. While some tasks may disappear, others may remain or arise; specifically, the creation, test, and interpretation of the outputs of ML techniques cannot be outsourced to ML themselves.

The chapter is organized as follows. In Sect. 15.2, we introduce the practice perspective that we use in the rest of the chapter. In Sect. 15.3, we describe machine learning techniques from this point of view, using natural language processing and ML in neuroscience as illustrations. We show that we need a more nuanced understanding of 'displacement of humans', meaning that scientists do not just 'vanish' completely. In Sect. 15.4, we continue the discussion of the potential displacement of humans by focusing on the outputs of ML models, and particularly by paying attention to questions of opacity and explainability and of bias. We conclude in Sect. 15.5 by returning to the concept of 'in-betweenness' and we argue that human scientists remain in the chain, in various capacities and roles, and foremost of the design choices that are needed in several moments of the practice of ML.

We write this contribution from the perspectives of philosophy of techno-science, philosophy of science in practice, and of empirical research in machine learning and neuroscience. We disclose our positionality to help the reader better understand not just the theoretical presuppositions of this chapter, but also its ambitions in terms of setting up a fruitful dialogue between philosophy and empirical research. Specifically, we embrace a 'practice perspective' and the emphasis on the intertwinement of science-technology and of humans-instruments, as is elaborated in the philosophy of techno-science (Russo, 2022). In this chapter, we specifically investigate how this approach helps shed light specifically on machine-learning practices, which is part of the research project "The Future of Creativity in Basic Research",[1] and how it can further shape empirical research ("How to bridge neurobiology and psycholinguistic theory by computational modelling",[2] "The biased reality of online media - Using stereotypes to make media manipulation visible"[3]). To do so, we combine expertise in philosophy of techno-science (Styger, Russo) and in machine learning (de Heer Kloots, van der Wal).

15.2 Machine Learning from a Practice Perspective

This chapter moves from a distinct approach in philosophy of science that gives prominence to the *practice of science*, in this case the practices of using machine learning techniques in scientific contexts. The methodology of a practice approach is the one broadly used in the circles of the philosophy of science in practice (Poliseli et al., 2022; Poliseli & Russo, 2022), and that has been specifically theorized by authors such as Chang (2014) in history and philosophy of science or as Pickering (1992) in STS (for a discussion, see the whole volume edited by Soler et al. (2014) and Russo (2022, Chap. 3).

In a practice approach, emphasis and attention is given to *how* science is carried out, over and above its finished products (e.g., established theories or successful explanations). Following Chang, it is important to note that in science we never deal with *one* well-defined practice, but rather with systems of practices, in which various elements of analysis play a role. Consider the following elements, (see Chang, 2014), especially Table 2 on p. 16):

- **Activity:** What is being done in the practice in question?
- **Aims:** What is the inherent purpose of this activity, and what external function does it serve?
- **Systematic context:** Does the activity constitute a part of a broader system of practices?
- **Agent:** Who is doing the activity?

---

[1] The Future of creativity in basic research: can artificial agents be authors of scientific discoveries?, funded by VW-Stiftung grant Az:97721 https://www.philosophie.uni-konstanz.de/ag-mueller/forschung/projekte/the-future-of-creativity/.

[2] Funded by the Netherlands Organization for Scientific Research (NWO), through a Gravitation Grant 024.001.006 to the Language in Interaction Consortium. https://www.dcc.ru.nl/languageininteraction/research-organization/big-questions/big-question-1/.

[3] Funded by the NWO as part of the Open Competition Digitalisation under project number 406.DI.19.059. https://www.nwo.nl/en/projects/406di19059.

- **The second person:** To/with whom?
- **Capabilities:** What must the agent be capable of, in order to carry out this activity?
- **Resources:** Which tools are necessary for this activity to be successful?
- **Freedom:** What kind of choices does the agent make?
- **Metaphysical principles:** What must we presume the world to be like, in order for this activity to be coherent?
- **Evaluation:** Who is judging the results, and by what criteria (in addition to coherence)?

A practice approach, at least in the version proposed by Russo (2022), does not prescribe to address *all* these aspects at once, but to specify, on any given occasion, the elements that are of interest or relevance. In our case, we are specifically interested in how human epistemic agents and artificial agents interact in scientific inquiries that employ ML-techniques, and we ask the question what their epistemic aims and purposes are. We emphasise this question because, following the 'in-betweenness' scheme, it looks like the application of ML techniques in science belongs to a third-order configuration of technologies, in which scientists are displaced and do not play a proper role in the scientific inquiry anymore. Is this really the case? As mentioned, we will show that scientists are *not* completely displaced. To answer this question, we embrace another important stronghold of a practice approach, namely its highly interdisciplinary endeavor, requiring an engagement between philosophers and scientists. The way in which such engagement happens is not predetermined and can take various forms.

With a practice approach, we wish to change the perspective from which ML models are typically analyzed. The usual narrative is about what an "algorithm does" and consequently, about what its output is. It is certainly true that technologies, and especially computational technologies in AI, have (degrees of) agency, as they have been described as intelligent agents (Russell & Norvig, 2021) or learning agents (Eva et al., 2023). However, it is important to consider how these artificial systems operate in an environment, which can be the space in which an algorithm works but also more broadly the "hybrid" network of humans and non-humans of socio-technical systems (Emery et al., 1960; Latour, 2005). In order to better understand what we mean by "environment", we borrow some intuitions from the work of Latour and Woolgar (1986), and notably the *Actor Network Theory*, that suggests analyzing any given case by mapping *all* the actors involved (humans and non-humans) that operate in an environment or, in the terminology of Chang, in a given context. That is certainly the specific computational framework for the case of ML, but also the physical, social, institutional setting in which ML practices are performed. The environment in which ML operates includes scientists as well as a whole range of infrastructure, from the institutional embedding to the material lab equipment, from the protocols of the experiment to the cultural and moral values of the human agents that design and use them. As scholars in Science and Technology Studies would put it, we are dealing with a *socio-technical system* and, from a practice perspective, human epistemic agents are just as important as the algorithms and ML models in such systems.

In the following, we reconstruct salient aspects of ML from a practice perspective, introducing two *episodes* that belong to said practices. We use the term episode here, rather than 'case study' or 'example' to convey the idea that we are not dealing with special, isolated cases, and that while they are relevant examples of ML in practice, they also belong to a wider and longer narrative, or are like a *series* of episodes (Russo, 2022, Chap. 3).

15.3 Machine Learning in Practice

15.3.1 The Episode of Natural Language Processing

Machine learning as a field is only one of many research endeavors focused on developing artificial intelligence, i.e., computer systems that perform intelligent behaviors. The AI research community has historically been divided into so-called *symbolic* AI and *subsymbolic* (or *nonsymbolic*) AI. As described by Mitchell (2020), symbolic AI used logic-based systems to formalize descriptions of conscious thought processes, resulting in programs operating on manually defined symbols and rules. On the other hand, subsymbolic AI took more inspiration from neuroscience and unconscious thought processes, resulting in programs essentially consisting of mathematical operations on numbers that do not necessitate humanly interpretable reasoning steps. The past decade of developments in AI technologies has been dominated by the latter, subsymbolic paradigm, and specifically by the rise of *deep neural networks* (DNNs) — a class of machine learning systems that learns from large amounts of data, by adjusting weights in layered configurations of computing units. All AI application areas, including for example the image, audio, and text processing domains, have by now transitioned to largely deep learning-based systems, with large-scale pre-trained 'foundation models' at the core of most applications (Bommasani et al., 2022).

Machine learning refers to a whole set of techniques that allow models to learn from a given data set (called a training data set). ML models are designed to make accurate inferences on new data with a similar structure and distribution to the training data, i.e., to learn generalizable solutions. Typical tasks performed by ML models are categorization and recognition (e.g., of objects in images or of sentiments in texts) and generation (e.g., new text, following provided previous context).

ML approaches can be divided into different classes of learning problems that vary in the extent to which the prediction target requires expert design (Bishop, 2006; Jurafsky & Martin, 2023). Whereas supervised systems are trained on data with human-annotated labels (for example pairs of images and annotations of the pictured objects), unsupervised systems are rather tasked with clustering data points into groups (e.g., customer buying behaviors) without labeled classes being defined beforehand. With the rise of deep representation learning approaches and internet-scale data, so-called self-supervised systems have gained traction, where the initial prediction target is contained within the available unannotated data (e.g., the next word in a sentence, given previous context). In the following, we go into more detail on the course of this transition and some of its consequences for scientific practice in the domain of natural language processing (NLP) specifically.

Natural language processing is an interdisciplinary field that is concerned with algorithms dealing with natural language related tasks. Examples of NLP domains are machine translation (e.g., Google translate), information retrieval (e.g., search engines), and chatbots (e.g., ChatGPT) (Jurafsky & Martin, 2023). The field of NLP has seen many changes in its research goals and epistemic activities. For instance, where 'Good Old-Fashioned AI' (GOFAI) focused on symbolic (explicit, rule-based) solutions, this was later abandoned in favor of statistical techniques that required fewer explicit rules to be hard-coded. This in turn, was largely replaced by subsymbolic (connectionist) techniques that relied on deep neural networks, which attempts at avoiding encoding any 'expert domain knowledge' in the instrument at all.

One trend, it seems, is that the development of computational methods is moving to that of increasingly *general* solutions (Sutton, 2019): where not so long ago, 'feature engineering' (the manually processing and selection of features from raw data) was a crucial part of the NLP modeling process, we now use self-supervised approaches to let these models learn such features from 'raw data'.[4]

At this point in time, language models based on deep neural networks—of which the Transformer (Vaswani et al., 2017) is currently the most popular one—have become the de facto NLP model. These models are trained to do a simple task—e.g., assign probabilities to sentences or predict the most likely next word in a given context (Jurafsky & Martin, 2023)—but have proven to be important building blocks for other tasks more 'downstream' (e.g., chatbots, like ChatGPT, are based on such language models). Some researchers refer to these language models as 'foundation models', since these are general purpose models that can be used in a broad range of applications (Bommasani et al., 2022). An even more recent development is that of 'prompt engineering', where the models are asked in natural language instructions (prompts) to perform specific tasks, without any explicit adaptation to a specific application (Liu et al., 2023). Interestingly, the adoption of these 'self-learning models' has given rise to a field of study (initially referred to as 'BERTology', now 'interpretability') that investigates the inner workings of transformer-based models: e.g., what features are found by the models (Rogers et al., 2021).

The practices in NLP have changed in other important ways, because of the technological changes it has relied on. The success of language models (and deep

[4] In a sense, one can argue that data is never a perfect and objective representation of the world. See for a discussion, for instance, Friedler et al. (2021). With this caveat, the data is less structured than it traditionally was (e.g., instead of syntactic and semantic features explicitly labeled, now the data may contain just textual data), which is why it is referred to as 'raw data'.

neural networks more generally) rests predominantly on the availability of (cheap) computation power (in the form of GPUs) and vast amounts of data ('big data'). Initially, abstractions of the NLP modeling process in the form of pretrained models (foundation models) and code libraries (e.g., those offered by HuggingFace, see Wolf et al. (2020)) have made it much easier for researchers to work with these language models and apply these to new contexts. However, now these models and datasets have become so large that it has become difficult to analyze, let alone train them without the right resources. State-of-the-art language models now comprise hundreds of billions of parameters, and training and analyzing models at that scale is substantially harder than 'smaller' models that fit on computer systems most researchers have access to. Limiting this access even further, many organizations only allow some access to these models via their own proprietary application programming interfaces (APIs). Most researchers are now confined to researching the models only 'downstream' in the modeling pipeline (i.e., studying the output of these models, a point raised in e.g., Talat et al., 2022). This asymmetry has consequences for the autonomy researchers and research institutions have in making epistemic design choices as well as the overall transparency of their research; a point we further discuss in Sects. 15.4 and 15.5.

The aims of researchers working in and with NLP are very diverse. These range from purely technical goals such as training and designing these models (including the collection of vast training datasets), evaluating and analyzing these models, to applying these to other fields. We also see how NLP has become an important factor in 'other sciences'. While NLP (and ML more generally) has clearly served as important inspirations and tools for other scientists, there has been critique on how it has constricted the epistemic practices of these researchers. For example, Gyllingberg et al. (2023) argue that the ML-based focus on fitting models to data draws attention away from the more fundamental scientific activities of formulating and analyzing the models yourself. In this light, we now turn to the epistemic use of ML generally, and NLP models more specifically, in cognitive neuroscience.

15.3.2 Machine Learning as a Modeling Practice: Episodes from Cognitive Neuroscience

The use of ML in cognitive neuroscience (see Chap. 20 by Crook and Kästner 2025) is particularly interesting to contextualize and conceptualize the use of ML as a *modeling practice*. Recalling Chang's element of 'activity', we here address the question of 'what is being done' when we model a phenomenon in science. By focusing on *modeling practices* that employ ML techniques, we can make explicit the epistemic activities, the epistemic aims and the epistemic agents involved in this techno-scientific practice. This is a first necessary step on our way to answer the question if human researchers are being displaced in scientific practice.

In the following episode from cognitive neuroscience, we contextualize the practice of modelling brain activity with ML-based systems. Which epistemic aims and agents are involved in studying the brain with ML? Neuroscientists have long employed statistical regression models in building predictive models of brain activity, which learn from data what stimulus features are most important for predicting activation in (particular areas of) the human brain. However, recent work in this area makes increasing use of features extracted from deep learning systems as the input to such predictive models (a configuration that prima facie fits third-order technologies mentioned above), rather than manually defined predictors designed by scientists. While brain activity of a participant watching an image can be predicted based on annotated features of the scientists' choice (e.g., image brightness, colors, contents), higher predictive accuracy is achieved when using the internal activations of a deep neural network processing the raw image pixels instead, even though such features are not easily interpretable for humans (for a discussion, see also Chap. 12 by Kieval 2025) and Chap. 13 by Freiesleben 2025). Hence, it might appear that the human scientist's role in the practice of modelling brain activity (choosing the relevant predictors) has been replaced by deep learning technologies. However, humans still play a crucial role in the *design* of experiments using these technologies, and thus in scientific progress, which underlines that scientists are not displaced. Here, the early design choices made by scientists on how to study brain activity by modelling brain activity with ML should therefore not be mistaken for displacement. It is also worth noting, as we also mentioned earlier, that human epistemic agents still have roles to play at different stages of the process, a point in line with 'computational reliabilism' as elaborated by Durán (2023).

Cognitive neuroscience is the scientific field that studies the neurobiological bases of mental processes such as language and visual perception, and it is one area of science where ML-based approaches have made a transformative impact. In the past decade, a new strand of research in this field has been developed that makes use of artificial intelligence technologies as scientific models of cognitive tasks (see i.a., (Kriegeskorte & Douglas, 2018; Cichy & Kaiser, 2019; Ma & Peters, 2020)). The general 'cognitive computational neuroscience', 'neuroconnectionist', or 'Neuro-AI' approach embraced in this line of work is to directly compare deep learning systems (for example an image classification or language model) with human or animal subjects in a neuroscientific experiment. This comparison involves the internal states generated inside the deep neural network while processing a stimulus on one side, and the experimental subjects' brain activity when processing the same stimulus on the other side (for example, observing an image or reading a piece of text). DNN models are then evaluated on their ability to capture neurally relevant stimulus properties, reflected in how well brain activity aligns with, or is predictable from, the models' internal states. Such DNN-based predictive models of brain activity based on internal states of AI systems first took off with computer vision systems and neural activity in visual perception (Yamins & DiCarlo, 2016), but have now also produced impressive results with NLP systems and neural activity related to language comprehension (Schrimpf et al., 2021; Caucheteux & King, 2022).

The epistemic aim of predicting neural activity from observed stimuli is not new, although it contrasts with more traditional research practices in neuroscience and psychology based on controlled lab-based experiments. Such studies generally aim to demonstrate relevant contrasts between two or more theory-based but usually quite artificial experimental conditions. In other words, psychological theories are tested by contrasting very specific subcomponents of the stimulus space against each other with the goal to confirm or reject hypotheses. One reason this ML-based approach to studying the brain is embraced is that it allows for more naturalistic experimental paradigms: Studying brain activity through predictive models arguably allows for theorizing about more naturalistic and ecologically valid human behaviors (Yarkoni & Westfall, 2017; Hasson et al., 2020).

However, the goal of predicting brain activity recorded in more naturalistic settings in itself does not require the use of DNN-based features as input to the prediction model. An example of predictive modelling in the language domain can be found in the work of Wehbe et al. (2014), who built a generative model of brain activity recorded during story reading. In their model, brain activity is predicted based on descriptive features of the text being read (e.g., the length of words or their grammatical category). Yet, recent work using models based on internal states of NLP systems achieve higher neural predictivity in language regions than encoding models based on lexical and grammatical feature annotations (Kumar et al., 2023). Hence, a second major reason for scientists' design choice to employ AI technologies as instruments for neuroscience is empirical: Predictive models using DNN-based features achieve much higher accuracy than encoding models using theory-based hand-engineered features.

In shifting from hand-engineered to DNN-based features, some interpretability and control over encoding model design is lost in exchange for predictive accuracy: The basis of encoding models are now features learned by a deep learning system to achieve a technological goal, rather than those formulated by scientific theories of particular cognitive phenomena. As mentioned above, this is an instance of 'outsourcing tasks' in the sense of bestowing 'agency' to ML techniques by means of epistemically using ML agents for our scientific goals. This trade-off is welcomed by researchers who emphasize that nature does not owe us easy answers, so to speak: There is no reason to assume that the inherent complexity of brain functioning can be captured in human linguistic terminology. For example, Doerig et al. (2023, 432) write: "(...) human-interpretable labels for neural activity are limited by the imagination of researchers, or simply by language. But natural mechanisms are not necessarily bounded within these constraints: neural selectivity can often rely on more complex features that only imperfectly map onto human-interpretable categories".

Nevertheless, the NeuroAI approach also receives criticism. Often, these critiques are grounded in a disagreement about the value of predictive models and emphasize the importance of explanation beyond prediction in scientific progress. Accurate prediction of neural activity is deemed useless when models are not grounded in explanatory cognitive theories; in this way, the practice of predicting neural activity using DNN features without such a theory has been likened to perfectly predicting the time on analog clocks with the digits displayed on digital clock, which is not an explanation of either clock's internal working (Guest & Martin, 2023). Neural activity may be well predicted by completely unrelated processes (Meijer, 2021), or by models that do not show human-like patterns in their output behavior (Bowers et al., 2022). A general point of criticism is that DNN-based predictions of brain activity are not explanations of brain functioning precisely because the internal states of DNN models are not directly interpretable by human scientists. Hence, the reason for high predictive accuracies remains uninterpretable as well. Here, the displacement of theory-based experimental control by DNN-discovered features seems to cross a line where the kind of scientific understanding that some neuroscientists strive for is out of reach.

However, the choice to prioritize predictive accuracy over human interpretability or explainability does not mean that human scientists are displaced (for a discussion, see Chap. 7 by Páez, 2025). Important choices in the brain activity prediction pipeline are still human-made: Human scientists choose stimuli, construct the setting in which brain activity is recorded, and make choices about what model architecture to use, what data to train it on, exactly what model-generated output to use for prediction. These choices allow scientists to use and investigate brain activity prediction models epistemically by seeking explanations for why particular modelling choices lead to better predictions than others. The choice of using deep learning technologies in cognitive modelling confines the possible space of scientific inquiry to those behaviors and cognitive phenomena that deep learning models can capture, rather than those that are most accessible to human imagination. But this also opens up new epistemic aims and activities, directed at improving and better understanding the modelling pipeline—an area of research where neuroscientists and artificial intelligence researchers are aligned and can mutually benefit from new discoveries (Ivanova et al., 2021; Kar et al., 2022).

15.4 Design Choices at the Core of ML Practice and the Case of Algorithmic Bias

As explained earlier in Sect. 15.2, a practice perspective gives emphasis to *how* science is carried out, over and above its finished products. It is however a good moment to consider the finished products of ML techniques, because it is precisely their opacity (see Chap. 1 by Beisbart (2025) and Chap. 2 by Formanek 2025) that has given rise to philosophical and empirical discussions about trust, reliability, and explainability (Humphreys, 2009; Durán & Jongsma, 2021; Langer et al., 2021; Durán & Formanek, 2018). Specifically, we ask the question: How should we consider results and evidence generated by ML, if we know little about *how* they are generated?

Addressing this question leads us to discuss algorithmic bias or data bias. These are well-known problems, not only for computer scientists, as many different forms of biases affect our research and ML practices. From a practice perspective, it is obvious that 'we' (those who do science) are responsible for many forms of biases ending up in our ML models, which means that our scientific results may be flawed as a consequence. However, as discussed in Sect. 15.3, even if researchers are able to understand their own positionality, the *asymmetry* in access to these models (i.e., researchers cannot train all models themselves and rely on other organizations to provide pre-trained models) makes it close to impossible to study all sources of biases in the ML models they use. Hence, the responsibility of human epistemic agents to remain actively in the loop to check and control our ML practices, for example, by applying fairness principles to the optimizations, by understanding how marginalized groups can be mis- or underrepresented in the results, and by generally being aware of the potential sources of bias.

Using the episode from cognitive neuroscience, it is still possible for scientists to manipulate models in relevant ways to successfully investigate research questions, even when model features are initially not transparently interpretable by humans. Still, the lack of interpretable reasoning steps in ML-based systems underscores the importance of investigating biases and finding ways to ensure e.g., reliability, trust, and explainability. We connect here to questions of explainability and of transparency that are widely discussed in the literature, as well as elsewhere (see Chap. 4 by Durán, 2025, Chap. 5 by Alvarado, 2025, Chap. 6 by Zednik and Verreault-Julien, 2025, Chap. 8 by Buijsman, 2025, and Chap. 9 by Räz 2025). Our intention is not to contribute to the conceptualization and/or operationalization and implementation of these concepts *directly*, but rather to reiterate a point made earlier: humans are only *allegedly* displaced, it is for us mainly and foremost a question of *design* that is at stake. We instead pose the question of transparency or explainability in the following terms: how much are design choices involved in requiring that certain standards of explainability and of transparency are met and ensured? This is why we think we should (also) focus on our choices about how much explainability or transparency is desirable or required, and depending on the context, and not (only) on what they are. Because ultimately it is the human epistemic agent such as the scientist engaging in ML-based scientific research, who wants and needs their ML models to be trustworthy, transparent, and reliable. ML models and their results should be transparent, reliable and explainable, both from a technological and a political policy-making point of view.

Furthermore, if we can establish a prominent role of humans via design choices, it is also important to note that ML tools (and any other technical artefacts) are not merely executors of some scientific protocol, in the sense that we might also obtain unexpected and difficult-to-explain outputs, in part because we don't have full understanding of the inner working of ML, and in part because novel outputs may emerge from machine-machine or human-machine interactions. Collectively, our considerations about the role of humans in the practice of ML and the conceptualization of (digital) technology as not being inert (i.e., there is a dynamic influence between epistemic practices and technologies) leads to abandoning a form of opposition between *us* (humans) and *them* (the technologies), and embrace a view according to which human and artificial agents are *in a partnership* in the process of knowledge production. We will specify this partnership in more detail with the idea of humans remaining in the loop in the next section. It also follows that in this partnership the terms are constantly negotiated, and so it must be a continuous question we ask about our role and that of the technologies, what we are prepared (not) to accept from the results generated by technologies (and ML specifically in this case). This argument has been made reflecting on techno-scientific practices in general (Russo, 2022), and we have particularized here some of those considerations for the case of ML.

We continue our reflections on the 'outputs' of ML in general and of NLP in particular by looking at the debates on bias. ML is an area in which questions of bias are becoming relevant and urgent, primarily because the community is increasingly acknowledging cases of harmful biases in ML outputs, such as in recommendations for health systems, where the implications of algorithmic bias are being extensively discussed (Panch et al., 2019; Aquino et al., 2023).

While the study of algorithmic bias is not new (see e.g., Friedman & Nissenbaum, 1996)), recent developments in ML models, especially those based on neural networks, and their increased adoption in real-world applications, has led to growing concerns about the social biases and harms that these models may propagate and even amplify (Zhao et al., 2017; Lloyd, 2018; Hooker et al., 2020; Hall et al., 2022).

To take the example of natural language processing (as discussed in Sect. 15.3.1), we see a trend of increasingly large language models (LLM), both in parameter size and the amount of textual training data, which poses many challenges for researchers to effectively study the problem of bias e.g., doing it qualitatively/manually becomes impossible. More so, because these biases may be introduced at different stages in the development cycle of LLMs (e.g., in the choice of training data, objective function, or task definition, see (Sun et al., 2019; Hovy & Prabhumoye, 2021). On top of that, researchers face many conceptual problems, because of the socio-technical nature of algorithmic bias: Rather than it being a purely statistical phenomenon, defining bias relies on the normative questions and socio-technical context at hand (e.g., linguistic, cultural, legal; see e.g., (Blodgett et al., 2020; Talat et al., 2022).

In fact, while it is tempting to view ML models as simply encoding pre-existing biases from the data, the adoption of ML models in real-world applications may be seen as part of socio-technical systems themselves, which introduce new biases and harms, and even reshape norms and values as society adapts to these changes (see e.g., the discussion by Kidd and Birhane, 2023). For instance, the use of applications like ChatGPT by medical practitioners may propagate harmful race-based medicine (Omiye et al., 2023). The problem of bias, therefore, requires an interdisciplinary approach to tackle these challenges. This also means that a purely black-box (e.g., evaluating the model's behavior through prompting) or mechanistic understanding (e.g., studying what kind of representations are encoded in the model's parameters) of the model is not sufficient for tackling the questions of bias and harms.

Besides technical work on detecting and mitigating bias, we need other approaches, based on ethnographic and social science methods, and including as well cultural analyses. Studies like these would allow approaching ML models as socio-technical systems, and treat them as part of a larger context: Who designs these systems? What are the social harms when using these? What are the power structures at hand? This is in line with a growing sentiment in the field of algorithmic bias, see e.g., (Margaret Mitchell et al., 2019; Blodgett et al., 2020; Bender et al., 2021; Talat et al., 2022). We believe this approach is equally important in understanding the biases of these models, which means that scientists do not only have a responsibility to work towards the transparency of their epistemic practices using ML models as we discussed in Sect. 15.3, but also to engage with a broader research and stakeholder community outside of their own field.

In sum, in our view, scientists have an even more important role in the scientific process now, since ML models are increasingly used in real-world systems affecting actual people. In combination with the increasing opacity of ML systems, this means that we have a high epistemic and moral responsibility and must play an active role in these practices.

15.5 The Place of Scientists in ML-Based Inquiries and the Missing Link in the 'in-betweenness' Chain

We build here on the previous section with some philosophical considerations about the possible displacements of humans in general and scientists in particular. Our arguments are partly descriptive and partly normative. On the descriptive side, detailed and informative reconstructions of ML practices show that humans are not entirely displaced. For one thing, this is not a question requiring a yes or no answer: Rather, the role of the human epistemic agent is changing rather than diminished. In certain tasks, humans are partly displaced (e.g., in feature engineering as discussed in Sect. 15.3), but other responsibilities and tasks have emerged as a consequence such as the interpretation of models (explainability) and ensuring the 'fairness' and reliability of the scientific practices. On the normative side, our point is that mere presence of humans is not enough to conclude that they are not displaced, because we should require that ML is practiced with a high-level of self-reflection related to design choices, as we now argue.

In the practice of machine learning, there are many translation steps that need to be performed by the researchers. For a certain task, data must be collected (often annotated as well), an objective function defined, a model designed and trained, and finally its performance evaluated. Often, researchers also need to fine-tune a pre-trained model to adapt it for a specific domain or task it was not originally designed for. At each of these steps, the human epistemic agents have to make numerous decisions about what is needed for their epistemic goals.

While in the early days of NLP with ML, researchers had to design the whole 'pipeline', they now often rely on 'pre-trained' language models. As also discussed in Sect. 15.3.1, others with often lots of resources train a model on an unsupervised language modeling task, which researchers can then use as the basis for their own specific use-case. NLP researchers are more and more relying on these and other 'abstractions', allowing them to focus on the particular research question at hand. However, this also means that researchers accept the design decisions others have made by proxy (including normative ones).

Yet, at the same time, these abstractions also serve as bottlenecks, and the engineering of most low-level computational aspects of the ML pipeline (e.g., writing GPU drivers, understanding the computational hardware; see also Anthony et al. (2024)) becomes important to facilitate the creation of large ML models. This is a skill that is not very prevalent, especially not among researchers. Together with the larger demand of (computational) resources, we find that institutions have come to play a central role. This is not to say that researchers have no choice in the face of technological determinism, as many institutions committed to 'open science' have sprung up to counter the initially closed-source tech-corporation dominated field (see e.g., BigScience,[5] EleutherAI,[6] Aya[7]). Hence, researchers have a choice to engage with the organizations that align with their own principles and choose their ML models accordingly. Institutions are now playing a crucial role in ML research, and while smaller labs and individual researchers cannot build these larger ML systems themselves anymore, they still have a say in which institutions and their models to engage with.

It is now useful to return to the idea of in-betweenness introduced in Sect. 15.1. We make two interrelated points. First, in the third order relations (technology-technology-technology), the absence of humans is not real but only apparent—this follows from applying a practice approach descriptively; second, the presence or absence of humans in ML practices can also be a *choice* made at design level, which is a normative consideration.

Descriptively, humans may be displaced in the practice of ML, for instance in the use of a ML algorithm for specific purposes. Not quite: somebody *has* designed this algorithm and is using it. At times, these human scientists are not very visible because they are part of large research consortia or institutions. A key feature of ML is the degree of autonomy and agency we wish to give such algorithms is in fact a *design choice*. So, not only are we still in the loop, but we can decide and control *how much* we want to be in the loop—all these design choices require a high level of self-reflection. To reconsider the schematization of third order relations, the situation would look like this (Fig. 15.1).

Our proposed schematization of the third order relations explicitly shows that human epistemic agents or scientists are still connected in various ways to technological artifacts, even to technologies that can be said to have certain aspects of agency and that are more and more autonomous from us.

By using episodes from NLP in cognitive neuroscience, we have been able to show that we need a more nuanced understanding of 'human displacement'.

---

[5] https://bigscience.huggingface.co/.

[6] https://www.eleuther.ai/.

[7] https://cohere.com/research/aya.

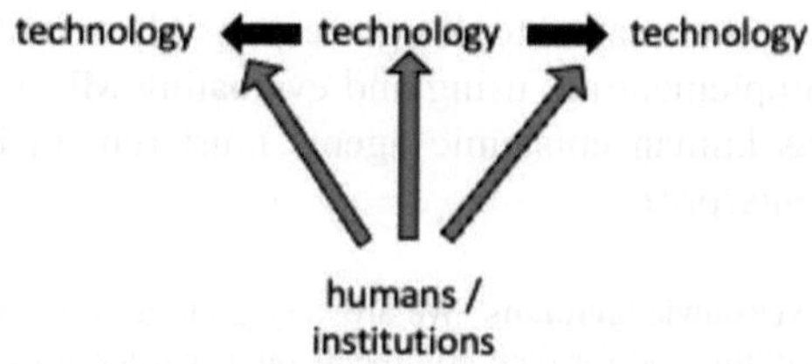

Fig. 15.1 Third-order technology relations including humans and institutions

Scientists do not simply 'disappear' when we engage more intensively with ML-based investigations in science. On the contrary, the NeuroAI episode illustrates how the epistemic use of ML technologies can be seen as a deliberate design decision by researchers that opens up new avenues for scientific inquiry. It is a modeling strategy and a deliberative decision by the scientists in charge, whose goal is to predict human brain activity. This can be described as a different approach to the problem of studying the human brain and demonstrates the potential epistemic benefits of using ML in science.

Also, when we look at the outputs of ML in ML-based scientific inquiries, we were able to emphasize where the place of humans is: Since there is social bias and scientific outputs can be influenced by it, humans are responsible for actively staying in the loop to check and control, e.g., by implementing fairness principles such as trust, reliability and explainability and investigating the bias of these models.

There is another important way in which human epistemic agents are not displaced, but play a key role and "new" places are created for them. Research and the use of ML techniques take place in institutional contexts where there are often protocols and guidelines for the correct, ethical and fair conduct of research. This is not so much about individuals, but rather about epistemic agents qua scientific communities, qua roles in institutions and organizations, and about the institutions and organizations themselves.

15.6 Concluding Remarks

In this chapter, we were tasked to explore the question whether the use of ML techniques displaces human epistemic agents in the process of scientific inquiry, thus leaving little or no roles to humans. We agree that ML does indeed change the mode of scientific inquiry. But the point is: This is not a change that we passively go along with. Rather, it is a change that we initiate and that we choose. Since we are designers and users of ML, we must ask ourselves the question: How many aspects of agency and autonomy do we want to grant to artificial systems? Our answers to this question will, in turn, determine the extent to which we are willing to be displaced in ML-based scientific practices. In other words, our displacement is not automatic or inevitable, but largely a choice. We need to stay in control over what we want ML to do for us, while accepting that we cannot control everything. This is another reason to design responsibly and to guide the entire process of developing, implementing, using and evaluating ML. Of course, this is all an iterative process, as human epistemic agents must remain in the loop of the ML-based epistemic enterprise.

Acknowledgements We are very grateful to the editors for the opportunity to contribute to this volume and for their comments on an earlier draft. Research for this paper has been supported by VW-Stiftung, 2021–2024, support by Volkswagen Stiftung grant Az:97721 (S.A. Styger), NWO Gravitation Grant 024.001.006 to the Language in Interaction Consortium (M.de Heer Kloots), NWO, Open Competition Digitalisation under project number 406.DI.19.059 (O. vd Wal).

References

Alvarado, R. (2025). Challenges for computational reliabilism: Epistemic warrants, endogeneity and error-based opacity in machine learning. In J. M. Durán & G. Pozzi (Eds.), *Philosophy of science for machine learning: Core issues and new perspectives*. Synthese Library, Springer.

Anthony, Q., Hatef, J., Narayanan, D., Biderman, S., Bekman, S., Yin, J., Shafi, A., Subramoni, H., & Panda, D. (2024). The case for co-designing model architectures with hardware. arXiv. https://doi.org/10.48550/arXiv.2401.14489.

Aquino, Y. S. J., Carter, S. M., Houssami, N., Braunack-Mayer, A., Win, K. T., Degeling, C., Wang, L., & Rogers, W. A. (2023). Practical, epistemic and normative implications of algorithmic bias in healthcare artificial intelligence: A qualitative study of multidisciplinary expert perspectives. *Journal of Medical Ethics, 51*(6), 420–428. https://doi.org/10.1136/jme-2022-108850

Beisbart, C. (2025). In which ways is machine learning opaque? In J. M. Durán & G. Pozzi (Eds.), *Philosophy of science for machine learning: Core issues and new perspectives*. Synthese Library, Springer.

Bender, E. M., Gebru, T., McMillan-Major, A., & Shmitchell, S. (2021). On the dangers of stochastic parrots: Can language models be too big? In *Proceedings of the 2021 ACM conference on fairness, accountability, and transparency* (FAccT ‘21) (pp. 610–623). Association for Computing Machinery. https://doi.org/10.1145/3442188.3445922

Bishop, C. M. (2006). *Pattern recognition and machine learning. Information science and statistics*. Springer. https://link.springer.com/book/9780387310732

Blodgett, S. L., Barocas, S., Hal Daumé, I. I. I., & Wallach, H. (2020). Language (technology) is power: A critical survey of “bias” in NLP. In *Proceedings of the 58th annual meeting of the Association for Computational Linguistics* (pp. 5454–5476). Association for Computational Linguistics. https://doi.org/10.18653/v1/2020.acl-main.485

Bommasani, R., Hudson, D. A., Adeli, E., Altman, R., Arora, S., von Arx, S., Bernstein, M. S., et al. (2022). On the opportunities and risks of foundation models. arXiv. https://doi.org/10.48550/arXiv.2108.07258.

Boon, M. (2020). How scientists are brought back into science—The error of empiricism. In M. Bertolaso & F. Sterpetti (Eds.), *A critical reflection on automated science* (Vol. 1, pp. 43–65). Springer International Publishing. https://doi.org/10.1007/978-3-030-25001-0_4

Bowers, J. S., Malhotra, G., Dujmović, M., Montero, M. L., Tsvetkov, C., Biscione, V., Puebla, G., et al. (2022). Deep problems with neural network models of human vision. *Behavioral and Brain Sciences, 46*, 1–74. https://doi.org/10.1017/S0140525X22002813

Buijsman, S. (2025). Machine learning models as mathematics. In J. M. Durán & G. Pozzi (Eds.), *Philosophy of science for machine learning: Core issues and new perspectives*. Synthese Library, Springer.

Caucheteux, C., & King, J.-R. (2022). Brains and algorithms partially converge in natural language processing. *Communications Biology, 5*(1), 1–10. https://doi.org/10.1038/s42003-022-03036-1

Chang, H. (2014). Epistemic activities and systems of practice: Units of analysis in philosophy of science after the practice turn. In L. Soler, S. Zwart, M. Lynch, & V. Israel-Jost (Eds.), *Science after the practice turn in the philosophy, history, and the social studies of science* (Routledge studies in the philosophy of science) (pp. 67–79). Routledge, Taylor & Francis Group.

Cichy, R. M., & Kaiser, D. (2019). Deep neural networks as scientific models. *Trends in Cognitive Sciences, 23*(4), 305–317. https://doi.org/10.1016/j.tics.2019.01.009

Crook, B., & Kästner, L. (2025). Don't fear the bogeyman: On why there is no prediction-understanding trade-off for deep-learning in neuroscience. In J. M. Durán & G. Pozzi (Eds.), *Philosophy of science for machine learning: Core issues and new perspectives*. Synthese Library, Springer.

Doerig, A., Sommers, R. P., Seeliger, K., Richards, B., Ismael, J., Lindsay, G. W., Kording, K. P., et al. (2023). The neuroconnectionist research programme. *Nature Reviews Neuroscience, 24*(7), 431–450. https://doi.org/10.1038/s41583-023-00705-w

Durán, J. M. (2023). Machine learning, justification, and computational reliabilism.

Durán, J. M. (2025). Beyond transparency: Computational reliabilism as an externalist epistemology for algorithms. In J. M. Durán & G. Pozzi (Eds.), *Philosophy of science for machine learning: Core issues and new perspectives*. Synthese Library, Springer.

Durán, J. M., & Formanek, N. (2018). Grounds for trust: Essential epistemic opacity and computational reliabilism. *Minds and Machines, 28*(4), 645–666. https://doi.org/10.1007/s11023-018-9481-6

Durán, J. M., & Jongsma, K. R. (2021). Who is afraid of black box algorithms? On the epistemological and ethical basis of trust in medical AI. *Journal of Medical Ethics, 47*(5), 329–335. https://doi.org/10.1136/medethics-2020-106820

Emery, F. E., Trist, E. L., Churchman, C. W., & Verhulst, M. (1960). Socio-technical systems. In C. W. Churman & M. Verhulst (Eds.), *Management science models and techniques* (Vol. 2, pp. 83–97). Pergamon.

Eva, B., Ried, K., Müller, T., & Briegel, H. J. (2023). How a minimal learning agent can infer the existence of unobserved variables in a complex environment. *Minds and Machines, 33*(1), 185–219. https://doi.org/10.1007/s11023-022-09619-5

Ezenkwu, C. P., & Starkey, A. (2019). Machine autonomy: Definition, approaches, challenges and research gaps. In K. Arai, R. Bhatia, & S. Kapoor (Eds.), *Intelligent computing* (Advances in intelligent systems and computing) (pp. 335–358). Springer International Publishing. https://doi.org/10.1007/978-3-030-22871-2_24

Floridi, L. (2016). *The 4th revolution: How the Infosphere is reshaping human reality*. Oxford University Press.

Formanek, N. (2025). How I stopped worrying and learned to love opacity. In J. M. Durán & G. Pozzi (Eds.), *Philosophy of science for machine learning: Core issues and new perspectives*. Synthese Library, Springer.

Freiesleben, T. (2025). Artificial neural nets and the representation of human concepts. In J. M. Durán & G. Pozzi (Eds.), *Philosophy of science for machine learning: Core issues and new perspectives*. Synthese Library, Springer.

Friedler, S. A., Scheidegger, C., & Venkatasubramanian, S. (2021). The (im)possibility of fairness: Different value systems require different mechanisms for fair decision making. *Communications of the ACM, 64*(4), 136–143. https://doi.org/10.1145/3433949

Friedman, B., & Nissenbaum, H. (1996). Bias in computer systems. *ACM Transactions on Information Systems, 14*(3), 330–347. https://doi.org/10.1145/230538.230561

Guest, O., & Martin, A. E. (2023). On logical inference over brains, behaviour, and artificial neural networks. *Computational Brain & Behavior, 6*(2), 213–227. https://doi.org/10.1007/s42113-022-00166-x

Gyllingberg, L., Birhane, A., & Sumpter, D. J. T. (2023). The lost art of mathematical modelling. arXiv. https://doi.org/10.48550/arXiv.2301.08559.

Hall, M., van der Maaten, L., Gustafson, L., Jones, M., & Adcock, A. (2022). A systematic study of bias amplification. arXiv. https://doi.org/10.48550/arXiv.2201.11706.

Hasson, U., Nastase, S. A., & Goldstein, A. (2020). Direct fit to nature: An evolutionary perspective on biological and artificial neural networks. *Neuron, 105*(3), 416–434. https://doi.org/10.1016/j.neuron.2019.12.002

Hey, A. J. G., Tansley, S., & Tolle, K. M. (Eds.). (2009). *The fourth paradigm: Data-intensive scientific discovery* (Vol. 1). Microsoft Research Redmond. https://www.microsoft.com/en-us/research/publication/fourth-paradigm-data-intensive-scientific-discovery/

Hooker, S., Moorosi, N., Clark, G., Bengio, S., & Denton, E. (2020). Characterising bias in compressed models. arXiv. https://doi.org/10.48550/arXiv.2010.03058.

Hovy, D., & Prabhumoye, S. (2021). Five sources of bias in natural language processing. *Language and Linguistics Compass, 15*(8), e12432. https://doi.org/10.1111/lnc3.12432

Humphreys, P. (2004). *Extending ourselves: Computational science, empiricism, and scientific method*. Oxford University Press.

Humphreys, P. (2009). The philosophical novelty of computer simulation methods. *Synthese, 169*(3), 615–626. https://doi.org/10.1007/s11229-008-9435-2

Ivanova, A. A., Hewitt, J., & Zaslavsky, N. (2021). Probing artificial neural networks: Insights from neuroscience. arXiv. https://doi.org/10.48550/arXiv.2104.08197.

Jumper, J., Evans, R., Pritzel, A., Green, T., Figurnov, M., Ronneberger, O., Tunyasuvunakool, K., et al. (2021). Highly accurate protein structure prediction with AlphaFold. *Nature, 596*(7873), 583–589. https://doi.org/10.1038/s41586-021-03819-2

Jurafsky, D., & Martin, J. H. (2023). *Speech and language processing*. (3rd ed.) https://web.stanford.edu/~jurafsky/slp3/

Kar, K., Kornblith, S., & Fedorenko, E. (2022). Interpretability of artificial neural network models in artificial intelligence versus neuroscience. *Nature Machine Intelligence, 4*(12), 1065–1067. https://doi.org/10.1038/s42256-022-00592-3

Kidd, C., & Birhane, A. (2023). How AI can distort human beliefs. *Science, 380*(6651), 1222–1223. https://doi.org/10.1126/science.adi0248

Kieval, P. H. (2025). Representation learning without representationalism. A non-representationalist account of deep learning models in scientific practice. In J. M. Durán & G. Pozzi (Eds.), *Philosophy of science for machine learning: Core issues and new perspectives*. Synthese Library, Springer.

Kriegeskorte, N., & Douglas, P. K. (2018). Cognitive computational neuroscience. *Nature Neuroscience, 21*(9), 1148–1160. https://doi.org/10.1038/s41593-018-0210-5

Kumar, S., Sumers, T. R., Yamakoshi, T., Goldstein, A., Hasson, U., Norman, K. A., Griffiths, T. L., Hawkins, R. D., & Nastase, S. A. (2023). Reconstructing the cascade of language processing in the brain using the internal computations of a transformer-based language model. bioRxiv. https://doi.org/10.1101/2022.06.08.495348.

Lam, R., Sanchez-Gonzalez, A., Willson, M., Wirnsberger, P., Fortunato, M., Pritzel, A., & Ravuri, S., et al. (2022). GraphCast: Learning skillful medium-range global weather forecasting. arXiv. https://doi.org/10.48550/arXiv.2212.12794.

Langer, M., Oster, D., Speith, T., Hermanns, H., Kästner, L., Schmidt, E., Sesing, A., & Baum, K. (2021). What do we want from explainable artificial intelligence (XAI)?—A stakeholder perspective on XAI and a conceptual model guiding interdisciplinary XAI research. *Artificial Intelligence, 296*(July), 103473. https://doi.org/10.1016/j.artint.2021.103473

Latour, B. (2005). *Reassembling the social: An introduction to actor-network-theory*. Oxford University Press.

Latour, B., & Woolgar, S. (1986). In J. Salk (Ed.), *Laboratory life: The construction of scientific facts*. Princeton University Press. https://press.princeton.edu/books/paperback/9780691028323/laboratory-life

Liu, P., Yuan, W., Jinlan, F., Jiang, Z., Hayashi, H., & Neubig, G. (2023). Pre-train, prompt, and predict: A systematic survey of prompting methods in natural language processing. *ACM Computing Surveys, 55*(9), 195:1–195:35. https://doi.org/10.1145/3560815

Lloyd, K. (2018). Bias amplification in artificial intelligence systems. arXiv. https://doi.org/10.48550/arXiv.1809.07842.

Ma, W. J., & Peters, B. (2020). A neural network walks into a lab: Towards using deep nets as models for human behavior. arXiv. https://doi.org/10.48550/arXiv.2005.02181.

Meijer, G. (2021). Neurons in the mouse brain correlate with Cryptocurrency Price: A cautionary tale. PsyArXiv. https://doi.org/10.31234/osf.io/fa4wz.

Mitchell, M. (2020). *Artificial intelligence: A guide for thinking humans*. Penguin. https://www.penguin.co.uk/books/294649/artificial-intelligence-by-mitchell-melanie/9780241404836

Mitchell, M., Wu, S., Zaldivar, A., Barnes, P., Vasserman, L., Hutchinson, B., Spitzer, E., Raji, I. D., & Gebru, T. (2019). Model cards for model reporting. In *Proceedings of the conference on fairness, accountability, and transparency* (FAT* '19) (pp. 220–229). Association for Computing Machinery. https://doi.org/10.1145/3287560.3287596

Nguyen, T., Brandstetter, J., Kapoor, A., Gupta, J. K., & Grover, A. (2023). ClimaX: A foundation model for weather and climate. arXiv. https://doi.org/10.48550/arXiv.2301.10343.

Omiye, J. A., Lester, J. C., Spichak, S., Rotemberg, V., & Daneshjou, R. (2023). Large language models propagate race-based medicine. *npj Digital Medicine, 6*(1), 1–4. https://doi.org/10.1038/s41746-023-00939-z

Páez, A. (2025). Axe the X in XAI: A plea for understandable AI. In J. M. Durán & G. Pozzi (Eds.), *Philosophy of science for machine learning: Core issues and new perspectives*. Synthese Library, Springer.

Panch, T., Mattie, H., & Atun, R. (2019). Artificial intelligence and algorithmic bias: Implications for health systems. *Journal of Global Health, 9*(2), 010318. https://doi.org/10.7189/jogh.09.020318

Pickering, A. (Ed.). (1992). *Science as practice and culture*. University of Chicago Press.

Poliseli, L., Coutinho, J. G. E., Viana, B., Russo, F., & El-Hani, C. N. (2022). Philosophy of science in practice in ecological model building. *Biology & Philosophy, 37*(4), 21. https://doi.org/10.1007/s10539-022-09851-4

Poliseli, L., & Russo, F. (2022). Philosophy of science in practice and weak scientism together apart. In M. M. Mizrahi (Ed.), *For and against scientism: Science, methodology, and the future of philosophy* (Ch.6. Collective studies in knowledge and society). Rowman & Littlefield.

Räz, T. (2025). From explanations to interpretability and Back. In J. M. Durán & G. Pozzi (Eds.), *Philosophy of science for machine learning: Core issues and new perspectives*. Synthese Library, Springer.

Rogers, A., Kovaleva, O., & Rumshisky, A. (2021). A primer in BERTology: What we know about how BERT works. *Transactions of the Association for Computational Linguistics, 8*(January), 842–866. https://doi.org/10.1162/tacl_a_00349

Russell, S., & Norvig, P. (2021). Artificial intelligence, global edition. https://elibrary.pearson.de/book/99.150005/9781292401171

Russo, F. (2022). *Techno-scientific practices: An informational approach*. Rowman & Littlefield.

Schrimpf, M., Blank, I. A., Tuckute, G., Kauf, C., Hosseini, E. A., Kanwisher, N., Tenenbaum, J. B., & Fedorenko, E. (2021). The neural architecture of language: Integrative Modeling converges on predictive processing. *Proceedings of the National Academy of Sciences, 118*(45), e2105646118. https://doi.org/10.1073/pnas.2105646118

Soler, L., Zwart, S., Lynch, M., & Israel-Jost, V. (2014). *Science after the practice turn in the philosophy, history, and social studies of science. Routledge studies in the philosophy of science*. Routledge, Taylor & Francis Group.

Sun, T., Gaut, A., Tang, S., Huang, Y., ElSherief, M., Zhao, J., Mirza, D., Belding, E., Chang, K.-W., & Wang, W. Y. (2019). Mitigating gender bias in natural language processing: Literature review. In *Proceedings of the 57th Annual Meeting of the Association for Computational Linguistics* (pp. 1630–1640). Association for Computational Linguistics. https://doi.org/10.18653/v1/P19-1159

Sutton, R. (2019). The bitter lesson. *Incomplete Ideas* (blog). 13 March 2019. http://www.incompleteideas.net/IncIdeas/BitterLesson.html

Talat, Z., Névéol, A., Biderman, S., Clinciu, M., Dey, M., Longpre, S., Luccioni, S., et al. (2022). You reap what you sow: On the challenges of bias evaluation under multilingual settings. In *Proceedings of BigScience Episode #5—Workshop on challenges & perspectives in creating large language models* (pp. 26–41). Association for Computational Linguistics. https://doi.org/10.18653/v1/2022.bigscience-1.3

Vaswani, A., Shazeer, N., Parmar, N., Uszkoreit, J., Jones, L., Gomez, A. N., Kaiser, Ł., & Polosukhin, I. (2017). Attention is all you need. In *Advances in neural information processing systems* (Vol. 30). Curran Associates, Inc. https://proceedings.neurips.cc/paper_files/paper/2017/hash/3f5ee243547dee91fbd053c1c4a845aa-Abstract.html

Wehbe, L., Murphy, B., Talukdar, P., Fyshe, A., Ramdas, A., & Mitchell, T. (2014). Simultaneously uncovering the patterns of brain regions involved in different story Reading subprocesses. *Edited by Kevin Paterson. PLoS ONE, 9*(11), e112575. https://doi.org/10.1371/journal.pone.0112575

Wolf, T., Debut, L., Sanh, V., Chaumond, J., Delangue, C., Moi, A., Cistac, P., et al. (2020). Transformers: State-of-the-art natural language processing. In *Proceedings of the 2020 Conference on empirical methods in natural language processing: System demonstrations* (pp. 38–45). Online: Association for Computational Linguistics. https://doi.org/10.18653/v1/2020.emnlp-demos.6

Yamins, D. L. K., & DiCarlo, J. J. (2016). Using goal-driven deep learning models to understand sensory cortex. *Nature Neuroscience, 19*(3), 356–365. https://doi.org/10.1038/nn.4244

Yarkoni, T., & Westfall, J. (2017). Choosing prediction over explanation in psychology: Lessons from machine learning. *Perspectives on Psychological Science, 12*(6), 1100–1122. https://doi.org/10.1177/1745691617693393

Zednik, C., & Verreault-Julien. (2025). Can explainable AI contribute to justification? In J. M. Durán & G. Pozzi (Eds.), *Philosophy of science for machine learning: Core issues and new perspectives*. Synthese Library, Springer.

Zhao, J., Wang, T., Yatskar, M., Ordonez, V., & Chang, K.-W. (2017). Men also like shopping: Reducing gender bias amplification using corpus-level constraints. In *Proceedings of the 2017 Conference on empirical methods in natural language processing* (pp. 2979–2989). Association for Computational Linguistics. https://doi.org/10.18653/v1/D17-1323

Sahra A. Styger has been an academic staff member and PhD candidate at the Department of Philosophy at the University of Konstanz, Germany, since 2021. As a focus for her PhD thesis, she studied epistemic agents and scientific products in ML-based science. Her research specifically addresses the foundations of ML-based science from a philosophy of science and epistemology perspective.

Marianne de Heer Kloots is a PhD candidate at the Institute for Logic, Language and Computation at the University of Amsterdam, the Netherlands, since 2021. Her research focuses on interpreting deep learning models for audio and text processing, and using them in models of human speech and language comprehension.

Oskar van der Wal is a PhD candidate at the Institute for Logic, Language and Computation at the University of Amsterdam, the Netherlands, since 2020. His research broadly focuses on understanding why and how language models exhibit social biases using interpretability tools. He is particularly interested in how we can reliably measure and mitigate bias, and the question of how to ground current evaluation practices in real-world harm. The views expressed in this work do not represent any undisclosed current or future affiliations.

Federica Russo is Professor of Philosophy and Ethics of Techno-Science and Westerdijk Chair at the Freudenthal Institute, Utrecht University. Her research concerns epistemological, methodological, and normative aspects they arise in the health and social sciences, with special attention to policy contexts and to the highly technologized character of these fields. She is the author of *Techno-Scientific Practices. An Informational Approach* (RLI, 2022).

Chapter 16
Values, Inductive Risk, and Societal-Epistemic Coupledness in Machine Learning Models

Milou Jansen and Koray Karaca

Abstract We offer a novel perspective on how to manage inductive risk in the context of machine learning (ML) models used for societal purposes. Since ML models can produce erroneous results to varying extents, whether to accept a given ML model requires taking account of the inductive risk associated with its use. Deciding which levels and types of inductive risks are acceptable in the application contexts is a value-laden activity. While ML models are increasingly used for societal purposes, their epistemic features are more strongly connected with their societal purposes and underlying value judgments. We suggest that the epistemic and societal value judgments should be understood as coupled in the sense of mutually influencing each other, rather than as two separate kinds of factors, in the construction and assessment of societal ML models. We provide a coupled societal-epistemic analysis of the inductive risk in the context of societal ML models. Based on a case study concerning the societal use of an ML model by the City of Amsterdam, we suggest that finding the right balance of acceptable inductive risk is a collective quest, in the sense that this risk is prioritized and interpreted differently by the relevant stakeholders of the model.

Keywords Machine learning · Error · Inductive risk · Social and epistemic values · Social-epistemic coupledness · Value-ladenness

16.1 Introduction

Governmental institutions are increasingly relying on machine learning (ML) models to optimize societal decision-making processes in the name of the public good (Veale & Brass, 2019). These models are implemented in a wide range of

M. Jansen (✉)
Amsterdam, Netherlands

K. Karaca
University of Twente, Enschede, Netherlands

J. M. Durán, G. Pozzi (eds.), *Philosophy of Science for Machine Learning*,
Synthese Library 527, https://doi.org/10.1007/978-3-032-03083-2_16

public domains for various optimization purposes, such as to optimize routes of garbage trucks and predict and distribute pedestrian and vehicle traffic intensity. The deployment of ML models in public domains indicates that the social and the technical are increasingly intertwined. What should be taken as optimal in those domains differs widely, and it is determined not only through technical and epistemic considerations but also through social considerations. For what is technically optimal may not always be normatively defensible in societally relevant terms, and the optimal performance of an ML model is ultimately bounded by its fit with society. Therefore, ML models and the practices surrounding their construction and deployment should not be understood as neutral, but rather laden to varying extents with both epistemic and societal value judgments. This not only raises the question of what kind of knowledge base we assign to these tools, but also what kind of values are embedded in them (Dotan & Milli, 2020; Karaca, 2021; Ratti & Graves, 2021). This in turn illustrates that modeling practices involve diverse choices resulting from human value judgments (Douglas, 2000; Winsberg, 2012; Diekmann & Peterson, 2013).

In this chapter, we will argue that epistemic and societal value judgments should not be seen as separate, but rather as coupled throughout the ML modelling practice, including not only their construction but also their validation. To this end, we will develop an analysis as a heuristic tool to jointly analyze the epistemic and societal dimensions of the ML modelling practice. Throughout this chapter, we use *societal values* to refer to considerations—including ethical, political, and pragmatic ones—that shape judgments about the social consequences of ML models. We avoid the term *social values* to maintain conceptual clarity and to highlight how societal framing reveals the coupling between a model's epistemic features and the specific features of its use context. Our analysis shares similarities with those developed in similar techno-science studies to characterize couplings of different kinds, e.g. between descriptive and evaluative, between facts and values, and between different philosophies. In recent years, it has been shown that the coupling approach has contributed to a better understanding of the interplay between the epistemic and ethical dimensions of knowledge-production practices in complex and social contexts (Katikireddi & Valles, 2015; Tuana, 2013; Valles et al., 2019; Russo et al., 2023), while the question of how the values embedded in these bridging concepts should be managed remains as a challenge (Alexandrova, 2015). In this chapter, we will also address this challenge by making use of the concept of inductive risk that has been fruitful in dealing with the management of value trade-offs (Elliott & Richards, 2017).

The plan of the present chapter is roughly as follows. In Sect. 16.2, we provide a characterization of the concept of inductive risk that is applicable to the context of ML modelling. In Sect. 16.3, we discuss the involvement of epistemic and societal values in ML modelling. Sections 16.4 and 16.5 are devoted to what we shall call the coupled societal-epistemic analysis of inductive risk, which is subsequently illustrated through a case study discussed in Sects. 16.6 and 16.7. In Sect. 16.8, we provide concluding remarks concerning the societal uses of ML models.

16.2 Error and Inductive Risk in ML Modeling

ML refers to mathematically modelling the mapping relationship that exists between a set of inputs $X = (x_1, \ldots, x_n)$ and a corresponding set of outputs $Y = (y_1, \ldots, y_m)$ with respect to a specified task, such as a classification task (Alpaydin, 2010). For instance, in the case of binary (two-class) classification which is widely used in societal applications, the inputs are assigned to the set of outputs consisting of two distinct classes denoted by a binary tuple (y_1, y_2). ML is based on the assumption that there exists a target function $f : X \rightarrow Y$ that correctly maps all the given inputs to the corresponding outputs for a given task. ML is used to find the function g that provides an optimal approximation to the unknown function f. The function g is a parametric function of the input variables and it is referred to as the ML model that optimally represents the foregoing mapping relationship. In the case of supervised learning, which we shall consider in this chapter, the training data contains examples of inputs together with the corresponding outputs.[1]

The essential elements of supervised ML include a training data set, learning algorithm, hypothesis set (H), and loss (error) function together with an optimization procedure. H contains the candidate functions, namely the functions chosen by a ML modeler to be suitable to represent the mapping relationship under consideration. The learning algorithm is used to find the most optimal set of parameters that will constitute the function g as an ML model. This is achieved through a training process that is basically an optimization process aiming at minimizing the loss (or cost) function that specifies the total cost incurred due to error. Here, error denotes the extent of the difference between the outputs drawn from the training data and the outputs provided by the function g. The chosen learning algorithm is implemented according to an optimization procedure in such a way that instances drawn from the training data, such as (x^1, y^1), (x^2, y^2), $\ldots (x^n, y^n)$, are introduced to the learning algorithm until the latter finds the set of optimal parameters (of the function g) that minimizes the total loss. As a result of this optimization process, the parameters of the function g are fixed by the learning algorithm, thus fixing this function as the ML model of the sought mapping relationship. The ML optimization process underlies the inductive character of ML model construction, in that it consists in generalizing from the input-output relations in the training data (found by the learning algorithm) to a model (namely an ML model) that can be used to determine the input-output relations in previously unseen data sets, which have the same characteristics as the training data.

Loss function is used to quantify the loss incurred due to incorrect predictions (such as misclassifications) made during the training process. What is called training error is defined to be the sum of losses over the individual input-output instances classified during the training. For example, in binary classification tasks, classification classes are usually marked as positive and negative. While true

[1] See Alpaydin, 2010 for other types of ML.

positives and negatives are correct outcomes, false positives and negatives are erroneous outcomes. Since training error is concerned with the construction of an ML model, it can be regarded as an indication of the error-proneness of an ML model based on its performance on the training data. In this sense, the extent of training error can be taken to indicate the extent of inductive risk built into the model during its construction. Here, inductive risk can be understood as arising from the possibility that the outcomes of the model could turn out to be erroneous to varying extents due to the inductive character of ML model construction (Douglas, 2000). This means that minimization of training error during ML optimization serves to minimize the inductive risk built into an ML model (Karaca, 2021).

16.3 Value-Ladenness in ML Modelling

Accepting a particular model also means accepting its potentially erroneous outcomes and thus facing the potential consequences of those errors. Securing an optimal balance in the performance of a model is an important aspect of model construction. In this quest, modelers are expected to take account of not only epistemic values but also ethical and other societal values in their practice concerning modelling (Winsberg, 2012; Karaca, 2021; Schmidt, 2025; Ratti, 2025). This is mainly because it is not possible for them to provide the relevant stakeholders with value-free information regarding the constructed models and let them deal with the social implications of these models (Parker & Winsberg, 2018). As a result, ML modelers are forced to move beyond the value-free ideal of knowledge production when engaging in model practices. What constitutes objectivity, according to the value-free ideal of knowledge discovery, is the active exclusion of social values from the inferential aspects of science (Reiss & Sprenger, 2020). This means that only epistemic values should be allowed when making inferences about which models to accept. In this regard, Martin Carrier (2013) suggested that epistemic values play both what he calls significance and confirmation roles in the process of knowledge production. In this section, we will suggest that this two-fold role equally holds for social values in the practice of societally relevant modelling.

Even in the idea of a value-free knowledge pursuit, societal values are accepted to play a role in the external significance requirements of modelling. First, both epistemic and societal values play important roles in the selection of modeling problems to pursue (Longino, 1990). The extent to which a societal goal is accurately reflected in a model depends on the influence of relevant societal values (Diekmann & Peterson, 2013). Therefore, some models are influenced, and ought to be influenced, by social values. Second, the way in which the knowledge is put to use in society requires the consideration of societal values. Third, both epistemic and societal values place direct limitations on methodological options. Societal ML models are implemented as solutions to directly address and analyze societal problems. This involves choices about how to define and frame a problem and thereby its solution.

As the previous section has shown, optimization is key to ML modeling. Optimization is generally regarded as something instrumentally valuable rather than inherently valuable. This means that optimization is not valuable in itself, but rather is valued for what it brings about. It is in this sense that modelers share an overall perspective on design as an exercise in optimization (Franssen et al., 2023). The technical requirements for most modelling problems do not explicitly prescribe what should be optimized. In the context of societal ML modelling, social and epistemic judgments are thus needed to justify the optimal performance of ML models (Basu, 2020; Dobbe et al., 2020; Wong, 2020). What to optimize for is a directly value-laden question when taking account of significance requirements. Social influences on goal and problem framing of a public issue have bearing on the assumptions underlying an ML model and directly influence the representation of public issues. While technical requirements on ML models cannot always be endorsed because of epistemic and social motivations (Ratti & Graves, 2021), these requirements do impact the outcomes of ML models and this way they have epistemic and social repercussions.

Confirmative requirements help in assessing how well a model can be confirmed by the evidence available given its underlying assumptions and uncertainties. In the light of error and associated inductive risk, there are several types of uncertainty to consider here, namely data uncertainty, structural model uncertainty, and parameter uncertainty (Winsberg, 2012). The informational content of data is often regarded as fixed and context independent. However, this is an oversimplification of the data component in scientific inquiry, as even what is called raw data is already processed and gathered with a specific goal in mind (Leonelli, 2019). To truly understand the representational capacities of data, it is important to distinguish data about a phenomenon from the phenomenon itself (Woodward & Bogen, 1988). While data does not provide direct access to or the representation of the phenomena, it sets the evidential space within which phenomena are identified (Leonelli, 2019). In this sense, it is unknowable if all data points and features are included, which causes data uncertainty in the modelling process. Data uncertainty may lead to under or non-representation of issues, characteristics, events, or groups of people.

Each model brings different auxiliary assumptions on the underlying relationships of the data. These assumptions lay in structural model choices, resulting in structural model uncertainty. The same model structure will make different predictions for different values of the parameters. Parameters and thresholds bear assumptions on the weight and significance of different features and set the boundaries that balance the outcome of a model. It is often highly uncertain about what the best values for many of these parameters are, thus resulting in parameter uncertainty. Data uncertainty bears on both parameter uncertainty and structural uncertainty, in that it acts as input and thereby affecting our ability to judge the accuracy of our parameters and our model structures. ML models can be seen as optimal estimation models that are aimed to map accurate relations onto social phenomena under investigation. ML involves the ability to generalize from past experiences to deal with new situations related to those experiences. As a result, ML models require various assumptions, including idealizations and simplifications.

The assumptions made in ML modelling processes can cause value-laden structural and paratemer uncertainties that manifest as uncertainties in model predictions. As in the context of scientific models, the evidence never suffices to completely prove or disprove an ML model. As a result, what serves as sufficient evidence as well as what seems to be the best ML model in a given societal context becomes a matter of judgment related to the uncertainties in model predictions. This is why modelers aim to quantify this type of uncertainty, and this can be seen as an effort to quantify inductive risk (Karaca, 2021). In this regard, model choices reflect balances of inductive risk, and models are being optimized to particular purposes and thus to particular metrics of success (Winsberg, 2012).

16.4 Ethical-Epistemic Coupledness in Modelling

The framework for coupled ethical-epistemic analysis is conceptually developed by Nancy Tuana (2013) to advance interdisciplinary collaboration in scientific knowledge production. Interdisciplinary practices give space for reflection on both epistemic and social issues that are typically intertwined. She states that this type of analysis should lead to trustworthy scientific practice and thus better science. First, because it advances knowledge by building on current knowledge. Second, because it leads to scientific research that better benefits society. Most concretely, it broadens our understanding of the types of value decisions that are entrenched in research models and methods, offering resources for identifying the ethical implications of research decisions, and providing a lens for identifying the questions that are ignored, under-examined, and rendered invisible.

The focus here is not only on the ethical aspects of the scientific research process itself, but rather on the ethical issues external to the production of scientific research as well as on the ethical values embedded in the production of scientific knowledge (Tuana, 2013). A coupled ethical-epistemic analysis will provide a more interdisciplinary approach that links procedural ethics, but especially extrinsic and intrinsic ethical issues closer together with the epistemic issues under investigation. Katikireddi and Valles (2015) reformulated Tuana's account through a set of questions concerning the choice of variables. They used this analysis to examine the interplay between the ethical and epistemic challenges related to the choice of variables in the practice of health public research. Valles et al. (2019) extended the scope of the foregoing framework's applicability so as to include "research practices" through the following set of questions:

Q1. What are the research practices' strengths and weaknesses for evidence gathering and analysis?
Q2. What are the research practices' ethical strengths and weaknesses?
Q3. How will the research practices' strengths and weaknesses for evidence gathering and analysis affect ethical issues?

Q4. How will the research practices' ethical strengths and weaknesses affect evidence gathering and analysis? (Ibid., 270–271)

In this analysis, Q1 and Q2 respectively concern epistemic and ethical features, while Q3 and Q4 jointly concern the interplay between epistemic and ethical features, namely how epistemic features affect ethical features (Q3), and vice versa (Q4). The "epistemic" concerns the requirements set for the knowledge-production processes as well as the judgments made concerning the quality of the produced knowledge. The coupledness of the ethical and epistemic makes it possible to conceptualize the epistemic and ethical value influences on models and the insights they produce, as being connected, rather than as being two separate factors that independently shape the model's capacity for knowledge production. They are linked and thus mutually shaping. Epistemic strengths and weaknesses give rise to ethical strengths and weaknesses and vice versa. Some of the dynamics therein are more conceptual than causal (Valles et al., 2019). Coupledness can then also be defined more philosophically, as it points out that in a conceptual sense, features of one type (epistemic or ethical) can be altered by features of the other type without being caused by them (Valles et al., 2019). In ML modelling, this primarily means combining the context of model use with the various aspects of modelling, including data-uncertainty, model performance, structural model uncertainty and parameter uncertainty, and thereby linking these aspects to the management of inductive risk.

16.5 Societal-Epistemic Coupled Analysis Applied to ML Modeling Practices

We will now turn to an elaboration of how the above coupledness framework can be specifically applied in the context of public ML modelling practices, while addressing concerns related to value-ladenness and management of inductive risk. A coupled ethical-epistemic analysis can be conducted in advance as well as in retrospective by answering the following questions as paraphrased based on Valles et al. (2019). To this end, we will first respectively look into the societal and epistemic dimensions of the ML modelling practices. Here again, we note that by societal we mean all kinds of social considerations including those that are ethical. In this context, model construction and model validation are two main epistemic activities whose strengths and weaknesses are to be assessed. Therefore, we shall adapt Q1 above by specifying "research practices" as "ML research practices" and accordingly by replacing "evidence gathering and analysis" with "model construction and model validation" as well as "ethical" with "societal." Moreover, since we suggest that not only ethical but also societal strengths and weaknesses need to be considered, we shall modify Q1 and Q2 in such a way that they read as follows:

Q1. What are the ML research practices' strengths and weaknesses concerning model construction and model validation?
Q2. What are the ML research practices' societal strengths and weaknesses?

Accordingly, Q3 and Q4 read as follows:

Q3. How do the ML research practices' strengths and weaknesses concerning model construction and model validation affect societal issues?
Q4. How do the ML research practices' societal strengths and weaknesses affect model construction and model validation?

Concerning Q1, we suggest that the epistemological weaknesses and strengths of ML research practices lie in the extent to which the methodological choices concerning model construction and validation are justified. Like other types of scientific models, ML models are constructed and validated based on certain methodological choices, which are in turn justified on the basis of certain methodological assumptions made by the developers of these models and on the basis of the existing limitations of the chosen models. This means that ML research practices concerning model validation and construction are epistemically strong or weak to the extent that their underlying methodological choices are justified by the modelers. The previous discussion indicates that the modelers need to make certain methodological choices to specify the following elements of ML model construction: training data, hypothesis set, learning algorithm, loss (error) function, optimization procedure (including utilities assigned to different types of errors). Similarly, they also need to make certain methodological choices to specify the following elements of model validation: validation data and procedure, and performance metric to be used to measure the predictive performance of the model.

Concerning Q2 above, we suggest that ML research practices' societal strengths and weaknesses are related to the social constraints imposed on ML models by the particular contexts in which these models are (to be) used. This means that the extent to which ML research practices are societally strong or weak depends on whether the constraints concerning the phases of model construction and validation are correctly identified by the modelers prior to these phases. Since Q2 is aimed to isolate social aspects from epistemic aspects, it is not directly concerned with the extent these methodological constraints are satisfied by the models. This issue will be addressed in what follows.

The coupledness of the epistemic and societal aspects of ML research practices can be analyzed by addressing Q3 and Q4 above. Q3 is concerned with how epistemically unjustified methodological choices concerning model construction and model validation can result in societal issues, such as unacceptably large errors in model predictions used in high-stake decision contexts, e.g. medical context, and biased or discriminatory decisions based on those predictions, e.g. credit applications. Q4 is concerned with the kinds of methodological constraints entailed by the social constraints imposed by the contexts in which ML models are to be used. The latter type of constraints need to be translated by the modelers into methodological constraints underlying ML models. These methodological constraints may concern the intended use of the model; the inclusion or exclusion of different types of data—such as datasets associated with underrepresented groups or personal characteristics; the choice of utilities assigned to false positives and false negatives for optimization purposes; and the selection of a metric for model validation.

Answering Q3 serves to identify the societal issues arising from unjustified methodological choices, while answering Q4 serves to identify the methodological constraints imposed by the contexts-of-use dependent social constraints. It is worth noting that answering Q3 and Q4 requires in the first place answering Q1 and Q2, as the foregoing methodological choices and constraints can be identified through answering the latter questions.

The foregoing societal-epistemic coupledness analysis can be used to carry out an inductive risk analysis. This analysis can be used by the relevant stakeholders of an ML model to effectively manage inductive risk associated with its practical applications, or to understand if the developers of the model have effectively managed the inductive risk built into the model during its construction.

Since ML models are typically used for societal applications, an inductive risk analysis in such contexts is concerned with the social consequences of the risk of erring in accepting and using these models for their intended uses (Rudner, 1953; Douglas, 2000). As has been argued above, their contexts of use impose social constraints on these models. There will be social consequences arising from the failure of the ML models to meet these methodological constraints. The extent of inductive risk will be mainly concerned with these social consequences in the sense that whether the ML models are to be accepted and uses for their intended purposes will depend on the extent to which these models satisfy the methodological constraints to be imposed by their social constraints. Therefore, the question that is relevant to determine the extent of inductive risk in a specific application of a ML model is the following:

> To what extent do the methodological choices underlying ML model construction and validation satisfy the methodological constraints related to inductive risk?

Since the methodological constraints and choices are identified through the coupledness analysis outlined in Q1–Q4, this analysis provides the basis for answering the foregoing question.

The above question serves to determine the extent of inductive risk. Its management is a related but different question. Users can manage inductive risk through the value judgments they make (Rudner, 1953; Douglas, 2000). That is to say that given some form of evidence, they make certain methodological choices (including accepting or rejecting a given model) when they decide that they can (not) afford taking the inductive risk, i.e., the risk that the model's predictions are false. This also indicates that management of inductive risk is mainly about assessing the severity of the inductive risk associated with the methodological choices. This assessment is always made (by users) with respect to both the epistemic and societal values that they hold. Therefore, the management of inductive risk requires in the first place determining those values and associated value judgments. This leads to the following question:

> What epistemic and societal values, and associated value judgments, held by users underlie the methodological choices involved in ML model construction and validation?

The severity of the inductive risk would differ with respect to different values hold by users. This is mainly due to the fact that different values may clash with each other about the severity of the inductive risk. In other words, the severity of the inductive risk may weigh differently for different values. This means that an effective management of inductive risk requires users to understand how values clash in the assessment of its severity. Therefore, the following question needs to be answered:

> What trade-offs exist among the different values relevant to methodological decisions in model construction and validation?

As pointed out above, users can manage inductive risk through value judgments they make. If different values (hold by users) are relevant to the methodological choices, then the following question needs to be answered to understand how the values of users bear upon the methodological choices:

> What value judgments do users make to guide their methodological choices in accordance with value trade-offs?

16.6 Applying the Societal-Epistemic Coupledness Framework to ML in Practice

In Amsterdam, citizens can report issues that take place in the public space of the city. These issues are so called micro-events that concern a small geographical area and only affect a limited number of citizens. They are linked to nuisance in public space, e.g. a loose tile, a streetlight that is not working, a non-functioning container, bars making too much noise or a fallen tree on the street. These cases fall under the responsibility of different municipal departments. An ML classifier model is used to detect the class of a report and route it to the correct department (Sukel et al., 2019). The ML model fulfills a small but crucial role in the process from report submission to report closing. After a citizen report has been submitted, the classifier ML model automatically classifies the report, while sending it to the right department. The officers of the specific department follow up on the report. This model is implemented by the city-government of Amsterdam, and it thereby plays an important public role in facilitating the correct and timely response towards issues in public space.[2] As such, its epistemic qualities are intertwined with a multitude of

[2] See the following URL through which the citizens can submit reports to be processed by the ML model in question: https://algoritmes.overheid.nl/nl/algoritme/gm0363/21579514/meldingen-openbare-ruimt The same model is currently used by other Dutch municipalities.

societal considerations as we will argue in this section by performing a societal-epistemic coupled analysis of this case.

16.6.1 Introducing the Case and Structuring the Coupledness Analysis

Societal-epistemic coupled analysis may give us a better grip on the impact of a model on its social context. A focus on societal or epistemic values alone cannot cover the inductive societal risks of ML models. Looking at how exactly both kinds of values are mutually influencing each other brings new tools to better understand the epistemic status attributed to these models in society. To this end, we will draw upon the analysis we presented earlier by revisiting each of its underlying questions in turn. It is worth noting that the discussion in what follows in part depends on personal communications with individuals who were involved in the design and management of the ML model in question.

Q1. What are the ML research practices' strengths and weaknesses concerning model construction and model validation?

Since Q1 is concerned with both the phase of model construction and the phase of validation, we will address this question for each phase separately.

Model Construction The ML model in question is a textual classification model that was constructed (or trained) by using citizen reports as training data (Sukel et al., 2019, p. 1460). Over 500,000 citizen reports were divided in two parts, i.e. data for training and data for validation separately. The use of such a training data makes the performance of the model for a specific issue highly dependent on previous representation of that issue in the training data. This is justified by the expected similarity of context in which the reports were produced as well as how they will be produced in the future. The reporting behavior in previous years as well as the recurring issues in public space are expected to be relatively similar.

For the training (or optimization) of the ML model, what is called *term frequency–inverse document frequency* technique was used (Ibid., p. 1458). This technique determines weights for each word. These weights show how unique they are for the specific citizen report compared to the overall collection. It aims to define how important a keyword is for a specific class (or category). This makes it suitable for classes that have very specific words describing them, as is the case with citizen reports. A word like "the" will get a low weight, and a word as "garbage" will get a higher weight. In addition, a multi-class probabilistic classification model based on logistic regression is used to classify these reports. This means that this classification model is a classifier with multiple classes. During its implementation in 2019, the classifier consisted of 8 classes and more than 50 sub-classes.[3]

The model was trained such that it can perform as good as possible on each class. On the other hand, a difference exists in how well the classifier works for each class. Words—understood here to include phrases—which are less precise will be less accurate in their classification. The more unique the word, the better, since every class has specific wordings. 'Speeding Boat' for example, does not have many previous reports, but very precise wording. New or underrepresented events will perform less. For example, 'processionary caterpillar in trees' and 'Covid19 house-party' are more difficult for the model to classify correctly, since not many previous examples have come across. The learning algorithm was tasked with finding the balance among which the overall accuracy across classes was the highest. This threshold was partly determined based on the notion that the threshold should enable the best possible average performance for each class. Probability proved optimal at a 40% threshold, meaning that there had to be at least 40% certainty on the assigned class being the correct class. If this percentage is less, the classification is being processed manually by employees of the Service Center.

Model Validation The model performs better on some classes than the others, because the words were represented more often in the training data or the wordings were very precise and therefore easy to be classified by the model. That is why an optimal balance between classes is important. Therefore, a balance of recall and precision is sought throughout all classes (Ibid., p. 1460). F1 score was used to evaluate the classification performance of the ML model, because the imbalance among the multiple classes had to be considered. F1 score is a metric (ranging between 0 and 1) that takes into account uneven class distributions and works well if false positives and false negatives do not have similar costs (Karaca, 2021). F1 score was calculated for every class as well as for the entire model performance (Sukel et al., 2019, Fig. 5).[4] F1 score of the entire model performance was found to be close to 0.7.

The issues related to the model's performance per class are taken into consideration through manual corrections and periodically retraining of the model. Citizens have provided the past report on which the model was trained, but they do not provide perfect reports as input. They submit reports that are incomplete, have spelling mistakes, lack language skills or provide excessive information. In addition, this input is also not perfectly distributed among societal groups and neighborhoods. The number of citizens that are more or less likely to submit reports

[3] See the URL: https://medium.com/maarten-sukel/how-to-use-machine-learning-for-the-classification-of-citizen-service-requests-b71159a85f36

[4] Precision is the ratio of correctly predicted positive observations to the total positives, this relates to a low false positive rate. It shows how many labelled reports are classified correctly. Recall is the ratio of all correctly predicted observations to the total of observation. It indicates how many reports belonging to a particular class were actually labelled correctly. F1 score is defined to be the harmonic mean of precision and recall.

to the municipality differs across neighborhoods and socio-economic factors in society. It is important to take account of this lack of diversity. Not doing so would give rise to a problem of lack of representativeness of certain groups within the training data, but also in the validation data.

Q2. What are the ML research practices' social strengths and weaknesses?

User Context of the Model At its core, the ML model is in place to ensure an optimal internal process for allocating reports to the right department. As such, the ML model seems to primarily facilitate the effective management of the public space, while also improving its livability. The main criteria for success here are improved service towards citizens, limiting follow-up time of reports and limiting the number of mistakes. All in all, the collaboration between departments had to be improved upon, resulting in a better service towards the city. The identified goals are social, as they the sense that they prioritize usefulness over knowledge discovery.

Changing Priorities Over Time and Across Users Civil servants came by with requests for the prioritization of specific keywords or classes related to their specific department. Ideally, finding the optimal balance between which words link to which classes is a task for the learning algorithm. Initially, the reporting system did not work well enough specifically for street lighting. Therefore, street-lighting reports had to be prioritized over other classes, in order to increase the feeling of safety in the city, which was a political choice. Reports concerning individuals do not occur often, so have less frequency, but are crucial to quickly follow up on. However, a report referring to confused, lost or harmed people must be classified correctly 100% of the time, in order to prevent dangerous situations for those involved.

Ambiguity in Purpose Citizen reports sometimes seem to function as indirect signals for civil servants across different domains. They help to identify urban issues and gain actionable intelligence about objects on the street (Sukel et al., 2020). The model implicitly seems to move from a classification tool to a signaling device. However, only a small number of citizens actually report issues in the public space. It should be safeguarded that the servicing of the entire city does not get customized solely to the needs of reporters. From a social perspective, this risks the exclusion of certain types of stakeholders and phenomena that are not included in the reports.

Q3. How do the ML research practices' strengths and weaknesses concerning model construction and model validation affect social issues?

There are several aspects that link to data, model and parameter uncertainties affecting social issues. Regarding the accuracy of classifications, there will always be some classes that perform slightly better than others. It was seen as the model's task to find an optimal balance among accuracy of keywords and classes, by setting the parameters accordingly. A modeler can prioritize the overall performance of the model or manually guide its performance towards specific classes within the limits of this model. This is where room for choice arises and where requirements of confirmation are relevant. While the ML model's calculations determined the balance among the classes, two were prioritized on normative grounds, as shall be discussed in Q4 below.

Training-Data and Biased Performance How well a specific class performs in terms of accuracy relies on the frequency of keywords in past reports, as was earlier explained. It is important to note here is that prior reports reflect a particular language and education-level. Foreign languages, lower language skills and individual levels of trust towards governments also cause limitations for equal representation. Moreover, some citizens seem more willing to submit a report to the municipality and submit requests sooner and more frequently. These characteristics can be traced back to over- and underrepresentation of certain groups within the training data.

Use of Output in Public Context The topics that find their way into statistical overviews often serve as indication for decision making across institutions and departments. The fact that reports concerning garbage on the street have decreased in a particular neighborhood does not necessarily mean that the amount of garbage on the street has actually decreased there. When fewer complaints come in, it could also mean that people are tired of complaining. No reports do not mean no issues. On the other hand, receiving more reports can also be a result of better service. When making a report is effective, people are likely to do it more. It is therefore important to consider reports as a question for help from citizens, not as an accurate representation of the factual state of events.

Q4. How do the ML research practices' social strengths and weaknesses affect model construction and model validation?

In the context of the current ML model, because of an erroneous classification, a wrong department will receive the public issue report, resulting in a delay of the issue being resolved. As mentioned, one can prioritize the overall performance of the model or manually guide its performance towards specific classes. This system was fully reliant on the self-learning capabilities of the ML model. It is the model's purpose to find an optimal balance among accuracy of keywords and classes. Two expectations were made based on social grounds, as shall be described.

Zero Uncertainty Tolerance The accuracy of reports involving people was prioritized over the overall performance of the model in order to ensure fast intervention. Reports concerning people do not occur often, so have less frequency, but are incredibly important. Reports referring to confused, lost or harmed people have to be classified with almost full accuracy in order to prevent dangerous situations for those involved. As mentioned, underrepresented events or new events (low frequency) have been harder to classify correctly. As a result, new trends might be missed in the beginning, resulting in a delayed action on new and upcoming public issues.

Political Priority Initially, the reporting system did not work well enough specifically for street lighting. Therefore, street lighting reports had to be prioritized over other classes, in order to increase the feeling of safety in the city. This was mentioned as a political priority that directly influenced the optimization priorities of model development, which may conflict with the developer's aim to reach a statistical optimum. Likewise, civil servants requested the prioritization of specific keywords or classes relevant to their departments. Ideally however, finding the optimal balance between which words link to which classes is a task for the learning algorithm. This is why no other interventions were made based on employee preferences.

16.6.2 Inductive Risk and Societal-Epistemic Trade-offs

The purpose of the ML model is to find an optimal balance among the accuracy of keywords and classes, by setting the parameters accordingly. This is where room for choice arises and where requirements of confirmation are relevant. Uncertainties that arise here indicate inductive risks associated with a model, in that uncertainties give rise to potential errors (false positive and false negatives) that have potential consequences.

A trade-off necessarily occurs between the entire model performance and between every separate class, which provides the opportunity to judge and balance the societal consequences of these errors. The F1 score was used for this evaluation to take account of the imbalances among multiple classes.

What is seen as most important within the scope of cleaner and safer streets more specifically, may differ from time to time. Sub-priorities could relate to societal problems concerning trash, streetlights, or tourism. Stakeholders have different perspectives on what is useful. In this sense, the model has multiple meanings as a tool for optimization.

With an overall F1 score that is close to 0.7, the textual classifier ML model works well for most issue-reports. The classification and the performances concerning each class bear coupled social-epistemic implications because their focus determines which sub-topics receive signaling and how departments act based on these signals. In what follows, we shall identify the trade-offs that exist among the different values. This provides further insight into the value judgments that are made by users to determine their methodological choices in accordance with the value trade-offs.

Overall Accuracy vs Care Care is a social value and refers to valuing what is necessary for the health, welfare, maintenance, and protection of someone or something. In the use-context of the ML-model, care requires a high degree of accuracy for one specific class of reports, namely that any report containing the word "person" or "confused person" needed to be classified almost fully correctly in order to ensure fast intervention. Reports concerning people do not occur often, so have less frequency, but are crucial in terms of its social consequences concerning care. As a result, the accuracy of the model concerning reports involving confused persons was prioritized over other classes and thus over the overall accuracy of the model.[5] This means that while the model performs with great accuracy on one specific class, its overall accuracy for all other classes is sacrificed.

Overall Accuracy vs Public Safety Due to public safety, which is a social value, the class of broken streetlight is also prioritized in such a way that a predictive accuracy that is less than the overall predictive accuracy of the model is not accepted for this class. Initially, the reporting system did not work well enough specifically for street lighting. Therefore, street-lighting reports had to be prioritized, to increase the feeling of safety in the city. This was mentioned as a political priority that directly influenced the optimization priorities of model development, which may conflict with the developers' aim to reach a statistical optimum.

Overall Accuracy vs Organizational Preferences In navigating the priorities between overall accuracy and organizational class preference, the main focus of the ML modelling process leaned towards optimizing the overall accuracy across classes, rather than facilitating specific optimizations regarding specific classes. Many specific requests or preferences were shared with the developing team. These requests came from a wide range of policy departments working with the ML-model. Given its operational context, the construction of the model underscored practical effectiveness as a priority. Many classes already existed for a long time within the process of the organization. However, this approach risks engendering a "class bias," where labeling tendencies conform to established classes and objectives (Barocas & Selbst, 2018, p. 677–678), potentially compromising result accuracy. Classes (labels) for the ML model were constructed in dialogue with the departments of the municipality as the users of the model. A designated class manager oversees this iterative process, aligning classes with existing municipal processes and departments. Nevertheless, what remained decisive in the model development and evaluation process was the overall accuracy of the model across all classes. This in turn limited the possibility for interventions based on department or class specific preferences in general. Exceptions were made only for the classes of 'confused people' and 'street lights.' For all other classes, the optimization of the overall accuracy was prioritized over class specific accuracy optimizations.

The aforementioned trade-offs need to be managed in ways acceptable by the relevant stakeholders. The current model is accepted as sufficiently optimal by the municipality, since the harm of errors for reports on people has been mitigated and the bias towards certain neighborhoods can be limited by other process interventions. The first one being a human in the loop approach, by having a helpdesk team ready to tackle any wrongly classified report, also available by phone. The second one being to also drive independent service routes with municipal vehicles, and not only respond to areas with a high reporting rate. The third one being a 'class-manager' that represents different departments when they would manually improve something about a class. This manager speaks to the ML modelers to see what is possible. Finally and importantly, special attention has been paid to training the

[5] This information is based on personal communications with the designers of the model.

ML model on different specific language levels of citizens to reduce unintentional performance bias towards different language levels. The classification primarily bears on social-epistemic implications because their focus determines which subtopics receive signaling and how departments act based on these signals and who benefits from this service.

The above analysis demonstrates that various relevant trade-offs exist between epistemic and societal values in the context of the present case study. This indicates the societal-epistemic coupledness aspects of the inductive risk and its management in this context. Here, the societal and epistemic dimensions are coupled in the trade-off between statistical accuracy and care through the justification that reports concerning confused, lost, or harmed individuals must be classified correctly 100% of the time. Similarly, the trade-off between accuracy and public safety represents the societal-epistemic coupledness of the acceptability of certain inductive risks over others. The trade-off between accuracy and practical effectiveness, the latter being a societal value, directly effects the overall understanding of public issues that can be produced by the model. Importantly, which classes are added determines what you optimize in the city, which becomes coupled when you let the managerial structure of an organization determine the labels that you use to classify issues in public space. Finally, due to variations in language use, language skills, and reporting behavior, groups submitting deviating reports are likely to consist of people who can often be linked to similar neighborhoods or backgrounds. Over time, particular classes will continuously be processed slower or without corrections. They will eventually perform less. The opposite is true for classes that are processed faster over time. Therefore, inductive risks and the potential consequences of error are unequally distributed among those affected by the model.

16.7 Stakeholder Perspectives and the Management of Inductive Risk

Uncertainties, which are inherent to ML modelling, have different levels of acceptability, giving rise to the need for responsible management of inductive risk in ML modelling. The goals we assign to ML models are shaped by societal and political perspectives that carry both epistemic and societal value judgments. Epistemic values are central to the assessment of how much uncertainty we can afford to take. Societal values concern the acceptability of such uncertainty, and this is where inductive risk considerations arise. It requires reflection on the level of uncertainty acceptable to stakeholders, as well as on which kinds of errors they are willing to tolerate. Constant negotiations concerning what is optimal manifest on the level of model features as well as on model outcomes through confirmative requirements that are related to inductive risks and associated possible errors. False negatives and false positives are unequally distributed among social groups, meaning that these errors do not necessarily entail equal costs for all stakeholders. As the case study has shown, this results in inductive risks and their potential consequences varying across stakeholder groups.

Stakeholders can co-determine decisions concerning specific trade-offs in managing inductive risk. Negotiations may also take place through the articulation of significance requirements in the context of use. The epistemic status of ML models as knowledge producers in public decision-making processes should not be regarded as merely instrumental, nor as objective statistical tools for automation. Since ML models—and the knowledge they produce—are societally and epistemically coupled, the inductive risks associated with their use will be prioritized differently by different stakeholders. This raises the question of who has—or should have—the authority to make such decisions.

We suggest that the inclusion of relevant stakeholders is necessary to justify the use of ML models in public domains. This is because stakeholders may evaluate inductive risks differently, as their potential consequences can vary significantly across social groups.

This asks for normative and value-laden decisions, as there is no common measure for weighting inductive risks and potential consequences. The second reason is that the moral, political and epistemic standards we set for ML models as knowledge producing tools are related to the collective meaning we assign to such methods. The present case study has demonstrated that ML models serve to implicitly negotiate relations among different stakeholder groups. Likewise, the framing of the optimization goals for both models has been subject to multiple formulations. The goals and perspectives that underlay these models is not static. When straightforward answers on how to arrive at acceptable principles of conduct are lacking, we speak of normative indeterminacies (Dobbe et al., 2020). When city governments, as authorities and democratic institutions, use ML models to produce insights there is a need for democratic dissent mechanisms for such indeterminacies (Daniels & Sabin, 2008; Binns, 2018; Wong, 2020). As such, public participation could strengthen reliable and responsible tools for public knowledge-production and decision-making.

What serves as sufficiently optimal for the performance of an ML model should be determined on both epistemic and societal grounds. Depending on which stakeholders are concerned, a less accurate model may better align with societal needs. Value-laden choices navigate how one handles the uncertainty of a model, its potential errors, the desirability and severity of such errors and the acceptance of the available evidence. Model-choices reflect particular balances of inductive risk and models are being optimized for particular purposes towards particular metrics of success (Winsberg, 2012). This means that while possible errors are unavoidable, we can influence both the types of errors that are likely to occur and their extent through our modeling decisions. These decisions represent a negotiation of different priorities to arrive at the most optimal tool needed for a particular goal or quest. Judging a model's optimality requires making it intelligible and acceptable to specific groups within particular contexts. There is not one stable, single, essential characteristic that brings a fixed predictable power or agency to these models. Rather, the focus should be on the set of relations that enable an

ML model to be brought into being (Neyland, 2016). The interpretations of a model will be different for different stakeholders across society (Seaver, 2017). This stems from varying interpretations of the data, the phenomena and the outcome among stakeholders (Erasmus et al., 2020). Therefore, ML models would need to fulfill diverse explanatory requirements, as different stakeholders are likely to pose different kinds of questions (Zednik, 2021).

16.8 Conclusions

In this chapter, we have shown that incorporating philosophy through the lens of coupled societal and epistemological questions cannot only deepen our understanding of the values embedded in ML models used in public decision-making, but also improve the management of inductive risks by involving diverse stakeholder perspectives. This approach will unveil the value and limits of ML modelling as a knowledge producing practice that is part of the public domain. The ML model in the presented case study is geared towards increased livability in Amsterdam's urban environment. This model is concerned with the objects in public space, which only indirectly affect people's lives in the city. The case study has shown that using ML models as tools for knowledge production and decision-making in governmental processes raises not only the question of what kind of knowledge base we embed in these tools in our cities, but also what kinds of moral norms or virtues should be programmed into them.

According to the concept of inductive risk, when accepting a ML model, its uncertainties, and thus its potential errors, are also indirectly accepted. These errors are likely to affect particular groups in the city differently. As such it will always be the same kinds of people, often linked to particular neighborhoods, who live through such systems without any problems, as in the presented case study. Simultaneously, it is the same kind of people that always fall in the potential error-zone of uncertainty, or that fail to comply with such a system. This is closely linked to coupled judgments of inductive risk. Specific potential errors may be judged as highly undesirable because of the social evaluation of their consequences. Rather than treating the epistemic model-performance and its social consequences as separate entities, the norms we hold for each of them should be seen as coupled. Our treatment of the case study, based on this analysis, demonstrates that like epistemic values, societal values play a key role in the production and evaluation of the model's epistemic qualities and potential consequences. The coupled societal-epistemic analysis presented in this chapter places societal values at the center of the evaluation of model performance and outcomes. It thus suggests that assessing optimal functioning of ML models requires epistemic and societal value judgments, as they are coupled and mutually influence each-other. This means that accepting ML model performance as sufficient means accepting their epistemic output and both direct and indirect societal consequences.

The societal-epistemic coupledness analysis shows that the knowledge generated by societal ML models is not neutral. Since moral norms are an essential part of a ML model's appropriate epistemic functioning, what counts as usable knowledge for public decision-making becomes highly political. As a result, the insights to be produced by these models should not be understood and perceived as purely epistemic, but as coupled from the start. Going beyond the epistemic-societal divide will provide new tools to discuss the status we assign to ML models as knowledge producers in public decision-making processes. In such contexts, societal and epistemic aspects are deeply intertwined. For instance, the case study presented in this chapter shows a sensitivity to the consequences of the model's potential errors, as the cost of error is not seen as neutral towards those affected. This suggests that involving citizens more in the goal and outcomes of ML models may reduce uncertainties concerning the moral and epistemic judgments of these model-practices. Without such collective insights about public issues and phenomena in the real-world, these interpretations and interventions are highly rhetorical. Hence, the current analysis offers fertile ground for more democratically epistemic practices that promote transparency and strengthen the reliability of knowledge production.

References

Alexandrova, A. (2015). Well-being and philosophy of science. *Philosophy Compass, 3*, 219–231.

Alpaydin, E. (2010). *Introduction to machine learning*. The MIT Press.

Barocas, S., & Selbst, A. D. (2018). Big data's disparate impact. *SSRN Electronic Journal, 671*, 671–732.

Basu, R. (2020). The specter of normative conflict: Does fairness require inaccuracy? An introduction to implicit bias: Knowledge, justice, and the social mind. In E. Beeghly & A. Madva (Eds.), *An introduction to implicit bias: Knowledge and the social mind* (pp. 1–19). Routledge.

Binns, R. (2018). Algorithmic accountability and public reason. *Philosophy and Technology, 31*(4), 543–556.

Carrier, M. (2013). Values and objectivity in science: Value-ladenness, pluralism and the epistemic attitude. *Science and Education, 22*(10), 2547–2568.

Daniels, N., & Sabin, J. (2008). *Setting limits fairly: Learning to share resources for health* (2nd ed.). Oxford University Press.

Diekmann, S., & Peterson, M. (2013). The role of social values in engineering models. *Science and Engineering Ethics, 19*(1), 207–228.

Dobbe, R. I. J., Gilbert, T. K., & Mintz, Y. (2020). Hard choices in artificial intelligence. *Artificial Intelligence, 300*, 103555.

Dotan, R., & Milli, S. (2020). Value-laden disciplinary shifts in machine learning. In *Proceedings of the 2020 conference on fairness, accountability, and transparency*, 294. Association for Computing Machinery.

Douglas, H. (2000). Inductive risk and values in science. *Philosophy of Science, 67*(4), 559–579.

Elliott, K., & Richards, T. (Eds.). (2017). *Exploring inductive risk: Case studies of values in science*. Oxford University Press.

Erasmus, A., Brunet, T. D. P., & Fisher, E. (2020). What is interpretability? *Philosophy and Technology, 33*(4), 541–546.

Franssen, M., Lokhorst, G.-J., & van de Poel, I. (2023). Philosophy of technology. In E. N. Zalta, & U. Nodelman (Eds.), *The Stanford encyclopedia of philosophy* (Spring 2023 Edition). Retrieved from https://plato.stanford.edu/archives/spr2023/entries/technology/

Karaca, K. (2021). Values and inductive risk in machine learning modelling: The case of binary classification models. *European Journal for Philosophy of Science, 11*, 102.

Katikireddi, S. V., & Valles, S. A. (2015). Coupled ethical-epistemic analysis of public health research and practice: Categorizing variables to improve population health and equity. *American Journal of Public Health, 105*(1), e36–e42.

Leonelli, S. (2019). What distinguishes data from models? *European Journal for Philosophy of Science, 9*(2).

Longino, H. (1990). *Values and objectivity in scientific inquiry*. Princeton University Press.

Neyland, D. (2016). Bearing accountable witness to the ethical algorithmic system. *Science Technology and Human Values, 41*(1), 50–76.

Parker, W. S., & Winsberg, E. (2018). Values and evidence: How models make a difference. *European Journal for Philosophy of Science, 8*(1), 125–142.

Ratti, E. (2025). Mostly harmless? Machine learning and the ethics of induction. In J. M. Durán, & G. Pozzi (Eds.), *Philosophy of science for machine learning: Core issues and new perspectives*. Synthese Library. Springer.

Ratti, E., & Graves, M. (2021). Cultivating moral attention: A virtue-oriented approach to responsible data science in healthcare. *Philosophy & Technology, 34*(4), 1819–1846.

Reiss, J., & Sprenger, J. (2020). Scientific objectivity. In E. N. Zalta (Ed.), *The Stanford encyclopedia of philosophy* (Winter 2020 Edition). Retrieved from https://plato.stanford.edu/archives/win2020/entries/scientific-objectivity/

Rudner, R. (1953). The scientist qua scientist makes value judgments. *Philosophy of Science, 20*(1), 1–6.

Russo, F., Schliesser, E., & Wagemans, J. (2023). *Connecting ethics and epistemology of AI*. AI & Society.https://doi.org/10.1007/s00146-022-01617-6

Schmidt, E. (2025). Stakes and understanding the decisions of artificial intelligent systems. In J. M. Durán, & G. Pozzi (Eds.), *Philosophy of science for machine learning: Core issues and new perspectives*. Synthese Library. Springer.

Seaver, N. (2017). Algorithms as culture: Some tactics for the ethnography of algorithmic systems. *Big Data and Society, 4*(2), 1–12.

Sukel, M., Rudinac, S., & Worring, M. (2019). Multimodal classification of urban micro-events. In *Proceedings of the 27th ACM Conference on Multimedia (MM'19)* (pp. 1455–1463). Association for Computing Machinery.

Sukel, M., Rudinac, S., & Worring, M. (2020). Urban object detection kit: A system for collection and analysis of street-level imagery. *Proceedings of the 2020 International Conference on Multimedia Retrieval (ICMR 2020)*, 509–516.

Tuana, N. (2013). Embedding philosophers in the practices of science: Bringing humanities to the sciences. *Synthese, 190*(11), 1955–1973.

Valles, S. A., Piso, Z., & O'Rourke, M. (2019). Coupled ethical-epistemic analysis as a tool for environmental science. *Ethics, Policy and Environment, 22*(3), 267–286.

Veale, M., & Brass, I. (2019). Administration by algorithm? Public management meets public sector machine learning. In K. Yeung & M. Lodge (Eds.), *Algorithmic Regulation*. Oxford University Press.

Winsberg, E. (2012). Values and uncertainties in the predictions of global climate models. *Kennedy Institute of Ethics Journal, 22*(2), 111–137.

Wong, P. H. (2020). Democratizing algorithmic fairness. *Philosophy and Technology, 33*(2), 225–244.

Woodward, J., & Bogen, J. (1988). Saving the phenomena. *The Philosophical Review, 97*(3), 303–322.

Zednik, C. (2021). Solving the black box problem: A normative framework for explainable artificial intelligence. *Philosophy & Technology, 34*, 265–288.

Milou Jansen, at the time of writing, worked as Lead Digital Rights and Ethics at the City of Amsterdam and was the Coordinator of the Cities Coalition for Digital Rights. She combines her local and international experience in the field combined with a Master's Degree in Philosophy of Science, Technology & Society, obtained at the University of Twente. Most recently, she worked as Project Coordinator of the 'International Guidelines on People-Centered Smart Cities' at UN Habitat (United Nations Human Settlements Programme).

Koray Karaca is an assistant professor in the Department of Philosophy at the University of Twente. He holds a Ph.D. in theoretical physics from the Middle East Technical University, and a Ph.D. in history and philosophy of science from Indiana University. Previously, he worked as a postdoctoral researcher on the interdisciplinary project *The Epistemology of the Large Hadron Collider*, based at the University of Wuppertal, and he taught in the Department of Philosophy at the University of South Florida. His current research interests lie in the epistemology of experimentation and simulation, as well as the epistemology and ethics of machine learning. He currently serves as the Ethics Coordinator for *LLMs4EU—Large Language Models for the European Union*, a project supported by the Digital Europe Programme.

Chapter 17
Machine Learning and the Ethics of Induction

Emanuele Ratti

Abstract This chapter analyzes the inferential structure of machine learning (ML) systems, and shows how these can be value-laden in unexpected ways. ML systems follow an inductive inferential strategy, which is based on two components. First, there is the basic assumption that we are entitled to predict future events on the basis of past occurrences because the world will not drastically change. This assumption is called 'uniformity of nature' (UoN). Second, 'canons of inductive inference' (CIIs) are required to narrow down the set of possible hypotheses that one can generate from UoN. Debates on the ethics of ML have focused on CIIs. Here I show that UoN plays an important ethical role, in particular in eroding human agency.

Keywords Machine learning · Induction · Ethics of AI · AI ethics

17.1 Introduction

In the past few years, there has been a growing concern about the problem of 'algorithmic bias' in machine learning (ML) tools. In a general sense, 'bias' is understood as a systematic error due to a deviation from a norm or standard, where norms or standards can be statistical, moral, epistemic, etc. (Danks & London, 2017). Research has shown that, because of the way they are designed, algorithmic systems incorporate plenty of biases—in particular social and moral biases—that cause harmful impacts, especially to minorities. Stemming from these works, several proposals on how to identify biases and design algorithmic systems that are less biased have been developed (Fazelpour & Danks, 2021). All in all, this rich literature is focused on how choices behind the design of algorithmic systems have repercussions on their outputs and impact individuals differently.

E. Ratti (✉)
University of Bristol, Bristol, UK
e-mail: mnl.ratti@gmail.com

J. M. Durán, G. Pozzi (eds.), *Philosophy of Science for Machine Learning*,
Synthese Library 527, https://doi.org/10.1007/978-3-032-03083-2_17

In this chapter, I want to show how investigating ML tools from the point of view of their inferential structure can be fruitful. In particular, I will show that (supervised) ML tools follow an inductive inferential structure, and that the debate on the ethical issues raised by ML systems has focused only on specific components of this inferential structure, at the expense of others. But these neglected aspects of the inferential structure, I claim, reveal different kinds of ethical problems.

The structure of the chapter is as follows. In Sect. 17.2, I will make the case that (supervised) ML tools follow an inductive strategy. In particular, this strategy is based on two components. First (Sect. 17.2.1), there is the uniformity of nature (UoN) component, according to which we are allowed to infer future observations on the basis of similar past observations because of assumed uniformity between the past and the future. This principle is insufficient to ground any reliable inference, as the past and the future may look alike in an indefinite number of ways. UoN is supplemented by non-evidential background assumptions (that is, the second component), which I refer to as 'canons of inductive inference[1]' (CII). CIIs make sure that the space of possible ways in which the past and future are similar is narrow enough to allow inference. I then show (Sect. 17.2.2) how ML follows this particular strategy, where CIIs take the form of different choices that shape the final ML model, and elucidate (Sect. 17.2.3) how debates in ethics of ML are mostly about CIIs. In Sect. 17.3, I turn to the analysis of UoN, first by describing the 'harmless' role that it plays in the natural sciences (Sect. 17.3.1), and then (Sect. 17.3.2) to show how, in the context where human agency is salient, UoN can actually erode human agency itself.

17.2 Induction and Machine Learning

Let me start by describing the main features of induction and inductive inferences, and in which sense ML systems are inductive.

17.2.1 Varieties of Induction and Machine Learning

Here I refer to 'inductive inference' in a somewhat broad way—"whenever we note that evidence lends support to an hypothesis (. . .) while not establishing it with deductive inference" (Norton, 2005, p. 1). Induction is said to be *ampliative* because, unlike deduction, the conclusion is not contained in the premises, or the conclusion is more than just a reformulation of the available evidence. According to these broad indications, there are several types of inferences that (at least at first) can count as

[1] This is, as I will specify, much broader than the way this term is used in the literature in philosophy of science and epistemology.

'inductive'. These include enumerative induction; abduction; eliminative induction (Earman, 1992; Norton, 1995); variational induction; material induction etc. Let me briefly illustrate them.

Enumerative induction—also known as induction by simple enumeration—is the most common form of induction, an archetype of inductive generalization (Norton, 2005). What enumerative induction is supposed to establish is a general dependency between two types of events. The idea is that if there is a sufficient number of co-occurrences of these two types of events, then we can infer that there is a general dependency between them (Canali & Ratti, 2024). It is obviously ampliative: if the co-occurring instances are seen as premises, and establishing a general dependency is the conclusion, then the conclusion is not contained in the premises.

Eliminative induction is also known as 'induction by means of deduction' (Hawthorne, 1993), 'strong inference' (Platt, 1964), or 'eliminative inference' (Forber, 2011). It works in the following way. First, a set of premises corresponds to a finite set of competing hypotheses. How to establish these premises is tricky, but the idea is that they are built from a prior state of individual practices (Kitcher, 1993; Ratti, 2015), which drive the selection of the competing hypotheses. In my past work (Ratti, 2015), I show how this is done in genomics. Next, other premises and new evidence stimulates a process where hypotheses are discarded until the 'true' hypothesis remains. Famously, this scheme captures Sherlock Holmes' strategy that can be summarized with the slogan 'when you have eliminated the impossible, whatever remains, however improbable, must be the truth'. Eliminative induction has a renowned tradition in science and philosophy, from Mill to Dretske. But whether it is really a kind of induction is an open question, as the means are in fact deductive, and its ampliative nature is dubious, given that the hypothesis ending up to be true is one of the premises that survive the process of elimination (so, in a sense, the conclusion is contained in the premises).

Variational induction is not as common as the previous two. A comprehensive characterization is provided by Pietsch (2021, 2022), who borrows the term from Russo (2009). Pietsch conceptualizes it as a method to infer causal relevance between a phenomenon and a set of circumstances, "by relying on variational evidence, i.e., on evidence that tracks changes in a phenomenon resulting from systematic variations of circumstances" (Pietsch, 2021, p. 29). A precursor of variational induction is, according to Pietsch, Mill's methods, in particular the method of difference and the method of agreement. In my understanding, variational induction is ampliative because the conclusion (i.e., establishing causal relevance) is not contained in the premises (i.e., variational evidence).

Abduction/inference to the best explanation is another form of inductive inference. According to Norton (2005), there is a family of inductive inferences known as 'hypothetical induction', whose archetype is the so-called 'saving the phenomena' strategy: "the ability of an hypothesis to entail deductively the evidence is a mark of its truth" (p. 5). Abduction is an extension of this strategy, where an additional constraint is put on the hypothesis. In addition to entail the evidence, the hypothesis must also explain it. This is considered to be ampliative, because the conclusion does not follow from the premises.

Now, it is usually said that ML works by adopting a form of inductive inference. Given that there are different types of inductive inference, which one does characterize induction? Because ML is a rather broad family of techniques and algorithms, we might expect that more than one type of inductive inference will fit it.

We can exclude two cases from the start. First, ML does not fit a type of 'inference to the best explanation' type of induction. This is because ML algorithms work by fitting a real function, and so in a sense they aim at 'saving the phenomena', but they do not pose explanatory constraints on the models themselves. To be more precise, what is learned by an algorithm is called a ML model (Facchini & Termine, 2021). This is a learned function, consisting in a mathematical representation of the statistical patterns that the algorithm has learnt from a data set. Sometimes it is said that the function or the model is selected, among the ones possible. But other than fitting the data in the right way, the model need not meet explanatory requirements or desiderata to be 'selected' by the algorithm.[2] Of course, scientists using ML models may opt for the model that better explains the data from a variety of different angles, but the algorithm *per se* is neutral with respect to this particular explanatory constraint. We can also exclude eliminative induction. Even if eliminative inferences characterizes disciplines making ample use of ML techniques like genomics, the eliminative inductive strategy is usually ascribed to the whole system of inquiry, rather than the ML algorithm (Ratti, 2015). In other words, scientists in genomics may use ML techniques in their own eliminative inductive investigations, but the algorithm *per se* does not proceed by eliminative inferences. This is because the algorithm does not start with any predefined hypothesis (Pietsch, 2022).

Let me now turn to the positive cases. According to Pietsch (2022), variational induction seems to characterize a significant number of strategies used in Big Data analytics. He notices that in many cases (e.g. decision trees, naïve Bayes, neural networks), ML algorithms are in fact trained by relying on variational evidence, where confirmation or performance increases with the variety of evidence. For instance, decision trees are constructed by taking into account both positive and negative instances, and the relevance of predictor variables is established in terms of the difference that these variables make for the accuracy of a classification—these can be seen as characteristics of variational induction. The same can be said, according to Pietsch, for neural networks. He also recognizes that simple enumerative induction characterizes some methods, including association rules. But it is possible to show that variational induction and enumerative induction are based on similar ideas, and that in a way variational induction can be translated in terms analogous to enumerative induction. One important aspect that the two have in common is that both enumerative and variational induction assumes that more data is generally better (Canali & Ratti, 2024). As mentioned above, in enumerative induction it is generally assumed that we should provide a sufficient number of co-occurrences between two events to make the case for their dependency, and in

---

[2] But, as we will see, they come with several other constraints.

absence of specific guidelines on how much is enough, then we should assume that the higher the amount, the better. But simple enumerative induction is agnostic with respect to the type of evidence of co-occurrence that one should use. Variational induction is more precise on this; we need to increase the number of features considered (or 'circumstances') in order to observe "as many different situations in terms of changing circumstances as possible" (Pietsch, 2021, p. 30). Under this lens, variational induction is a more precise version of enumerative induction when it comes to the kind of evidence required for inference. Sure, one can say (as Pietsch does) that variational induction establishes causal relations, while enumerative induction does not, but whether variational induction is legitimate for causal inference—especially when understood in the context of ML, where causal inference is rare—is contentious (Canali & Ratti, 2024). In other words, it is often possible to reconstruct cases of variational induction-based ML as specific cases of enumerative inductions. In fact, I agree with Harman and Kulkarni (2007) when they point out that enumerative induction "applies to many examples of machine learning, including perceptron and feed-forward neural net learning" (p. 25)

17.2.2 Enumerative Induction in Machine Learning

Now it is time to show more precisely the inductive structure of ML. Let me go back to enumerative induction.

In (Johnson, 2023), Johnson explores the relationship between methods like ML and the value-free ideal in science. She does this by tracing back the problematic nature of this relation to Hume's problem of induction. Typically, Hume's problem is understood as being about the justification of induction. In deduction, true premises will always guarantee true conclusions, which is not what happens in induction. Consider for instance a case where you have a set of premises arguing that many apples have been found to have seeds and no apples have been found to be seedless. From these premises, the hypothesis/conclusion/prediction that the next apple will have seeds can turn out to be false, even if the premises are true. One may say that the more apples with seeds, the more likely that the next apple will have seeds. Psychologically, this is what we would expect, but whether we are justified in having this expectation is a different matter. In fact, it does not matter how much support a hypothesis has; there is nothing that can justify why known instances would provide any degree of support to predict unknown instances: there is "nothing logically at odds with the world suddenly becoming drastically different" (Johnson, 2023, p. 6). This means that we cannot justify (without further constraints) the practice of using patterns identified in evidence to predict novel cases. However, this practice is common in the sciences, and in ML in particular.

This problem is typically overcome by making one or more non-evidential assumptions. The most basic assumption that one can do—and that is used by Hume himself—is assuming that nature is uniform. Given that the world will not be drastically different tomorrow, we can use the past to reliably predict what there will be in the future on the basis of similarities between the world as it is now, and the world as it will be tomorrow. Let us call this UoN (i.e., Uniformity of Nature). The problem with UoN is that, as noted by Johnson and many others, there are many ways in which the world of today can be similar to the world of tomorrow. Today's world may have apples with seeds and blue skies, but tomorrow's world—while still having blue skies—may have apples without seeds. This implies that there is not much guidance in restricting the set of possible 'similarities' between past and future worlds (which then translate into possible hypotheses/predictions about novel instances) to a tractable size.

Therefore, UoN can be a ground norm used for inductive inferences in the sciences, but we need also other assumptions, which will provide additional "non-evidential way[s] of limiting the hypothesis space to a tractable size" (Antony, 2016, p 161). Johnson calls these assumptions 'biases'. In the history of science, philosophy of science, and epistemology, assumptions vary, and they are known under many names, such as cognitive heuristics (Bechtel & Richardson, 2010), theoretical/epistemic virtues (Kuhn, 1977; McMullin, 1983), non-cognitive values (Douglas, 2009), canons of inference (Levi, 1960), etc. I prefer to use 'canons of inductive inference' (CII), and my use of the term will be very broad, to include also non-epistemic and non-canonical criteria. In this way, my use of CII leaves open the question of whether these assumptions are epistemic or not. Moreover, unlike the use of 'biases', using CII is not normatively charged (even though Johnson means 'bias' without normative connotations). So the idea is that we are entitled to infer from known instances the prediction of novel instances because of UoN *plus* a host of other assumptions that would constrain the way patterns should look like. Certain kinds of CIIs (most notably, epistemic and non-epistemic values) are also invoked to solve old problems in philosophy of science, such as underdetermination (Longino, 1990)—an example being the so-called 'gap feminist empiricism' (Solomon, 2012), where the impasse of theory choice is solved by values.

By following Harman and Kulkarni (2007), it is possible to show that typical problems in statistical learning theory—whose applications are embedded in many ML systems—have an inductive structure of the kind just shown. In particular, consider classificatory problems. At a basic level, this is a problem about reaching conclusions about the next X on the basis of observing prior Xs, where the number of correct conclusions should be as high as possible. We want to find a method for doing this—in this context, the method is sometimes called 'a rule', or 'model'. The rule is a classification rule because its goal is to help us to classify something as X, on the basis of observed characteristics of previous Xs.

This way of framing the problem seems misleading though, since it reflects old problems. Consider the question 'what kind of classification rule do we want?' in the context of, for instance, distinguishing between French Bulldogs and Boston Terriers (Castro, 2022). The rule we are looking for will be in the form of "if you run into a dog that is either a French Bulldog or a Boston Terrier, take measurements x and y (e.g. height and weight); if x is below z, and y is above z, then it is a French Bulldog; otherwise it is a Boston Terrier". But how do we establish if this is the best rule? One may say that the best classification rule is the one with least errors.

But, as Harman and Kulkarni note, this is not a precise way of identifying the best rule. If we think about the rules to select as forming a set, then an unconstrained classification problem like this one will have a set that includes all possible rules, and as a consequence there will be also many rules with similar error rates (as well as different types of errors), which will classify new cases differently. To avoid this problem, there should be restrictions on the rules that can be selected. This is exactly the same issue motivating the use of CII as a way to constrain UoN. Harman and Kulkarni (2007) call these restrictions 'inductive biases'. They say that the framing of a classificatory problem,

> "must prefer some rules over others. It must be biased in favor of some rules and against others. If the method is the sort of enumerative induction which selects a rule from C [i.e. rules for classifying] with the least error on the data, there has to be a restriction on what rules are in C. Otherwise, we will not be able to use data in that particular way to restrict the classifications of new cases" (p. 23)

This reflects typical problems in underdetermination and induction. There can be different hypotheses that make sense of the same data (in our case, there can be different rules with the least errors based on the same observations). Because of this, we need to restrict the range of possible hypotheses: there can be different ways in which the present and future worlds are similar, and these will grant different predictions. In the case discussed by Harman and Kulkarni, restricting the possible rules is necessary, and this is done by adopting an inductive bias. In the context illustrated by Johnson, in order for theory/hypothesis choice to overcome impasse of the vagueness of UoN or the underdetermination problem, we need CIIs (e.g. theoretical/epistemic virtues; non-epistemic values; biases; cognitive heuristics; etc.). But while the contexts are slightly different, they are pointing to the same thing—in fact, the inductive bias Harman and Kulkarni talk about is just one instance of CIIs. It is also important to notice that CIIs do not do all the work; in the case of statistical learning theory and classificatory problems, UoN plays a substantial role too: it is only under the assumption that the world will not drastically change that the claim 'given these past Xs, here is a novel instance of X' makes sense—without UoN, there would be no starting point.

This framing of classificatory problems in statistical learning theory represents exactly the kind of problem solved by most supervised ML systems. Rather than using the word 'rule', several uses the word 'model' (i.e. it is the model of the data that is selected [James et al., 2013]) or simply function.[3] The *x*s that are used

[3] 'Function', 'model', 'rule' (in case one refers to a classificatory problem), are used interchangeably in this literature. This emerges by considering different introductory texts. For instance, James et al. (2013) set the goal of statistical learning as the estimate of a function f mapping input variables into output variables. In discussing the trade-off between prediction accuracy and interpretability of the function $\hat{f}$ which estimates the real function f, they refer to 'model interpretability', and 'model predictive accuracy', making clear that they are referring to the function $\hat{f}$. Burkov (2019) identifies the goal of ML as the construction of statistical models, and then defines the model explicitly as a function (p. 5). Jiang (2021) also goes back and forth using 'model' and 'function' interchangeably. Harman and Kulkarni (2007) talk about 'rules of

to select the best model are typically training data sets (i.e., data used to train the algorithm) and test data sets (i.e. data used to test the accuracy of the model selected as a result of first training). These data sets are 'labelled', in the sense that there is a label attached to different data points indicating what the data is about (i.e. a target variable such as a cancer sample; a dog; etc.). After a model is 'selected', it is applied to unlabeled data sets to automatically classify new cases. In this process, CIIs play substantial roles. I take canons in ML to mean something broad, namely anything that constrains or shapes the set of possible models that will be selected: *CIIs in ML are any criteria guiding choices made in building these algorithmic systems that constraint the 'selection' of the final model.* For instance, in another article (Ratti & Graves, 2021) I decompose the ML construction process into various phases, and show how technical and micro-incremental choices in how to prepare data—even before the actual training—can have important consequences for the final model selected. Johnson focuses on similar things. In fact, it is a truism that predictive models can be built in different ways, and these will reflect different 'patterns' found in the data. But norms in the ML community will restrict the range of acceptable 'models' or 'methods'—such as setting hyperparameters, best practices for when to use certain algorithms rather than others on the basis of the problem at hand, different measures of performance (e.g. accuracy, precision, sensitivity, specificity, etc.).

To sum up, many supervised ML systems proceed by induction. What they do is solve classification problems typical of statistical learning theory. These are conceptualized as problems of identifying the next x on the basis of observation of former xs. The basis for the prediction is a rule R. Such a rule is selected by choosing among alternatives within a set of rules. This set is limited and constrained by what are called inductive biases. An additional and upstream assumption is that it makes sense to identify new xs on the basis of past observed xs only if UoN is true. ML systems follow this structure, with 'models' usually standing for rules, and 'inductive biases' as being any constraint that is posed on the functioning of algorithms that impacts model selection. This structure is a typical inductive inference strategy, where UoN makes sure that the world of past observations will not look drastically different from the world of new instances, and CIIs constrain the set of competing hypotheses that can be inferred from the past observations.

classification' as rules "for using observed features of items to classify them" (p. 29). Rules are found by using data, where data are thought to be a random sample coming out of a background probability distribution, about which nothing is assumed. The best classification rule is the one that makes the least errors—in other words, the one that estimates as much as possible the background probability distribution. But the background probability distribution is a function, so the rule will be f' as meant by James et al. and, ultimately, a model.

17.2.3 AI Ethics Is About Canons of Inductive Inference

I have shown that ML systems are typically designed to solve problems formulated within the remit of statistical learning theory, which in turn can be characterized as going through inductive procedures. In induction, two components stand out in making inferences possible. First there is UoN, namely the assumption that the world will not drastically change, thereby making it easier for us to predict the novel on the basis of past observations. The second component are CIIs, which are constraints posed on UoN, which facilitate the restriction of the space of hypotheses/predictions/rules to consider.

Debates on the ethical, societal, and political dimensions of ML tools are mostly about CIIs. The large literature on algorithmic bias (Fazelpour & Danks, 2021) focuses on the different aspects of ML systems that may be salient or relevant for ethical, societal, and political issues. It has been shown how different choices in designing ML systems can be motivated by values whose nature is non-epistemic (Ratti & Graves, 2022; Johnson, 2023; Biddle, 2020) [see Jansen & Karaca, 2025]. For instance, Biddle (2020) reconstructs the systematic presence of value-laden decisions in designing algorithmic systems in the justice system. In the typical way scholars of the inductive/epistemic risk literature proceeds, Biddle shows how at each step of the ML pipeline more than one decision is possible, and each choice comes with related risks. In deciding which risks are acceptable and which are not, ML practitioners use value-judgement, which often involves endorsing certain ethical, social, and political values. All these choices motivated and/or justified with values (Ward, 2021) can be conceptualized as choices pertaining to CIIs. For instance, Biddle argues that in collecting data for building a tool to predict recidivism, questions about the baseline population or about the features that correlate with crimes come up—and choosing certain features rather than others is based on values and has consequences for values. Or another typical problem is about the evaluation of the performance of ML systems. In choosing the relevant metric (be it accuracy, precision, sensitivity, specificity or something else), the importance of increasing or decreasing false positives and false negatives is fundamental, and choosing one over the other reflects certain constraints that we want ML system to act within—and these are, ultimately, CIIs. These values may be 'technical', social, moral, political, etc., and they all constitute forms of CIIs that shape choices behind ML systems. One can also formulate the problem not by relying on 'risks', but rather on 'contingency'. This is the 'science and value' approach developed by Brown (2020). Scientific activities like building a ML system requires practitioners to execute certain actions. These actions are contingent, in the sense that alternative actions are always possible. This is the case also in ML (Ratti, 2020a, 2020b). *Practical reasons* resolve the impasse of alternative actions. Pragmatic choices based on practical reasons means taking into account various CIIs in order to orient the design of the ML tool in a direction rather than another. Problems of bias, fairness, privacy, safety, etc., can all be formulated as problems concerning the risks of (or contingent actions behind) constraining the design of ML systems via specific CIIs rather than others.

These observations are relevant to make a specific point: the ethics of ML is focused on a specific component of ML inferential structure, which is CII. However, focusing only on CII overlooks an important ethical dimension of ML tools, which is related to UoN. In the next section, I will show what this means.

17.3 Machine Learning and Uniformity of Nature

In the previous section, I have shown in which sense common supervised ML tools follow an inductive inferential structure, by drawing from well-known literature in statistical learning theory. In this section I will delineate some consequences of this. In particular, I will show how in cases where ML is applied to the natural sciences, the inductive structure does not pose any significant ethical problems. But in Sect. 17.3.2 I will show how, in the context where human agency is involved, that same inductive structure is problematic from an epistemic point of view and, most important, ethically controversial. But this is not just because of the typical ethical issues investigated by the ML ethics literature concerning CIIs; rather, the ethical problems are about UoN.

17.3.1 ML, Induction, and the Natural Sciences

Let us see this inferential structure in action in the natural sciences. Let's consider, for instance, how supervised ML tools are used in biology, in particular in genomics. In this field, biologists have been using these tools for more than 20 years. Recently, genomics has benefited from the use of deep learning systems (Eraslan et al., 2019), and the first steps have been made in adapting large language models to its context (Consens et al., 2023). But these are complicated cases, and we do not need to consider them to show the kind of inferential structure in place in genomics when problems are approached from a ML perspective. For this reason, I will refer to simpler examples of the 'pre-deep learning' era.

Let me start with a basic example of a classificatory task in genomics, such as identifying transcription start sites (TSS) in genomics data (Libbrecht & Noble, 2015; Ratti, 2020a, 2020b). Using ML to classify TSSs is a representative example of what people in this context are after. In genomics, the goal is to identify the genetic basis (that is, at the DNA level) of biological phenomena. It is common to gather data about genomes, and then look for biological entities that are biochemically active at the DNA level, including genes, DNA mutations, TSSs, transcription factors, etc. To get a preliminary understanding of TSS, consider the first phases of the mechanism of protein synthesis. Very briefly put, in eukaryotic cells the mechanism starts in the nucleus, where a given portion of DNA molecules (i.e. a

DNA sequence known as 'gene') is copied (i.e. transcribed) into a RNA nucleotide sequence, which is known as pre-messenger RNA (pre-mRNA). Pre-mRNA as template-molecule is typically subjected to a process of chemical modification known as *splicing*, where the template is modified by removing some bases (*introns*) and the remaining (*exons*) are assembled together. The new molecule/template is known as messenger RNA (mRNA), and it is moved into the cytoplasm, where other processes (beyond the scope of this brief exposition) will lead to the synthesis of the primary structure of proteins. In this context, TSSs are DNA sequences indicating where the transcription (i.e. the copying mechanism) of a gene should approximately start. We know a great deal about TSSs. For instance, they are usually associated with the so-called TATA boxes, namely sequences of Thymine-Adenine-Thymine-Adenine where transcription factors bind. This is taken to indicate where transcription should approximately start, around 25 nucleotides upstream TSSs themselves. Anytime there is a TATA box, then it is likely to have a TSS nearby. Similarly, TSSs are usually located downstream of CpG islands, which are regions where cytosine and guanine are separated by only one phosphate group. The interesting thing is that, in building a ML system to automatically predict the presence of TSSs, you do not need to use any of this theoretical information about TSSs.[4]

Typically, the ML model is built in this way. An algorithm is trained on data sets to find 'rules' or 'models' or 'functions' that can be used to classify TSS sequences in new data. The data collected for the task are 'labeled', in the sense that it is specified whether given sequences contain a TSS. The data set is usually divided into training and test data sets. The training data set is provided as input to the algorithm, in order to produce a model that can be used to classify TSSs. This model is then 'tested' against the test data set. The training set with labeled TSS is the set of known instances, while the test data set is the set of unknown instances. But the latter are 'unknown instances' only for the sake of training; in fact this data set is labeled, but it is used only for testing the initial model, in the sense that the model will predict the labels for the testing data set, and then these are compared to the actual labels of the data. The model/rule/function is what we use to infer unknown instances, also beyond testing data sets. The way the model is built reflects various CIIs, in the sense that it is possible to have an indefinite number of models/rules/functions fitting the initial data sets, and so we need to restrict the space of possible models/rules/functions. Typical CIIs are decisions taken at each step of the ML pipeline. When deciding which data to collect to use as input for training, several choices must be made. First, one would have to choose the data themselves. In a pioneering case of building a classifier for TSSs, Down and Hubbard (2002) extracted mammalian promoters from the EPD database, but they decided to discard those on human chromosome 22, and those with less than 500 bases of upstream sequences available. While they do not

[4] Even though you may need to use your knowledge about TSSs to 'externally' validate the final ML model (see Ratti, 2020a, 2020b).

explain why they do this, this is certainly a case where a CII (that is, a choice in building the system) influences how the final model will look like. Another common CII is the so-called ‘feature selection’, which is how one reduces the dimensions of feature vectors by selecting only those features that are seen as relevant for the problem at hand. For instance, in building a support vector-machine (SVM) to classify polymorphisms (that is, DNA and amino acids mutations) for cancer genomics, Capriotti and Altman (2011) end up choosing 51 different features for building the input vector, where these features include components about local sequence environment, sequence profile, gene ontology (GO) terms associated with the relevant DNA region, etc. Another important choice or ‘canon’ is the choice of the algorithm itself (Jiang, 2021). Whether one will choose a discriminative algorithm (e.g. nonlinear kernels, decision trees, convolutional neural networks, etc.) or a generative algorithm (e.g. transformers, variational autoencoders, etc.) will depend on many factors, one of which is the data modality (e.g. image, text, etc.). And each algorithm will also come ‘pre-packaged’ with a number of assumptions about parameters, the kind of function family characterizing the model, etc. Moreover, even evaluating the performance of the model relies on an important choice as to which metrics to use (e.g. sensitivity, accuracy, specificity, etc.). Finally, it is also important to note that in many cases there is also a ‘domain knowledge consistency check’, namely that the model is not evaluated only via quantitative measures, but also on the basis of whether what it has learnt is consistent with domain knowledge. In the case of TSSs, Down and Hubbard (2002) notice *en passant* that what the model has learnt was consistent with what we know about TSS. In the case of polymorphisms/mutations, Capriotti and Altman (2011) even construct a new quantitative metric called GO log-odds (LGO) score. This measures the correlation between the effect of a mutation and a related gene ontology (GO) term. As shown in the literature (Leonelli, 2016), GO terms are supposed to refer to specific biological entities and activities, By establishing the connection between a mutation and a GO term, one can get a grasp on the strength of the association between a mutation and typical biological processes. In this way, Capriotti and Altman can establish whether their model is consistent with the known underlying biology.

The use of (supervised) ML in this context reflects the kind of inductive strategy described in Sect. 17.2. Problems are typical classificatory problems of statistical learning theory. These are conceptualized as problems of identifying the next X on the basis of observation of former Xs. In the genomics case we have just reviewed, the problem is to identify a TSS on the basis of data sets about TSS, or to identify certain mutations on the basis of data sets of mutations. The basis of the identification is a rule, which in the case of TSS or somatic mutation is a model. Such a rule/model is selected from a (metaphorical) set of alternatives rules/models, where the set is limited and constrained by what are called inductive biases in statistical learning theory, or in general by ‘canons of induction’ (again, understanding this term very broadly). These ‘canons’ are much wider than the typical list of CIIs in philosophy, and in the case of genomics and ML include all choices that have an impact on the final model. But in these procedures to build genomics ML models, an important role is played by UoN. In fact, it makes sense to identify new TSSs on the basis of past observed TSSs, or new mutations on the basis of past somatic mutations, only under the assumption of UoN. Therefore, ML systems follow an inductive strategy by grounding their inferences on UoN, plus a number of CIIs.

The role that UoN plays should not be overlooked. As an assumption, it plays an important normative role. I take 'normative' in the way discussed by Hans Radder in his work about the normativity of artifacts (Radder, 2019). According to Radder, an artifact is normative when its realization implies that in the context in which an artifact used, people ought to follow certain (implicit and/or explicit) norms (where a norm is a directive about what to say, do, or be), where these norms will facilitate and not disturb the functioning of the artifact itself. The nature of norms vary, as they can be social, political, moral, epistemic, technical, cognitive, etc. Background assumptions in science can play an analogous 'normative' role. For instance, theoretical virtues are background assumptions on how to choose theories. Committing to the theoretical virtue of simplicity means that, in theory choice, you *ought to* choose the simplest theory among the alternatives, because simplicity is seen as a virtue/value, namely something desirable for scientific theories. UoN is a basic background assumption, and it can be interpreted as an epistemic norm. In fact, *via* UoN, you are entitled to use the past to infer about the future, on the basis of similarities between the past and the future. But you are not just entitled: in a sense it creates an epistemic obligation to do so. This can be put as a conditional: if one wants to predict what will happen, then the past *ought to* be used as a basis for inference. In a context like genomics, UoN is also a *desirable* epistemic norm (which, in a sense, is already implicit in the fact that it is a norm). Through UoN, we have a basis for controlling natural phenomena. From the fact that we can predict the future status of a biological phenomenon on the basis of its past observations, we can also vary the conditions that allow us to predict future occurrences, and we can control the phenomenon itself. This work of 'varying conditions' is done by selecting the right CIIs. But the very possibility of having a conversation about CIIs – that is, about how to restrict the space of possible similarities between the past and the future - is made possible by UoN, as it is only through UoN that it makes sense to take the past as a basis for predicting the future.

17.3.2 ML, Induction, and Human Agency

UoN is considered mostly harmless, and in fact there is not much debate about it. It is of course important in the natural sciences, as it sets the starting point for many investigations, and it is certainly not considered controversial from other points of view. In particular, it is not considered ethically problematic, and as such it should not be the focus of scholars working in e.g., the ethics of ML. However, here I claim that, in cases where ML is used to make decisions concerning human agency, UoN is not just an epistemic norm—it also has an ethical dimension. This has some interesting consequences. In discussions on ethics of ML, ethical issues are usually identified as problems with CIIs. The idea is that a particular combination of CIIs generate unfair outcomes, e.g. biased data sets, or the wrong performance metrics, or a problematic feature selection process, etc. Under this lens, there will always be, at least in principle, a way to shape CIIs to make the algorithmic system more 'ethical'. However, sometimes we might just want to argue that we *should not use* ML in a certain context—and this is what our ethical analysis of UoN can provide, unlike the literature focusing on CIIs. Let me unpack this.

In the context of natural sciences such as genomics, UoN is an assumption that is epistemically salient. The conditional statement put forth by UoN in this case is: if you want to investigate biological phenomena and be able to predict and control them, then you *ought to* use the past as a basis for your inferences. What UoN is saying, is that there are ways in which TSSs or somatic mutations of the past are indeed very similar to TSSs or somatic mutations of the present, and even of the future. But in the context where human agency is involved, acting under the assumption that there are strong reasons to take the future as similar to the past is not just a harmless epistemic norm.

First, it is epistemically problematic, unlike in the natural sciences. One may just notice that predicting human behavior is notoriously difficult. For instance, in political science (Larsen, 2018) there have been discussions about this very issue for a long time. On the one hand, the side of predictivists claims that we can indeed discover patterns and regularities in human behavior, and these will form the basis of reliable predictions. On the other hand, anti-predictivists are quick in pointing out that the track-record of predictivism is quite poor, and includes some spectacular failures like the inability to see 2016 Trump's election coming; Obama's election; the 2008 economic crisis; etc. This is, in part, an epistemic problem, in the sense that "every future state [about human behavior] comes about as a unique result of a unique interplay of many factors—a one-of-a-kind situation for which there is no normal distribution". Moreover, the unpredictability of human behavior is exacerbated by the fact that humans will respond unpredictably to policies, which may then backfire "or produce new problems elsewhere in the policy space" (Larsen, 2018, p. 318). This does not mean that nothing at all can be predicted; in fact, one can take Hayek's distinction between point and pattern predictions,[5] and claim that maybe pattern predictions are possible. I am not taking any side here, but I just want to show how controversial the idea of predicting is when it comes to human agency.

Second, in the context of human agency, assuming that we have strong reasons to believe that the future will be similar to the past is ethically controversial. Here, I define 'ethical' as pertaining to views on how one ought to live their own life. In particular, 'ethics' is about choices and actions unfolding with respect to explicit or implicit conceptions of the so-called 'Good Life'. This is a typical virtue-theory

[5] Point predictions are about a specific element of a given phenomenon, while pattern predictions are about "some of the general attributes of the structures that will form themselves, but not containing specific statements about individual elements" (Hayek, 1974)

or Neo-Aristotelian way of defining ethics, which has found a recent revival in the philosophy and ethics of technology (Vallor, 2016; Ratti & Stapleford, 2021; Ferdman & Ratti, 2024). According to this view, something is ethically salient when it shapes not just human lives concretely (as many CIIs do by shaping the final ML model), but also conceptions or views that individuals hold on how they ought to live their lives, as well as their perceptions on whether their 'aspirations' can be fulfilled. For instance, assaulting individuals in a neighborhood is not just ethically controversial because you are actually harming other individuals, but also because you are shaping individuals' perception of how they ought to live their lives (e.g. living in fear; not going out when it is dark; etc). Norms that impinge on these aspects are normative from an ethical point of view. Given this way of understanding 'ethical', the ethically controversial aspects of UoN in ML is two-fold:

1. The past is usually subjected to a constant ethical evaluation by individuals, e.g. how they wish they could change or not; how it is seen as valuable or desirable or not; how the future can be shaped on the basis of the desirability of past experiences; etc. This can include very mundane things like wanting to change route to go to work to avoid rush hour or traffic, or more important things like wishing to have done a certain degree instead of another. But saying that the future has to be similar to the past makes any reflective exercise on the past quite useless, as there is little point to reflect on the past if you are just doomed to repeat it, other than just gaining a better understanding of one's predicament. Therefore, we might say that something like UoN *bypasses* a moral evaluation of the past.
2. The future is indeed the realm of agency and, as such, an important focus of ethics. Again, ethics is about how one ought to live their own life, and one's conception of living well. Whatever 'good life' one sees fit for him/herself, it will consist in certain plans and goals, and it will be realized by instantiating certain patterns of actions or behaviors. But by saying that the future will be similar to the past, we are bypassing those fundamental ethical and forward-looking considerations that any human being will gauge in order to exercise substantial freedoms. In other words, UoN can potentially *erode* human agent's substantial freedom to an open future by establishing how the future ought to look like

This commitment to erode the freedom to an open future is relevant to the ML context. Consider the well-discussed case of the use of ML tools to predict recidivism in the justice system (Biddle, 2020; Pruss, 2021). In the specific case of COMPAS, the debate between ProPublica vs Northpoint/Equivant was all about CIIs. One crux of the controversy was that, according to ProPublica, COMPAS was racially biased because it did score poorly on a measure called 'predictive equality' (Castro, 2022), and as a consequence blacks were more likely to be falsely classified as 'high risk', while whites were more likely to be falsely classified as 'low risk'. Northpointe/Equivant argued that that the tool was not racially biased at all, because it scored well on another measure called 'calibration', where risk group defendants were found to reoffend at similar rates, regardless of race. This debate, and the analysis under the lens of frameworks such as epistemic risk (Biddle, 2020), sees the ethical issue as a problem of how we align the science to the right values. On this analysis, it is always possible (at least in principle) to adopt a different combination of CIIs, and make sure that a tool like COMPAS scores well in predictive equality, or any other measure.[6] But consider the role that UoN is playing. What a tool like COMPAS does because of UoN, is to estimate what an individual's conduct in the future will be on the basis of what other individuals (similar to the individual in question) have done in the past. In a sense, UoN is establishing the fact that we should expect individuals' agency to unfold exactly as observed in the past (where the specific range is later established by CIIs). But one may use other considerations as bases for the inference. For instance, one may just base the estimation of future conduct on whether a defendant does indeed regret past conduct, or on the basis of how sincere the defendant's plans for their future life are, or on the basis of the conditions that led the defendant to commit a certain crime (e.g. poverty). Therefore, one might make a strong claim, and say that our future agency should not be evaluated on the basis of how similar agents have acted in the past. This has been expressed in the literature as saying that we should not use 'bare statistical evidence' in court proceedings (Schmidt et al., 2023). This, I think, may potentially apply even to 'individualized evidence' if this is the *only* criterion considered. One may argue, for instance, that agents should have the opportunity to reflect on their own past, and be evaluated on the basis of a combination of considerations about their own past actions, their intentions and plans about the future. In other words, one can argue that you ought to evaluate future agency on the basis of something different than just the past track-records of individuals or past track-records of individuals with a similar agency.

Debating the specific similarities between past and present agents (e.g. race, educational background, etc.) as the basis through which the evaluation takes place is already framing the debate in terms of CIIs (most notably, in terms of feature selection). By addressing UoN, one can claim that setting up the evaluation of individual agents from the point of view of past observations as a ground criterion to predict the future is akin to denying agency to individuals in the first place, as their future actions are not determined by their own plans and goals, but *solely* by what other similar agents (or themselves) have done in the past. One may say that it is not the case that these decisions are taken on the basis of *only* considerations of past track-record. However, this does not consider that the use of ML—based on UoN—tends to promote the consideration of factors pertaining to past track-records as the 'objective' factors to follow.

These considerations about UoN in the ML context have interesting consequences. By shifting the focus from CIIs to UoN, one can argue that we ought not to use ML tools in specific contexts where human agency is salient. The view can be expressed as a conditional: if one wants to promote human agent's substantial freedom to an open future, then one ought to promote opportunities for agents to

[6] It should be noted that ML tools cannot be tweaked to achieve a high score for all relevant measures, as various impossibility theorems have shown.

plan their lives on the basis of their own goals, and views; but ML tools, because of the focus of UoN, will consider agents' future action as a predictive problem to be solved on the basis of past instances, rather than as a problem concerning what agents plan to do on the basis of their own goals. This means that, from a ML perspective, the agent's future is not open; rather, it is similar to what has already happened. Therefore, ML tools, because of UoN, tend to distort human agency, and ought not to be used by those who want to promote agents' substantial freedom to an open future. To put it more bluntly, focusing on UoN is one way to motivate the claim that we should not use ML tools in certain contexts.[7] By claiming that we do not want to frame the problem as a predictive problem where the past works as a basis to infer the future when human agency is involved, we immediately exclude the use of ML tools, which employ an inductive inferential structure. By focusing only on CIIs, we do not have this opportunity: we run the risk of giving the impression that it is possible, at least in principle, to come up with a combination of CIIs that will make the ML tool shape human agency in an uncontroversial ethical way. Another interesting consequence is that UoN, in a context like COMPAS, will necessarily frame the problem to solve as a predictive problem of crime-control, rather than as a problem of what a defendant might deserve on the basis of culpability and other factors (including defendant's willingness to pursue a different, good life). This has been seen as a 'domain distortion' by Pruss[8] (2021), as well as a value-promoting[9] aspect of ML tools (Ratti & Russo, 2024).

17.4 Conclusion

In this chapter, I have shown how analyzing the inferential structure of ML can be useful to distinguish different classes of ethical problems. ML tools (at least supervised tools) follow an inductive strategy based on UoN and various CIIs. The literature on ethics of ML has focused especially on CIIs. Here, I have shown how UoN raises issues about human agency—in fact, ML tools used in the context where human agency is involved are based on the assumption that the future must be similar to the past, thereby undermining the importance of human agency and open future. I have ended with some consequences concerning this analysis of UoN, especially in the kind of ethical arguments that focusing on UoN vs focusing on CIIs can provide.

---

[7] As noted by a Reviewer, legal constraints may deliver this, even though it would be strictly a legal reason and not an ethical motivation.

[8] By 'domain distortion' Pruss (2021) means how ML can influence the concepts and assumptions of a given context of application

[9] In a co-authored work with Federica Russo (2024), we refer to 'value-promoting' science to the phenomenon when either scientific methodologies or concepts promote certain values at the expense of others because of their own identifiable characteristics, independently of the value endorsed by those using the methods or the concepts

References

Antony, L. (2016). Bias: Friend or Foe? In M. Brownstein & J. Saul (Eds.), *Implicit bias and philosophy, Volume 1: Metaphysics and epistemology* (pp. 157–190). Oxford University Press.

Bechtel, W., & Richardson, R. (2010). *Discovering complexity – Decomposition and localization as strategies in scientific research*. The MIT Press.

Biddle, J. B. (2020). On predicting recidivism: Epistemic risk, tradeoffs, and values in machine learning. *Canadian Journal of Philosophy*, 1–21. https://doi.org/10.1017/can.2020.27

Brown, M. (2020). *Science and moral imagination*. University of Pittsburgh Press.

Burkov, A. (2019). *The hundred-page machine learning book*.

Canali, S., & Ratti, E. (2024). Between quantity and quality: Competing views on the role of Big Data for causal inference. In P. Illary & F. Russo (Eds.), *The Routledge handbook of causation and causal methods*. Routledge.

Capriotti, E., & Altman, R. B. (2011). A new disease-specific machine learning approach for the prediction of cancer-causing missense variants. *Genomics, 98*(4), 310–317. https://doi.org/10.1016/j.ygeno.2011.06.010

Castro, C. (2022). Just machine. *Public Affairs Quarterly, 36*(2).

Consens, M. E., Dufault, C., Wainberg, M., Forster, D., Karimzadeh, M., Goodarzi, H., Theis, F. J., Moses, A., & Wang, B. (2023). To transformers and beyond: Large language models for the genome. http://arxiv.org/abs/2311.07621

Danks, D., & London, A. J. (2017). Algorithmic bias in autonomous systems. *IJCAI International Joint Conference on Artificial Intelligence, Ijcai*, 4691–4697. https://doi.org/10.24963/ijcai.2017/654

Douglas, H. (2009). *Science, policy, and the value-free ideal*. University of Pittsburgh Press.

Down, T. A., & Hubbard, T. J. P. (2002). Computational detection and location of transcription start sites in mammalian genomic DNA. *Genome Research, 12*, 458–461. https://doi.org/10.1101/gr.216102

Earman, J. (1992). *Bayes or bust? A critical examination of Bayesian confirmation theory*. The MIT Press.

Eraslan, G., Avsec, Ž., Gagneur, J., & Theis, F. J. (2019). Deep learning: new computational modelling techniques for genomics. In *Nature reviews genetics* (Vol. 20, Issue 7, pp. 389–403). Nature Publishing Group. https://doi.org/10.1038/s41576-019-0122-6

Facchini, A., & Termine, A. (2021). Towards a taxonomy for the opacity of AI systems. In *Conference on philosophy and theory of artificial intelligence* (pp. 73–89). Springer.

Fazelpour, S., & Danks, D. (2021). Algorithmic bias: Senses, sources, solutions. *Philosophy Compass, 16*(8). https://doi.org/10.1111/phc3.12760

Ferdman, A., & Ratti, E. (2024). What do we teach to engineering students: Embedded ethics, morality, and politics. *Science and Engineering Ethics, 30*(1), 7. https://doi.org/10.1007/s11948-024-00469-1

Forber, P. (2011). Reconceiving eliminative inference. *Philosophy of Science, 78*(2), 185–208.

Harman, G., & Kulkarni, S. (2007). *Reliable reasoning – induction and statistical learning theory*. The MIT Press.

Hayek, F. A. (1974). *The Pretence of Knowledge*. Lecture to the memory of Alfred Nobel, December 11.

Hawthorne, J. (1993). Bayesian induction is eliminative induction. *Philosophical Topics, 21*(1).

James, G., Witten, D., Hastie, T., & Tibshirani, R. (2013). *An introduction to statistical learning*. Springer. https://doi.org/10.1016/j.peva.2007.06.006

Jansen, M., & Karaca, K. (2025). Values, inductive risk and epistemic-ethical coupledness in machine learning models. In J. M. Durán, & G. Pozzi (Eds.), *Philosophy of science for machine learning: Core issues and new perspectives*. Synthese Library. Springer.

Jiang, H. (2021). *Machine learning fundamentals – A concise introduction*. Cambridge University Press.

Johnson, G. M. (2023). Are algorithms value-free? Feminist theoretical virtues in machine learning. *Journal of Moral Philosophy, 21*, 1–35.
Kitcher, P. (1993). *The advancement of science*. Oxford University Press.
Kuhn, T. (1977). Objectivity, value judgement and theory choice. In *The essential tension: Selected studies in the scientific tradition and change* (pp. 356–367). University of Chicago Press.
Larsen, P. (2018). When human behavior enters the equation. *Critical Review, 30*(3–4), 316–324. https://doi.org/10.1080/08913811.2018.1565729
Levi, I. (1960). Must the scientist make value judgments? *The Journal of Philosophy, 57*(11), 345.
Leonelli, S. (2016). *Data-centric Biology*. University of Chicago Press.
Libbrecht, M. W., & Noble, W. S. (2015). Machine learning applications in genetics and genomics. *Nature Reviews Genetics, 16*(6), 321–332. https://doi.org/10.1038/nrg3920
Longino, H. (1990). *Science as social knowledge: Values and objectivity in scientific inquiry*. Princeton University Press.
McMullin, E. (1983). Values in science. *PSA: Proceedings of the Biennial Meeting of the Philosophy of Science Association, 2*, 686–709.
Norton, J. (1995). Eliminative induction as a method of discovery: How Einstein discovered general relativity. In J. Leplin (Ed.), *The creation of ideas in physics*. Kluwer Academic Publishers.
Norton, J. (2005). A little survey of induction. In P. Achinstein (Ed.), *Scientific evidence – Philosophy theories and applications*. The Johns Hopkins University Press.
Pietsch, W. (2021). *Big data*. Cambridge University Press. https://doi.org/10.1017/9781108588676
Pietsch, W. (2022). *On the epistemology of data science – Conceptual tools for a new inductivism*. Spring. https://link.springer.com/bookseries/6459.
Platt, J. R. (1964). Strong inference. *Science, 146*(3642).
Pruss, D. (2021). Mechanical jurisprudence and domain distortion: How predictive algorithms warp the law. *Philosophy of Science, 88*(5), 1101–1112.
Radder, H. (2019). *From commodification to the common good: Reconstructing science, technology, and society*. The University of Pittsburgh Press.
Ratti, E. (2015). Big data biology: Between eliminative inferences and exploratory experiments. *Philosophy of Science, 82*(2), 198–218.
Ratti, E. (2020a). Phronesis and automated science: The case of machine learning and biology. In M. Bertolaso & F. Sterpetti (Eds.), *A critical reflection on automated science – Will science remain human?* Springer.
Ratti, E. (2020b). What kind of novelties can machine learning possibly generate? The case of genomics. *Studies in History and Philosophy of Science Part A, 83*, 86–96. https://doi.org/10.1016/j.shpsa.2020.04.001
Ratti, E., & Graves, M. (2021). Cultivating moral attention: A virtue-oriented approach to responsible data science in healthcare. *Philosophy and Technology, 34*(4), 1819–1846. https://doi.org/10.1007/s13347-021-00490-3
Ratti, E., & Graves, M. (2022). Explainable machine learning practices: Opening another black box for reliable medical AI. *AI and Ethics*. https://doi.org/10.1007/s43681-022-00141-z
Ratti, E., & Russo, F. (2024). Science and values: A two-way direction. *European Journal for Philosophy of Science, 14*(1). https://doi.org/10.1007/s13194-024-00567-8
Ratti, E., & Stapleford, T. A. (Eds.). (2021). *Science, technology, and virtues: Contemporary perspectives*. Oxford University Press.
Russo, F. (2009). *Causality and causal modelling in the social sciences. Measuring variations*. Springer.
Schmidt, E., Sesing-Wagenpfeil, A., & Köhl, M. A. (2023). Bare statistical evidence and the legitimacy of software-based judicial decisions. *Synthese, 201*(4). https://doi.org/10.1007/s11229-023-04141-2
Solomon, M. (2012). The web of valief: An assessment of feminist radical empiricism. In S. Crasnow & A. Superson (Eds.), *Out from the shadows: Analytical feminist contributions to traditional philosophy* (pp. 435–450). Oxford University Press. https://doi.org/10.1093/acprof:oso/9780199855469.001.0001

Vallor, S. (2016). *Technology and the virtues – A philosophical guide to a future worth wanting*. Oxford University Press.
Ward, Z. B. (2021). On value-laden science. *Studies in History and Philosophy of Science, 85*, 54–62. https://doi.org/10.1016/j.shpsa.2020.09.006

Emanuele Ratti Senior Lecturer in Philosophy of Science and Artificial Intelligence in the Department of Philosophy of the University of Bristol, UK. His area of research and teaching is Ethics and Philosophy of Science and Technology (with a focus on the life sciences and data science). He is particularly interested in the aspects of the natural sciences and data science that stand at the intersection of ethical and epistemic questions.

Part VII
ML in the Particular Sciences

Chapter 18
Beyond Classification and Prediction: The Promise of Physics-Informed Machine Learning in Astronomy and Cosmology

Helen Meskhidze

Abstract Though the use of machine learning (ML) is ubiquitous in astrophysics and cosmology, many still see the opacity of ML algorithms as a major issue to their scientific utility. One way of addressing this opacity is through an emerging trend in ML research: "teaching" ML algorithms physical laws and domain-specific knowledge. "Physics-informed machine learning" (PIML), as this methodology is called, promises to produce better predictions and yield more interpretable algorithms. It does so by using physical principles in the training process and/or by using physical principles to guide the development of the neural network architecture. In this chapter, I investigate two uses of PIML in astronomy/cosmology, each a representative example of the two PIML methods. In both cases, PIML provides improvements in terms of the predictions and efficiency of ML algorithms. However, I argue that only in the second case does PIML offer improvements to the interpretability of the algorithms.

Keywords Opacity · Interpretability · Understanding · Machine learning · Neural network · Astronomy · Cosmology

18.1 Introduction

"Physics-informed machine learning" (PIML) has been called "the next generation of artificial intelligence" by popular media (Andrzejczuk, 2023) and the scientists who use it claim that it can "transform our modeling, simulation, and understanding of complex physical systems in various science and engineering disciplines" (Chen et al., 2021). PIML aims to incorporate physical laws and domain-specific knowledge into machine learning in order to produce better predictions and yield

---

H. Meskhidze (✉)
Departments of Philosophy and Physics, University of Cincinnati, Cincinnati, OH, USA
e-mail: helen.meskhidze@uc.edu

J. M. Durán, G. Pozzi (eds.), *Philosophy of Science for Machine Learning*,
Synthese Library 527, https://doi.org/10.1007/978-3-032-03083-2_18

more interpretable algorithms. Domain-specific knowledge, in this context, can include symmetry laws, conservation laws, or even the governing dynamics of a system.

Two means have been proposed for the incorporation of such physical laws and domain-specific knowledge: by "teaching" machine learning algorithms this information and/or by designing specialized network architectures. In either case, physics-informed machine learning is argued to yield better predictions in the presence of imperfect or noisy data because the algorithms are more robust to any small irregularities in the data (Karniadakis et al., 2021, p. 423). More importantly, though, PIML is thought to be more interpretable because our knowledge of the physical world is incorporated from the start.

Since their introduction in Raissi et al. (2019), PIML methods have been developed and used in numerous domains across the sciences. Some examples of projects include performing parameter estimation for applications in systems biology (Daneker et al., 2023), modeling fluid dynamics (Cai et al., 2021), characterizing a crack in the surface of a material (Shukla et al., 2020), predicting the many-electron wave equation for applications in quantum chemistry (Pfau et al., 2020), and forecasting weather/climate processes (Kashinath et al., 2021). In astronomy and cosmology, PIML has been used to, e.g., model the formation of molecular clouds in the interstellar medium (Branca & Pallottini, 2023), solve the radiative transfer equation for supernova simulations (Chen et al., 2022; Mishra & Molinaro, 2021), investigate astrophysical shocks (Moschou et al., 2023), and model the gravitational fields around small astrophysical objects (Martin & Schaub, 2022a). These last two projects will be the case studies discussed in detail in the present chapter.

Given the rising use of PIML in various domains, questions about the prospects of the methodology and whether it fulfills its promise of increased scientific understanding are pressing. Are physics-informed ML algorithms truly novel in methodology? In what sense are they the "next generation of artificial intelligence"? Are they able to overcome the issues of opacity faced by standard ML algorithms that worry scientists and philosophers alike? These questions will be the focus of this chapter.

I will begin by reviewing some standard uses of machine learning in astronomy and cosmology (Sect. 18.2). I then turn to philosophical analyses of the issues raised by ML (Sect. 18.3). With these issues in mind, I present two case studies (Sect. 18.4). These are representative of the two methods of PIML: incorporating physical principles into the training of the ML algorithm and designing a specialized network architecture. In both cases, PIML improves the predictions and efficiency of ML algorithms. However, I argue that improvement in the interpretability of the algorithms only occurs in the second case. This is because the inclusion of physical principles as part of the training of the ML algorithm is insufficient to ground claims of increased transparency.

18.2 Machine Learning in Astronomy and Cosmology

The use of machine learning has a long history in astronomy and cosmology with some of the first uses dating back to the 1980s (see Miller, 1993 for a review). Given the large amounts of data astronomers typically collect, early projects primarily used neural networks (NNs) for object classification on existent data sets. Occasionally, astronomers took advantage of the computational speeds offered by ML for real-time applications. For instance, another early application of ML in astronomy was for detector event filtering in high-energy telescopes (a kind of classification but used in real-time to filter events; see Meetre & Norris, 1991). Another real-time application of ML during this time was for adaptive telescope optics, to quickly adjust telescope mirrors in response to atmospheric distortions (Angel et al., 1990). While projects today have broadened the scope of the use of ML/NNs, many still use NNs for image classification-related tasks. Below, I review a few significant, contemporary uses of ML in astronomy and cosmology. Though the projects detailed below constitute only a small sample of the various uses of ML in astronomy and cosmology today, they provide a sense of the scope of the use of such methods.[1] Understanding the applications of ML in astronomy/cosmology will help us understand what issues are relevant to these fields. Then, we will be able to situate PIML in this context and assess whether it resolves the pressing issues.

18.2.1 Imaging and Data Processing

Applications of machine learning to imaging are the most common use of ML in astronomy, especially to identify individual objects. For instance, ML has been used for the detection of strong gravitational lenses (Metcalf et al., 2019). These are astrophysical objects (like galaxies) that cause the light from even more distant objects to be bent as it travels to the Earth. Such objects can be used to probe dark matter or various cosmological parameters of interest. ML has also been used to comb through big data sets in search of regularities that indicate the presence of pulsars (a spinning neutron star emitting radiation from its poles; Lin et al., 2020) or even exoplanets (planets that are part of other solar systems; Jin et al., 2022). Such projects involve sifting through a large amount of astronomical data to look for regularities, a task that ML is particularly well-suited for.

A more complex application of machine learning to imaging has to do with astronomical catalogs. These catalogs are created by taking images of different parts of the sky and matching them up to create a larger image, a catalog. Typically, astronomers use a "best-fit" model of the constituents of each image to synthesize them into a catalog. This process involves a kind of source matching: a source that

[1] See George Stein's GitHub page (2023) for a more comprehensive listing of various contemporary projects using ML in astronomy/cosmology.

is identified in one image is found in another, allowing these images to be lined up. Machine learning has the potential to allow probabilistic cataloging of such images, allowing one to consider a distribution across the possible constituents of an image (Brewer et al., 2013). When images are crowded and noisy, this kind of probabilistic cataloging can be very valuable.

Along similar lines, machine learning has been used for data processing. Data processing is a complex undertaking for astronomers, especially given the large amount of atmospheric interference many telescopes are subject to. Some of the uses of ML for data processing in astronomy are similar to applications of ML for image analysis. For instance, algorithms like Source Extractor have been developed to detect and extract foreground sources in an image (i.e., sources that are between the detector and the object of interest in the image; Bertin & Arnouts, 1996). Other algorithms have been developed to remove sensor artifacts, cosmic rays, etc. These kinds of algorithms often operate in the data processing "pipeline" between when data is collected by the instrument and when the scientist can use the data.

One emerging use of ML in the image analysis context is with multi-messenger astronomy. Here, astronomers conduct wide-field surveys, sweeping the sky for any objects of interest. If such an object is detected, the goal is then to use another instrument (often in another frequency) to analyze the object in more detail. Thus, speed is critical. One must be able to sift through the data collected by the wide-field survey quickly to then point another instrument at the object of interest (see, e.g., Narayan et al., 2018 for an example of real-time classification of data from the Large Synoptic Survey Telescope). Much of the research in this area uses supervised or semi-supervised ML, but there is space to employ unsupervised methods to detect rare events (see e.g., Li et al., 2021).

18.2.2 Simulations

A particularly interesting use of machine learning that bridges data analysis and simulations is in cosmic parameter estimation. Here, the task is to determine the value for various cosmological parameters, usually by comparing simulations to observations of the large-scale structure of the universe. In particular, astronomers and cosmologists require a huge suite of simulations to perform parameter estimation. This is because one needs to compare the collected data with possible values of the various cosmological parameters. Running the suite of simulations is very computationally costly. Therefore, some astronomers and cosmologists have instead run a small suite of such simulations to train an "emulator," a machine learning algorithm aimed at recreating these results for many different parameter combinations but for a fraction of the computational cost (see, e.g., Heitmann et al., 2009). One then uses machine learning to pick out what combination/values of parameters best capture the data. In short, machine learning is being used to interpolate amongst the various combinations of parameters and the results are then compared to observations for scientifically interesting results.

In a recent project along these lines, Villaescusa-Navarro and collaborators investigate the scales of data required to extract cosmological information (2022). They study whether one can use ML methods to extract cosmological information from data collected on the scale of a single galaxy. To do so, they train an NN on the output of many simulations, produced using various values of cosmological parameters. Then, they provide the NN with data from a single galaxy and ask it to predict the cosmological model that is consistent with that particular data set. Given the limited data that Villaescusa-Navarro and collaborators provide to the trained NN, it must be very sensitive to the impacts of changes in the various cosmological parameters.

18.2.3 Methods of ML

Given the numerous astronomical projects that use ML for imaging-related tasks, convolutional NNs tend to be the most suitable and commonly used architecture.[2] However, in a review of the state of the field, Dvorkin and collaborators critique such methods. They note that "...many network architectures are sequential and recursive, which does not easily allow parallel computation (to account for the size of cosmological data sets) or they assume a Markov process, which means that they cannot easily learn long-range dependencies" (Dvorkin et al., 2022, p. 13). Thus, they note, there is an opportunity to develop architectures specifically geared towards the context of cosmology. In particular, they highlight PIML, noting that such methods "have shown promise in reducing the dimensionality of the underlying latent space of a network with an associated reduction in the size of data sets needed to train the network" (Dvorkin et al., 2022, p. 13). Besides offering advances in efficiency, PIML is also thought to improve the interpretability of ML algorithms. Thus, before considering PIML in more detail, I will first review the issues of interpretability/opacity that arise with standard uses of ML.

18.3 Interpretability of ML

Given the importance of machine learning methods for scientific progress, many have argued that greater attention needs to be given to better understanding these methods. Philosophers, scientists, and computer scientists alike have turned their attention to this task. The fields of explainable artificial intelligence (XAI) and interpretable machine learning (IML) both aim to tackle questions along these

[2] An explanation of why CNNs are well-suited for imaging-related tasks generally is beyond the scope of this chapter (see Stewart, 2019 for such an explanation). Put very briefly, one can imagine that the architecture of a CNN respects the relationships among the pixels in an image.

lines.[3] Some propose novel ML algorithms for the task while others propose frameworks in which to situate such questions. Much progress has been made in clarifying what is at issue, outlining frameworks to think within, and defining key terms in the debate. I outline some themes from this literature next in order to ultimately ask whether physics-informed machine learning is responsive to the demands being made.

18.3.1 Interpretability Through Transparency

In Lipton's influential paper, "The Mythos of Model Interpretability," he remarks

> ML-based systems do not know why a given input should receive some label, only that certain inputs are correlated with that label... As ML penetrates critical areas such as medicine, the criminal justice system, and financial markets, the inability of humans to understand these models seems problematic. (2018, p. 3)

Lipton reflects that though many propose "interpretability" as a remedy, few articulate precisely what they mean by interpretability.[4] With this in mind, he outlines five goals one might hold when demanding interpretability: increasing our trust in an ML model, helping support causal reasoning based on the model, transferring/generalizing the findings of the ML model to nearby contexts, being an informative component in human decision-making, and helping facilitate fair and ethical decision making (2018, pp. 9–12).

Lipton then turns to considering how we might achieve these goals. The first method he considers for achieving interpretability is transparency.[5,6] Transparency, he argues, can be evaluated at various levels: at the level of the model, at the level of individual components, and at the level of the training algorithm (2018, p. 12). These levels each correspond to their own notion of transparency: simulatability, which captures the sense in which a model can be understood as a whole; decomposability, which captures the sense in which one can understand the various components of the model individually; and algorithmic transparency, which captures the sense in which one can understand the learning algorithm itself. Ultimately, Lipton argues that whether one algorithm is more interpretable than another will turn on what notion of interpretability one subscribes to and what kind of transparency that might demand. For instance, deep neural networks (DNNs) may be algorithmically

[3] See Chap. 7 by Páez (2025) in this volume for more on XAI.

[4] See also Chap. 10 by Sullivan and Kasirzadeh and Chap. 11 by Schmidt in this volume for more on interpretability.

[5] Lipton also discusses post-hoc methods for increasing interpretability, but they are not important for the present purposes.

[6] See Chap. 1 by Beisbart and Chap. 2 by Formanek in this volume for a discussion of issues of opacity. For an alternative to transparency—computational reliabilism—see Chap. 4 by Duran in this volume.

more complex than linear models, but the high-dimensional or heavily engineered features of linear models mean that they may lose simulatability or decomposability respectively. Thus, in those domains, DNNs may exhibit more transparency and thus better interpretability.

The importance of specifying what kind of transparency one is interested in has also been highlighted in the recent philosophical literature. Creel (2020) argues that transparency is an important goal when considering opaque, complex computational systems. She presents three avenues for pursuing transparency: functional, structural, and run. Functional transparency has to do with the algorithmic functioning of the whole, i.e., understanding the high-level logical rules governing the system. Structural transparency recognizes that algorithms can be multiple realized so it asks how the algorithm is realized in the code. Finally, run transparency asks whether one has knowledge of the program as it was actually run in a particular instance (Creel, 2020, pp. 572–582). With this framework in mind, let us now turn to evaluating the uses of ML in astronomy and cosmology and the kinds of transparency they may or may not exhibit.

18.3.2 Transparency of ML in Astronomy and Cosmology

ML algorithms are routinely referred to as "black boxes." This might lead us to think that the prospects for transparency when using ML are rather bleak. However, recall that in many of the uses of ML in astronomy/cosmology described above, the ML algorithms are just being used as statistical tools. It is precisely because they are used as statistical tools that we need not be worried about issues of transparency. Consider, for instance, the algorithms used to emulate the results of large-scale structure simulations. As mere interpolation devices, their results are neither surprising nor mysterious. They take a coarsely sampled parameter space as input and fill it in. In cases like this, a particular step in the process is being black-boxed using machine learning. However, since there is a well-understood, physically-motivated methodology one can always refer back to, the scientists could easily open up the black box. This means the machine learning algorithms can be made to have functional, structural, and run transparency. The overall functioning of the algorithm follows clear statistical rules appropriate for interpolation, the structure of the algorithm is relatively straightforward, and the result of the algorithm in any particular instance can be investigated. If one were to ask for a more physical explanation of any particular result, the simulations used to train the emulator can be appealed to. Put differently, we can distinguish between the "black-boxing" occurring in this instance—an (unproblematic) methodological step—and using a black box to provide understanding. This is a distinction I have argued for in more detail elsewhere (Meskhidze, 2023). There, I argue that the case of emulators is an unproblematic use of methodological black-boxing.

Another way of distinguishing problematic from unproblematic uses of ML is through Sullivan's notion of "link uncertainty." In "Understanding from Machine

Learning Models" (2022), Sullivan argues that the complexity or black-box nature of a model need not undercut the understanding that the model can provide. Instead, she argues that this understanding is undercut when "[t]here is a high level of *link uncertainty*, that is, a lack of scientific and empirical evidence supporting the link that connects the model to the target phenomenon" (2022, p. 6). To illustrate this point, she contrasts a DNN that identifies cases of melanoma from images of moles on skin to a DNN that uses facial recognition to identify an individual's sexual orientation. Since, as she argues, "The level of scientific justification and background knowledge linking the appearance of moles to instances of melanoma is extensive," (2022, p. 23) the link uncertainty is greatly reduced. But the link uncertainty between facial features and sexual orientation is much higher. The transparency of the DNNs in these two cases is largely the same and irrelevant to understanding. Thus, what matters for assessing understanding, on Sullivan's view, is the degree of link (un)certainty.

In their review of the recent philosophical discussion about the interpretability of machine learning, Beisbart and Räz discuss Sullivan's argument. They note the following consequence of her view: "Black boxes need not compromise the scientists' abilities to use a model if they merely black-box the implementation of the fulfillment of a known task (e.g. the calculation of a factorial)" (Beisbart & Räz, 2022, p. 5). I agree and would also refer to this process as unproblematic "black-boxing." Indeed, in the case of emulators, the link certainty is well-established; one can always run a simulation to check that the link between a particular cosmological parameter and the resultant large-scale structure of the universe accords with the predictions of the emulator. Therefore, we can conclude that issues of interpretability are not troubling in many of the above-described applications of machine learning in astronomy and cosmology.

There are some cases in which we do lack the necessary interpretability. Consider a simple case of image analysis. We may want to know, for instance, why the ML algorithm predicts an exoplanet from a set of input data. In this case, the ML algorithm is "making a decision" with respect to the image so the kind of statistical explanation proposed above for emulators might not be satisfactory. On Sullivan's account, this case seems akin to using moles for melanoma detection: we have clear reason to think that various aspects of the image correlate with the properties of interest. We have link certainty. Nonetheless, one might still worry whether the ML has selected some set of *real* indicators for the context/object of interest or whether it has captured some spurious correlations. In other words, though the link may be well-established, unless we are confident that the ML has captured that link, we might still worry about the interpretability and robustness of the results. If, following Lipton, we want our ML algorithms to support causal reasoning and allow generalizations, we need enough transparency to ensure that whatever "link" our ML algorithm is using to make predictions is the "link" we believe exists in the data. Put differently, while a lack of link certainty does undermine interpretability, having link certainty (without transparency) does not immediately grant interpretability.

The limitations of using link certainty to guide understanding can be made even more pressing with the following considerations. In "Two Dimensions of Opacity and the Deep Learning Predicament," Boge introduces a distinction between "h-opacity" and "w-opacity" where the former concerns *how* the machine is learning while the latter concerns *what* is learned by the machine. On his account, w-opacity is what sets deep learning apart from other scientific modeling. The features of the data picked out by the deep learning model correspond to "automatically discovered insights; complex, non-obvious features that can be abstracted from the data and allow the machine to discriminate" (Boge, 2022, p. 61). This framing is helpful for what I am highlighting above: though some features of interest may exist in the data and the link may exist in the world, w-opacity undermines our ability to say whether these are the features the algorithm is leveraging. This is, of course, worrisome: there may or may not be link certainty but we have no way of determining which is the case. However, Boge presents a more worrisome possibility: that the ML leverages real links in the data that are, at present, inconceivable to humans (see also Boge, 2023). In such a case, there would be link certainty in the sense that the ML would be leveraging real links. But we would not be able to say this as we could not conceive of the very link being exploited. These are cases in which, perhaps if we had greater transparency, we would be able to use the ML to make novel discoveries. The implication is that, in some cases, requiring link certainty would undermine our ability to make novel discoveries with ML.[7]

Creel's discussions of opacity/transparency can help pinpoint where we want greater clarity. It may not completely resolve the more troubling cases Boge is worried about, but it will assist in the cases of interest for the present chapter. On Creel's taxonomy, what we are worried about is the functional transparency of the algorithm. An example Creel gives nicely parallels this discussion. She writes:

> ...in all but the smallest networks it would be difficult to predict the outcome without tracing each step or to understand the behavior of the network, especially if the network includes feedback loops. More importantly, without further analysis it would be unclear to the observer why this neural net successfully classified an image and to what extent each of the neurons contributed to the result or why different neural nets might have different patterns of classification. In this sense, although we know how the learning algorithm works and what formal guarantees (if any) we have about its performance, we do not know how the learned 'algorithm' brings about the classification result. Thus, we lack functional transparency. (2020, pp. 579–580)

To summarize: there are many frameworks for discussing interpretability, some highlight transparency while others highlight link certainty. Here, I have shown that, while appealing, link certainty is insufficient for interpretability. This is because, without transparency, we cannot be sure that the link captured by the ML algorithm is the actual link we expect to exist. Though many uses of ML in astronomy/cosmology are unproblematic because the ML algorithm is just being used as a statistical tool, some uses still highlight the need for interpretability. With

[7] My worries in the present chapter are confined to the first category. For discussion of the second (i.e., cases when scientists do not have the resources to conceptualize the link being exploited), see Chap. 19 Boge and de Regt in this volume. There, Boge and de Regt discuss the prospects of using ML for discovering novel concepts and phenomena in particle physics and the problems that arise.

all this in mind, we can now turn to physics-informed ML in order to ultimately ask whether such methods yield more interpretable ML.

18.4 Physics-Informed Machine Learning

18.4.1 General Methodology of PIML

The first projects and papers introducing physics-informed machine learning claim that the methodology is an improvement over standard ML approaches. When standard approaches are trained on huge collections of data, they may make physically inconsistent or implausible predictions due to biases that might be in the data or because the ML is being asked to extrapolate beyond the domain represented by the data. Authors advocating for PIML claim,

> ... there is a pressing need for integrating fundamental physical laws and domain knowledge by 'teaching' ML models about governing physical rules, which can, in turn, provide 'informative priors'—that is, strong theoretical constraints and inductive biases on top of the observational ones. (Karniadakis et al., 2021, p. 423)

To understand this claim it will be essential to understand the methodology of such algorithms. In general, there are two methods for introducing physics into ML. The first is by including physical constraints or principles into the learning process, creating a learning bias. Projects in this vein might train the NN on both the data and a partial differential equation (PDE) known to model the situation. This is accomplished by having terms in the loss function for both the alignment of the NN with the data and the PDE. More specifically, the general form of the solution is fed into the network and soft penalties ensure that the network finds the relevant parameters. The second method uses the architecture of the network itself to incorporate the physical principles. Projects in this vein tend to be more diverse than those in the previous group. One salient example involves using graph neural networks to learn the structure of chemical bonds. There, the success of the architecture for the particular purpose relies on "their ability to pick up on structures in the graph at multiple different scales, while satisfying the crucial requirement that the output be invariant to permutations of the vertices" (Hy et al., 2018, p. 1). To better understand these methodologies, I now turn to describing two case studies, each of which highlights the use of one of the two methods of PIML in astronomy/cosmology.

18.4.2 Astrophysical Shocks

PIML methods have been used to study astrophysical shocks. Here, I consider the use of PIML to study the solar termination shock in particular. The solar termination shock occurs at the edge of a solar system (for our solar system, this is around 100 times the distance between the Earth and Sun). There, streams of charged particles being released from a sun (i.e., solar winds) interact with the matter and radiation that exists between star systems in a galaxy (i.e., the interstellar medium). This causes the particles to slow down suddenly. The termination shock is formed as the solar wind goes from super to subsonic speeds causing compression and heating in the plasma.

In a recent project, Moschou and collaborators use a physics-informed neural network (PINN)[8] for modeling the solar termination shock in the presence of a gravitational field (2023). A nice feature of this context is that a simulation code already exists (PLUTO) that effectively models astrophysical gas dynamics. The computational expense of using PLUTO for the full analysis, however, is quite high. Therefore, the authors turn to machine learning, using PLUTO to generate a synthetic database that the PINN is trained on.

The inclusion of physical principles in this context comes with the form of the loss function used. The researchers add two terms to the standard loss function. One of the new terms captures how close the model is to satisfying the partial differential equation thought to describe the system. The other new term corresponds to the initial and boundary conditions. The PINN is then trained on the synthetic data created by PLUTO and put to work. Generally, the PINN performs very well and models shocks effectively. Additionally, it requires only a fraction of the data a standard NN would (Moschou et al., 2023, see §3.2).

The capability of the PINN goes beyond merely speeding up computations though, as the authors argue. They claim that the PINN can be "used to discover the underlying physics from data" (Moschou et al., 2023, p. 6). But what do they mean by this? The "underlying physics" learned by the PINN in this context is the prediction of a particular parameter in the PDE being used to model the system. This parameter is known as the effective polytropic index. The polytropic index describes the exchange of energy between a gas and the environment.[9] In this case, the researchers are using an effective polytropic index that "mimics the effects of adding heating in the system" (Moschou et al., 2023, p. 3) but avoids the computational expense of actually calculating the heating and cooling. The "ground truth" value of this parameter is found in the PLUTO simulation and the value predicted by the PINN is then compared to this "ground truth" value. In sum, the discovery of underlying physics by the PINN amounts to being able to predict the value of an effective parameter occurring in the PDE that is used as part of the loss function used to train the PINN.

Despite their success, PINNs still face challenges. One discussed by Moschou and collaborators has to do with the convergence of the PINN. The inclusion of

[8] I will be using PIML and PINN interchangeably going forward; PINNs are a subset of PIML just as NNs are a subset of ML.

[9] For natural fluids, the polytropic index (γ) falls between 1 and 5/3 (where 5/3 is the index for an ideal gas). Here, they use an effective polytropic index $\gamma < 5/3$.

two additional terms in the overall loss function used to train the PINN makes the convergence of the NN trickier. Whereas the convergence of standard NN is well-studied, the convergence of these PINNs with multifaceted loss functions is not. As Moschou and collaborators write, "The neural network optimizer might have to deal with losses that differ by several orders of magnitude which makes the minimization task and reaching a unique solution challenging" (2023, p. 15).[10] Issues with the convergence of the cost function and questions about what to include in the cost function will arise with the next case study as well.

18.4.3 Gravitational Fields Around Small Bodies

Another context in which PINNs have been applied is to study gravitational fields around small astronomical bodies. Here, I consider a three-part project by Martin and Schaub. They began by first applying these methods to predict the gravitational field around the Earth and Moon (Martin & Schaub, 2022b) before turning to small astrophysical bodies (Martin & Schaub, 2022a, 2023). Their approach goes beyond the previous example of PINN: it alters the network architecture as part of the "physics-informed" aspect of the model. I begin by detailing the problem at issue and then outline the implementation of the PINN.

Suppose you want to take soil samples from the surface of an asteroid. Doing so requires that you land a spacecraft on its surface. Landing a spacecraft on the surface of an asteroid, though, is no easy feat. It requires being able to accurately model the gravitational field around an irregularly-shaped object, a field whose potential is described by Laplace's equation. One common method of modeling the gravitational field around objects like asteroids and comets—of solving Laplace's equation—is to use spherical harmonic functions (see, e.g., Bucha & Sanso, 2021). While spherical harmonics may provide an exact, analytic solution to the problem, they are not suitable when one is close to the body of interest (e.g., approaching the surface) because they diverge. Indeed, as noted by Martin and Schaub, the divergence of the spherical harmonics "poses a problem for irregularly shaped

[10] They note that some (Jin et al., 2021, in particular) have tried to address this issue by using adaptive weighting algorithms to more effectively minimize the loss function and find convergence. Adaptive weighing does introduce additional model hyper-parameters that have to be set up by the researchers. However, these hyper-parameters have to do with the convergence of the network, not the physical problem at issue. Nonetheless, Moschou and collaborators are not convinced that this method will be sufficient. As evidence, they cite another study (Fuks & Tchelepi, 2020) that aimed to solve a non-linear hyperbolic equation. Fuks and Tchelepi explored many NN algorithms and architectures for their PINN but could not find convergence until they switched the equation form to a parabolic one. This leads Moschou and collaborators to believe that no matter what algorithm is being used to minimize the loss function, it cannot "address the fundamental problems in the PINNs optimization procedure" (2023, p. 15).

objects, like asteroids, where spacecraft may operate at substantially lower radii than the Brillouin sphere" (2022a, p. 46).[11,12]

Traditional ML, NNs in particular, has been applied to this problem. However, the nature of the problem makes the use of NNs difficult: one would need a large amount of data at various altitudes for each astronomical body to capture any surface irregularities. Additionally, there are various physical constraints the researchers know that could be used to guide solutions to the problem. For instance, any derived gravitational potential must be a solution to Laplace's equation. However, there is no way of incorporating this kind of knowledge into a traditional NN. As Martin and Schaub put it,

> Traditional neural networks are not trained with these physics properties in mind. Instead, they prioritize predicting an accurate acceleration from a position vector, irrespective of the more fundamental properties. In this sense, the network will be trained agnostic to the fact that the gravity field it represents produces conservative forces, and the underlying potential must be sufficiently smooth and continuous for sensible dynamics. (2022a, p. 9)

With these kinds of considerations in mind, Martin and Schaub turn to a physics-informed approach. As mentioned above, they begin by applying their methods to the gravitational field around the Earth and Moon (Martin & Schaub, 2022b) and then generalize (Martin & Schaub, 2022a, 2023). They begin with a relatively simple model that is similar to the first study. Their methodology evolves over the course of the three papers though, allowing this to serve as an example of how to incorporate physics-informed principles in the architecture of the NN.

Martin and Schaub begin with minimal physical constraints: they only include one additional term into the PINN loss function. This term corresponds to the equation describing the relationship between the gravitational potential and the acceleration ($\mathbf{a} = -\nabla U$). They limit the modifications they make to the loss function for two reasons. First, further constraints undermine the convergence of the PINN. Second, other constraints would likely not be of the same order-of-magnitude and would require rescaling to be incorporated into the loss function (discussed in more detail below).

In the second paper of the series, Martin and Schaub aim to incorporate further physical principles in the PINN. To do so, they adopt a modified network architecture. This change also helps with the issue regarding the convergence of the loss function mentioned in the previous case study. Instead of a fully connected network (in which every neuron in one layer is connected to every neuron in the previous and the next layer), Martin and Schaub adopt an architecture proposed by Wang et al. (2021) that uses recent developments in neural attention mechanisms. Essentially, they introduce transformers that project the inputs into a higher-

[11] The "Brillouin sphere" is the sphere encompassing all the field-generating mass. It is within this sphere that the spherical harmonic functions diverge.

[12] At low altitudes, such as those used for touch and go or landing maneuvers, one can use alternative representations (like the polyhedral gravity representation). However, these have their own issues (e.g., they assume uniform density and are typically computationally expensive).

dimensional feature space. As Wang, Teng, and Perdikaris note, this: “(i) explicitly accounts for multiplicative interactions between different input dimensions, and (ii) enhances the hidden states with residual connections” (2021, A3069). These architectural modifications alleviate issues of convergence that otherwise arise with multi-faceted loss functions.

The final iteration of Martin and Schaub’s PINN is presented in Martin and Schaub (2023). There, they reflect that previous iterations of their algorithm “use a cost function which inadvertently leads models to prioritize low-altitude field points, where the accelerations are largest” (2023, p. 2). In order to address this issue and various other performance issues, they have again redesigned the architecture of the network, rescaling the output of the NN, changing the loss function, and including more physically-informed principles. I detail some of these changes below. The important takeaway here is the degree of control they have over the components of the PINN architecture and how this enables them to develop understanding.

One of the reasons low-altitude field points were prioritized is because the gravitational field is small at high altitudes, so small that the field values encroach on machine precision. To address this issue and deprioritize the low-altitude predictions, Martin and Schaub rescale the output of the NN before it is used in the loss function (2023, p. 10). Another effect of the gravitational field values being so much larger at low altitudes than at high altitudes is that the error values are also much larger. Therefore, the standard error used in these contexts (a percentage error) prioritizes low-altitude predictions. To account for this, they switch from using mean percent error to using root mean squared error (2023, p. 7). Finally, Martin and Schaub re-introduce the spherical harmonic functions for performance, “fus[ing] an a-priori gravity model with the neural network solution” (2023, p. 9). In terms of the final performance, the first iteration of the PINN already performed better than the spherical harmonic representation. However, it still required a large data set. In comparison, the third generation converges reliably and quickly with a small training set (2023, pp. 13–14).

In sum, this series of three papers introduces and develops a PINN for predicting the gravitational field around small astrophysical bodies. As Martin and Schaub improve their PINN, they increase the number of physical principles incorporated in their PINN, alter the network architecture to better suit the problem at hand, and reintroduce the form of the solution they expect.

18.5 Discussion

The two case studies presented above are characteristic examples of the two approaches taken in physics-informed machine learning: incorporating physical principles into the training of the ML algorithm and using the network architecture to capture some physical properties of the system. We can now ask whether these algorithms are in fact more interpretable than standard ML methods.

Let us begin with the first case study. One can imagine training a standard NN to learn on the same synthetic data that Moschou and collaborators used to train their PINN. Such an NN would be able to similarly predict the polytropic index of some new data. Now, the question is whether the PINN is in a substantially better position than such an NN.

Most simply, one might argue that, because of the input physics, the improved efficiency of the PINN ought to correspond to an improvement in interpretability. In particular, one might claim that because PINN requires fewer nodes, hidden layers, etc. compared to NN for the same performance, it is easier to interpret. I would argue, however, that when we have tens of thousands of nodes or more, even an order of magnitude fewer does not seem significant to transparency. More importantly, recalling Creel's taxonomy of the types of transparency from Sect. 18.3.1, it is clear that the kind of transparency at stake when considering the number of nodes, the weights, etc. of these algorithms is run transparency. Though run transparency may be important for some purposes, it does not help with the overall interpretability of the network, i.e., its functional transparency.

Do the physical principles used to train the algorithm themselves provide improvements to the functional transparency? While the addition of physical principles improves the performance of the algorithm, I argue that they have no effect on the functional transparency, and thus interpretability, of the algorithm. To understand why, consider the difficulties both teams of researchers encountered when adding terms to the loss functions that represented the physical principles. They worried about the weighting of the various terms and the convergence (or lack thereof) of the network. If one wanted to argue that the functional transparency is increased with the addition of physical principles, understanding how those principles are being leveraged would be critical. However, the developers of these algorithms do not (yet) have a clear sense of how the algorithm is prioritizing the various components of the loss function. Though improvements are being made with further research, it seems too early to say whether PINNs are leveraging and respecting these physical principles.

Despite this, I do not believe the situation is worrisome. We may not have the transparency required to tell whether PINNs are levering these physical principles in the way we expect, but we also do not need it. To understand this claim, recall the discussion of emulators presented in Sect. 18.3.2. There, a particular step of the methodology was being "black-boxed" when scientists used ML methods like emulators. The first case study of PIML is akin to the use of emulators. Here, like in the case of emulation, a particular step of the method is being "black-boxed." In particular, one knows the physical principles underlying the problem and, if needed, could use a first-principles-style simulation for the same goal. Put differently, one has an interpretable simulation to fall back on for explanatory purposes. We need not require the ML algorithm to serve explanatory purposes; like emulators, its purpose is to make computations faster and more efficient.

We can also analyze the situation with respect to link certainty. Should we be worried about whether the PINNs are using the "right" link? Since what we are predicting is the value of the effective polytropic index—not some further detail of the underlying physics—, it does not matter whether our ML algorithm is leveraging the right link or not. The PINN is just a statistical tool. And, again, because the background physical principles are well-understood, if needed, we could always revert to more of a first-principles kind of analysis.

In sum, the methodology of the PINN in this first case does not present any novel strides in interpretability/transparency. But this is in part because one need not be worried about the interpretability/transparency of the ML algorithm in the first place. If one treats the original ML algorithms akin to how one treats emulators—as statistical tools—such questions about interpretability do not arise. Of course, there may be other reasons to adopt the PIML methods (e.g., they are better suited to noisy data and more efficient), but these reasons are irrelevant to considerations of the interpretability of the algorithms.

Let us now consider the second case study. I argue that this case study is an example in which the physics-informed nature of the ML helps with its interpretability. This is because (1) the goal is different and (2) the methodology the researchers adopt is different. Martin and Schaub's overall goal is to predict the gravitational field close to an astronomical body. In this case, at least in the first two generations, the PINN is not bound by any particular physics. Though a physical principle is introduced, it is not predicting some particular parameter (like the polytropic index). Instead, it is predicting the gravitational field across the surface of the body. While the spherical-harmonic representation might be the best for analytic calculations, their PINN is not constrained to this representation. As Martin and Schaub write,

> Such an approach allows the PINN to efficiently learn custom and physically motivated basis functions which represent the natural features in the gravitational potential of a planetary body (e.g., craters on the Moon or mountain ranges on the Earth) rather than imposing basis functions like spherical harmonics which are inefficient at capturing these idiosyncratic and often discontinuous features. (2022a, p. 3)

Clearly, this is a different goal from the previous case study. This different goal guides the methodology. Like in the first case study, Martin and Schaub introduce a physical principle in their loss function. However, their iterative methodology—the fact that they consider how the loss function behaves, adjust the architecture to ensure convergence, and later, incorporate further physical principles—raises the likelihood that the PINN is leveraging that loss function and respecting the physical principles incorporated. And, as Martin and Schaub input more physical principles (a low-fidelity analytic model, boundary conditions, adjust the loss function to allow the network to be more sensitive at high-altitudes), the interpretability of the PINN improves further. Some of these changes are of course possible with standard NN models. However, the physics-informed NNs and the architectural flexibility they provide allow the researchers to have a better handle on the performance of the PINN overall.

We can again consider the situation in terms of link certainty and whether the PINN has leveraged the right link. In the previous case, the physical principles did not meaningfully impact the interpretability of the model. However, there, it did not matter whether the PINN leveraged the link we knew to exist between the data and the polytropic index. Here, there is a link between the data and the gravitational field and we want the NN to leverage that link. In adjusting the architecture of the network, our belief that the algorithm is leveraging that link increases. This is because our understanding of the algorithm is increasing as we investigate various components of it and its performance. Thus, the PIML methods, and the ability to iteratively adjust them, do present an improvement in transparency. This is important because, unlike in the first case, we do not have a first-principles style argument to fall back on. Our analytic models fail in the domains that the PIML is being used. To summarize, this second case study offers a case in which we have link certainty and we want functional transparency to ensure our ML algorithm is indeed leveraging that link. PIML methods—the inclusion of physical principles and iterative adjustment of the network architecture—increase our confidence that the ML algorithm is indeed leveraging that link.

18.6 Conclusion

Is PIML really the "next-generation of artificial intelligence"? Is it as "transformative" as proponents would have us believe? PINN algorithms outperform standard ML techniques by introducing physical principles into their loss functions. In terms of their interpretability, there are many contexts in which we need not be worried about the interpretability of the algorithms in the first place; in such contexts, PINNs seem to be in just the same situation as traditional NN. In cases where greater transparency is desired, PINN methods—ones that adjust the network architecture to reflect physical principles—can offer improvements.

Acknowledgments My thanks to Florian Boge and Giorgia Pozzi for their comments on earlier drafts of this chapter.

This work was supported by the Black Hole Initiative, which is funded by grants from the John Templeton Foundation (Grant #62286) and the Gordon and Betty Moore Foundation (Grant GBMF-8273). The opinions expressed in this work are those of the author and do not necessarily reflect the views of these Foundations.

References

Andrzejczuk, D. (2023). Physics informed machine learning. *Medium: Technology, Invention, App, and More*. https://medium.com/the-quantum-data-center/physics-informed-machine-learning-the-next-generation-of-artificial-intelligence-solving-89ca4bb2e05b

Angel, J. R. P., Wizinowich, P., Lloyd-Hart, M., & Sandler, D. (1990). Adaptive optics for array telescopes using neural-network techniques. *Nature*, *348*(6298), 221–224.

Beisbart, C. (2024). In which ways is machine learning opaque? In J. M. Duran & G. Pozzi (Eds.), *Philosophy of science for machine learning: Core issues and new perspectives*. Synthese Library. Springer.

Beisbart, C., & Räz, T. (2022). Philosophy of science at sea: Clarifying the interpretability of machine learning. *Philosophy Compass*, *17*(6), e12830.

Bertin, E., & Arnouts, S. (1996). Sextractor: Software for source extraction. *Astronomy and Astrophysics Supplement Series*, *117*(2), 393–404.

Boge, F. J. (2022). Two dimensions of opacity and the deep learning predicament. *Minds and Machines*, *32*(1), 43–75.

Boge, F. J. (2023). Functional concept proxies and the actually smart hans problem: What's special about deep neural networks in sciences. *Synthese*, *203*, 16.

Boge, F. J., & de Regt, H. W. (2024). Machine learning discoveries and scientific understanding in particle physics: Problems and prospects. In J. M. Duran & G. Pozzi (Eds.), *Philosophy of science for machine learning: Core issues and new perspectives*. Synthese Library. Springer.

Branca, L., & Pallottini, A. (2023). Neural networks: Solving the chemistry of the interstellar medium. *Monthly Notices of the Royal Astronomical Society*, *518*(4), 5718–5733.

Brewer, B. J., Foreman-Mackey, D., & Hogg, D. W. (2013). Probabilistic catalogs for crowded stellar fields. *The Astronomical Journal*, *146*(1), 7.

Bucha, B., & Sanso, F. (2021). Gravitational field modelling near irregularly shaped bodies using spherical harmonics: A case study for the asteroid (101955) Bennu. *Journal of Geodesy*, *95*, 1–21.

Cai, S., Wang, Z., Fuest, F., Jeon, Y. J., Gray, C., & Karniadakis, G. E. (2021). Flow over an espresso cup: Inferring 3-d velocity and pressure fields from tomographic background oriented schlieren via physics-informed neural networks. *Journal of Fluid Mechanics*, *915*, A102.

Chen, X., Jeffery, D. J., Zhong, M., et al. (2022). Using physics informed neural networks for supernova radiative transfer simulation. Preprint. arXiv:2211.05219.

Chen, Z., Liu, Y., & Sun, H. (2021). Physics-informed learning of governing equations from scarce data. *Nature Communications*, *12*(1), 6136.

Creel, K. A. (2020). Transparency in complex computational systems. *Philosophy of Science*, *87*(4), 568–589.

Daneker, M., Zhang, Z., Karniadakis, G. E., & Lu, L. (2023). Systems biology: Identifiability analysis and parameter identification via systems-biology-informed neural networks. In *Computational modeling of signaling networks* (pp. 87–105). Springer.

Duran, J. M. (2024). Beyond transparency: Computational reliabilism as an externalist epistemology for algorithms. In J. M. Duran & G. Pozzi (Eds.), *Philosophy of science for machine learning: Core issues and new perspectives*. Synthese Library. Springer.

Dvorkin, C., Mishra-Sharma, S., Nord, B., Villar, V. A., Avestruz, C., Bechtol, K., Ciprijanovic, A., Connolly, A. J., Garrison, L. H., Narayan, G., & Villaescusa-Navarro, F. (2022). Machine learning and cosmology. Preprint. arXiv:2203.08056.

Formanek, N. (2024). How I stopped worrying and learned to love opacity. In J. M. Duran & G. Pozzi (Eds.), *Philosophy of science for machine learning: Core issues and new perspectives*. Synthese Library. Springer.

Fuks, O., & Tchelepi, H. A. (2020). Limitations of physics informed machine learning for nonlinear two-phase transport in porous media. *Journal of Machine Learning for Modeling and Computing*, *1*(1), 19–37.

Heitmann, K., Higdon, D., White, M., Habib, S., Williams, B. J., Lawrence, E., & Wagner, C. (2009). The Coyote Universe. II. Cosmological models and precision emulation of the nonlinear matter power spectrum. *The Astrophysical Journal*, *705*, 156–174.

Hy, T. S., Trivedi, S., Pan, H., Anderson, B. M., & Kondor, R. (2018). Predicting molecular properties with covariant compositional networks. *The Journal of Chemical Physics*, *148*(24), 241745-1–241745-11.

Jin, X., Cai, S., Li, H., & Karniadakis, G. E. (2021). NSFnets (Navier-Stokes flow nets): Physics-informed neural networks for the incompressible Navier-Stokes equations. *Journal of Computational Physics*, *426*, 109951.

Jin, Y., Yang, L., & Chiang, C.-E. (2022). Identifying exoplanets with machine learning methods: A preliminary study. *International Journal on Cybernetics & Informatics*, *11*, 109951-1–109951-26.

Karniadakis, G. E., Kevrekidis, I. G., Lu, L., et al. (2021). Physics-informed machine learning. *Nature Reviews Physics*, *3*(6), 422–440.

Kashinath, K., Mustafa, M., Albert, A., Wu, J. L., Jiang, C., Esmaeilzadeh, S., Azizzadenesheli, K., Wang, R., Chattopadhyay, A., Singh, A., & Manepalli, A. (2021). Physics-informed machine learning: Case studies for weather and climate modelling. *Philosophical Transactions of the Royal Society A*, *379*(2194), 20200093.

Li, X., Ragosta, F., Clarkson, W. I., & Bianco, F. B. (2021). Preparing to discover the unknown with Rubin LSST: Time domain. *The Astrophysical Journal Supplement Series*, *258*(1), 2.

Lin, H., Li, X., & Luo, Z. (2020). Pulsars detection by machine learning with very few features. *Monthly Notices of the Royal Astronomical Society*, *493*(2), 1842–1854.

Lipton, Z. C. (2018). The mythos of model interpretability: In machine learning, the concept of interpretability is both important and slippery. *Queue*, *16*(3), 31–57.

Martin, J., & Schaub, H. (2022a). Physics-informed neural networks for gravity field modeling of small bodies. *Celestial Mechanics and Dynamical Astronomy*, *134*(46), 1–28.

Martin, J., & Schaub, H. (2022b). Physics-informed neural networks for gravity field modeling of the earth and moon. *Celestial Mechanics and Dynamical Astronomy*, *134*(2), 13.

Martin, J., & Schaub, H. (2023). The physics-informed neural network gravity model revisited: Model generation III. In *33rd AAS/AIAA Space Flight Mechanics Meeting, Austin, United States*.

Meetre, C., & Norris, J. (1991). Recognition of gamma-ray events in egret using neural networks. *Bulletin of the American Astronomical Society*, *23*, 905 pp.

Meskhidze, H. (2023). Can machine learning provide understanding? How cosmologists use machine learning to understand observations of the universe. *Erkenntnis*, *88*(5), 1895–1909.

Metcalf, R. B., Meneghetti, M., Avestruz, C., et al. (2019). The strong gravitational lens finding challenge. *Astronomy & Astrophysics*, *625*, A119.

Miller, A. (1993). A review of neural network applications in astronomy. *Vistas in Astronomy*, *36*, 141–161.

Mishra, S., & Molinaro, R. (2021). Physics informed neural networks for simulating radiative transfer. *Journal of Quantitative Spectroscopy and Radiative Transfer*, *270*, 107705.

Moschou, S., Hicks, E., Parekh, R., et al. (2023). Physics-informed neural networks for modeling astrophysical shocks. *Machine Learning: Science and Technology*, *4*(3), 035032.

Narayan, G., Zaidi, T., Soraisam, M. D., Wang, Z., Lochner, M., Matheson, T., Saha, A., Yang, S., Zhao, Z., Kececioglu, J., & Scheidegger, C. (2018). Machine-learning-based Brokers for Real-time Classification of the LSST Alert Stream. *Astrophysical Journal Supplement Series*, *236*(1), 9.

Páez, A. (2025). Axe the X in XAI: A plea for understandable AI. In J. M. Duran & G. Pozzi (Eds.), *Philosophy of science for machine learning: Core issues and new perspectives*. Synthese Library. Springer.

Pfau, D., Spencer, J. S., Matthews, A. G., & Foulkes, W. M. C. (2020). Ab initio solution of the many-electron Schrödinger equation with deep neural networks. *Physical Review Research*, *2*(3), 033429.

Raissi, M., Perdikaris, P., & Karniadakis, G. E. (2019). Physics-informed neural networks: A deep learning framework for solving forward and inverse problems involving nonlinear partial differential equations. *Journal of Computational Physics*, *378*, 686–707.

Schmidt, E. (2024). Stakes and understanding the decisions of artificial intelligent systems. In J. M. Duran & G. Pozzi (Eds.), *Philosophy of science for machine learning: Core issues and new perspectives*. Synthese Library. Springer.

Shukla, K., Di Leoni, P. C., Blackshire, J., Sparkman, D., & Karniadakis, G. E. (2020). Physics-informed neural network for ultrasound nondestructive quantification of surface breaking cracks. *Journal of Nondestructive Evaluation*, *39*(3), 1–20.

Stein, G. (2023). https://github.com/georgestein/ml-in-cosmology

Stewart, M. (2019). https://towardsdatascience.com/simple-introduction-to-convolutional-neural-networks-cdf8d3077bac

Sullivan, E. (2022). Understanding from machine learning models. *The British Journal for the Philosophy of Science*, *73*, 109–133.
Sullivan, E. and Kasirzadeh, A. (2024). Explanation hacking: The perils of algorithmic recourse. In J. M. Duran & G. Pozzi (Eds.), *Philosophy of science for machine learning: Core issues and new perspectives*. Synthese Library. Springer.
Villaescusa-Navarro, F., Ding, J., Genel, S., et al. (2022). Cosmology with one galaxy? *The Astrophysical Journal*, *929*(2), 132.
Wang, S., Teng, Y., & Perdikaris, P. (2021). Understanding and mitigating gradient flow pathologies in physics-informed neural networks. *SIAM Journal on Scientific Computing*, *43*(5), A3055–A3081.

Helen Meskhidze is an Assistant Professor in the Departments of Philosophy and Physics at the University of Cincinnati. She was previously a Postdoctoral Fellow at the Black Hole Initiative and holds a Ph.D. in Philosophy from the Department of Logic and Philosophy of Science at the University of California, Irvine. Her research concerns the epistemology of large-scale simulations and uses of machine learning in astronomy/cosmology. She is also interested in the foundations of spacetime theories.

Chapter 19
Machine Learning Discoveries and Scientific Understanding in Particle Physics: Problems and Prospects

Florian J. Boge and Henk W. de Regt

Abstract Particle physicists have been among the early adopters of Machine Learning (ML) methods, the most notable ML systems being Deep Neural Networks (DNNs). Today, ML's use in Particle Physics (PP) ranges from the reconstruction of signals inside the detector to the simulation of events and the determination of statistical ratios in the final analysis. Most intriguingly, there is some evidence which suggests that DNNs might be able to independently acquire complex physical concepts—concepts that are relevant for the discovery and understanding of new particles and phenomena. We here argue that these two possibilities, that of discovering novel concepts per se, and that of discovering novel phenomena by means of them, pose epistemic challenges for particle physicists. In turn, we will analyse ways of mitigating these challenges, both actual and at present merely possible.

Keywords Particle physics · Scientific discovery · Concepts · Phenomena · Understanding · Deep learning

19.1 Introduction

Particle physicists face a tremendous amount of data in their experiments. Inside detectors at the Large Hadron Collider (LHC) at CERN in Geneva, numbers can ramp up to 1 billion collisions per second, leading to an incredible amount of some

F. J. Boge (✉)
Institute for Philosophy and Political Science, TU Dortmund University, Dortmund, Germany

Lamarr Institute for Artificial Intelligence and Machine Learning, Dortmund, Germany
e-mail: florian-johannes.boge@udo.edu

H. W. de Regt
Institute for Science in Society, Faculty of Science, Radboud University, Nijmegen, The Netherlands
e-mail: henk.deregt@ru.nl

J. M. Durán, G. Pozzi (eds.), *Philosophy of Science for Machine Learning*, Synthese Library 527, https://doi.org/10.1007/978-3-032-03083-2_19

petabyte of collision-data per second.[1] These data are filtered down by sophisticated 'trigger systems' to just some hundreds of megabytes per second, but of course this still means a huge amount of data to be stored and analysed. Furthermore, the response of these triggers must be designed so that indeed only 'uninteresting' events get discarded. Finally, due in part to the nature of the underlying theory, the connection between theoretical predictions and recorded data is highly mediated (Morrison & Morgan, 1999), and analyses leading to discovery and measurement claims are intricate (Boge & Zeitnitz, 2021; Boge, 2021).

Given the sheer amounts of data and the complexities thus involved in data-generation, management and analysis, it is no wonder that "[p]article physicists began fiddling with artificial intelligence (AI) in the late 1980s, just as the term 'neural network' captured the public's imagination." (Cho, 2017). Over these 40 or so years, there has been a steady co-evolution of particle physics (PP) and Machine Learning (ML) methods.[2] Today, applications of ML in PP are vast and their number is ever-growing.[3] What we here, in this chapter, will focus on are a few developments which suggest that genuine discoveries in PP seem possible with ML, and that these may outstrip physicists' present-day *understanding* of the sub-atomic domain. Thus, we will here confront the possibility of discovery without (human) understanding; a phenomenon which bears close connections to the pertinent topic of scientific understanding's role in scientific progress (De Regt, 2017; De Regt et al., 2009; Grimm et al., 2017) and to the question of whether explanation falls behind in ML (Boge et al., 2022).[4]

In particular, we will here discuss the potential discovery of novel *concepts* by ML in PP,[5] and of novel *phenomena*. Prima facie, in either of these events, particle physicists will be at a loss regarding understanding while facing novelty in a discovery-related sense: Understanding relies on theories and models, which in turn rely on concepts. If an ML system thus conceptualizes a targeted domain in a novel way, it will be at a relative advantage compared to the scientists using the ML system (and not, or not yet, in possession of the relevant concepts). Similarly, if a phenomenon is discovered in the absence of theory, then it will not be well-understood.[6]

---

[1] See, e.g., https://home.cern/news/news/computing/cern-data-centre-passes-200-petabyte-milestone (checked 05/23).

[2] See, e.g., https://physicsworld.com/a/ai-and-particle-physics-a-powerful-partnership/ for the reciprocal aspects of this relationship.

[3] See, e.g., the living review of ML methods in PP, https://iml-wg.github.io/HEPML-LivingReview/.

[4] In this connection, see also Chap. 7 by Páez, Chap. 8 by Buijsman, and Chap. 10 by Sullivan and Kasirzadeh in this volume.

[5] In this connection, see also Chap. 12 by Kieval and Chap. 13 by Freiesleben in this volume.

[6] Of course, this can also happen in the absence of ML, but with the vast and unwieldy data-sets of PP and like fields, the gap between discovery and understanding will certainly become significantly larger, and the challenges for overcoming it quite different.

Indeed, it seems very much plausible that ML-based discoveries can inspire new physics-concepts (Krenn et al., 2022; Barman et al., MS; Iten et al., 2020), and this has already somewhat happened in certain areas of physics (see Ananthaswamy, 2021). To connect these issues specifically to *particle* physics, we will use two sets of case studies, the first being concerned with the potential discovery of complex PP-concepts by Deep Neural Networks (DNNs) which have not in any way been directly given to the DNN during training (Baldi et al., 2014; Chang et al., 2018). The second will be concerned with the possibility of discovering novel phenomena by means of unsupervised learning, i.e., without specific, targeted outputs to be computed by the DNN at any given data instance (Farina et al., 2020; Heimel et al., 2019; Finke et al., 2021; Fraser et al., 2022; Dillon et al., 2022).

Assuming that understanding is one major goal of science, how will scientists remedy these sorts of situations? In this connection, we will discuss two recently proposed frameworks for 'learning from the machine' (Krenn et al., 2022; Barman et al., MS). These proposals are aimed at mitigating the effects of novel concept-discoveries by ML (but not by humans), and hence the advantage in understanding that the machine may be said to have over human researchers. Furthermore, we will also address current limitations to discovering novel phenomena with ML, as a robust performance in this connection is actually unlikely to obtain *without* significant reliance on theory (Boge et al., MS). Thus, hybrid approaches, integrating theory and ML, are presumably the more realistic road to an ML-assisted discovery of phenomena. At the same time, these would lessen the scope and impact of 'discovery without understanding' through ML.

19.2 ML's Discovery Potential in Particle Physics

19.2.1 Discovery of Novel Concepts by DNNs

The maybe most intriguing aspect of the use of advanced ML systems like DNNs in fields such as PP is that they appear to be able to acquire complex physical concepts without being instructed about them. One might be skeptical about applying such a mentalistic vocabulary to ML systems which, on a more sober view, are just complex functions on high-dimensional spaces realized on a computer and optimized over a large space of parameters.[7] Indeed, while some are optimistic that this is possible under a modest reading of 'concept' (Räz, 2023), one of us (Boge, 2024) has recently argued that it is sufficient (and indicated) to assume that DNNs can merely *emulate* concept-acquisition, and that this will even *amplify* the issues discussed below. In this chapter, however, we will act as if talk of concept-possession in DNNs

[7] See Radder (2006, Ch. 5) as well as Chap. 13 by Freiesleben in this volume and references discussed therein for similar skepticism.

was fully appropriate, and avoid deeper discussion of the thorny philosophy of mind-issues.

First note that this purported concept-acquisition is a general feature that has nothing to do with PP per se. But it acquires a special relevance therein, as we shall argue. In the philosophical literature, such an acquisition of concepts by DNNs has been most prominently acknowledged by Buckner (2018), but for our purposes, it will be most instructive to consider some actual examples from the ML and PP literature.

Let's begin with the works by Bau et al. (2017, 2018), which fall outside the scope of specifically scientific applications of ML. In a first study, Bau et al. (2017) introduced *annotation masks* for a data set of scenery-images, which contained labels for elements of the scenery on various levels (from pixel to whole image). These masks were intended to isolate conceptually meaningful patterns (say, sets of pixels representing dogs), and their accuracy was tested for on human test-subjects. Additionally, the top activated nodes in an image-classifying DNN—i.e., a DNN that is fed with an image and then spits out a label—were used to define an *activation map* over the images pixels, by blackening everything but those pixels for which a given set of nodes exceeded a very high threshold of activation. Using the matching-percentage between annotation and activation, Bau et al. (2017) could identify nodes that were reasonably specialized to concepts such as DOG—*even though the network had only been trained to classify whole sceneries.*

This is certainly impressive, as it seems as though the DNN had inevitably acquired a concept DOG as a means for classifying typically dog-containing sceneries (see also López-Rubio, 2020). A second study by Bau et al. (2018) is even more impressive in this respect. Here, Bau and colleagues investigated a *generative* DNN, which takes in white noise-images and spits out photo-realistic ones after proper training. Said generative DNN was trained to produce, e.g., sceneries involving churches, which of course also included things like trees. Walking through a similar procedure as above, Bau et al. (2018) were here able to identify, say, tree-specialized units, even though the DNN had not been specifically trained to produce trees. The novelty of this second study was that Bau et al. (2018) successively set the activations of tree-correlated nodes to zero by fiat, which led to a successive decrease of the trees in generated images. In this way, Bau et al. (2018) could show that these nodes and their activations indeed functioned *causally* in the ways that concepts supposedly do for human beings (see also Boge, 2024).

Thus, there is evidence from within the science of ML that DNNs may be said to 'possess (and even develop) concepts', or at least to be able to mimic concept-possession and -development by means of activation patterns correlated with data-elements that human beings would consider conceptually meaningful. However, certainly the most impressive point here is that these were concepts the DNN *had not even been 'educated' on.*

Evidence suggestive of this very same fact has been found within various physics-related applications as well. An impressive set of examples has been presented by Iten et al. (2020): A DNN with a specific architecture, called an 'autoencoder' (Fig. 19.1 for illustration), has been shown to plausibly acquire

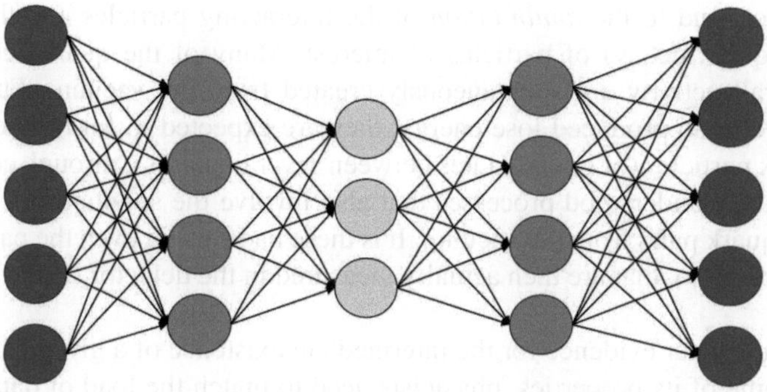

Fig. 19.1 Generic depiction of the 'autoencoder' architecture. The yellow, slim layer is called the 'bottleneck'; the layers preceeding it an 'encoder', those succeeding it a 'decoder'

concepts that are crucially important for understanding the underlying physics. The autoencoder has a very 'slim' intermediate layer, with only a few nodes (say two or three), so that it can store only the most salient information for the sake of predicting a particular output. When Iten et al. (2020) trained (a variant of) such an autoencoder to predict the positions of a simulated pendulum (a damped harmonic oscillator) based on past positions, they could show that the most condensed layer had specialized to the two constants determining the underlying equation (the reduced spring and damping constant), *without ever having 'seen' said equation*. Similarly, when said DNN was trained to predict the positions of planets in our solar system and of the sun from a geocentric perspective, they found that nodes in the condensed layer had specialised to angles describing the trajectories in a *heliocentric* view.

These futuristic-sounding observations have a correlate also within *particle* physics. In order to expound on this, let us briefly recapture the very basic ideas surrounding particle physics events, such as those going on at the LHC in Geneva. In order to be able to detect the particles they are interested in, particle physicists smash together other particles that they know how to isolate and exert control over. At the LHC these are protons, accelerated to peak collision energies of almost 14 TeV. The interaction between these is conceived of in terms of interactions between the quarks and gluons, collectively referred to as 'partons', that the protons are 'made of'.

Typically, multiple partonic interactions will take place at once, and there may even be several proton-proton collisions taking place in close proximity at roughly the same time. This leads to additional activity, sometimes collectively called an 'underlying event'. Strictly speaking, all these 'events' are of a quantum-mechanical nature, which means that there can be interference between several distinct processes of partonic interaction (Schwartz, 2021; Passon, 2019). However, as a matter of fact, there are theoretical reasons to think that, at very high energies, such interactions can be treated in close analogy to statistical ensembles of 'classical' interactions between tiny particles (Schwartz, 2014). Another aspect of the quantum field-theoretical nature of the underlying theory is the fact that such events correspond to the *annihilation* of the interacting particles and the *creation* (and subsequent decay) of particles of interest. Many of the quarks contributing to the overall activity are spontaneously created from the vacuum. Furthermore, as the particles so produced lose energy, they are expected to form 'hadrons', i.e., multi-quark particles (or bound-states between several quarks), through complicated and only partly understood processes that also involve the spontaneous creation of quark-antiquark pairs from the vacuum. It is these hadrons, or even the particles they quickly decay into, that are then actually measured in the detector (Boge & Zeitnitz, 2021).

To provide clear evidence for the intermediate existence of a given particle, or to measure some of its properties, physicists need to match the load of data produced in multiple events to theoretical models and their predictions. Typically, they do not use the data directly 'given' to them by the detector for this. As mentioned above, electric currents received from the detector will first be filtered down by means of a sophisticated 'trigger system' (Karaca, 2018), intended to store only events that are potentially of interest, so that there even is a manageable amount of data left. The data that survive this procedure are stored as 'event records', specifying the activity in the detector in terms of lists of numbers (Delfino, 2020; Albertsson et al., 2018). They are also referred to as the 'raw data', and these will be processed further into 'low level' data specifying the momenta, scattering angles, and some further identifying properties of the particles actually measured in the detector.

However physicists usually even go one step further and reconstruct, say, the (invariant) masses of intermediate particles from these 'low level data', by appeal to physical laws such as the relativistic energy-momentum relation. Data so derived, through these and further intermediate steps, are then referred to as 'high level data' (Delfino, 2020). The key reason for using high level data is that this representation is most discriminating between a null hypothesis, typically called the 'background only' hypothesis in the jargon, and the alternative (that there is something unexpected there). The classic image is that of a 'bump', as displayed in Fig. 19.2, indicating the intermediate production of Higgs bosons, decayed into two photons, in addition to a 'background' of pairs of photons to be expected also in the absence of Higgs bosons.

Remarkably, some DNNs trained on low level data to classify events as either 'signal' or 'background', and so to help even determine whether there is an excess of 'signal' data, did not profit from being handed also the high level information (Baldi et al., 2014). Even more impressively, Chang et al. (2018) found that, when the data were prepared so that the information on higher level variables such as the invariant mass was removed, an initially very successful DNN started to fail miserably. In just a little more detail, consider the histogram displayed in Fig. 19.3. In the two left most panels, one can see two histograms displaying the frequency of events with a specific reconstructed invariant mass. As is readily seen, the top histogram here features the characteristic bump indicative of a new particle. However, manipulating the data, by weighting each column in this histogram inversely by its height, one

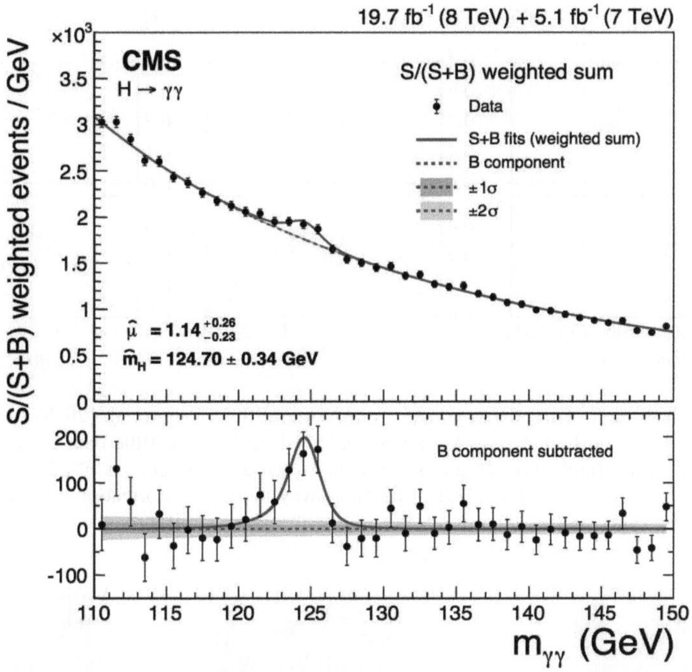

Fig. 19.2 Statistical displays of the Higgs-observation in 2012. Taken from Khachatryan et al. (2014) under a CC By 4.0 license. Color available online

obtains the lower of the two histograms.[8] Obviously, no 'bump-information' for the mass distribution is contained in the data anymore. This procedure was called 'data-planing' by Chang et al. (2018), as the result is a flat, uniform distribution.

As a result of this planing-procedure the DNN lost its ability to discriminate signal from background. However, the crucial point is only seen by considering the remaining histograms (middle and right), where the top one always corresponds to the original histogram, *before* the planing, and the bottom one to the result of planing. These are histograms for *different* quantities and, as is readily seen, the changes in these histograms are rather swift. Hence, it is unclear whether one would even *notice* a specific loss of information *unless one was already in possession of the relevant concept* (invariant mass). In other words: This study suggested quite vividly that the relevant DNN had recovered the mass information on its own, without having been instructed about masses at all.

These fascinating observations hint at the possibility that DNNs and other complex ML systems may acquire concepts on their own. However, combined with another observation, this even yields the possibility that DNNs may develop

[8] Actually, $10^{-x} \times 10^{x} = 1$, so the final histogram also includes a normalization-step.

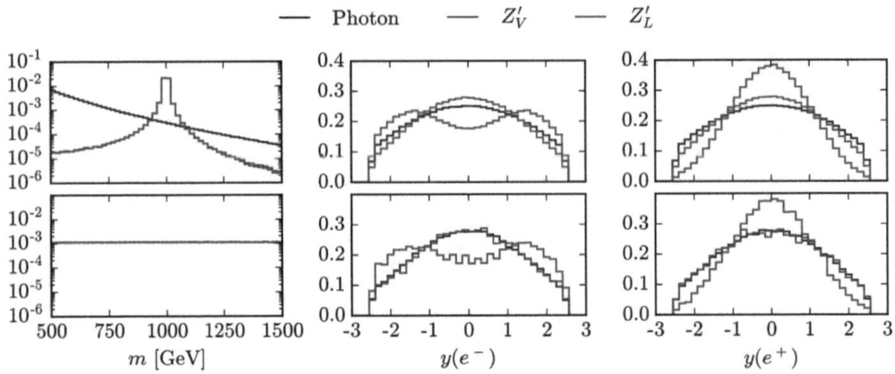

Fig. 19.3 Illustration of the data-planing procedure used by Chang et al. (2018) to rob a DNN of the information on higher level variables contained in the data. Blue data correspond to the 'background only' simulation, red and green data to two distinct versions of a hypothetical new Z'-particle. As can be seen, the effect of planing away the mass information only leads to swift changes in the other plotted quantities ($y(e^{\pm})$) for the Z'_L-case. Reproduced from Chang et al. (2018) under a CC BY 4.0 license. Colour available online

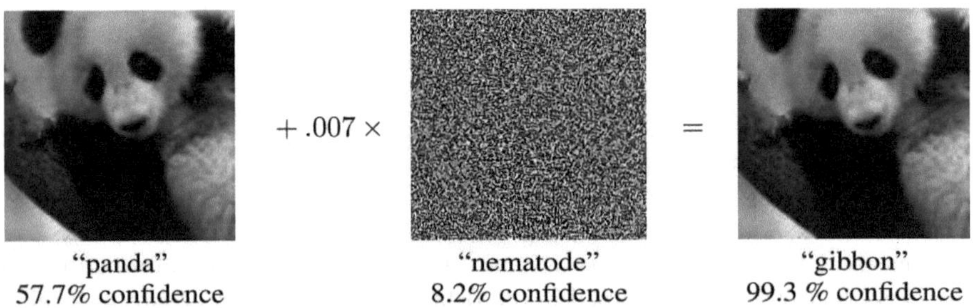

Fig. 19.4 Famous example of an adversarial, taken from Goodfellow et al. (2014). Color available online

(or maybe 'emulate') concepts that we, as human researchers, are not even in possession of.

In this connection, consider the prominent case displayed in Fig. 19.4, in which a DNN is made to fail by adding some dedicated noise to an originally well-classified image of a panda. After the addition of said noise, the DNN recognises the image as displaying a gibbon, and even more confidently than was the case with the panda before. Images (or more generally: data) of this sort are usually called *adversarials*. Next to data consciously created to fool the DNN, as in the panda case, a more permissive reading of the term comprises any sort of image (or data-point) that is easily recognizable for a human but surprisingly misidentified by a given DNN. Such a situation may be brought about, say, by including an unusual background to an object (Hendrycks et al., 2021), or by displaying it in a particular pose (Alcorn et al., 2019).

Crucial for us here is the reason *why* DNNs fail on such images or data. As Szegedy et al. (2013) noticed early on, the presence of adversarials seems to be fairly general across DNN architectures and data-sets used for training. This suggests that there might be generic reasons related to the very functioning of DNNs that lead to the possibility of finding or generating adversarials. Indeed, recent work by Zhang et al. (2019) suggests that adversarials are closely connected to what is atypical *for* the DNN, relative to its training, in the sense of being 'distant' from the training set in a specific metric. This metric was defined by comparing what a DNN encoded in its hidden layers when confronted with a new example to what it so encoded on the training data. Thus, if $h(x_{\text{test}})$ is the activation of some specific hidden layer of a DNN on a test-image, x_{test}, and $h(x_{\text{train}})$ is the corresponding activation on some training image, then the distance in question was defined as the average distance between $h(x_{\text{test}})$ and the k nearest $h(x_{\text{train}})$ for x_{train} from the training set in an ℓ_p metric. Images 'far from' the training data in this sense were, however, usually not in any specific sense atypical for humans. When such DNN-atypical images were just slightly distorted, this resulted in an adversarial, while still looking perfectly good to humans.

This last point is key here: Computer scientists refer to the functions DNNs compute within their hidden nodes as 'features'; the suggestion being that the given number computed corresponds to some property the DNN has managed to abstract from the data (also Buckner, 2018). And it thus seems that the features recognized by DNNs may differ markedly from those recognized by humans: An image that is atypical in terms of the features recognized by the DNN is easily turned into something it misidentifies, but neither the atypicality nor the strong response to these swift changes are correlated with corresponding human responses.

Furthermore, a study by Ilyas et al. (2019) recently suggested that, while these features so abstracted by DNNs may not be very robust in the sense that a small amount of noise can often destroy them, they may still be well-generalizing across various data-sets. This was shown by Ilyas et al. (2019) in the following way: Defining the *utility* of a certain feature by the degree to which it is positively correlated with a desired output in classification, and its *robustness* by the degree to which it remains useful under perturbations, Ilyas et al. (2019) constructed a data-set in which only *non-robust*, but still *useful* features remained. Then, training a DNN on such a data set and testing its performance on a standard test set (with robust features included), the DNN could be shown to exhibit a very good performance. So it seems that the DNN recognizes, and exploits, the very same (non-robust) features in the test-set.

Assuming that it is, hence, meaningful to talk as if DNNs develop concepts for the features they 'recognize', a fascinating possibility (recently discussed by Buckner, 2020, in only slightly different terms) arises: DNNs could develop concepts that, while being fairly incomprehensible for humans, are nevertheless scientifically productive.

To understand this suggestion a little better, consider Goodman's (1955) bleen and grue. To us, being 'green and observed before t or else not so observed and blue' does not seem like a meaningful property. But remember that green and blue can be similarly defined by means of bleen and grue. Hence, the switch between the two sets of concepts has much in common with a change of coordinate system, or frame of reference. It thus requires additional justification to argue for the objective preferredness of one frame of reference over the other.

Now, as Buckner (2020, p. 3) points out,

> Quine[9] famously suggested that [...] evolution has shaped our perceptual and cognitive faculties [in such a way that] certain features [...] jump out at us as natural candidates for investigation [...].

Combined with Ilyas et al.'s (2019) observation that the 'non-robust' features perturbed in the generation of adversarials are nevertheless well-generalizing, this suggest that DNNs and like ML systems rely on features, or even entertain concepts, that we would consider 'non-natural'—somewhat like bleen and grue. However:

> If scientific investigation would become more productive by tracking non-natural features [...] then even Quine would be likely to embrace this alternative route to scientific progress. (Buckner, 2020, p. 3)

In other words: It seems very much possible that DNNs and similar ML systems might have an advantage on us in conceptualizing vast and unwieldy data sets, not just despite the fact that they seem to conceptualize very differently, but even *because* they do so: Maybe the 'features' with the greatest predictive impact are very hard for humans to even grasp as features (i.e., as properties of the data or of the underlying objects), but can easily be utilized by a DNN or other ML system. However, assuming that such a situation actually occurs—which could in principle very well happen in the exploratory PP-research to come in the near future (Boge, 2022)—where would that leave us as human researchers in the quest for scientific understanding?

19.2.2 Discovery of Novel Phenomena by DNNs

The discovery of novel concepts by ML is certainly one interesting path to discovery that might well be around the corner in PP, but there are also other (not entirely unrelated) paths. Duede (2023) has recently investigated the discovery of novel theories with the help of DNNs, which seems at least partly connected to the discovery of concepts discussed above: If the DNN discovers relevant concepts, we might treat it as a source of inspiration for our own theorising (more on that in Sect. 19.3.4). However, we here want to turn to the discovery of novel *phenomena*, in the *absence* of theory. As a matter of fact, this issue is also partly connected to the above discussion on the possibility of novel concepts, but in rather subtle ways.

To unfold these issues, consider the autoencoder DNN depicted in Fig. 19.1 again. Using such an autoencoder, particle physicists have devised a basic method

[9] See Quine (1969).

for finding anomalies without building (strongly) on theory. However, to understand this suggestion, we first need to consider what an 'anomaly' actually is. First, recall the incredible amounts of data in PP and the highly mediated connection between these and the theory. Could a single, unexpected data-point tip us off as an 'anomaly' in PP? Certainly not. In PP at least, but presumably in science more generally, a recognizable anomaly will have to be of a statistical nature, reflecting an excess of data over the theory-guided expectation. Anomalies in PP are thus scientific *phenomena*: "something public, regular, possibly law-like, but perhaps exceptional." (Hacking, 1983, p. 222).

Furthermore, the word 'anomaly' suggests a deviation from a law of nature; so anomalies should be recognized as something that is inconsistent with accepted theory (containing the candidate laws for a given domain of inquiry). Hence, it seems that the very idea of recognizing anomalies in a theory-independent way is flawed: What counts as an anomaly is *defined* by means of existing theory. This is certainly correct in principle, but it need not imply any particularly *interesting* sense of theory-dependence: If the DNN is trained on data that are assumed to be well-understood but then indicates something odd on new data, this means that the oddity only relies on the acknowledged theory, not on any rivals.

Such a methodology is certainly at odds with standard methodology in PP. Recall how PP's Standard Model (SM) comes with certain limitations: It does not feature a model of gravity or, relatedly, dark matter (Martens & Lehmkuhl, 2020), and has many suspicious fine-tuning properties (Rosaler & Harlander, 2019). Hence, a major task in present-day PP is to find signs of physics 'beyond the SM' (BSM). Yet, "the search for many of the favoured BSM scenarios has been unsuccessful" (Butterworth et al., 2017, p. 2), and so new strategies are indicated. Strategies that particle physicists have explored comprise probing the room left for favoured BSM scenarios in the space spanned by their parameters by measurement uncertainties around the SM predictions (Butterworth et al., 2017), using simplified models that include certain BSM properties (McCoy & Massimi, 2018), and extending the SM by exploring rather arbitrary new combinations of its operators (Bechtle et al., 2022). However, these approaches clearly presuppose not only the SM but also some candidate, or preliminary, *rivals*. In contrast, using a DNN to probe the data promises a much greater extent of theory-freedom.

So far, this is all certainly consistent with the traditional ways in which anomalies have promoted progress: Remember how Kuhn (1970, pp. 52–53) held them responsible for paradigm-shifts:

> Discovery commences with the awareness of anomaly [and] continues [...] until the scientist has learned to see nature in a different way [...].

Clearly, one need not be a Kuhnian to acknowledge that anomalies matter to progress though: More moderate thinkers like Lakatos (1970, 1976) also reserved an important role in theory change for them, and one of us (De Regt, 2020) has recently shown how the resolution of an anomaly in the kinetic theory of gases led to an increase in understanding—and thus to scientific progress.

However, as was explained above, even recognizing an anomaly becomes very difficult when the data are as vast, and the theory is as strongly mediated by models, as is the case in PP. Thus, the crucial point is that theory may even be insufficient to say when something is *not* consistent with accepted laws.

So DNNs promise as theory-independent a method for recognizing anomalies as possible, and in domains where this would otherwise hardly be humanly feasible. Let us see in more detail how it works. In two benchmark studies, Farina et al. (2020) and Heimel et al. (2019) could demonstrate some astonishing successes with autoencoders on simulated data. Focusing on the study by Farina et al. (2020), the task faced by the autoencoder was to recognize anomalous 'jet images', which were chosen to stem from top-quark events.

To unpack the physics a little here, recall how quarks form hadrons (multi-quark particles like protons and neutrons) at lower energies—a phenomenon known as 'quark confinement'. The related process is complicated, and involves particles spontaneously created from the vacuum, as well as decays of semi-stable particles that form over short times (Boge & Zeitnitz, 2021). However, what one measures in the detector are actually the products of this 'hadronization process'. The distribution of tracks of such particles in the detector from events involving the annihilation of (almost) free quarks is identifiable by its characteristic, conical shape. Such a cone of tracks is referred to as a 'jet'. Furthermore, jets from distinct types of quark-events may exhibit different sub-structures (i.e., different internal groupings of tracks jointly forming a jet). The identification of both jets and their sub-structures is non-trivial. However, jet-substructures can be made visible by 'looking into the jet from above', i.e., displaying an image of the distribution of energy-momentum inside the detector when looked onto from above the center of the jet, down into the jet's core. For humans to recognize characteristic structures in this way, it is necessary to average over some tens or even hundreds of thousands of such images. Obviously, this requires first being able to *sort* them according to the particle-type they stem from.

Clearly, such an approach doesn't make much sense if one is searching for an anomaly. Hence, the approach pursued with autoencoders is slightly different. Instead of averaging over images, the autoencoder is trained to reconstruct individual jet-images which can be assumed to be consistent with the known physics. If after this training, the autoencoder is unable to reconstruct a given jet-image well, this is an indication that it contains anomalous features and thus may be indicative of an as yet unknown ('anomalous') physics process.

Surprisingly, in benchmark studies with simulated data wherein top-quark jets were used as the anomaly and the remaining jets from quantum chromodynamics as the 'background' (i.e., the data consistent with known physics), an autoencoder trained on a data set that was contaminated with up to 10% of anomalous jet-images could be successfully used for detecting anomalies after training (Farina et al., 2020). Hence, on the face of it, this seems like a very promising approach to finding anomalies in as theory-free ways as possible, and in domains where one could otherwise hardly find them.

There is an important catch with the theory-independence of this method, but before we discuss it, first consider the connection to concepts. Recall that the concepts discovered by DNNs (if any) may actually not be closely aligned with human concepts, and that this could actually be the factor that gives them an edge in scientific discovery (Buckner, 2020). As a matter of fact, it appears that this is actually what happens in unsupervised anomaly detection (see Boge et al., MS): The autoencoder reacts to subtle changes in the image that are barely even recognizable to humans. Because in the condensed, bottleneck-layer, it can only focus on a small number of salient features which it abstracts from the data—but which typically won't correspond to anything meaningful for humans—, a failure to reconstruct anomalous images well indicates a difference in features *recognized by the DNN*. As we will explain in more detail in Sect. 19.3.2 the recognition of certain features (or even objects) defining a new phenomenon requires the presence of appropriate concepts, which may then give rise to new theories. But if this is correct, research into the discovery of phenomena by DNNs and into the concepts developed by them might ultimately co-align.

Here comes the catch though: It is far from clear that this method for recognising anomalies in such a theory-free way can be made robust (Boge et al., MS). The evidence for this comes, again, from PP. In particular, in a study by Finke et al. (2021), the autoencoder was first trained in the same basic way as in the study by Farina et al. (2020), though without contamination by top-quark anomalies. However, Finke et al. (2021) then turned things around and trained their autoencoder to reconstruct top-images (making them the 'background') and to recognize the remaining quantum chromodynamics-jets as anomalies. Surprisingly, in this latter set-up, the autoencoder performed "worse than picking anomalies randomly." (Finke et al., 2021, p. 3).

Indeed, it turned out that the excellent performance in the original setup was probably due to the fact that top-jets are simply *more complex*, in the sense that they contain more pixels that are important to their structure, and that they simply have more structure in the first place (ibid.). Hence, the seemingly successful autoencoders in the studies by Farina et al. (2020) and Heimel et al. (2019) may have actually *only learned to reconstruct simple images well*, while remaining unable to reconstruct complex (and *coincidentally*: anomalous) ones.

Again, the most crucial point for us here must be extracted with some additional effort. To this end, first note that there was a certain dependency of the observed effect on the training regimen. In most of these studies, the 'loss' encountered by the DNN at any given image was quantified by the average number of pixels it failed to reconstruct correctly. However, it is not difficult to see that this can be related to bright pixels becoming more important: The brightness of certain pixels corresponds to them having 'high values', and so reconstructing these can quickly raise the average. However, given the capacity constraints imposed by the bottleneck, it will be difficult for the autoencoder to reconstruct images with 'too many' bright (and thus: salient) pixels.

Second, Finke et al. (2021) tried out several ways of mitigating the problem that were in no obvious way theory-dependent. For example, they tried to 'smear out' certain pixels, so that the autoencoder would be less focused on individual, salient pixels prominent in the simpler images. However, such strategies turned out to be quite unsuccessful (ibid., 18).

Third, there were additional studies done in which similar observations could be made, but with slightly different upshots. In particular, Fraser et al. (2022) tried out different combinations of loss-function (determining the training) and anomaly-measure (determining how badly a DNN would be able to reconstruct an image *after* training, i.e., indicating how anomalous it was). As it turned out, though, different combinations were here favourable for different types of (simulated) background and anomalous data. Hence Fraser et al. (2022) concluded that "without a signal model in mind, optimizing analysis strategies is hard to do in a principled manner" (13), or even that "one cannot optimize without a signal model in mind" (8).

One might suspect that all this is due to the architectural constraints imposed by the autoencoder and that a different architecture—without a bottleneck-layer—might fare better as an anomaly detector. However, note that the very idea of anomaly detection with ML is predicated on the presence of a bottleneck: Any sufficiently rich DNN would be trivially able to reconstruct any image, as it could learn the identity mapping. Hence, the bottleneck is *necessary* for at least the traditional ways of performing anomaly detection with ML. Furthermore, a larger bottleneck might be necessary to reconstruct complex anomalies, but figuring this out would require *preconceptions* of the type of anomaly in question on the side of the researcher—exactly our point.[10]

There appears, in other words, to be a certain trade-off between the robustness of the method and its theory-independence (Boge et al., MS): Methods which start from very innocent ideas that require little input from theory will turn out to be sensitive to many changes in, say, the types of data they encounter, the ratios between different such types, and the correlation between choices of loss-function for the training and the type of data they will be confronted with. They can hence not be said to perform robustly. As it turns out, this situation can be remedied (or at least mitigated) by appeal to expectations of what potential anomalies might look like—thus re-introducing a fairly strong theory-dependence.

In sum, it seems that the possibility of discovering of novel phenomena with DNNs must, at least at present, be taken with a grain of salt. Whether scientists come up with methods that are not subject to the same robustness-theory trade-off only time can tell (however, see also Boge et al., MS, for a principled argument exposing the obstacles). We will return to this issue below.

[10] There are several other, less popular methods (e.g. Mattia et al., 2021), but at the level of loss functions and training data, it remains unclear whether they can do without significant input from theory.

19.3 Understanding (and) ML's Impact on Particle Physics

19.3.1 Scientific Understanding: A Philosophical Account

Above, we have indicated tentative problems for gaining scientific understanding in PP when discoveries—of either concepts or phenomena—are partly ML-driven therein. However, what do we mean by 'understanding' here, and what are the connections to phenomena and concepts, respectively?

The nature of scientific understanding has been a central topic of debate in philosophy of science since the turn of the millennium. In contrast to traditional philosophical work on scientific explanation (a notion that is obviously related to understanding), the increasing philosophical attention for understanding is motivated by the practice-turn in the philosophy of science. Study of scientific practice (and its history) clearly shows that understanding plays a key role in the process of science, contrary to what traditional philosophers of explanation believed. An influential example of such a practice-oriented analysis is De Regt's (2017) contextual theory of scientific understanding, which is based on historical case studies of scientific development, especially physics, and on recent insights that philosophers of science have derived from studying scientific practice in general. In our chapter we will use this theory as a starting-point for our analysis of the role of ML in discovery and understanding in PP.

De Regt's core idea is the thesis that scientists achieve understanding of a phenomenon P if they construct an appropriate model of P based on a theory T. If it is empirically adequate, the resulting model constitutes an explanation that provides understanding of P. While this may not yet sound very different from traditional accounts, De Regt's claim is that study of the history and practice of science teaches us that only if theory T is intelligible to scientists is it in fact possible to construct a successful model-based explanation. Hence, the following Criterion for Understanding Phenomena (CUP; De Regt, 2017, p. 92):

CUP A phenomenon P is understood scientifically if and only if there is an explanation of P that is based on an intelligible theory T and conforms to the basic epistemic values of empirical adequacy and internal consistency.

The key term in this criterion is 'intelligible', but what exactly is intelligibility? De Regt (2017, p. 40) defines it as follows:

Intelligibility: the value that scientists attribute to the cluster of qualities of a theory T (in one or more of its representations) that facilitate the use of T.

Note that intelligibility is not an intrinsic property of theories, but a context-dependent value: whether a theory is intelligible to scientists depends on contextual factors such as their skills and background knowledge. Why do scientific theories need to be intelligible to the scientists who use them? Again, this follows from the study of scientific practice, especially from the work of philosophers like Cartwright (1983) and Morrison and Morgan (1999), who highlighted the crucial role of modelling in explanatory practices. On their model-based account of scientific explanation scientists acquire understanding of the phenomena by constructing models, which 'mediate' between relevant theories and the phenomenon-to-be-explained. Constructing such mediating models involves pragmatic judgments and decisions, since models do not follow straightforwardly from theories and neither from the empirical data. In particular, suitable idealizations and approximations need to be made, which cannot be deduced from theory or data. De Regt argues that the construction of such models—which provide explanatory understanding of phenomena—requires theories that are intelligible in the sense defined above. Only if scientists' skills to work with the theory allow them to make suitable pragmatic judgments, will they succeed in constructing explanatory models. In short, understanding a phenomenon on the basis of T depends on an appropriate combination of skills of the scientist S and qualities of T.

A test for intelligibility can be described by the following criterion (De Regt, 2017, p. 102):

CIT$_1$ A scientific theory T (in one or more of its representations) is intelligible for scientists (in context C) if they can recognize qualitatively characteristic consequences of T without performing exact calculations.

This is a sufficient but not a necessary condition for intelligibility. This is why there is a subscript 1 in CIT$_1$, since there might be other criteria for intelligibility. CIT$_1$ holds primarily for theories with a mathematical formulation and is thereby particularly suitable for physics. The key aspect of this condition is the ability to derive (qualitative) consequences.

Although De Regt's contextual theory is not the only one on the market, we believe it is a good starting-point for analyzing the prospects and limitations for using ML methods to achieve understanding in particle physics. First, it is an influential account that is widely accepted as capturing basic intuitions regarding scientific understanding, and second, it has been geared towards the practice of physics. It is immediately obvious that the contextual theory applies specifically to human scientific understanding: it has been developed via analysis of how human scientists achieve understanding in practice, and its emphasis on the role of skills and context highlights the characteristically human nature of understanding. This is captured by its central notion of intelligibility (of theories or models), which relates to human scientists (with particular skills), in particular historical, disciplinary and social contexts.[11]

Given the fundamentally human nature of scientific understanding the question arises whether ML-driven science can conform to extant criteria for understanding, especially the intelligibility criterion. This is prima facie not the case. To begin

[11] It is not a coincidence that philosophical interest in the notion of scientific understanding emerged only after the traditional objectivist approaches in the philosophy of science (and in the debate about scientific explanation in particular) had been abandoned and the subjective and pragmatic nature of understanding was not regarded as detracting from its epistemic importance.

with, according to the contextual theory, (human) scientific understanding requires theories as a basis for constructing models of the phenomena to be explained. However, both theory and model are representational devices that are allegedly completely absent in ML systems. To be sure, such systems (e.g. DNNs) are called models as well, but they are models in a specific sense that does not conform to the idea of models as mediators between theory and phenomena; instead they are complex mathematical representations that learn from input data in order to make predictions, but remain black boxes that are opaque and thereby unintelligible to humans (Facchini & Termine, 2021; Boge, 2022). Hence, DNNs seem to be incapable of generating scientific understanding in the sense of CUP, because their predictions are not based on scientific theories (indeed, they are data-driven rather than theory-driven) and they do not provide intelligible models of the phenomena.

This conclusion is not merely a peculiar feature of De Regt's theory: it resonates with the views of many philosophers and scientists. The fact that DNNs are unintelligible black boxes has generally been regarded as an obstacle to their capacity for providing understanding. Thus, Chirimuuta (2021, p. 787) argues that in computational neuroscience the use of artificial neural networks for modeling neuroscientific phenomena entails "a trade-off between predictive accuracy and the ability of models to confer understanding". She suggests that intelligibility—which she rightly claims to be a "human-relative virtue"—should not be given up in favor of mere prediction and control. In a similar vein, Greif (2022, p. 130) evaluates DNNs with regard to 'model intelligibility' and concludes that "DNNs, like ML approaches more generally, are neither designed for [scientific explanation and understanding], nor can they be recruited for it in a similar way to analogue or more traditional computer models. Their ability to master complex cognitive tasks in ways that are in part beyond human comprehension actually testifies to that".[12]

To be sure, some have argued that one can acquire scientific understanding of phenomena without model intelligibility. Kästner and Crook (MS, ch. 19, this volume), for example, claim that the arguments above rest on a misguided conflation of understanding models and 'model-induced understanding' of phenomena. They adopt a conception of scientific understanding that incorporates elements of De Regt's contextual theory, namely that understanding consists in a qualitative grasp of the target phenomenon P, involving the abilities to make coarse-grained predictions and to control and intervene on P. However, they apply this directly to the phenomena rather than to representations thereof by means of theories and models, arguing that understanding a phenomenon cannot be reduced to understanding a model, since "real-world scientific understanding of complex phenomena frequently depends upon facts and details extrinsic to any individual model". Subsequently, Crook and Kästner compare simple models with complex ones and claim that

[12] Both Chirimuuta and Greif adopt De Regt's notion of intelligibility. Ultimately, we think that even considering ML models as 'models' in a very literal sense is a non-starter in fields outside computational neuroscience. Hence, some of the positions discussed here may actually rest on a confusion. We will make all this clearer in Sect. 19.3.3.

"partial understanding of a complex but accurate model can lead to greater understanding of a phenomenon than complete understanding of a simple but inaccurate model". However, even if we grant this much, we submit that in the limit where complex models become completely unintelligible black boxes (as appears to be the case with DNNs), understanding of the phenomena has vanished for human scientists because they have lost their qualitative grasp in terms of abilities to predict, intervene and control on the phenomenon.

Crook and Kästner are surely right that model intelligibility and model-induced understanding differ and should not be conflated. However, interaction between the two is essential: it constitutes a feedback mechanism between constructing and using explanations that produces new scientific knowledge and thereby advances scientific understanding. This mechanism shows that understanding, as De Regt (2017, pp. 44–47) states, is both "a means and an end". Rather than a conflation, it contains a 'virtuous circularity': scientists employ their initial understanding of theories (read: model intelligibility) to construct explanations that, if empirically successful, produce understanding of phenomena. Next, they may apply, extend and refine their knowledge of the target systems using the same skills that were needed to construct the original explanations (the skills associated with model intelligibility).

We conclude that scientific understanding requires theories on the basis of which scientists construct (model-based) explanations of target phenomena that generate understanding. At first sight it seems that while DNNs may excel in predictive power, their understanding-providing power is low because they lack intelligible theories. In the next subsection we will examine this tentative conclusion more closely by focusing on the role of concepts and the nature of phenomena.

19.3.2 Concepts, Phenomena, and Scientific Understanding

Above, we have argued that (intelligible) theories are required for achieving scientific understanding, but we have not yet given a characterization of what a theory is. De Regt (2017, pp. 30–32) adopts Giere's (2006) notion of theory as a collection of principles that provide the basis for the construction of models of parts/aspects of the real world. This is a rather minimal characterization that does not put strong constraints on the notion of theory. But it does entail that the fundamental building blocks of theories are *concepts*. At this point there seems to be an opening for DNNs to develop theories after all, since—as we have seen in Sect. 19.2.1—DNNs abstract 'features' from data, a process that can be regarded as the acquisition of corresponding concepts. However, these concepts can differ wildly from ordinary (human) concepts and may therefore be unintelligible to humans. Hence, even if DNNs develop theories on the basis of their 'concepts' and acquire understanding via these theories, this understanding may not be accessible to humans. In order to explore the ways in which DNNs' concepts (and hence theories) agree and/or differ from that of human concepts and theories, let us first consider human scientists' concepts and their relation to theories, phenomena and understanding. In Sect. 19.3.4 we will then examine whether, and if so how, conceptualizations by DNNs can be made intelligible to humans.[13]

Concepts are a thorny subject in the philosophy of mind and cognitive science. While there appears to be vast agreement that concepts are central to philosophy and human cognition in general, there is little agreement on the details: a variety of accounts exists of the nature and structure of concepts, of concept acquisition, and of the dynamics of conceptual change. We cannot possibly cover this debate here and will hence restrict ourselves to discussing a few insights that are relevant to our project.

To begin with, there are at least three views regarding the nature of concepts, regarding them as either (1) mental representations, (2) abilities (peculiar to cognitive agents), or (3) abstract objects (see Margolis & Laurence, 2023). For our purposes, (1) and (2) are most relevant. If concepts are assumed to be *mental* representations, DNNs by definition cannot possess them. However, DNNs can mimic concept possession by extracting features from data sets, which can be regarded as representing target phenomena. Structuring observation, detecting empirical regularities, and classifying objects are representational roles that DNNs may fulfill. The roles of enabling inferences and explanations better fits the ability view of concepts.

As regards the structure of concepts the classical view is that concepts involve necessary and sufficient conditions. This strict definitional view is challenged by the looser 'prototype theory', inspired by Wittgenstein's idea of 'family resemblance' (see Hampton, 2006). Prototype theory is congenial to the ability view of concepts insofar as it seems to reduce concept-possession to "having a bundle of inferential capacities" (Fodor, 1994, p. 108). However, a general worry raised by Fodor (1994, p. 109)—one that we share—is that the *identification* of concepts with (inferential) abilities goes too far, as this would likely defy the compositional structure of concepts.

Prototypes, furthermore, are not to be misconstrued as 'typical examples', but are "the centers of clusters of similar objects", which "allows for the representation of different possible values of relevant features such as that apples can be red, green, brown, or yellow [...]." (Hampton, 2006, p. 80). It is hence difficult to pin them down as mental representations.

In response to the classical and the prototype theory, a third view has been advanced that solves problems of its predecessors and is therefore gaining popularity in cognitive science: the 'theory theory' of concepts (Carey, 2011). On this view, concepts should not be considered in isolation but as part of more general theories that relate various concepts: by being part of a coherent theory concepts acquire their meaning and function, both in science and in everyday life. Hence, the theory theory is fully in line with our claim above that scientific theories are networks of concepts.

---

[13] For related discussions, see also Chap. 12 by Kieval and Chap. 13 by Freiesleben in this volume.

As mentioned, we cannot settle these debates here. For our purposes three conclusions are relevant. First of all, concepts are crucial for, and perhaps even inherently intertwined with, *theories*. Second, concepts can fulfill *representational* roles. Finally, concepts appear to *provide* (though they are not to be identified with) certain *abilities*, such as "recognition, naming, inference, and language understanding." (Piccinini, 2011, p. 179) Turning to the role of concepts in science, important abilities that concepts may provide are: enabling the recognition of phenomena, as well as enabling inferences and explanations, and thereby—as we will argue below—achieving *understanding* of those phenomena. These conclusions align with Arabatzis (2019, p. 86), who lists a variety of roles that concepts can play in scientific practice. For our purposes, the following roles are relevant: they can structure scientific observation; enable the detection of regularities and empirical laws; go hand in hand with the classification of objects; enable inferences about the objects they refer to; and enable the explanation of phenomena via hidden entities and mechanisms.

That concepts are crucially involved in the recognition of phenomena has been acknowledged since logical empiricism was challenged by philosophers such as N.R. Hanson and Thomas Kuhn. They famously argued that (scientific) observation is always 'theory-laden', because it is in part determined by the conceptual framework of the observer. In this vein, Koningsveld (1973, p. 5, emph. altered) writes: "*what* we observe, the empirical datum, is co-constituted *as that datum* by the theory or the concepts through which we observe." In these classic debates no distinction was made between observations, data and phenomena: they were simply lumped together as the empirical basis for scientific theorizing.

More recent discussions, however, make a threefold distinction between data, phenomena and theory. This started with the work of Bogen and Woodward (1988), who argue that phenomena rather than data are the object of scientists' activities of explanation and prediction, where data can serve as evidence for the existence of phenomena (since they are caused by them). Phenomena, then, are regularities that can be either observable or unobservable (cf. Hacking, 1983). While data are "idiosyncratic to particular experimental contexts", phenomena have "stable, repeatable characteristics which will be detectable by means of a variety of different procedures, which may yield quite different kinds of data" (Bogen & Woodward, 1988, p. 317).

Bogen and Woodward's view of phenomena is uncompromisingly realist and seems to imply that phenomena are independent of our conceptualizations. Their account has triggered extensive debate and inspired more elaborate analyses of the nature of phenomena. Thus, Hans Radder (2006), Mieke Boon (2020) and Michela Massimi (2022) have developed accounts with a Kantian flavour, in which conceptual frameworks or 'perspectives' play a crucial role in the conception of phenomena. Hasok Chang (2022, pp. 73–75) defends a similar view on phenomena, arguing that phenomena are "realities, or attributes of realities" that are "mind-framed but not mind-controlled". He rejects the traditional realist idea "that reality has well-defined parts and properties that exist independently of all conceptualization", which he coins the *fallacy of pre-figuration*.

The fact that concepts form the basis of scientific theories, and are also crucial for the constitution (or at least the recognition) of phenomena, has important implications for the nature of scientific understanding. As we have seen above in Sect. 19.3.1, on De Regt's view scientific understanding of phenomena requires intelligible theories, and hence concepts. This idea aligns with the generally accepted view that concepts are crucial for understanding. Morrison and Morgan, for instance, already stated that it "is th[e] process of interpreting, conceptualising and integrating that goes on in model development which [...] provides the starting point for understanding" (Morrison & Morgan, 1999, pp. 31–33; see also Elgin, 2017, p. 120). At the same time, the phenomena that are the object of understanding are also partly dependent on conceptualizations.[14] This fundamental role for concepts in the process of achieving scientific understanding of phenomena raises the question of whether, and if so how, human scientists will always be able to understand the results of ML-driven science. For if DNNs acquire concepts (or functional proxies) that humans have no grasp of, they can perhaps recognize phenomena and develop theories that are unintelligible to humans. In the next sections we will discuss this issue in more detail and consider ways in which the situation might be remedied.

19.3.3 Paths to the Future (I): Integrating Theory and Machine Learning

A notable response to the challenges raised by ML in PP, as discussed in this chapter, connects to the theory-robustness trade-off mentioned in Sect. 19.2.2. Recall that, at least for the studies discussed therein, a certain trade-off between theory-freedom and robustness was noted: Training an autoencoder on data contaminated to a small extent with anomalous data (Farina et al., 2020; Heimel et al., 2019) appeared to allow the finding of excesses of such anomalies, as the autoencoder remained unable to successfully reconstruct them after training. However, as shown in the study of Finke et al. (2021), this apparent ability actually depended on the complexity of the 'anomaly' in question. Furthermore, no generally successful method for finding anomalies could be determined, *regardless* of their complexity (Finke et al., 2021), and *without* appeal to some sort of physics-model that *provided* (information about) the (simulated) anomalous data (Fraser et al., 2022).

However, from the vantage point of scientific understanding, it seems this vice might be turned into a virtue. Recall that an automated discovery of anomalies, as envisioned by particle physicists, means a *loss* of understanding: According to CUP,

[14] The concepts on which theories are based need not be the same as those related to phenomena, so there is not necessarily a circularity involved. In particular, the concepts developed by a DNN in order to recognise a novel phenomenon need not be identical to concepts contained in any humanly known theory.

we need an intelligible theory T to build models that provide us with understanding of phenomena, where intelligibility may be evidenced by the fact that scientists can recognize qualitatively characteristic consequences of T without performing exact calculations, according to CIT_1. However, as was mentioned already in Sect. 19.3.1, DNNs do not seem to be the right sorts of models to provide us with understanding. Let us briefly explain this in more detail.

First off, one could be misled to think that DNNs *are* understanding-promoting, as they are constructed under the auspices of an intelligible theory, i.e., deep learning theory (Goodfellow et al., 2014; Shalev-Shwartz & Ben-David, 2014, Ch. 20). Furthermore, successful architectures are often constructed in fairly heuristic ways that are only qualitatively sanctioned by the fundamental theory, rather than corresponding to rigorously proven properties. Hence, following CIT_1, we may take this to show that deep learning theory is intelligible to ML scientists.

However, note that this does not at all establish how DNNs can (and should) themselves promote understanding *about the relevant kind of target*: At best, it shows that, even though many details are presently ill-understood, we have a kind of general grasp of *how DNNs work*. Since their functioning is generally modelled on selective aspects of cognition, neuroscience, and statistical learning, this sort of intelligibility might help us towards an understanding of some aspects of human (or biological) learning—if ever so crudely and selectively (Buckner, 2018; Chirimuuta, 2021). But this, emphatically, doesn't imply that we thereby also obtain an understanding of the *subject matter analysed with* DNNs—in our case: particle physics.

Furthermore, we mentioned the fact that ML 'models' in general are not the sorts of models that mediate between theories and phenomena. Rather, they are similar to purely predictive models, as used, e.g., in descriptive statistics (Boge, 2024). The crucial point to recognize is that DNNs are by and large not *designed to represent* anything about a targeted system or phenomenon: Their elements (weights, biases, activation functions) are not ab initio interpreted as referring to a system's states and properties. These elements are actually devoid of any content at the outset and merely employed for the purpose of getting predictions right; something that Boge (2022) has coined being 'instrumental qua devoid of content'. At best, typical ML models are used to, say, represent the statistical distribution of certain features across data, and nothing more.

It is a common, empirically supported assumption, though, that DNNs *develop* representations (or models) during their training (e.g. Goodfellow et al., 2016; López-Rubio, 2020; Buckner, 2018, and above), which then may be understanding-promoting. In fact, we have tacitly bought into this by claiming that DNNs might develop novel concepts, and by making a connection between concepts, theorizing, modelling, and understanding.

However, it is important here to take into account also the specific sort of *opacity* that accompanies DNNs (and possibly other, similarly complex and flexible ML

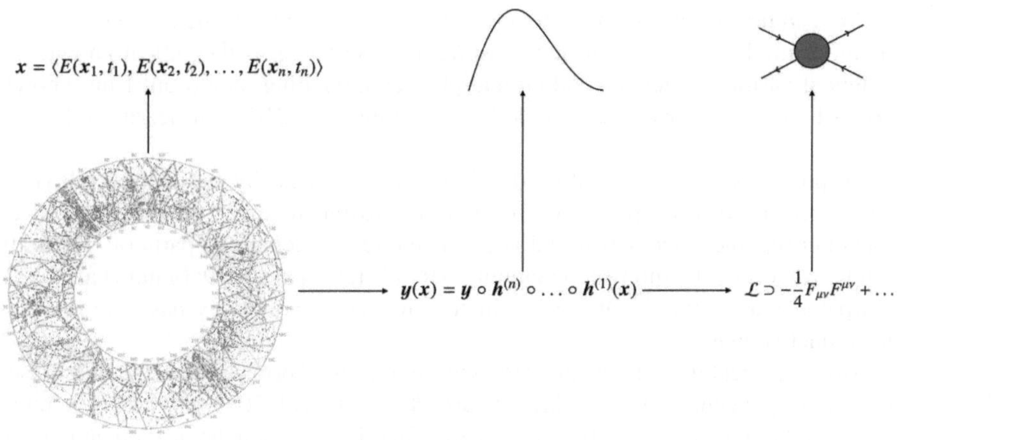

Fig. 19.5 Modeling steps that can lead from ML successes to scientific understanding. The first step means interpreting certain data, which are just currents received from a detector, in terms of energy-deposits from particle-interactions. The second step means (e.g.) running a DNN-classifier that distinguishes signal- from background-events on this data and interpreting it as providing a statistical relation, such as a likelihood ratio. The third step then means connecting this likelihood ratio back to explanatory physics-models, which might be a formidable task if the underlying processes are entirely unknown. Taken from Boge (2022) under a CC BY 4.0 license. Color available online

systems). As argued at some length in Boge (2022),[15] it is not only opaque *how* the machine learns (and functions), but also *what* it learns, i.e., what hidden pieces of information *in the data* drive the parameter-updating that we call 'learning'. As the study by Chang et al. (2018) discussed in Sect. 19.2.1 shows, these pieces of information can be highly relevant for our understanding of the subject matter: Having the mass-information in stock clearly alters the range of models we could employ to render the observed tracks in the detector understandable.

This 'what-opacity' (Boge, 2022), conjoined with the specific instrumental character of ML models in PP, implies that it will usually require additional effort to get from the recognition of what the ML model itself is used to represent (say, a statistical ratio or distribution) to an understanding of the subject matter: The model is *used* to represents fairly little (only said distribution or ratio), and what it might 'represent *to itself*' will be difficult to identify.[16] In particular, connecting a successful DNN to more understanding-promoting representations, such as theory-bound physics models, will require additional modelling (see Fig. 19.5).

[15] Beisbart (Chap. 1, this volume) offers an account of opacity that is phrased in different terms but roughly consistent with Boge's; Formanek (Chap. 2, this volume) instead offers a deflationary take on opacity, though he does not engage with Boge's arguments in detail. See also (Boge & Mosig, 2025) for an extended approach to the use of ML for understanding in science.

[16] We will turn to a specific class of proposals on how this might be done below.

We can now easily see how the vice of theory-presence in ML-based discovery-methods might become a virtue regarding understanding: If theoretical knowledge enters at various stages, this additional physics-modelling may profit from knowledge about how theory is used to delimit the things the DNN can 'learn' in the first place.

Approaches which have this idea built into them go by the name of 'hybrid ML'. Various distinct proposals have been subsumed under this label, such as patching together theory-inspired and ML models or replacing parts of DNNs by known, theoretically motivated mappings (Reichstein et al., 2019; Maier et al., 2022; Karpatne et al., 2017).[17] Of greatest interest to us here are theory-based approaches to 'inductive bias'.

Broadly speaking, this means "the incorporation of prior knowledge that biases the learning mechanism." (Shalev-Shwartz & Ben-David, 2014, p. 3). A little more specifically, it means restricting the class of functions that can be instantiated by the given DNN or ML system (ibid., 16). As has been pointed out by Sterkenburg and Grünwald (2021), it is possible to interpret the so called 'no free lunch theorems' (Wolpert, 1996), which say that under certain modest-seeming assumptions, "every learning algorithm can be expected to do no better (or worse) than random guessing" (Sterkenburg & Grünwald, 2021, p. 9982), as suggesting that inductive bias is *needed* to make ML work.

Insofar as this interpretation is correct, there is hardly anything puzzling about the above findings on autoencoders and their model-dependence:[18] Some idea about what a given anomaly looks like is needed; otherwise we don't stand a chance of *reliably* detecting it.

Furthermore, from the vantage point of scientific understanding, the situation is as follows: Insofar as a given implementation of inductive bias is motivated by theory—say, using a specific loss-function whose shape is guided by a physics model (Fraser et al., 2022)—, we can conclude that a successful autoencoder trained in such a way will find signals consistent with our theoretical expectations. This immediately allows us to understand the phenomena discovered, but of course it won't be anomaly-detection anymore: This would require the discovery of phenomena *inconsistent* with any theoretical expectations.

One might try to mediate between this loss in theory-freedom and gain robustness by employing an extended scheme that uses a broad range of differently trained autoencoders and thus looks for signals in the data consistent with various theoretical scenarios. This would certainly maintain a respectable increase in discovery potential by means of ML while also making the gap to theoretical

[17] For an appraisal in the context of astronomy and cosmology, see also Meskhidze (Chap. 17, this volume).

[18] Note that this is meant in a purely intuitive sense: Autoencoders trained as described above instantiate unsupervised learning, so they are not subject to the theorems mentioned here.

understanding smaller. But, admittedly, it would certainly also mean losing our handle on automated, largely theory-free discovery of phenomena.

19.3.4 Paths to the Future (II): Learning from the Machine?

Above we have seen that hybrid approaches, combining data-driven methods with theory-driven ones, can improve robustness and at the same time allow (at least in principle) for scientific understanding in the sense of CUP. However, the question remains whether novel conceptualizations developed by DNNs, that form the basis of their modeling and theorizing, can be made intelligible to human scientists. In this section we address this question by discussing two recent proposals for transferring understanding generated by AI, esp. DNNs, to humans.

In their paper 'On scientific understanding with artificial intelligence' Krenn et al. (2022) analyze how AI may contribute to achieving scientific understanding, by reviewing the current use of AI in science and discussing future prospects. They submit that AI might foster scientific understanding in three ways: (1) as a "computational microscope"; (2) as a "resource of inspiration"; and (3) as an "agent of understanding". (1) comprises uncovering patterns in complex data sets, and is already widely used. (2) concerns identifying surprises in the data or in the scientific literature, which may require that novel concepts are developed. Currently, interpretation and conceptualization has to be done by human scientists, but Krenn et al. (2022, pp. 754–756) do not preclude the possibility that future AI will be able to come up with novel concepts by itself. The third case goes one step further. Here AI generates new explanations and has thereby acquired novel understanding on its own: the AI is an "agent of understanding". Also this future is still ahead of us but not unlikely according to Krenn et al. In light of what we have discussed above, development of novel concepts and of novel explanations and understanding is closely related, and in any case raises the question of whether, and if so how, human scientists can learn from AI. If the newly developed concepts and understanding are not intelligible to humans, the result is a divide between human and artificial scientific understanding, which would be an obstacle to making progress in scientific understanding as a whole.

So, the crucial question is whether, and if so how, the AI can transfer its understanding to humans. To answer this question, we start with a suggestion made by Krenn et al.: they propose a 'scientific understanding test' for determining whether an AI has scientific understanding. They start with De Regt's above-mentioned criterion for intelligibility CIT_1 and suggest the following analogous condition: "An AI gained scientific understanding if it can recognize qualitatively characteristic consequences of a theory without performing exact computations and use them in a new context" (Krenn et al., 2022, p. 767). By itself this is not yet sufficient for actually assessing whether an AI has understanding. Therefore, Krenn et al. add a second condition, according to which "an AI gained scientific understanding if it can transfer its understanding to a human expert". Combining the two conditions, they formulate their 'scientific understanding test' for judging whether an AI has in fact gained understanding:

> A human (the student) interacts with a teacher, either a human or an artificial scientist. The teacher's goal is to explain a scientific theory and its qualitative, characteristic consequences to the student. Another human (the referee) tests both the student and the teacher independently. If the referee cannot distinguish between the qualities of their non-trivial explanations in various contexts, we argue that the teacher has scientific understanding. (Krenn et al., 2022, p. 767)

This obviously resembles a Turing test (as Krenn et al. readily admit). However, rather than using it as such, it can also serve a different purpose, namely as providing a link between human and artificial understanding and as a means to examine the extent to which humans can learn from AI. But the test does not yet tell us *how exactly* the referee measures the quality of the alleged understanding transfer from teacher to student. Krenn et al. do not elaborate on this issue, but Barman et al. (MS) provide an answer. They argue that evaluation of understanding in both humans and other agents (e.g. AIs) should be based on their *abilities* to perform relevant tasks, an approach that clearly aligns with the ability view of concepts. Barman et al. present a general framework *measuring* an agent's scientific understanding:

AUP The degree to which agent A scientifically understands phenomenon P can be determined by assessing the extent to which (i) A has a sufficiently complete representation of P; (ii) A can generate internally consistent and empirically adequate explanations of P; (iii) A can establish a broad range of relevant, correct counterfactual inferences regarding P. (Barman et al., MS, 6)

The idea behind this proposal is that scientific understanding is not an all-or-nothing affair (captured by necessary and sufficient conditions) but a matter of degree, comprising three levels of increasing value. The framework allows for different ways to measure the degree to which A has (i), (ii) and/or (iii). Barman et al. focus on Large Language Models (LLMs), for which they suggest the following specification: "[AUP](i–iii) can be measured, given a certain context (series of prompts) via what-, why-, and what-if-questions respectively." (ibid.) The ability to correctly answer what-questions reflects possession of relevant information about P (i), while an ability to answer why-questions regarding P reflects the capability to produce an explanation of it (ii). Finally, the ability to answer what-if-questions reflects competency at establishing counterfactual inferences about P (iii). This requires not only having a good explanation but also knowing how to use it, and can accordingly be linked to an agents' breadth and depth of understanding. Barman et al. (MS) argue that application of this approach to the scientific understanding test of Krenn et al. (2022) allows for measuring the increase of understanding in the student, and hence the understanding transfer from AI to humans. Extending this approach to AI in general would require integration of LLMs with other types of DNNs, a goal the pursuit of which is currently still in its infancy (e.g. Singh et al., 2022; Barman et al., forthcoming) but will plausibly be realized in the near future.

19.4 Conclusions

In this chapter we have explored the possibility that DNNs in particle physics can (I) autonomously discover novel, physically meaningful concepts, and (II) recognize anomalous phenomena that would otherwise be concealed from human sight. Both (I) and (II) would prima facie create obstacles for gaining understanding for the following reasons. First, concepts are the building blocks of theories that are required for understanding, and hence the unintelligibility of ML-generated concepts entails a failure of theoretical understanding. Second, anomalies by definition constitute ill-understood phenomena whose resolution would require extraordinary effort in the case of ML-based discovery.

As regards (I), we have argued that it is not implausible that DNNs can possess concepts (or at least be able to mimic concept-possession) and perhaps also develop novel concepts. However, it is not guaranteed that such novel concepts are intelligible to humans, and if they are not, humans cannot use them directly for acquiring scientific understanding. We have suggested that this problem might be resolved by invoking two recent proposals for establishing and measuring a transfer of scientific understanding between artificial agents, such as DNNs, and humans.

As it turns out, the impression that (II) is indeed the case is deceptive. Existing methods require theoretical input to become robust, that is, input on what the 'anomalies' to be discovered look like, so these would actually not be anomalies anymore. From the vantage-point of understanding, this seems like a virtue though, as the gap between discovered phenomena and our understanding of these is immediately reduced or even closed: If we know the potential new phenomena that a DNN can recognize through theories, we may already understand them fairly well. Disappointingly, this also means that the discovery-potential associated with DNNs in PP is currently less impressive than one might have initially thought.

Acknowledgments We thank Helen Meskhidze for a number of comments. FJB acknowledges generous funding by the German Research Foundation (DFG), for his Emmy Noether grant *UDNN: Scientific Understanding and Deep Neural Networks* (grant number 508844757).

References

Albertsson, K., Altoe, P., Anderson, D., Andrews, M., Espinosa, J. P. A., Aurisano, A., Basara, L., Bevan, A., Bhimji, W., Bonacorsi, D., Calafiura, P. (2018). Machine learning in high energy physics community white paper. *Journal of Physics: Conference Series, 1085*(2), 022008.

Alcorn, M. A., Li, Q., Gong, Z., Wang, C., Mai, L., Ku, W.-S., & Nguyen, A. (2019). Strike (with) a pose: Neural networks are easily fooled by strange poses of familiar objects. In *Proceedings of the IEEE/CVF Conference on Computer Vision and Pattern Recognition* (pp. 4845–4854).

Ananthaswamy, A. (2021). AI designs quantum physics experiments beyond what any human has conceived. *Scientific American*, July 2. https://www.scientificamerican.com/article/ai-designs-quantum-physics-experiments-beyond-what-any-human-has-conceived/

Arabatzis, T. (2019). What are scientific concepts? In K. McCain & K. Kampourakis (Eds.), *What is scientific knowledge?* (pp. 85–99). Routledge.

Baldi, P., Sadowski, P., & Whiteson, D. (2014). Searching for exotic particles in high-energy physics with deep learning. *Nature Communications, 5*, 4308.

Barman, K. G., Caron, S., Claassen, T., & De Regt, H. W. (2024). Towards a benchmark for scientific understanding in humans and machines. *Minds & Machines, 34*(6), 1–16. https://doi.org/10.1007/s11023-024-09657-1

Barman, K. G., Caron, S., Sullivan, E., De Regt, H. W., et al. (forthcoming). Large physics models: Towards a collaborative approach with large language models and foundation models. *The European Physical Journal C*, Preprint: arXiv:2501.05382.

Bau, D., Zhou, B., Khosla, A., Oliva, A., & Torralba, A. (2017). Network dissection: Quantifying interpretability of deep visual representations. Preprint. arXiv:1704.05796.

Bau, D., Zhu, J.-Y., Strobelt, H., Zhou, B., Tenenbaum, J. B., Freeman, W. T., & Torralba, A. (2018). Gan dissection: Visualizing and understanding generative adversarial networks. Preprint. arXiv:1811.10597.

Bechtle, P., Chall, C., King, M., Krämer, M., Mättig, P., & Stöltzner, M. (2022). Bottoms up: The standard model effective field theory from a model perspective. *Studies in History and Philosophy of Science Part A, 92*, 129–143. https://doi.org/10.1016/j.shpsa.2022.01.014

Beisbart, C. (MS). In which ways is machine learning opaque? In G. Pozzi & J. M. Durán (Eds.), *Philosophy of science for machine learning*. Springer.

Boge, F., Grünke, P., & Hillerbrand, R. (2022). Minds and machines special issue: Machine learning: Prediction without explanation? *Minds & Machines, 32*, 1–9.

Boge, F. J. (2024). Why trust a simulation? Models, parameters, and robustness in simulation-infected experiments. *The British Journal for the Philosophy of Science, 75*(4), 843–870.

Boge, F. J., & Mosig, A. (2025) Put it to the test: Getting serious about explanation in explainable artificial intelligence. *Minds & Machines, 35*, 26. https://doi.org/10.1007/s11023-025-09724-1

Boge, F. J. (2022). Two dimensions of opacity and the deep learning predicament. *Minds and Machines, 32*(1), 43–75.

Boge, F. J. (2024). Functional concept proxies and the actually smart hans problem: What's special about deep neural networks in science. *Synthese, 203*(16), 39 pp. https://doi.org/10.1007/s11229-023-04440-8

Boge, F. J., Krämer, M., & Zeitnitz, C. (MS). Deep learning robustness for scientific discovery: The case of anomaly detection. Unpublished.

Boge, F. J., & Zeitnitz, C. (2021). Polycratic hierarchies and networks: What simulation-modeling at the LHC can teach us about the epistemology of simulation. *Synthese, 199*(1–2), 445–480.

Bogen, J., & Woodward, J. (1988). Saving the phenomena. *The Philosophical Review, XCVII*(3), 303–352.

Boon, M. (2020). The role of disciplinary perspectives in an epistemology of scientific models. *European Journal for Philosophy of Science, 10*, article 31.

Buckner, C. (2018). Empiricism without magic: Transformational abstraction in deep convolutional neural networks. *Synthese, 195*(12), 5339–5372.

Buckner, C. (2020). Understanding adversarial examples requires a theory of artefacts for deep learning. *Nature Machine Intelligence, 2*(12), 731–736.

Buijsman, S. (MS). Machine learning models as mathematics. In J. M. Durán & G. Pozzi (Eds.), *Philosophy of science for machine learning: Core issues and new perspectives*. Synthese Library. Springer.

Butterworth, J. M., Grellscheid, D., Krämer, M., Sarrazin, B., & Yallup, D. (2017). Constraining new physics with collider measurements of Standard Model signatures. *Journal of High Energy Physics, 2017*(3), 78.

Carey, S. (2011). *The origin of concepts. Oxford Series in Cognitive Development*. Oxford University Press.

Cartwright, N. (1983). *How the laws of physics lie*. Oxford University Press.
Chang, H. (2022). *Realism for realistic people: A new pragmatist philosophy of science*. Cambridge University Press.
Chang, S., Cohen, T., & Ostdiek, B. (2018). What is the machine learning? *Physical Review D*, *97*(5), 6.
Chirimuuta, M. (2021). Prediction versus understanding in computationally enhanced neuroscience. *Synthese*, *199*(1–2), 767–790.
Cho, A. (2017). Ai's early proving ground: The hunt for new particles. *Science*, *357*(6346), 20.
De Regt, H. W. (2017). *Understanding scientific understanding*. Oxford University Press.
De Regt, H. W. (2020). Understanding, values, and the aims of science. *Philosophy of Science*, *87*, 921–932.
De Regt, H. W., Leonelli, S., & Eigner, K. (2009). *Scientific understanding: Philosophical perspectives*. University of Pittsburgh Press.
Delfino, M. (2020). Distributed computing. In C. W. Fabjan & H. Schopper (Eds.), *Particle physics reference library* (Vol. 2, pp. 613–644). Springer.
Dillon, B. M., Favaro, L., Plehn, T., Sorrenson, P., & Krämer, M. (2022). A normalized autoencoder for LHC triggers. *ArXiv:2206.14225*, [hep-ph]. https://doi.org/10.48550/arXiv.2206.14225
Duede, E. (2023). Deep learning opacity in scientific discovery. *Philosophy of Science, 90*(5), 1089–1099.
Elgin, C. Z. (2017). *True enough*. MIT Press.
Facchini, A., & Termine, A. (2021). Towards a taxonomy for the opacity of ai systems. In V. C. Müller (Ed.), *Conference on philosophy and theory of artificial intelligence* (pp. 73–89). Springer.
Farina, M., Nakai, Y., & Shih, D. (2020). Searching for new physics with deep autoencoders. *Physical Review D*, *101*(7), 075021.
Finke, T., Krämer, M., Morandini, A., Mück, A., & Oleksiyuk, I. (2021). Autoencoders for unsupervised anomaly detection in high energy physics. *Journal of High Energy Physics*, *2021*(6), 161.
Fodor, J. (1994). Concepts: A potboiler. *Cognition*, *50*(1–3), 95–113.
Formanek, N. (MS). How I stopped worrying and learned to love opacity. In G. Pozzi & J. M. Durán (Eds.), *Philosophy of science for machine learning*. Springer.
Fraser, K., Homiller, S., Mishra, R. K., Ostdiek, B., & Schwartz, M. D. (2022). Challenges for unsupervised anomaly detection in particle physics. *Journal of High Energy Physics*, *2022*(3), 1–31.
Freiesleben, T. (MS). Artificial neural nets and the representation of human concepts. In J. M. Durán & G. Pozzi (Eds.), *Philosophy of science for machine learning: Core issues and new perspectives*. Synthese Library. Springer.
Giere, R. (2006). *Scientific perspectivism*. University of Chicago Press.
Goodfellow, I., Bengio, Y., & Courville, A. (2016). *Deep learning*. The MIT Press.
Goodfellow, I. J., Shlens, J., & Szegedy, C. (2014). Explaining and harnessing adversarial examples. Preprint. arXiv:1412.6572.
Goodman, N. (1955). *Fact, fiction, and forecast*. Harvard University Press.
Greif, H. (2022). Analogue models and universal machines. Paradigms of epistemic transparency in artificial intelligence. *Minds and Machines*, *32*(1), 111–133.
Grimm, S. R., Baumberger, C., & Ammon, S. (2017). *Explaining understanding*. Routledge.
Hacking, I. (1983). *Representing and intervening*. Cambridge University Press.
Hampton, J. A. (2006). Concepts as prototypes. *Psychology of Learning and Motivation*, *46*, 79–113.
Heimel, T., Kasieczka, G., Plehn, T., & Thompson, J. M. (2019). QCD or what? *SciPost Physics*, *6*(030), 21.
Hendrycks, D., Zhao, K., Basart, S., Steinhardt, J., & Song, D. (2021). Natural adversarial examples. In *Proceedings of the IEEE/CVF Conference on Computer Vision and Pattern Recognition* (pp. 15262–15271).

Ilyas, A., Santurkar, S., Tsipras, D., Engstrom, L., Tran, B., & Madry, A. (2019). Adversarial examples are not bugs, they are features. Preprint. arXiv:1905.02175.

Iten, R., Metger, T., Wilming, H., Del Rio, L., & Renner, R. (2020). Discovering physical concepts with neural networks. *Physical Review Letters, 124*(1), 010508.

Karaca, K. (2018). Lessons from the large hadron collider for model-based experimentation. *Synthese, 195*(12), 5431–5452.

Karpatne, A., Atluri, G., Faghmous, J. H., Steinbach, M., Banerjee, A., Ganguly, A., Shekhar, S., Samatova, N., & Kumar, V. (2017). Theory-guided data science: A new paradigm for scientific discovery from data. *IEEE Transactions on Knowledge and Data Engineering, 29*(10), 2318–2331.

Kästner, L., & Crook, B. (MS). Don't fear the bogeyman: On why there is no prediction-understanding trade-off for deep learning in neuroscience. In G. Pozzi & J. M. Durán (Eds.), *Philosophy of science for machine learning*. Springer. Online: http://philsci-archive.pitt.edu/22344/

Khachatryan, V., Sirunyan, A. M., Tumasyan, A., Adam, W., Bergauer, T., Dragicevic, M., Erö, J., Fabjan, C., Friedl, M., Fruehwirth, R., & Ghete, V. M. (2014). Observation of the diphoton decay of the Higgs boson and measurement of its properties. *The European Physical Journal C, 74*, 3076. https://doi.org/10.1140/epjc/s10052-014-3076-z

Kieval, P. H. (MS). Representation learning without representationalism: A non-representational account of deep learning models in scientific practice. In J. M. Durán & G. Pozzi (Eds.), *Philosophy of science for machine learning: Core issues and new perspectives*. Synthese Library. Springer.

Koningsveld, H. (1973). *Empirical laws, regularity and necessity*. H. Veenman & Zonen B.V.

Krenn, M., Pollice, R., Guo, S. Y., Aldeghi, M., Cervera-Lierta, A., Friederich, P., dos Passos Gomes, G., Häse, F., Jinich, A., Nigam, A., Yao, Z. (2022). On scientific understanding with artificial intelligence. *Nature Reviews Physics, 4*(12), 761–769.

Kuhn, T. (1970). *The structure of scientific revolutions* (2nd enlarged ed.). University of Chicago Press.

Lakatos, I. (1970). History of science and its rational reconstructions. In *Proceedings of the Biennial Meeting of the Philosophy of Science Association* (Vol. 1970, pp. 91–136).

Lakatos, I. (1976). Falsification and the methodology of scientific research programmes. In S. G. Harding (Ed.), *Can theories be refuted?* (pp. 205–259). D. Reidel Publishing Company.

López-Rubio, E. (2020). Throwing light on black boxes: Emergence of visual categories from deep learning. *Synthese*. https://doi.org/10.1007/s11229-020-02700-5

Maier, A., Köstler, H., Heisig, M., Krauss, P., & Yang, S. H. (2022). Known operator learning and hybrid machine learning in medical imaging—A review of the past, the present, and the future. *Progress in Biomedical Engineering, 4*(2), 022002.

Margolis, E., & Laurence, S. (2023). Concepts. In E. N. Zalta & U. Nodelman (Eds.), *The Stanford encyclopedia of philosophy* (Fall 2023 ed.). Metaphysics Research Lab, Stanford University.

Martens, N. C. M., & Lehmkuhl, D. (2020). Dark matter = modified gravity? Scrutinising the spacetime–matter distinction through the modified gravity/dark matter lens. *Studies in History and Philosophy of Modern Physics, 72*, 237–250.

Massimi, M. (2022). *Perspectival realism*. Oxford University Press.

Mattia, F. D., Galeone, P., Simoni, M. D., & Ghelfi, E. (2021). A survey on GANs for anomaly detection. arXiv, cs.LG(1906.11632).

McCoy, C. D., & Massimi, M. (2018). Simplified models: A different perspective on models as mediators. *European Journal for Philosophy of Science, 8*(1), 99–123.

Meskhidze, H. (MS). Beyond classification and prediction: The promise of physics-informed machine learning in astronomy and cosmology. In G. Pozzi & J. M. Durán (Eds.), *Philosophy of science for machine learning*. Springer.

Morrison, M., & Morgan, M. S. (1999). Models as mediating instruments. In M. Morrison & M. S. Morgan (Eds.), *Models as mediators* (pp. 10–37). Cambridge University Press.

Páez, A. (MS). Axe the x in XAI: A plea for understandable AI. In J. M. Durán & G. Pozzi (Eds.), *Philosophy of science for machine learning: Core issues and new perspectives*. Synthese Library. Springer.

Passon, O. (2019). On the interpretation of Feynman diagrams, or, did the LHC experiments observe $h \rightarrow \gamma\gamma$? *European Journal for Philosophy of Science*, *9*(2), 20.

Piccinini, G. (2011). Two kinds of concept: Implicit and explicit. *Dialogue*, *50*(1), 179–193.

Quine, W. (1969). Natural kinds. In N. Rescher (Ed.), *Essays in honor of Carl G. Hempel* (pp. 5–23). Springer.

Radder, H. (2006). *The world observed/the world conceived*. University of Pittsburgh Press.

Räz, T. (2023). Methods for identifying emergent concepts in deep neural networks. *Patterns*, *4*(6), 100761.

Reichstein, M., Camps-Valls, G., Stevens, B., Jung, M., Denzler, J., Carvalhais, N., & Prabhat, F. (2019). Deep learning and process understanding for data-driven earth system science. *Nature*, *566*(7743), 195–204.

Rosaler, J., & Harlander, R. (2019). Naturalness, Wilsonian renormalization, and 'fundamental parameters' in quantum field theory. *Studies in History and Philosophy of Modern Physics*, *66*, 118–134.

Schwartz, M. (2014). *Quantum field theory and the standard model*. Cambridge University Press.

Schwartz, M. D. (2021). Modern machine learning and particle physics. Preprint. arXiv:2103.12226.

Shalev-Shwartz, S., & Ben-David, S. (2014). *Understanding machine learning: From theory to algorithms*. Cambridge University Press.

Singh, C., Morris, J. X., Aneja, J., Rush, A. M., & Gao, J. (2022). Explaining patterns in data with language models via interpretable autoprompting. Preprint. arXiv:2210.01848.

Sterkenburg, T. F., & Grünwald, P. D. (2021). The no-free-lunch theorems of supervised learning. *Synthese*, *199*(3–4), 9979–10015.

Sullivan, E., & Kasirzadeh, A. (MS). Explanation hacking: The perils of algorithmic recourse. In J. M. Durán & G. Pozzi (Eds.), *Philosophy of science for machine learning: Core issues and new perspectives*. Synthese Library. Springer.

Szegedy, C., Zaremba, W., Sutskever, I., Bruna, J., Erhan, D., Goodfellow, I., & Fergus, R. (2013). Intriguing properties of neural networks. Preprint. arXiv:1312.6199.

Wolpert, D. H. (1996). The lack of a priori distinctions between learning algorithms. *Neural Computation*, *8*(7), 1341–1390.

Zhang, H., Chen, H., Song, Z., Boning, D., Dhillon, I. S., & Hsieh, C.-J. (2019). The limitations of adversarial training and the blind-spot attack. Preprint. arXiv:1901.04684.

Florian J. Boge is Junior-Professor for Philosophy of Science with a focus on Artificial Intelligence at TU Dortmund University and leader of the DFG-funded Emmy Noether research group *UDNN: Scientific Understanding and Deep Neural Networks*. He holds a PhD from the University of Cologne, for a thesis on the interpretation of quantum physics. Boge is associate editor for the *European Journal for Philosophy of Science* and was a postdoc in the subproject *The Impact of Computer Simulations and Machine Learning on the Epistemic Status of LHC Data* of the interdisciplinary DFG research unit *The Epistemology of the Large Hadron Collider*.

Henk W. de Regt is Professor of Philosophy of Natural Sciences at the Institute for Science in Society, Radboud University, The Netherlands. He holds an MSc in foundations of physics (Utrecht University) and a PhD in philosophy (Vrije Universiteit Amsterdam). His research centers on scientific understanding and explanation, which he has recently extended to questions regarding the role of AI in science. His monograph *Understanding Scientific Understanding* (Oxford University Press, 2017) won the 2019 Lakatos Award. De Regt co-founded the *Society for Philosophy of Science in Practice* (SPSP) and the *European Society for the Philosophy of Science* (EPSA).

Chapter 20
Don't Fear the Bogeyman: On Why There Is No Prediction-Understanding Trade-Off for Deep Learning in Neuroscience

Barnaby Crook and Lena Kästner

Abstract Machine learning models, particularly deep artificial neural networks (ANNs), are becoming increasingly influential in modern neuroscience, often being employed as models of cognitive and perceptual systems. However, these models are often complex and opaque, leading some to worry that, by utilizing ANNs, neuroscientists are trading one black box for another. On this view, despite increased predictive power, ANNs effectively hinder our scientific understanding of the brain. We think these worries are unfounded. While ANNs are difficult to understand, there is no fundamental trade-off between the predictive success of a model and how much understanding it can confer. Thus, utilizing complex computational models in neuroscience will not generally inhibit our ability to understand the (human) brain. Rather, we believe, deep learning is best conceived as offering a novel and unique epistemic perspective for neuroscience. As such, it affords insights into the operation of complex systems that are otherwise unavailable. Integrating these insights with those generated by traditional neuroscience methodologies bears the potential to propel the field forward.

Keywords Deep learning · Neuroscience · Machine learning · Trade-off · Accuracy · Prediction · Scientific understanding · Explainability · Epistemic perspective

20.1 Introduction

Machine learning in general, and deep artificial neural networks (ANNs) in particular, have become an increasingly influential methodological tool in the arsenal of modern computational neuroscience. Because of their flexibility, ANNs are used

B. Crook (✉) · L. Kästner
University of Bayreuth, Bayreuth, Germany
e-mail: barnaby.crook@uni-bayreuth.de; lena.kaestner@uni-bayreuth.de

J. M. Durán, G. Pozzi (eds.), *Philosophy of Science for Machine Learning*,
Synthese Library 527, https://doi.org/10.1007/978-3-032-03083-2_20

in multiple distinct ways, from the identification of predictive variables to serving as models of cognitive and perceptual faculties (Glaser et al., 2019). As a result of their expressive power, ANNs have achieved unmatched success when it comes to task performance, prediction of neural activity, and replication of human-like behavioral patterns (Cao & Yamins, 2021; Kanwisher et al., 2023; Schrimpf et al., 2020). However, despite their undeniable predictive power, deep learning-based approaches to investigating the brain are not without limitations. One such limitation is that systems designed through deep learning are often highly complex and opaque to researchers. Because their functional structure is generated through an automated learning procedure, precisely *how* deep learning-based systems solve a given set of computational problems often remains unclear, even to their designers (Creel, 2020; Zednik, 2021; Samek & Müller, 2019). This applies not only to the use of deep learning in neuroscience but to deep learning systems at large. Many AI researchers worry that there is a "tension between machine learning performance (predictive accuracy) and explainability" (Gunning & Aha, 2019, p. 45; see Crook et al., 2023 for discussion). This general worry is also echoed with respect to neuroscience research specifically. For instance, Lindsay argues that "to have a model complex enough to perform real-world tasks, we must sacrifice the desire to make simple statements about how each stage of it works" (2021, p. 2024). Building on such worries, Chirimuuta portrays the situation as one where computational neuroscience faces "a trade-off between predictive accuracy and the ability of [its] models to confer understanding" (2021, p. 787).

While we agree with Lindsay's statement, we believe that the fear expressed in Chirimuuta's is misguided. There might indeed be a trade-off between the predictive success of individual models and the *ease* of understanding them; but this does not imply that utilizing complex computational models in neuroscience will generally inhibit our ability to understand the (human) brain. On the contrary: We firmly believe that rather than worrying about sacrificing understanding by utilizing deep learning in neuroscience, we should welcome ANNs into the neuroscientists' toolkit.[1] They bear the potential to generate new insights about the brain from a yet unavailable *epistemic perspective* (see Kästner, 2018) that is complementary to the perspectives provided by established neuroscientific methodologies. That is, deep learning contributes to, and does not detract from understanding neuroscientific phenomena.

We shall begin our exposition by diagnosing the source of trade-off worries in a misconception of scientific understanding (Sect. 20.2). The reasoning behind the notion that researchers are paying a high price for predictive success crucially depends on weighing models' predictive power against their intelligibility. While it is widely agreed that the aim of science is to understand *phenomena*,[2] the argument

[1] For ease of reading, we shall use the terms "deep learning models" and "ANNs" synonymously.

[2] There is a rich debate in contemporary philosophy of science about what precisely phenomena are and how to best characterize them (e.g., Colaco, 2019; Craver & Darden, 2013; Feest, 2017; Shagrir & Bechtel, 2017). Here we simply use the term "phenomenon" to refer to any explanandum

for a trade-off depends on equating this understanding with the intelligibility of particular models of those phenomena (see, e.g., Bokulich, 2017; Elgin, 2017).[3] However, we think this is mistaken in two ways. First, while we agree that models play a key role in the production of scientific understanding, we think such understanding depends upon integration of numerous distributed sources of explanatory information. Second, the intelligibility of a model is not a reliable guide for how much understanding it can ultimately confer. While interpreting complex models may be challenging, so long as it is feasible, even relatively unintelligible models can induce scientific understanding of phenomena.

Once we recognize this distinction between model understanding and *model-induced understanding of phenomena*, it becomes evident that the prediction-understanding trade-off is a bogeyman we need not fear: Deep learning can contribute to a progressive computational neuroscience that yields both improved predictive performance *and* deeper scientific understanding (Sect. 20.3). This can be achieved by integrating insights from traditional neuroscientific research with those delivered by ANNs and systematic post-hoc interpretability strategies. Insights from these different epistemic perspectives will contribute to a store of distributed knowledge that neuroscientists can draw on to explain and understand phenomena.

Because debates about what it takes to explain and understand complex systems have a particularly rich history in neuroscience, we focus our discussion on this field. Still, we think our insights will apply not only to neuroscience; but are likely to generalize to relevantly similar domains where deep learning is being utilized to study complex systems.

20.2 Trade-Off Worries

In this section, we discuss why trade-off worries come about. We shall first review arguments to the effect that a trade-off exists between the predictive success of a model and how much understanding it can confer (Sect. 20.2.1). Against this backdrop, we shall explain why we think these arguments are flawed: because they mistake model understanding for model-induced understanding of phenomena (Sect. 20.2.2).

investigated by neuroscientists. Prototypical examples include such things as color vision, face recognition, or cortical language processing.

[3] We use the term *intelligible* to capture how easy it is to understand a given model.

20.2.1 Why Worry?

A common conception is that models inducing understanding should be simple and intelligible, abstracting away from or usefully idealizing the complexities of the real world (cf., Bokulich, 2017; Elgin, 2017; Potochnik, 2016). Against this backdrop, it seems natural to expect that scientific understanding is best conferred by simple transparent models that domain experts understand perfectly. In traditional scientific modelling, scientists' domain knowledge is encoded by the construction of a model of a target phenomenon (Craver, 2001; Frigg & Hartmann, 2020). These models can frequently be represented by intuitive box-and-arrow diagrams or simple equations experts can grasp just by eyeballing them. The components of these models are often taken to represent relevant features of target phenomena, though these may be individuated functionally rather than structurally (Bennett et al., 2019; Craver & Kaplan, 2020). The construction, evaluation, and manipulation of models that represent phenomena is often taken to be central to the production of scientific understanding (de Regt, 2009; Elgin, 2007; Morgan & Morrison, 1999).

Deep learning models are different on various scores.[4] First, because they are trained with large datasets and consist of many parameters, they can be exceedingly complex. As such, they break with the expectation of being simple. Second, and relatedly, unlike with traditional modelling, domain knowledge typically only constrains choices about the initial *architecture* and training data of deep learning-based models (Goodfellow et al., 2016). The architecture consists of the number of layers, the number and types of neurons in each layer, connectivity profiles of each neuron, and learning rules of the network (Chollet, 2021; Richards et al., 2019). While these features are specified by researchers, the rest of what a trained ANN model encodes (e.g., what representations the system learns and how its functional units interact to solve the computational problem) is learned through an automated training procedure, typically involving backpropagation and some form of gradient descent (LeCun et al., 2015). On the one hand, this autonomy from human design is the central strength of deep learning. It is what gives ANNs the flexibility to learn subtle, task-relevant patterns that elude traditional modelling and thus to expand the range of domains in which models can achieve predictive success (Boge, 2022; Jumper et al., 2021; Zhuang et al., 2021). On the other hand, especially when ANNs have many parameters, it is often unclear both i) how the internal structure of a trained model implements the mapping from input to output (see Chirimuuta, 2021; Creel, 2020), and ii) what relation this internal

[4] The word 'model' has multiple meanings. In statistics, any deep learning system is called a model, regardless of the purpose for which it is constructed. However, in the philosophy of science, a model is any structure, mathematical, physical, or graphical, that is constructed to represent (some aspect of) a target system. In neuroscience, statistical models based on deep learning are used both as tools (e.g., for data pre-processing; cf. Glaser et al., 2019) *and* as models that represent target systems (Cichy & Kaiser, 2019). In what follows, we focus on the use of ANNs as *scientific* models (i.e., models constructed to represent).

structure bears to the target phenomenon (this is what Sullivan, 2022 calls 'link uncertainty').[5] As such, deep learning models can be predictively successful without necessarily illuminating how exactly the target phenomenon is produced; that is, without conferring understanding (See Páez, 2025; Sullivan & Kasirzadeh, 2025).

As a result of these unique properties, several authors writing about the increasing prevalence of deep learning in science have worried that the advantages we accrue in terms of predictive accuracy are offset by a decrease in explanatory power and understanding (Boon, 2020; Chirimuuta, 2021; Lindsay, 2021; López-Rubio & Ratti, 2021; Srećković et al., 2022). For example, Srećković and colleagues suggest that increased use of deep learning models may "lead us away from the explanatory aspect of science" (2022, p. 172). Along similar lines, Boon imagines a future in which "scientific researchers and scientific knowledge become superfluous as learning from large data sets, algorithms and data-models will be developed at a degree of complexity and adequacy far beyond the capacity of the human intellect" (2020, pp. 49–50). Finally, López-Rubio and Ratti state that, "when predictive performances increase, the possibility of elaborating a (mechanistic) explanation necessarily decreases" (2021, p. 1332). While the exact claims and arguments of these authors differ slightly, the overall perspective they present is clear: we must choose between increasing the predictive adequacy of our models and gaining understanding from them; we cannot have both.

Worries about this hypothesized trade-off between predictive success and understanding have touched many areas of science, from protein-folding (Bouatta et al., 2021), to physics (Boge, 2022; see also Boge & de Regt, 2025), to molecular biology (López-Rubio & Ratti, 2021). In what follows we concentrate on neuroscience, the discipline with which we are most familiar. Though we spell out our response with this specific focus, we suggest that the broad shape of our argument is likely to carry over to relevantly similar scientific domains, even if the details may differ.

Perhaps due to the intertwined history of neuroscience and AI research (Hassabis et al., 2017; McCulloch & Pitts, 1943), deep learning has had a particularly large and rapid influence on the study of the brain and nervous system (Cichy & Kaiser, 2019; Marblestone et al., 2016; Richards et al., 2019). As well as being used to identify predictive features and solve engineering problems (Glaser et al., 2019), deep learning has been employed in computational neuroscience to produce ANN models of cognitive and perceptual systems (Doerig et al., 2023; Kell et al., 2018; Schrimpf et al., 2020; Yamins & DiCarlo, 2016). It is in this capacity that deep learning's contributions to neuroscience have prompted worries of a trade-off. While ANNs have proven successful at implementing solutions to ethologically relevant tasks, predicting neural activity, and recapitulating human-like error profiles, critics have raised concerns that we are simply replacing one black box with another (Charles Leek et al., 2022; Chirimuuta, 2021; Thompson, 2021). This argument

[5] There is a vast body of literature on opacity. For a classic see Burrell (2016), for a recent comprehensive taxonomy see Mann et al. (2023), for further discussion see Beisbart (2025) and Formanek (2025).

has been made most explicitly by Mazviita Chirimuuta. We shall briefly reiterate her argument before explaining why we do not agree with its conclusion.

Chirimuuta directs her argument at models "intended to represent the functions computed by neural systems" (2021, p. 772).[6] As she details, traditional models, such as Georgopoulos and colleagues' (1986) *Population Vector Algorithm* for estimating arm movements from primary motor cortex activity, are fairly simple. Although such models knowingly encode false assumptions, in Chirimuuta's estimation they end up being "highly intelligible, representationally inaccurate but surprisingly useful" (2021, p. 774). At the other end of the spectrum, models based on deep learning, such as Sussillo et al.' (2012) recurrent neural network for decoding arm kinematics from neural activity, are complicated beasts. As Chirimuuta puts it, such models are "non-linear, opaque but predictively accurate" (2021, p. 776). This contrast recapitulates the distinction between traditional modelling strategies and deep learning that we have already discussed; and it leads directly to the suggestion that "there is a trade-off between a model's predictive power and its ability to increase the scientist's understanding of a neural response" (Chirimuuta, 2021, p. 768). However, we think this is too quick. To reach this conclusion, Chirimuuta uses the notion of 'model intelligibility' to capture not just the idea that models are easy to understand but *also* their overall ability to confer understanding. It is the conflation of these two notions that we take issue with.[7]

At this point, it is important to introduce a subtle but important distinction concerning the source of the unintelligibility or opacity of ANNs.[8] On the one hand, ANNs are trained via automated learning procedures. This is what makes it possible for researchers to develop models that solve certain tasks without understanding exactly how they do so. On the other hand, ANNs are typically very complex, consisting of many tunable parameters which interact in subtle ways to produce the behavior of the whole system. While these two sources of opacity are intimately related in the methodology of deep learning, they are conceptually distinct. The first problem is that an explanation is not *given* (i.e., automatically conferred through construction of the model), the second is that an explanation is not *simple*. In principle, the first source of opacity might be overcome through systematic post-hoc analysis and interpretation of a trained ANN (as we will discuss in Sect. 20.3).

---

[6] Chirimuuta argues for a principled distinction between *efficient coding* models which aim to capture the encoding functions used by the brain and *mechanistic* models which (also) encode anatomical isomorphisms, intending her argument to apply only to the former. However, in our view, the properties of ANNs that Chirimuuta invokes to make her case creates a structurally analogous problem for both kinds of model. As such, we take our response to address a generalised version of Chirimuuta's argument. This argument covers a broad class of ANN models in neuroscience, including those with mechanistic explanatory goals.

[7] We wish to stress that this is not a semantic dispute about how *intelligibility* should be defined, but a conceptual dispute about whether 'ease of understanding' and 'ability to confer understanding' ought to be equated with one another.

[8] We take opacity to mean "barrier to conferring understanding". Notice that opacity is at least partially a subjective phenomenon. What is opaque to a layperson may not be opaque to an expert (Humphreys, 2009).

Assuming we might reverse engineer the functional structure of ANNs, however, we are still left with the second source of opacity, viz. the structure we uncover might be quite complex. Chirimuuta states clearly that in her view this second source is sufficient to ensure the trade-off obtains:

> [...] even if we could write down by hand the equations embedded in the trained ANN's [...], those models would still be far less intelligible than their low-tech predecessors, because they would be nonlinear and contain very many more terms than the ones occurring in the traditional models. (2021, p. 782)

So, even if an explanation was given, it would not necessarily alleviate opacity because it might lack the simplicity required for humans to understand what a system is doing. In other words, even sophisticated interpretability methods may be powerless to overcome the trade-off between the predictive success of deep learning models and their potential to confer understanding. With these arguments in view, let us turn to our own assessment of the trade-off story.

20.2.2 Why Not to Worry!

We grant that there might be a trade-off between predictive success and ease of understanding for individual models. However, we take issue with the notion that this implies neuroscientists necessarily sacrifice understanding by utilizing ANNs in their research. This is because what they are after, we think, is not intuitive understanding of models of complex systems but understanding the *phenomena* that these systems exhibit with the help of models. Thus, there is no requirement for models to be simple. What matters is that they confer *understanding of the target phenomenon*, even if that understanding requires significant interpretative work. Indeed, as we will show below, there is nothing in principle stopping complex models from conferring greater understanding than simple ones.

Before we continue, a note on what we take (scientific) understanding to be is in order. Generally speaking, we are sympathetic to Elgin's suggestion that "understanding is a grasp of a comprehensive body of information that is grounded in fact, is duly responsive to evidence, and enables non-trivial inference, argument, and perhaps action regarding that subject the information pertains to" (2007, p. 39). Put simply, we take understanding to be having some kind of *qualitative grasp* of a target phenomenon (de Regt, 2009; Elgin, 2007; Strevens, 2013), some kind of *insight* into how a phenomenon of interest is exhibited by a system. It is a widely acknowledged feature of grasping that it is gradable, i.e., it comes in degrees (Elgin, 2007; Hills, 2016). With this in mind, we can say a subject S grasps a phenomenon P to the extent that they are able to (i) make coarse-grained predictions about P without performing exact calculations (cf., de Regt, 2009), (ii) give rich, detailed, and productive answers to a broad range of questions concerning P (cf., Chirimuuta, 2021), and (iii) intervene on and control P (or at least to describe such interventions, even if they would be impractical) (cf., de Regt, 2009). To serve these purposes, S might utilize some kind of (mental) representation of P where this representation might be considered an explanation for P or, alternatively, enable S to explain P.[9] Note that (i)–(iii) are also features commonly associated with good scientific explanations (see e.g., Craver, 2007; Pearl, 2009; Salmon, 1984; Woodward, 2003). This tight connection between understanding and explanation is no accident. If Alice has a better grasp of P than Bob (in the sense of (i)–(iii)), then she is likely to not only have a better understanding of P but to also be able to *explain* it better.

Admittedly, this characterization of understanding in terms of grasping is fairly minimal. What we say about the relation between (i), (ii), and (iii), about the role of (mental) representations, and about the precise relation between understanding and explanation remains intentionally sketchy. But since the argument we make in what follows is compatible with a range of more articulated accounts of understanding (see de Regt, 2009; Elgin, 2007; Hills, 2016; Strevens, 2013 for various detailed views; see Grimm, 2021 for an overview), we think this is a merit rather than a shortcoming. What is important for the current discussion is merely that scientific understanding is conferred by the very kind of information we utilize in scientific explanations. And that the targets of understanding and explanation in science are *phenomena* (e.g., Bechtel & Richardson, 2010; Colaco, 2019; Craver, 2007; Feest, 2017; Glennan, 2017; Hochstein, 2016; Kästner, 2018).

Let us now turn to our assessment of the trade-off arguments presented above (Sect. 20.2.1). In those arguments, a trade-off between model complexity and intelligibility is taken to imply a trade-off between model complexity and the potential of a model to confer understanding of phenomena. But understanding models is not the same as understanding phenomena. And the ease of understanding a model is not the same as how much understanding that model can confer. Granted, all of the tell-tale signs of grasping just described can be straightforwardly applied to models and their behavior (i.e., by casting them as target phenomena). However, while models play a crucial instrumental role in our understanding of phenomena, they are not typically the ultimate *targets* of inquiry in neuroscience or any other natural science; models are usually *mediators* (cf., Morgan & Morrison, 1999), not explananda. Thus, we must take care not to mistake model understanding for phenomenon understanding.[10]

The issue we are getting at is nicely captured by Lawler and Sullivan's (2021) distinction between *model explanation* and *model-induced explanation*.[11] In model

[9] Given grasping is a gradable notion, (i)–(iii) can be performed with various degrees of correctness or success.

[10] This is not to say model understanding cannot contribute to understanding phenomena. Indeed, model understanding plays a crucial role. However, the relationship between understanding models and understanding phenomena is complex. We will elaborate on this in Sect. 20.3.2.

[11] Note that Lawler and Sullivan are talking about explanation rather than understanding. As we stated above, we see an intimate link between the two. Depending on how exactly the relation between understanding and explanation is construed, the distinction might even collapse (e.g., when understanding is conceptualised as the possession of a mental representation of the right kind of explanation) (see Strevens, 2013 for an example of such a view).

explanation, the content of the model simply is the explanation. In model-induced explanation by contrast, "the relevant explanatory information is independent of the model but only closely intertwined with the model due to the history of obtaining the information" (Lawler & Sullivan, 2021, p. 1069). Talking about understanding we can reframe the point thusly: Whenever our understanding of a phenomenon includes information extrinsic to any particular model, model-induced understanding of a phenomenon is not reducible to model understanding. We think this is actually very commonplace in scientific practice (cf., Hochstein, 2016; Kästner, 2018; Mitchell, 2002, 2019). Thus, reducing understanding of phenomena to what is encoded in a model is both descriptively and normatively inadequate. It is descriptively inadequate because real-world scientific understanding of complex phenomena frequently depends upon facts and details extrinsic to any individual model. It is normatively inadequate because focusing on understanding models—rather than model-induced understanding—artificially prevents us from satisfying scientific desiderata that might otherwise be satisfied (i.e., increasing both predictive accuracy *and* understanding of phenomena).

With these distinctions in mind let us return to Chirimuuta's argument. Recall her claim that "models of neural systems are either very intelligible, or predictively accurate, but not both" (2020, p. 781). In a certain sense, we agree with this claim. It is obviously *easier* to understand simple models than it is to understand complex models (perhaps tautologically so). However, that does not imply that we obtain a greater understanding of phenomena through constructing and using simple models than we do through constructing, using, and interpreting complex ones. In fact, we believe complex models will often confer higher degrees of understanding of phenomena in the sense described above: They enable us to (i) make better predictions, (ii) answer more questions about a phenomenon, and (iii) allow for more intervention and control.

But what about the opacity of ANNs—does it not interfere with (i)–(iii)? To address this question, recall our distinction between whether explanatory information encoded in an ANN is *given* and whether it is *simple* (Sect. 20.2.1). Suppose that the *givenness* issue is overcome and we obtain an equation describing the function the model has learned, e.g., through interpretability methods (we will address the feasibility of this in Sect. 20.3.3). Chirimuuta argues that, even so, "eyeballing an equation of such complexity would not give the neuroscientist the same qualitative sense of how adjustment of parameters or variables would make a difference to the behavior of the system [as traditional, simpler models]" (2021, p. 782). Here, we think Chirimuuta conflates two things: the ease of coming to understand something is not the same as the degree of understanding that one might obtain from it. To make the point clear, recall Lindsay's claim that the operation of models of sufficient complexity to perform real-world tasks cannot be captured with simple statements (2021, p. 2024). Lindsay is probably right. However, we do not think that understanding of phenomena needs to consist in being able to make *simple* statements. Deep ANNs might be challenging to understand. It might be laborious to gain insights about how they work, those insights might not be simple, and establishing their relationship to target phenomena may be difficult. But they might still enable a deeper scientific understanding than highly intuitive but overly simple equations or box-and-arrow models. How easy it is to understand a model does not strictly determine how instrumentally valuable that model will be to understanding and explaining a phenomenon all things considered.

Importantly, partial understanding of a complex but accurate model can lead to greater understanding of a phenomenon than complete understanding of a simple but inaccurate model. To see this, let us focus on prediction for the moment (we will return to the other criteria for understanding in Sect. 20.3). Consider the following toy example: a system S elicits a behavior P (the phenomenon of interest). S consists of 5 interacting factors (f1, ..., f5) and P is described by a single scalar value y that can be obtained by simple mathematical operations performed on values taken by f1, ..., f5. Bob constructs a simple model to help him predict P. Bob's model consists of two predictors, x1 and x2, which could (but need not) approximate f1 and f2, respectively. To make predictions about P, Bob calculates $x1 + x2$ in his head, which he can do flawlessly. As such, we can say that Bob understands his model perfectly; but (due to the simplicity of his model) his predictions about the phenomenon are very inaccurate. Meanwhile, Alice constructs a complex model to help her understand P. Alice's model consists of five predictors $x'1, \ldots, x'5$ which, with accurate coefficients, precisely capture the factors of the target system: $0.83{*}x'1 + 0.96{*}x'2 + 0.76{*}x'3 + 0.22{*}x'4 + 0.01{*}x'5$. To make predictions about P without running the model (as demanded by the *qualitative* prediction criterion on understanding), Alice estimates $0.8{*}x'1 + x'2 + 0.8{*}x'3 + 0.2{*}x'4$ in her head. That is, she calculates using imprecise multiplier values and omits $x'5$ completely.[12] Thus, we might say her grasp of the model is partial or incomplete, even though the model itself is perfect. Still, Alice's predictions about P are reasonably accurate. So, despite Alice's model understanding being partial where Bob's is perfect, Alice's predictions about P are still better than Bob's. As such, all else being equal, we attribute greater *model-induced understanding* of P to Alice than to Bob.

This toy example makes it plain that there is an important conceptual distinction between model understanding and model-induced understanding. Even if we can understand a simple model perfectly, we may gain less scientific understanding about a phenomenon P than we can from a complex model of P we only partially understand. Whether we *actually* gain a deeper understanding of P by utilizing complex models remains an empirical question. Still, if our diagnosis is correct there is no *principled* trade-off between predictive accuracy and the potential for models to confer scientific understanding. There is no reason to fear the bogeyman! Utilizing deep learning in neuroscience need not systematically inhibit our understanding of the brain. Rather, we think, it can support our understanding

[12] We should think of Alice omitting $x'5$ both because it becomes too hard for her to perform a fifth multiplication in her head and because she recognizes it will not have a huge effect on the system's behavior. This detail about $x'5$ also mirrors cases where parts of the internal structure of ANNs remain opaque.

of neuroscientific phenomena in previously unavailable ways. This takes us to the next section.

20.3 Understanding the Brain with Deep Learning

So far, we suggested that scientific understanding consists in having a qualitative grasp of target phenomena and that this involves being able to predict, control, and answer questions about the phenomena in question. Further, we claimed that such understanding is often model-induced. That is, scientists come to understand phenomena through the construction, use, and analysis of models; this applies to simple equations or box-and-arrow-style models as well as to contemporary deep learning-based models. In this section, we elaborate on how ANNs may contribute to an improved understanding of neuroscientific phenomena. First, we argue that deep learning-based models offer a novel *epistemic perspective* which may render accessible previously inaccessible aspects of neuroscientific phenomena (Sect. 20.3.1). Second, we argue that understanding complex phenomena typically requires integrating *distributed explanatory information* produced from multiple epistemic perspectives (Sect. 20.3.2). With this conceptual machinery in place, we sketch a specific research strategy by which ANN-driven neuroscience may contribute to researchers' understanding of biological brains (Sect. 20.3.3).

20.3.1 A Novel Epistemic Perspective

Epistemic perspectives are best described as ways in which researchers approach a target phenomenon—through the use of tools, skills, and theoretical assumptions—to yield explanatory information about how it is elicited by a system. Figuratively speaking, we may think of an epistemic perspective as the outlook or view a scientist takes on a phenomenon; it is constrained by their specific knowledge and capacities, as well as their research questions. More specifically, epistemic perspectives can be characterized with respect to the following dimensions: (i) spatiotemporal granularity, (ii) specificity, i.e., which kinds of entities and dependencies are detected and described, (iii) point of view, including ontological and theoretical assumptions, (iv) sensitivity to different factors, and (v) scope, i.e., class of phenomena which are investigated (Kästner, 2018, p. 74). Crucially, different epistemic perspectives need not be conceived as competing with each other. Although research from a particular perspective is sometimes sufficient for answering specific questions, deeper understanding of complex systems requires figuring out how insights from multiple epistemic perspectives fit together (Mitchell, 2002, 2019). Before we consider the question of how multiple perspectives can be integrated (Sect. 20.3.2), let us briefly sketch the core features of the epistemic perspective a deep learning approach to computational neuroscience provides. Needless to say, this will be unavoidably simplistic. In reality, practitioners constructing ANN models of neural functions may endorse various theoretical commitments. Still, it is valuable to see where the contrast lies to more biologically-grounded neuroscience research.

Spatiotemporal Granularity Traditional neuroscience methods are constrained with respect to both spatial and temporal resolution, though in various ways (Churchland & Sejnowski, 1988; Grinvald & Hildesheim, 2004). Similarly, the deep learning approach to computational neuroscience can be applied at numerous spatial scales, e.g., by targeting neural populations in specific anatomical regions (Sussillo et al., 2012) or distributed processing along pathways like the ventral visual stream (Yamins et al., 2014). What is novel, though, is that *the same model* can shed light on neural phenomena across a wide range of temporal scales. For example, one model can be taken to represent processes extending all the way from a single perceptual judgement (i.e., the inference of the trained model) to evolutionary learning (i.e., the learning trajectory the system undertakes during training) (Cao & Yamins, 2021; Zador, 2019).

Specificity Being based on automated learning procedures, the deep learning approach has the potential to pick up on patterns in data and dependencies between factors that a human researcher will not usually see or hypothesize. A case in point are functional groups of neurons in V4 we discuss below (Sect. 20.3.3). For a detailed discussion of how deep learning-based models can help uncover structures inaccessible from other epistemic perspectives see Kästner and Crook (2024).

Point of View The deep learning approach casts the brain as an optimisation machine, with specialised modules solving individual computational problems (Cao & Yamins, 2021; Marblestone et al., 2016). This can be contrasted with more biologically-grounded approaches to neuroscience which reject some of these theoretical assumptions (Buzsáki, 2006; Pessoa, 2023; Pulvermüller et al., 2021).

Sensitivity Unlike biologically-grounded neuroscience, which may treat any and all features of neuroanatomy as relevant to understanding the brain (including, e.g., glial cells or endocrine signals), the deep learning approach is primarily sensitive to features which are necessary for solving the computational task at hand. Further, because practitioners are interested in how *biological brains* solve these tasks, they especially value models which replicate characteristic error profiles in behavioral tasks (say, object recognition) and display similar patterns of neural activity (as measured by, e.g., fMRI) (Bowers et al., 2022; Schrimpf et al., 2020).

Scope The flexibility of the deep learning approach renders it amenable to shedding light on a broad range of cognitive and perceptual phenomena, including vision (Zhuang et al., 2021), audition (Kell et al., 2018), language comprehension (Schrimpf et al., 2020) working memory (Kozachkov et al., 2022), and many more besides (see Cichy & Kaiser, 2019; Doerig et al., 2023 for discussion). Unlike biologically-grounded neuroscience, this research is not constrained by ethical concerns or the availability of model organisms. Thus, the deep learning approach has a broader scope allowing practitioners to study, for instance, the "evolutionary goals and historical or developmental constraints that are responsible for shaping a system" (Cao & Yamins, 2021, p. 2; see also Lillicrap & Kording, 2019). These dependencies can be illuminated by focusing on the *designed* components of ANNs—objective functions, architecture, and learning algorithms—which do not suffer from problems of opacity (Richards et al., 2019; Thompson, 2021). In addition, we think deep learning can also help uncover the functional and computational structure of (biological as well as artificial) neural networks (Sect. 20.3.3).

In summary, our exposition makes it plain that the deep learning approach provides a novel epistemic perspective on biological neural processing which renders accessible information unavailable from other epistemic perspectives (i.e., those typically taken by more biologically-grounded neuroscientists). As such, we think it crucially complements existing research practices aiming to understand the brain. To unlock its full potential, however, insights from the deep learning approach must likely be combined with those gained from traditional modelling and biologically-grounded neuroscience.

20.3.2 Distributed Explanation, Integration, and Constraints

We already hinted that the relationship between understanding models and understanding phenomena is complex. One major reason this is so, we take it, is that *in practice* a subject S's understanding a phenomenon P will not usually rely on just a single model. According to Hochstein, "an individual model is rarely applied in isolation, and is often used to complement a huge body of background information and pre-existing models about the target system" (2016, p. 1401). Recall that for S to understand P involves the abilities to (i) make coarse-grained predictions about P, (ii) give rich, detailed, and productive answers to a broad range of questions concerning P, and (iii) intervene on and control P (see Sect. 20.2.2). To achieve all of this, it will usually be required for S to draw on a *distributed body of information* about P and the system eliciting it.

To clarify, it will be useful to briefly talk about distributed explanations—where the idea of a distributed body of information has been explored. Distributed explanations consist, at least partially, in sets of models, where each model provides different explanatory information about a given phenomenon by illuminating it from different epistemic perspectives (cf., Craver & Kaplan, 2020; Hochstein, 2016; Veit, 2020). If we are correct that understanding of phenomena is routinely supported by distributed explanations, philosophers limiting their analyses to particular models are making a mistake (Veit, 2020, p. 108).[13] Individual models only ever yield

[13] Naturally, focusing on particular models is only a mistake for cases in which the aims of the analysis *demand* a more holistic view. We think the debate over deep learning's contribution to scientific understanding is such a case.

partial understanding of complex phenomena. While we agree that sets of models are key constituents of distributed explanations, we see no reason to stop there. The *explanatory store* which facilitates scientific explanation and—as far as we are concerned—understanding should also admit data graphs, diagrams, explanatory writing in natural language, and any other representational vehicles that support understanding (Burnston, 2016; Kästner, 2018; Kohár & Krickel, 2021).[14] Thus, the items that support scientific understanding (viz. the elements of the explanatory store) may be of very different kinds and, just as models, may come in varying degrees of complexity.

Despite the looming messiness of the overall situation, we think it is important to acknowledge both the distributed character of the information which supports scientific understanding and the varying contributions that different elements in the explanatory store can make to it. First and foremost, this is because it accords with scientific practice. For instance, Craver and Kaplan observe that "many models are involved, explicitly or implicitly, in most explanations" and that we use various models to "gesture towards explanations that invariably have more content about the causal structure of the world than any single, useful model can express" (2020, pp. 309–310). Besides, once we acknowledge that the information that supports scientific understanding is distributed, we can circumvent undesirable trade-offs inherent to model building (Matthewson & Weisberg, 2009). There is no principled reason we cannot enjoy predictive success *and* improved understanding if we pool insights from multiple models in a common explanatory store.

Unfortunately, these benefits do not come for free. Reaping them requires at least a provisional account of *how* different items in the explanatory store (viz. information generated from multiple epistemic perspectives) can support greater understanding than mastery of particular models (cf., Chang, 2012, Chap. 5). The answer, we think, is that many items in the explanatory store not only contribute explanatory information but also *constraints* (Lawler & Sullivan, 2021; Sullivan, 2022). Based on these constraints, insights provided by different models (or other items) may be integrated with one another (see also Kästner, 2018).

Focusing on models specifically, Mitchell states that each model "provides a partial grasp of the phenomenon, and each requires input and ongoing engagement with the other perspectives" (2019, p. 179). In our view, the engagement Mitchell describes often consists of *constraining interpretation*. The interpretation of each source of explanatory information is constrained by the rest of the explanatory store (or, for cognitive tractability, a small but relevant subset). The exact way this plays out in any single case is subtle, depending crucially on the evidentiary support different sources of explanatory information enjoy, their perceived relevance to the phenomenon at hand, and what aspects of the phenomenon scientists are particularly interested in. For illustration, we briefly present two cases that highlight ways in which the deep learning approach can support scientific understanding

---

[14] The term *explanatory store* is originally due to Kitcher (1981) and is also used by Craver and Kaplan (2020).

of biological neural processing. Though the cases are different, they each involve contributing insights from a novel epistemic perspective and integrating them with other evidence.

Case 1: Vision Bowers et al.' (2022) make the case that the deep learning-based approach to vision neuroscience has paid inadequate attention to research documenting the idiosyncratic features of biological vision. The failure to adapt computational models to these constraints has resulted in the development of models which fail to replicate observed experimental effects in human vision. For instance, while humans show a clear preference for shape over texture, vision ANNs show the opposite pattern (Geirhos et al., 2018). As a result, Bowers and colleagues argue, we cannot (yet) utilize these models to gain insights into the algorithmic structure of biological vision. However, in light of evidence showing similar functional specialisation of single neurons in ANNs and biological visual systems (Pospisil et al., 2018; Willeke et al., 2023), we think a more nuanced conclusion should be drawn. Our understanding of human vision should provisionally integrate *some* properties of vision ANNs (e.g., the hierarchical composition of representations and the functional specialisation of units), while leaving others out (e.g., early layer neurons' overly strong preference for texture). Then, by systematically varying ANN architectures (e.g., by introducing recurrence, Kar et al., 2019) and training procedures (e.g., switching from supervised to unsupervised learning, Zhuang et al., 2021) and comparing the resulting models with diverse sources of neuroscientific data, researchers can clarify which features of ANNs are really (i) biologically plausible, and (ii) functionally relevant. This way, neuroscientists can refine predictions about how neural populations will respond to certain stimuli and target loci of control (Bashivan et al., 2019). This effectively means integrating research from the epistemic perspectives of traditional vision neuroscience and the deep learning approach to increase our understanding of biological vision.

Case 2: Working Memory Kozachkov et al. (2022) use recurrent neural networks (RNNs) to explore the hypothesis that short-term synaptic plasticity (STSP) plays a functional role in working memory (Mongillo et al., 2008). The authors trained RNNs both with and without an STSP mechanism to maintain items in working memory over a delay period. They then assessed them with respect to three criteria: (1) the robustness of their performance in the face of distractors, (2) graceful degradation to synaptic loss, (3) similarity to the neural dynamics of the prefrontal cortex. Kozachkov and colleagues found that both types of RNN were robust to distractors, but that those with the STSP mechanism degraded more gracefully and exhibited dynamics closer to those of the (primate) prefrontal cortex. In other words, the epistemic perspective researchers adopted by applying the deep learning approach allowed them to reveal previously unknown dependencies between architectural and functional properties of neural structures. Put more generally, we might say that the deep learning approach lent credence to particular computational and mechanistic hypotheses about human brain function (e.g., Miller et al., 2018; Mongillo et al., 2008; see also Doerig et al., 2023). Crucially, in this case, too, it is the integration of insights from different epistemic perspectives that helps researchers advance their understanding of complex phenomena such as working memory.

Thus far, we argued that deep learning-based models offer an additional epistemic perspective for brain researchers that can complement and constrain insights from traditional modelling or biologically-grounded neuroscience by contributing to a distributed store of knowledge. An argument to the same effect could have been made about other innovations in research, e.g. the invention of fMRI scans or the availability of even simple machine learning methods. Against this backdrop readers may wonder: is there anything really *special* about ANN-driven neuroscience?

20.3.3 Understanding Biological Brains Through Re-engineering

We already argued that we should not fear the bogeyman, but is there reason to welcome him with open arms? We think there is! By combining the strengths of deep learning with post-hoc interpretability techniques and biologically-grounded neuroscience, researchers can access the patterns embedded in trained ANNs and use them to gain insight into the functional and computational structure of neurobiological systems. Notably, the mechanistic description of biological systems this enables is precisely what critics fear might go missing when researchers utilize ML-based models to try and understand the brain (Chirimuuta, 2021; López-Rubio & Ratti, 2021; Thompson, 2021). We do not think that the deep learning approach sacrifices functional and computational structures; rather, it sheds a new light on them.

To see why this is and how it might work, recall that deep learning models are usually opaque. While the learned parameters of the models are accessible, understanding how they represent information and implement the mapping from input to output requires a dedicated research effort (i.e., this information is—unlike with traditional scientific models – not *given*, cf. Sect. 20.2.1). To uncover the functional structure and computational properties of trained ANNs, the application of systematic post-hoc interpretability techniques is needed (e.g., Bau et al., 2017; Cammarata et al., 2020; Geiger et al., 2021, 2022; Nanda et al., 2023; Olah et al., 2018). As of yet (and as far as we are aware), the research area that most vigorously employs post-hoc interpretability techniques is explainable AI (XAI), not neuroscience (for theoretical discussion of XAI see Buijsman, 2025; Räz, 2025). While we are certainly not the only ones to think that XAI presents opportunities for scientific discovery generally (see Zednik & Boelsen, 2022), we believe that applying post-hoc interpretability techniques to trained ANN models provides a unique opportunity for neuroscience by facilitating an *iterative re-engineering process*.

The process we have in mind involves the following key steps. First, train an ANN to implement an ethologically relevant task (as described in Yamins &

DiCarlo, 2016). Second, probe the behavioral characteristics of the trained ANN and evaluate the results with respect to explanatory information distributed across the explanatory store (as in Bowers et al., 2022).[15] Third, employ systematic post-hoc interpretability techniques to uncover the functional and computational structure of the trained ANN (as in Cammarata et al., 2020). Fourth, use these findings to develop hypotheses about the target biological system (Willeke et al., 2023). Fifth, use a variety of more traditional neuroscientific methods to test whether those hypotheses apply to biological target systems. Each step in the process may need to be refined and repeated multiple times in light of the constraints provided by the other steps.

The greatest source of uncertainty over the promise of the iterative re-engineering process is the plausibility of the third step, viz. the usefulness of post-hoc interpretability techniques (Cearns et al., 2019). However, recent work by a group around Chris Olah (Cammarata et al., 2020, 2021; Elhage et al., 2021; Olah et al., 2018) serves as a proof of principle that the kind of strategy we have in mind is applicable in practice. The researchers set out to characterise the functional structure of the image classification ANN *InceptionV1* (note that this ANN was not developed explicitly as a model of biological vision). They applied systematic post-hoc interpretability techniques to reverse engineer the opaque system. Focusing on curve detection specifically they found that:

> although curve detection involves more than 50,000 parameters, those parameters actually implement a simple algorithm that can be read off the weights and described in just a few English sentences. (Cammarata et al., 2021)

Indeed, the detailed understanding of the system's functional structure enabled the researchers to re-design (i.e., hand-code) this part of the system (i.e., a curve circuit) from scratch. Subsequently, they investigated—with research strategies familiar from biologically-grounded neuroscience—to what extent this re-engineered system actually displayed the same properties as the original system (for a detailed discussion see Kästner & Crook, 2024). This way, they validated their interpretable model of inceptionV1's curve detectors. Though this kind of research is still in its infancy, it clearly illustrates that re-engineering these ANNs through the application of post-hoc interpretability techniques *is possible*. As such, we think doing so on ANNs trained on ethologically relevant tasks might provide a fruitful way to uncover functional and structural features that can be converted into hypotheses for brain research.

Indeed, we know of one case in which this research has already inspired hypothesis generation. After synthesizing images (MEIs) to maximally excite V4 neurons in the macaque visual system, Willeke et al. (2023, p. 2) observed that units were arranged into functional groups selective for various visual features including curves, oriented textural patterns, and eye-like shapes. Referring to Olah's

[15] Obtaining promising results in this step may be a pre-requisite for engaging in the laborious and challenging process of reverse-engineering an ANN. As such, steps 1 and 2 already form a mini-loop which may be iterated over numerous times before step 3 is reached.

work, these researchers noted a "striking similarity between [the] single cell MEIs of V4 neurons and single units in the InceptionV1 architecture" (2023, p. 10).[16] Building on these similarities, Willeke et al. derived novel hypotheses about feature selectivity in the visual processing areas of the primate brain. More specifically, they suggest that ANNs can be used to "derive predictions about colour boundary encoding in monkey V4 functional groups, which could subsequently be verified in in vivo experiments" (2023, p. 7). These descriptions capture steps 3–5 of our iterative re-engineering process precisely.

Naturally, which precise models, tools and theories will be utilized most effectively throughout such a process obviously remains an empirical question. Indeed, it will likely vary from case to case. Still, we believe that we have shown that the iterative re-engineering research strategy has great potential. It may improve scientists' ability to make coarse qualitative predictions about brain behavior, answer questions about biological neural processing, and design effective interventions on the brain. That is, it provides an opportunity to foster our understanding of the brain in yet unimagined ways.

20.4 Conclusion

The aim of (neuro)science is to foster understanding of phenomena in the world. Machine learning approaches, particularly deep learning-based ANNs, can contribute to this endeavour by offering a novel and unique epistemic perspective affording insights into the operation of complex systems that are otherwise unavailable. While some have been worrying that this strategy implies trading one black box for another and might actually hinder scientific understanding, we argued these worries are unfounded. There is no trade-off between the predictive success of a model and how much understanding it can confer. Thus, utilizing complex computational models in neuroscience will not generally inhibit our ability to understand the (human) brain. Quite to the contrary, we believe, deep learning is best conceived as complementary to established neuroscientific methodology. It offers a powerful addition to neuroscientists' toolkit enabling constraint-based integration of numerous complementary epistemic perspectives. We thus conclude that deep learning contributes to—rather than detracts from—understanding the phenomena neuroscientists investigate. While our argument has focused on the domain of neuroscience specifically, we suggest that it is likely to carry over to relevantly similar scientific domains, even if the details may differ.

Acknowledgements Work on this chapter has been supported by the project "Explainable Intelligent Systems (EIS)" founded by the Volkswagen Foundation (Az 9B830). The authors wish

[16] Note that Willeke and colleagues performed psychophysics experiments and quantitative analyses to verify the robustness of their qualitative similarity judgements.

to thank all project members for their discussion of an early version of this chapter. We are especially indebted to Astrid Schomäcker, Sara Mann and Andreas-Sesing-Wagenpfeil for ample constructive feedback. We also thank Gregory Lusk for reviewing the document and pressing us to be clearer on several issues.

References

Bashivan, P., Kar, K., & DiCarlo, J. J. (2019). Neural population control via deep image synthesis. *Science (New York, NY), 364*(6439), eaav9436. https://doi.org/10.1126/science.aav9436

Bau, D., Zhou, B., Khosla, A., Oliva, A., & Torralba, A. (2017). *Network dissection: Quantifying interpretability of deep visual representations*, 3319–3327. https://doi.org/10.1109/CVPR.2017.354.

Bechtel, W., & Richardson, R. (2010). *Discovering complexity: Decomposition and localization as strategies in scientific research*. https://doi.org/10.7551/mitpress/8328.001.0001

Beisbart, C. (2025). In which ways is machine learning opaque? In Durán, J. M., & Pozzi, G. (Eds.), *Philosophy of science for machine learning: Core issues and new perspectives*. Synthese Library. Springer.

Bennett, D., Silverstein, S. M., & Niv, Y. (2019). The two cultures of computational psychiatry. *JAMA Psychiatry, 76*(6), 563–564. https://doi.org/10.1001/jamapsychiatry.2019.0231

Boge, F. J. (2022). Two dimensions of opacity and the deep learning predicament. *Minds and Machines, 32*(1), 43–75. https://doi.org/10.1007/s11023-021-09569-4

Boge, F. and de Regt, H. W. (2025). Machine learning discoveries and scientific understanding in particle physics: Problems and prospects. In Durán, J. M., & Pozzi, G. (Eds.), *Philosophy of science for machine learning: Core issues and new perspectives*. Synthese Library. Springer.

Bokulich, A. (2017). Models and explanation. In L. Magnani & T. Bertolotti (Eds.), *Springer handbook of model-based science* (pp. 103–118). Springer. https://doi.org/10.1007/978-3-319-30526-4_4

Boon, M. (2020). How scientists are brought back into science—The error of empiricism. *A Critical Reflection on Automated Science: Will Science Remain Human?*, pp. 43–65. https://doi.org/10.1007/978-3-030-25001-0_4

Bouatta, N., Sorger, P., & AlQuraishi, M. (2021). Protein structure prediction by AlphaFold2: Are attention and symmetries all you need? *Acta Crystallographica. Section D, Structural Biology, 77*(Pt 8), 982–991. https://doi.org/10.1107/S2059798321007531

Bowers, J. S., Malhotra, G., Dujmović, M., Montero, M. L., Tsvetkov, C., Biscione, V., Puebla, G., Adolfi, F., Hummel, J. E., Heaton, R. F., Evans, B. D., Mitchell, J., & Blything, R. (2022). Deep problems with neural network models of human vision. *The Behavioral and Brain Sciences*, 1–74. https://doi.org/10.1017/S0140525X22002813

Buijsman, J. (2025). Machine learning models as mathematics. In: Durán, J. M., & Pozzi, G. (Eds.), *Philosophy of science for machine learning: Core issues and new perspectives*. Synthese Library. Springer.

Burnston, D. C. (2016). Data graphs and mechanistic explanation. *Studies in History and Philosophy of Science Part C: Studies in History and Philosophy of Biological and Biomedical Sciences, 57*, 1–12. https://doi.org/10.1016/j.shpsc.2016.01.002

Burrell, J. (2016). How the machine 'thinks': Understanding opacity in machine learning algorithms. *Big Data & Society, 3*(1). https://doi.org/10.1177/2053951715622512

Buzsáki, G. (2006). *Rhythms of the brain*. Oxford University Press. https://doi.org/10.1093/acprof:oso/9780195301069.001.0001

Cammarata, N., Carter, S., Goh, G., Olah, C., Petrov, M., & Schubert, L. (2020). Thread: Circuits. *Distill, 5*(3). https://doi.org/10.23915/distill.00024

Cammarata, N., Goh, G., Carter, S., Voss, C., Schubert, L., & Olah, C. (2021). Curve circuits. *Distill, 6*(1), e00024.006. https://doi.org/10.23915/distill.00024.006

Cao, R., & Yamins, D. (2021). Explanatory models in neuroscience: Part 2 – Constraint-based intelligibility. *ArXiv:2104.01489 [Cs, q-Bio]*. http://arxiv.org/abs/2104.01489

Cearns, M., Hahn, T., & Baune, B. T. (2019). Recommendations and future directions for supervised machine learning in psychiatry. *Translational Psychiatry, 9*(1), 1. https://doi.org/10.1038/s41398-019-0607-2

Chang, H. (2012). *Is Water H_2O? Evidence, realism and pluralism*. Boston Studies in the Philosophy and History of Science.

Charles Leek, E., Leonardis, A., & Heinke, D. (2022). Deep neural networks and image classification in biological vision. *Vision Research, 197*, 108058. https://doi.org/10.1016/j.visres.2022.108058

Chirimuuta, M. (2021). Prediction versus understanding in computationally enhanced neuroscience. *Synthese, 199*(1), 767–790. https://doi.org/10.1007/s11229-020-02713-0

Chollet, F. (2021). *Deep learning with python* (2nd ed.). Simon and Schuster.

Churchland, P. S., & Sejnowski, T. J. (1988). Perspectives on cognitive neuroscience. *Science (New York, NY), 242*(4879), 741–745. https://doi.org/10.1126/science.3055294

Cichy, R. M., & Kaiser, D. (2019). Deep neural networks as scientific models. *Trends in Cognitive Sciences, 23*(4), 305–317. https://doi.org/10.1016/j.tics.2019.01.009

Colaco, D. (2019). *An investigation of scientific phenomena* [PhD Thesis]. University of Pittsburgh.

Craver, C. F. (2001). Role functions, mechanisms, and hierarchy. *Philosophy of Science, 68*(1), 53–74.

Craver, C. F. (2007). *Explaining the brain: Mechanisms and the mosaic unity of neuroscience*. Oxford University Press.

Craver, C. F., & Darden, L. (2013). *In search of mechanisms: Discoveries across the life sciences*. University of Chicago Press. https://press.uchicago.edu/ucp/books/book/chicago/I/bo16123713.html

Craver, C. F., & Kaplan, D. M. (2020). Are more details better? On the norms of completeness for mechanistic explanations. *The British Journal for the Philosophy of Science, 71*(1), 287–319. https://doi.org/10.1093/bjps/axy015

Creel, K. A. (2020). Transparency in complex computational systems. *Philosophy of Science, 87*(4), 568–589. https://doi.org/10.1086/709729

Crook, B., Schlüter, M., & Speith, T. (2023). Revisiting the performance-explainability trade-off in explainable artificial intelligence (XAI). In *2023 IEEE 31st International Requirements Engineering Conference Workshops (REW)* (pp. 316–324). IEEE.

Doerig, A., Sommers, R. P., Seeliger, K., Richards, B., Ismael, J., Lindsay, G. W., Kording, K. P., Konkle, T., van Gerven, M. A. J., Kriegeskorte, N., & Kietzmann, T. C. (2023). The neuroconnectionist research programme. *Nature Reviews Neuroscience*, 1–20. https://doi.org/10.1038/s41583-023-00705-w

Elgin, C. (2007). Understanding and the facts. *Philosophical Studies, 132*(1), 33–42. https://doi.org/10.1007/s11098-006-9054-z

Elgin, C. (2017). *True enough*. MIT Press.

Elhage, N., Nanda, N., Olsson, C., Henighan, T., Joseph, N., Mann, B., Askell, A., Bai, Y., Chen, A., Conerly, T., DasSarma, N., Drain, D., Ganguli, D., Hatfield-Dodds, Z., Hernandez, D., Jones, A., Kernion, J., Lovitt, L., Ndousse, K., ... Olah, C. (2021). A mathematical framework for transformer circuits. *Transformer Circuits Thread*.

Feest, U. (2017). Phenomena and objects of research in the cognitive and behavioral sciences. *Philosophy of Science, 84*(5), 1165–1176.

Formanek, N. (2025). How I stopped worrying and learned to love opacity. In: Durán, J. M., & Pozzi, G. (Eds.), *Philosophy of science for machine learning: Core issues and new perspectives*. Synthese Library. Springer.

Frigg, R., & Hartmann, S. (2020). Models in science. In E. N. Zalta (Ed.), *The Stanford Encyclopedia of Philosophy* (Spring 2020). Metaphysics Research Lab, Stanford University. https://plato.stanford.edu/archives/spr2020/entries/models-science/

Geiger, A., Lu, H., Icard, T., & Potts, C. (2021). Causal abstractions of neural networks. *Advances in Neural Information Processing Systems, 34*, 9574–9586. https://proceedings.neurips.cc/paper/2021/hash/4f5c422f4d49a5a807eda27434231040-Abstract.html

Geiger, A., Wu, Z., Lu, H., Rozner, J., Kreiss, E., Icard, T., Goodman, N., & Potts, C. (2022). Inducing causal structure for interpretable neural networks. *Proceedings of the 39th International Conference on Machine Learning*, 7324–7338. https://proceedings.mlr.press/v162/geiger22a.html

Geirhos, R., Rubisch, P., Michaelis, C., Bethge, M., Wichmann, F. A., & Brendel, W. (2018, December 21). ImageNet-trained CNNs are biased towards texture; increasing shape bias improves accuracy and robustness. *International Conference on Learning Representations*. https://openreview.net/forum?id=Bygh9j09KX

Glaser, J. I., Benjamin, A. S., Farhoodi, R., & Kording, K. P. (2019). The roles of supervised machine learning in systems neuroscience. *Progress in Neurobiology, 175*, 126–137. https://doi.org/10.1016/j.pneurobio.2019.01.008

Glennan, S. (2017). *The new mechanical philosophy* (1st ed.). Oxford University Press.

Goodfellow, I., Bengio, Y., & Courville, A. (2016). *Deep learning*. The MIT Press.

Grimm, S. (2021). Understanding. In E. N. Zalta (Ed.), *The Stanford encyclopedia of philosophy* (Summer 2021). Metaphysics Research Lab, Stanford University. https://plato.stanford.edu/archives/sum2021/entries/understanding/

Grinvald, A., & Hildesheim, R. (2004). VSDI: A new era in functional imaging of cortical dynamics. *Nature Reviews. Neuroscience, 5*(11), 874–885. https://doi.org/10.1038/nrn1536

Gunning, D., & Aha, D. (2019). DARPA's Explainable Artificial Intelligence (XAI) Program. *AI Magazine, 40*(2), 2. https://doi.org/10.1609/aimag.v40i2.2850

Hassabis, D., Kumaran, D., Summerfield, C., & Botvinick, M. (2017). Neuroscience-inspired artificial intelligence. *Neuron, 95*(2), 245–258. https://doi.org/10.1016/j.neuron.2017.06.011

Hills, A. (2016). Understanding Why. *Noûs, 50*(4), 661–688. https://doi.org/10.1111/nous.12092

Hochstein, E. (2016). One mechanism, many models: A distributed theory of mechanistic explanation. *Synthese, 193*(5), 1387–1407. https://doi.org/10.1007/s11229-015-0844-8

Humphreys, P. (2009). The philosophical novelty of computer simulation methods. *Synthese, 169*(3), 615–626. https://doi.org/10.1007/s11229-008-9435-2

Jumper, J., Evans, R., Pritzel, A., Green, T., Figurnov, M., Ronneberger, O., Tunyasuvunakool, K., Bates, R., Žídek, A., Potapenko, A., Bridgland, A., Meyer, C., Kohl, S. A. A., Ballard, A. J., Cowie, A., Romera-Paredes, B., Nikolov, S., Jain, R., Adler, J., et al. (2021). Highly accurate protein structure prediction with AlphaFold. *Nature, 596*(7873), 7873. https://doi.org/10.1038/s41586-021-03819-2

Kanwisher, N., Khosla, M., & Dobs, K. (2023). Using artificial neural networks to ask 'why' questions of minds and brains. *Trends in Neurosciences, 46*(3), 240–254. https://doi.org/10.1016/j.tins.2022.12.008

Kar, K., Kubilius, J., Schmidt, K., Issa, E. B., & DiCarlo, J. J. (2019). Evidence that recurrent circuits are critical to the ventral stream's execution of core object recognition behavior. *Nature Neuroscience, 22*(6), 6. https://doi.org/10.1038/s41593-019-0392-5

Kästner, L. (2018). Integrating mechanistic explanations through epistemic perspectives. *Studies in History and Philosophy of Science Part A, 68*, 68–79. https://doi.org/10.1016/j.shpsa.2018.01.011

Kästner, L., Crook, B. (2024) Explaining AI through mechanistic interpretability. *European Journal for Philosophy of Science, 14*, 52. https://doi.org/10.1007/s13194-024-00614-4

Kell, A. J. E., Yamins, D. L. K., Shook, E. N., Norman-Haignere, S. V., & McDermott, J. H. (2018). A task-optimized neural network replicates human auditory behavior, predicts brain responses, and reveals a cortical processing hierarchy. *Neuron, 98*(3), 630–644.e16. https://doi.org/10.1016/j.neuron.2018.03.044

Kitcher, P. (1981). Explanatory unification. *Philosophy of Science, 48*(4), 507–531.

Kohár, M., & Krickel, B. (2021). Compare and contrast: How to assess the completeness of mechanistic explanation. In F. Calzavarini & M. Viola (Eds.), *Neural mechanisms: New challenges in the philosophy of neuroscience* (pp. 395–424). Springer. https://doi.org/10.1007/978-3-030-54092-0_17

Kozachkov, L., Tauber, J., Lundqvist, M., Brincat, S. L., Slotine, J.-J., & Miller, E. K. (2022). Robust and brain-like working memory through short-term synaptic plasticity. *PLoS Computational Biology, 18*(12), e1010776. https://doi.org/10.1371/journal.pcbi.1010776

Lawler, I., & Sullivan, E. (2021). Model explanation versus model-induced explanation. *Foundations of Science, 26*(4), 1049–1074. https://doi.org/10.1007/s10699-020-09649-1

LeCun, Y., Bengio, Y., & Hinton, G. (2015). Deep learning. *Nature, 521*(7553), 7553. https://doi.org/10.1038/nature14539

Lillicrap, T. P., & Kording, K. P. (2019). *What does it mean to understand a neural network?* (arXiv:1907.06374). arXiv. https://doi.org/10.48550/arXiv.1907.06374

Lindsay, G. W. (2021). Convolutional neural networks as a model of the visual system: Past, present, and future. *Journal of Cognitive Neuroscience, 33*(10), 2017–2031. https://doi.org/10.1162/jocn_a_01544

López-Rubio, E., & Ratti, E. (2021). Data science and molecular biology: Prediction and mechanistic explanation. *Synthese, 198*(4), 3131–3156. https://doi.org/10.1007/s11229-019-02271-0

Mann, S., Crook, B., Kästner, L., Schomäcker, A., & Speith, T. (2023, September). Sources of opacity in computer systems: Towards a comprehensive taxonomy. In *2023 IEEE 31st International Requirements Engineering Conference Workshops (REW)* (pp. 337–342). IEEE.

Marblestone, A. H., Wayne, G., & Kording, K. P. (2016). Toward an integration of deep learning and neuroscience. *Frontiers in Computational Neuroscience, 10*. https://www.frontiersin.org/article/10.3389/fncom.2016.00094

Matthewson, J., & Weisberg, M. (2009). The structure of tradeoffs in model building. *Synthese, 170*(1), 169–190. https://doi.org/10.1007/s11229-008-9366-y

McCulloch, W. S., & Pitts, W. (1943). A logical calculus of the ideas immanent in nervous activity. *The Bulletin of Mathematical Biophysics, 5*(4), 115–133. https://doi.org/10.1007/BF02478259

Miller, E. K., Lundqvist, M., & Bastos, A. M. (2018). Working Memory 2.0. *Neuron, 100*(2), 463–475. https://doi.org/10.1016/j.neuron.2018.09.023

Mitchell, S. D. (2002). Integrative pluralism. *Biology and Philosophy, 17*(1), 55–70. https://doi.org/10.1023/A:1012990030867

Mitchell, S. D. (2019). Perspectives, representation, and integration. In *Understanding perspectivism* (pp. 178–193). Routledge. https://doi.org/10.4324/9781315145198-11

Mongillo, G., Barak, O., & Tsodyks, M. (2008). Synaptic theory of working memory. *Science, 319*(5869), 1543–1546. https://doi.org/10.1126/science.1150769

Morgan, M. S., & Morrison, M. (1999). *Models as mediators: Perspectives on natural and social science*. Cambridge University Press.

Nanda, N., Chan, L., Lieberum, T., Smith, J., & Steinhardt, J. (2023). *Progress measures for grokking via mechanistic interpretability* (arXiv:2301.05217). https://doi.org/10.48550/arXiv.2301.05217

Olah, C., Satyanarayan, A., Johnson, I., Carter, S., Schubert, L., Ye, K., & Mordvintsev, A. (2018). The building blocks of interpretability. *Distill, 3*(3). https://doi.org/10.23915/distill.00010

Páez, A. (2025). Axe the X in XAI: A plea for understandable AI. In: Durán, J. M., & Pozzi, G. (Eds.), *Philosophy of science for machine learning: Core issues and new perspectives*. Synthese Library. Springer.

Pearl, J. (2009). *Causality: Models, reasoning, and inference* (2nd ed.). Cambridge University Press.

Pessoa, L. (2023). The entangled brain. *Journal of Cognitive Neuroscience, 35*(3), 349–360. https://doi.org/10.1162/jocn_a_01908

Pospisil, D. A., Pasupathy, A., & Bair, W. (2018). 'Artiphysiology' reveals V4-like shape tuning in a deep network trained for image classification. *eLife, 7*, e38242. https://doi.org/10.7554/eLife.38242

Potochnik, A. (2016). *Scientific explanation: Putting communication first*, pp. 721–732.

Pulvermüller, F., Tomasello, R., Henningsen-Schomers, M. R., & Wennekers, T. (2021). Biological constraints on neural network models of cognitive function. *Nature Reviews Neuroscience, 22*(8), 8. https://doi.org/10.1038/s41583-021-00473-5

Räz, T. (2025). From explanations to interpretability and back. In: Durán, J. M. and Pozzi, G. (Eds.), *Philosophy of science for machine learning: Core issues and new perspectives*. Synthese Library. Springer.

de Regt, H. (2009). The epistemic value of understanding. *Philosophy of Science, 76*(5), 585–597. https://doi.org/10.1086/605795

Richards, B. A., Lillicrap, T. P., Beaudoin, P., Bengio, Y., Bogacz, R., Christensen, A., Clopath, C., Costa, R. P., de Berker, A., Ganguli, S., Gillon, C. J., Hafner, D., Kepecs, A., Kriegeskorte, N., Latham, P., Lindsay, G. W., Miller, K. D., Naud, R., Pack, C. C., et al. (2019). A deep learning framework for neuroscience. *Nature Neuroscience, 22*(11), 1761–1770. https://doi.org/10.1038/s41593-019-0520-2

Salmon, W. C. (1984). *Scientific explanation and the causal structure of the world*. Princeton University Press.

Samek, W., & Müller, K.-R. (2019). Towards explainable artificial intelligence. In W. Samek, G. Montavon, A. Vedaldi, L. K. Hansen, & K.-R. Müller (Eds.), *Explainable AI: Interpreting, explaining and visualizing deep learning* (Vol. 11700, pp. 5–22). Springer. https://doi.org/10.1007/978-3-030-28954-6_1

Schrimpf, M., Kubilius, J., Lee, M. J., Ratan Murty, N. A., Ajemian, R., & DiCarlo, J. J. (2020). Integrative benchmarking to advance neurally mechanistic models of human intelligence. *Neuron, 108*(3), 413–423. https://doi.org/10.1016/j.neuron.2020.07.040

Shagrir, O., & Bechtel, W. (2017). Marr's computational level and delineating phenomena. In D. M. Kaplan (Ed.), *Explanation and integration in mind and brain science* (p. 0). Oxford University Press. https://doi.org/10.1093/oso/9780199685509.003.0009

Srećković, S., Berber, A., & Filipović, N. (2022). The automated Laplacean Demon: How ML challenges our views on prediction and explanation. *Minds and Machines, 32*(1), 159–183. https://doi.org/10.1007/s11023-021-09575-6

Strevens, M. (2013). No understanding without explanation. *Studies in History and Philosophy of Science Part A, 44*(3), 510–515. https://doi.org/10.1016/j.shpsa.2012.12.005

Sullivan, E. (2022). Understanding from machine learning models. *The British Journal for the Philosophy of Science, 73*(1), 109–133. https://doi.org/10.1093/bjps/axz035

Sullivan, E., & Kasirzadeh, A. (2025). Explanation hacking: The perils of algorithmic recourse. In: Durán, J. M., & Pozzi, G. (eds.), *Philosophy of science for machine learning: Core issues and new perspectives*. Synthese Library. Springer.

Sussillo, D., Nuyujukian, P., Fan, J. M., Kao, J. C., Stavisky, S. D., Ryu, S., & Shenoy, K. (2012). A recurrent neural network for closed-loop intracortical brain-machine interface decoders. *Journal of Neural Engineering, 9*(2), 026027. https://doi.org/10.1088/1741-2560/9/2/026027

Thompson, J. A. F. (2021). Forms of explanation and understanding for neuroscience and artificial intelligence. *Journal of Neurophysiology, 126*(6), 1860–1874. https://doi.org/10.1152/jn.00195.2021

Veit, W. (2020). Model pluralism. *Philosophy of the Social Sciences, 50*(2), 91–114. https://doi.org/10.1177/0048393119894897

Willeke, K. F., Restivo, K., Franke, K., Nix, A. F., Cadena, S. A., Shinn, T., Nealley, C., Rodriguez, G., Patel, S., Ecker, A. S., Sinz, F. H., & Tolias, A. S. (2023). *Deep learning-driven characterization of single cell tuning in primate visual area V4 unveils topological organization* (p. 2023.05.12.540591). bioRxiv. https://doi.org/10.1101/2023.05.12.540591

Woodward, J. F. (2003). *Making things happen: A theory of causal explanation*. Oxford University Press.

Yamins, D. L. K., & DiCarlo, J. J. (2016). Using goal-driven deep learning models to understand sensory cortex. *Nature Neuroscience, 19*(3), 356–365. https://doi.org/10.1038/nn.4244

Yamins, D. L. K., Hong, H., Cadieu, C. F., Solomon, E. A., Seibert, D., & DiCarlo, J. J. (2014). Performance-optimized hierarchical models predict neural responses in higher visual cortex. *Proceedings of the National Academy of Sciences, 111*(23), 8619–8624. https://doi.org/10.1073/pnas.1403112111

Zador, A. M. (2019). A critique of pure learning and what artificial neural networks can learn from animal brains. *Nature Communications, 10*(1), 3770. https://doi.org/10.1038/s41467-019-11786-6

Zednik, C. (2021). Solving the black box problem: A normative framework for explainable artificial intelligence. *Philosophy & Technology, 34*(2), 265–288. https://doi.org/10.1007/s13347-019-00382-7

Zednik, C., & Boelsen, H. (2022). Scientific exploration and explainable artificial intelligence. *Minds and Machines, 32*(1), 219–239. https://doi.org/10.1007/s11023-021-09583-6

Zhuang, C., Yan, S., Nayebi, A., Schrimpf, M., Frank, M. C., DiCarlo, J. J., & Yamins, D. L. K. (2021). Unsupervised neural network models of the ventral visual stream. *Proceedings of the National Academy of Sciences, 118*(3), e2014196118. https://doi.org/10.1073/pnas.2014196118

Barnaby Crook is a PhD student at the University of Bayreuth and part of the research project "Explainable Intelligent Systems". He has an M.A in philosophy from the Berlin School of Mind and Brain. Barnaby's research addresses questions about human understanding of intelligent systems, both artificial and biological.

Lena Kästner is professor for philosophy, computer science and AI at the University of Bayreuth. She has a background in Cognitive Science and Cognitive Neuroscience and received her PhD in philosophy from Ruhr-University Bochum. Prof. Kästner's research focuses on explanations, intelligence, and causation. Currently, she is also head-PI of the project "Explainable Intelligent Systems (EIS)".

Chapter 21
Artificial Intelligence in Climate Science: From Machine Learning to Neural Networks

Greg Lusk

Abstract This chapter analyzes the extent to which the introduction of NNs into climate science might change its methodology and the way we understand climate science epistemology. It proceeds by examining areas of climate research already heavily influenced by artificial intelligence in the form of machine learning, and then considers areas of research where the introduction of neutral networks would be novel. The goal of this chapter is to bring out the structure of the extant debates in the philosophy of climate science and demonstrate how they might change with the introduction of neural nets. Overall, this chapter shows that if there is a revolution coming anytime soon, it is unlikely to take the form of a wholesale replacement of theory-based climate models by data-driven neural nets. Rather, the analysis here demonstrates that there is likely to be further hybridization of theory-driven models with data-driven NNs in a way that trades some intelligibility or representational accuracy for computational efficiency.

21.1 Introduction

Climate science has not been spared the enthusiasm for artificial intelligence (AI) exhibited in public discourse since the launch of ChatGPT in 2022. In fact, some pronouncements indicate that the impact on climate science could be drastic. For example, Google's ex-CEO Eric Schmidt has declared that "With the advent of AI, science is about to become much more exciting—and in some ways unrecognizable," and specifically discusses AI's impact on climate science, claiming "The hoped-for revolution in climate modeling" is already underway (Schmidt, 2023). While it is unclear who exactly was doing the hoping that Schmidt refers to, climate scientists have been examining the extent to which artificial intelligence could be used to further their scientific goals. If the tone of recent publications is indicative

G. Lusk (✉)
Department of Philosophy, Durham University, Durham, UK
e-mail: greg.lusk@durham.ac.uk

J. M. Durán, G. Pozzi (eds.), *Philosophy of Science for Machine Learning*,
Synthese Library 527, https://doi.org/10.1007/978-3-032-03083-2_21

of the field (e.g. Chantry et al., 2021; Ebert-Uphoff & Hilburn, 2023), there is a high level of excitement that these methods of research can be fruitfully applied to problems in weather and climate research.

Whether this excitement will translate into a revolution is unclear, but there are reasons why such a claim looks hyperbolic. After all, since its early days, climate science has been inventive in the way it produces and manages data (Edwards, 2010), and in its modern form, embraces its own distinctive style of big data analysis (see Lloyd et al., 2022). Climate disciplines embraced machine learning (for example, kernel and 'random forest' techniques) beginning in the 1990's to help with classification tasks arising in the context of remote sensing and the assimilation of observations by numerical weather models. Since then, machine learning (ML) has become "a universal approach" in many areas of climate research (Reichstein et al., 2019, 196). So, if by "revolution" one means integrating AI in the guise of ML into everyday climate research and modeling, then this revolution is 20 or 30 years in the making.

But it is because climate research has long embraced AI in the form of machine learning that it is interesting in the present moment. The recent clamor over AI is not due to the kind of ML techniques that climate science has long employed, rather, the excitement is a product of the success that neural networks (NNs) have shown in accomplishing predictive tasks (see Buckner, 2019).[1] Such NNs have not been widely used in climate science, and thus it makes a good case to examine how NNs might alter research practices. And because climate science has long been a discipline that is comfortable with a broadly empirical-reliabilist epistemic stance, and has continually sought novel ways to leverage data-driven techniques to overcome theoretical and computational shortcomings, it seems ripe for utilizing these NNs.

This chapter analyzes the extent to which the introduction of NNs into climate science might change its methodology and the way we understand climate science epistemology. It proceeds by examining areas of climate research already heavily influenced by AI in the form of ML, and then considers areas of research where the introduction of NNs would be novel. In each area of research, I engage the relevant philosophical analyses of these practices and assess how NNs might impact scientific understanding, explanation, representational transparency, or other epistemically relevant properties. The goal of this chapter is to bring out the structure of the extant debates in the philosophy of climate science and demonstrate how they might change with the introduction of NNs. Overall, this chapter shows that if there is a revolution coming anytime soon, it is unlikely to take the form of a wholesale replacement of theory-based climate models by data-driven NNs. Rather,

[1] The terms "artificial Intelligence" (AI), "machine Learning" (ML) and "neural network"(NN) are often vague and used imprecisely in recent discourse. In this chapter, NN is the most specific category of the three and I use it to refer to all kinds of neural nets (e.g. convoluted, recurrent, perceptron). I use the term ML to refer to methods of learning that are not NNs, this includes decision trees and Bayesian statistical inference or optimization. I use AI as the broadest category to cover ML, NNs, and other instances of automated pattern recognition when used for inference.

the analysis here demonstrates that there is likely to be further hybridization of theory-driven models with data-driven NNs in a way that trades some intelligibility or representational accuracy for computational efficiency.

21.2 Climate Observations and Pseudo-Observations

When it comes to climate science, models get all the attention. However, climate science relies on a whole host of observations, and turning these observations into usable information is one potential area in where NNs might be helpful. The observations that climate science relies upon are diverse. Observations can be gathered from ground-based stations that provide information about temperature or precipitation, but information can also be acquired from ships, ocean buoys, aircraft, and earth-orbiting satellites.[2] Scientists can also learn about conditions of the past from so-called "proxy" sources, like ice cores, tree rings, and pollen found in sediments.

Though there is a wide variety of data available, making that data useful can be difficult. After all, there are certain characteristics that the data have that are not aligned with the information researchers often want to know. For example, researchers are interested in quantities that represent certain regions of the globe, or the entire globe itself, often over long periods of time. Observations, however, are local, only representing one spatial area at one particular point in time. Thus, one may have significant gaps in their data; there may be geographic areas where observations are simply lacking. Furthermore, observations are often made with different instruments with different characteristics that go in and out of service on different lifecycles and have different error profiles. To see the difficulty that climate scientists often face, and how they can leverage AI to overcome it, it is worth examining two methods of obtaining observations in climate science where philosophers have shown particular interest: remote sensing and reanalysis using data assimilation.

21.2.1 Observations: Remote Sensing

Oftentimes, scientists are interested in observing areas of the planet that are not easy to access, for example, areas near the poles or areas over remote ocean. To observe these areas over extended periods of time, climate researchers employ various remote sensing devices. These devices can be ground based—positioned in the arctic for example—or placed on satellites, and monitor the atmosphere on a consistent and (if possible) continuous basis.

[2] See Parker (2025) for a more detailed overview.

However, remote sensing is itself a complex procedure, one that has already taken advantage of AI. Take, for example, the detection of atmospheric trace gases—which include greenhouse gases and water vapor—in an area. It is often difficult to obtain trace gas measurements through relatively direct measurements. Scientists are interested in the vertical profile of gases, that is, how the gas concentrations change with altitude. To measure this relatively directly, one might use a weather balloon or airplane. But these methods of detection are troublesome: both are expensive, weather dependent, and require significant manual labor. Instead, much of the data on these gases come from remotely installed devices that measure the electromagnetic signature of the gas to infer their vertical concentrations. This method of remote sensing has the advantage of being cheaper, automated, and consistently return results over long periods of time. For example, a spectrometer can be placed in an arctic environment and return trace gas profiles on sunny days for years, often without the intervention of researchers. Such information is crucial for studying air quality, trace gas movement, and for providing estimates of water vapor that are used in research on cloud dynamics and in some climate models.

Extracting information about the profile of trace gases from the readings of devices like spectrometers is surprisingly difficult and computationally expensive. The difficulty is that the electromagnetic information received by many remote sensing devices underdetermines the vertical profile of the gases. Gases have relatively unique electromagnetic signatures, but the atmosphere can influence the way such signatures are detected by the remote sensing device, permitting the signatures run together. Methane, for example, has eight other chemical species whose signature can overlap with its own, for instance, species such as ozone and water vapor. It is thus possible for there to be two or more vertical profiles of trace gases that could have resulted in the same reading by a remote sensing spectrometer. Thus, there is no equation that can use as input the quantified observations of the remote sensing device and return a unique profile of trace gas concentrations. This is called an "inverse problem" by those in the climate science and remote sensing communities.

The solution to the inverse problem of trace gas detection is to forego attempts to find a mathematical relationship between the output of the device and the trace gas profile, and instead, attempt to recreate the observed profile from existing knowledge. This process is called retrieval, and it has long relied on ML algorithms. To perform retrieval, one needs a forward model, that is, a mathematical description of how the spectral information received by the detector is produced. This forward model includes a radiative transfer model—that specifies how radiation related to the gases propagates through the atmosphere—combined with a model of how the detection instrument functions and its error characteristics. Using this forward model, scientists can then produce synthetic instrument readings; what the instrument would have seen given a particular distribution of trace gases. Scientists then initiate a choice procedure. Starting with background knowledge regarding the most likely trace gas profiles, they repeatedly simulate what the detection device would have seen given certain distributions of gases, and then try to match the simulated profile to the one actually detected. This search procedure is often formulated in Bayesian terms and represents the ML component.[3] Essentially, the procedure is seeking to calculate the best estimate of trace gas concentrations, which is defined as the maximum a posteriori solution. To find this solution, scientists define a cost function which seeks to find the minimal difference between the simulated and the actual instrument readings. The cost function is explored algorithmically until the best match is located. Essentially, the profile that results in the best match (i.e. the maximum a posteriori solution) is the reported result of the observation. The algorithm qualifies as ML in the sense that it "learns" what parameters produce the best fitting profile (see Rodgers, 2000; Lusk, 2021).

21.2.2 Pseudo-Observations: Data Assimilation

Before discussing how NNs may change the way scientists approach climate observations, it is worth examining another practice central to climate science that relies on very similar algorithms. While remote sensing via retrieval helps provide scientists with a means to obtain data in hard to access areas, from a global perspective, the area that observations cover is still limited. Measuring instruments are, after all, not uniformly located across the globe and have gone in and out of service over the timescales of interest for climate. For climate scientists this presents some difficulty, as their models give global results on a uniform grid that covers the globe. Thus, they would like to have some way to make a like-to-like comparison between their model outputs and observations. There are other modeling needs as well. Weather forecasters use numerical models to analyze future atmospheric states. Their models, however, require initial conditions for each grid point, and the results are relatively sensitive to these conditions. Traditional observations do not typically have the coverage to provide the needed information to initialize the weather forecasting model. What is needed are datasets with uniform global coverage.

To produce these datasets, scientists employ a ML-based technique called data assimilation, which combines model output and traditional observations to produce a global gridded dataset. Essentially, scientists use a weather model to fill in information that cannot be detected by the observation system and attempt to reduce error by ensuring that data for one grid point is consistent with others nearby that should be causally related to it. To do this, forecasters employ their last or previous forecast as the first guess for the future time they are interested in. They then "assimilate" the observational data that has been gathered since their forecast or more plainly, the model is free to use the information from observations to check and adjust the first guess. This is done using an algorithmic process where the observations are compared to the first guess produced by the forecast. Although there are several different data assimilation techniques, the general approach is

[3] See Rodgers (2000) for an explanation of the theoretical basis of retrieval.

much like the one found in retrieval: there is an iterative search for the value of an atmospheric quantity in a particular location that is consistent with nearby observations, observations made before and after, the physics of the model, and minimizes error in some way. This is precisely why it is a ML product: the assimilation process "learns" the atmospheric parameters that respect the given observational and theoretical constraints and result in minimal error. One virtue of this process is that it can produce globally gridded datasets that are uniform in the way climate models are. For areas where observations are sparse, the model can be used to extract information from the nearest observations to produce physically consistent data.

The results of data assimilation are often thought of as pseudo observational, and may be called "data products" or "observationally-based" (Parker, 2025). The employment and refinement of data assimilation has resulted in increased skill in medium to long range weather forecasting. When it comes to climate timescales, scientists have assimilated historical observations for most of the twentieth century, essentially replaying this time period in a model constrained by the observations. For example, the European Center for Medium-Range Weather Forecasts (ECMWF) has reanalyzed 1940 to the present using a data assimilation system to produce a dataset (i.e. ERA5) with gridded global coverage that is physically consistent (Setchell, 2020). Datasets like these are crucial for assessing the performance of climate models.

21.2.3 Impact of NNs on Observations and Pseudo-Observations

NNs are poised to have some impact on the way that scientists approach observational data. Though still in their infancy, techniques to use artificial NNs within retrieval and data assimilation are in development. For example, researchers have developed NN-based retrieval processes for aerosols (e.g. see Di Noia et al., 2015; Di Noia & Hasekamp, 2018) and proof-of-concept approaches for atmospheric CO_2 column retrievals (Zhao et al., 2022). In the latter, they demonstrate that it is possible to train a multi-level NN with synthetic data produced using the radiative transfer forward model (Zhao et al., 2022). Though these approaches are still new, they seem to demonstrate promising reliability. When it comes to data assimilation, researchers have shown that they can reproduce the main results from one of the EMCWF's operational forecasting models (Bonavita & Laloyaux, 2020). At present, these NN implementations are not operational; they are for research purposes only. But they show promise that the kinds of problems that climate scientists face can be addressed using NN techniques.

The question is, what should we make of the shift to such techniques? Philosophically, retrieval and data assimilation techniques have been used to help understand the nature of measurement. For example, Lusk (2021) argues that retrievals produce a form of explanation that is unique to certain measurement contexts. Lusk is interested in the way phenomena are detected and the role of data in that process, an issue that was brought to philosophical prominence by Bogen and Woodward (1988). Bogen and Woodward argue that detecting phenomena is a process that proceeds mostly without appeals to theory, that it happens "bottom-up", starting with data and reasoning to phenomena, and that the crux of reasoning in these cases is establishing the reliability of data. Data, for Bogen and Woodward, are the publicly available records of experiments or measurements. Data are idiosyncratic in the sense that they are the product of specific experimental contexts and typically cannot occur outside those contexts. They are also the result of multiple causes, for example, data are often produced by the target of a measurement, the measurement device, and a whole host of error causing processes. A paradigmatic example of data would be one of Eddington's photographic plates of the eclipse used to measure the deflection of starlight in 1919 to test the theory of general relativity. Phenomena, on the other hand are (usually unobservable) effects or processes that are stable enough to reoccur under different sets of conditions. Phenomena are the result of fewer causes than data. A paradigmatic phenomenon would be the amount of starlight deflection that Eddington measured. Such a phenomenon would be explained "top-down" by theory (in this case general relativity).

Bogen and Woodward describe the process of phenomena identification as proceeding largely without appeals to theory; instead, what a scientist attempts to do is find reasonable ways of handling and reducing error. For example, scientists may attempt to ensure that their observational devices are screened from error producing sources. This helps ensure the data produced are reliable and are amenable to statistical analysis in ways that will help control or compensate for the sources of error that do exist. Crucially, they claim that scientists do not attempt to explain data: theory-based explanations of data are too difficult to produce and would be scientifically useless even if such explanation could be given. So, while theory might tell us the kind of thing to screen off, one never tries to give an account of exactly how the causes of error impacted the observational report.

Lusk (2021) takes issue with this picture of data production and handling. He argues that the retrieval process as it is employed for atmospheric trace gases constitutes an attempt to explain data. That is, the process is essentially setup in a way that takes the readings produced by a remote located spectrometer, and attempts to explain them by providing a model-based account of their production. Lusk claims "In the case of retrieval... scientists attempt to derive the characteristics of spectra from theory and other assumptions, in order to make inferences to trace gas phenomena" (Lusk, 2021 p294). This detailed account of the data is given by the aforementioned forward model and is a necessary component of retrieval. Essentially, Lusk sees scientists as attempting to simulate what the measurement instrument would produce using the forward model of the measurement process and knowledge about the measurement instrument and the likely distribution of trace gases. At least when it comes to certain notions of explanation, particularly Woodward's (2003) counterfactual notion, this process appears to constitute a search for an explanation of the instrument reports that can answer crucial what-if-things-had-been-different questions. Only with these explanations does it seem we can identify phenomena like trace gas distributions and movement patterns.

However, what happens when the retrieval process, and the forward model, are replaced by NNs; are scientists still providing, or at least aiming to provide, an explanation of data? NNs are generally acknowledged to be black boxes: once trained, it is difficult to tell what relationships the NN has discovered, and thus, it is difficult to know what inference the NN is making on our behalf (see Creel, 2020; Zednik, 2021; Sullivan, 2022a, 2022b). The worry is that by employing NNs, scientists are black boxing a crucial component of the process that does the explanatory work. For example, Zhao et al. (2022) demonstrate the viability of performing satellite retrievals of atmospheric CO_2 columns using NNs. Their retrieval system employs a Multi-Layer Perceptron (MLP) NN, which is not only one of the most popular forms of NNs, it is one known to lead to the black boxing of inferential steps. In MLP, there is an input layer, and output layer, and then several hidden layers placed in between. Each layer has one or several neurons that serve as nodes that connect to neurons in the next layer. The connections between neurons have weights and bias, and the object is to train the network "by repeatedly updating the weights and bias between neurons to minimize the loss function so that the output values generated by the network are matched with the actual corresponding values" (Zhao et al., 2022, p. 6). It is often the case, when trained, that the weights and biases in the hidden layer connections will not be known to researchers, and thus, exactly what function the MLP-NN is estimating will also not be known. This is when the NN becomes a black box; researchers simply do not know how the MLP-NN is producing the outputs given certain inputs. Zhao et al. employ two linked MLP-NNs that each have two hidden layers. As we will see, there are benefits to employing techniques like MLP-NNs in retrieval, but doing so partially obscures central aspects of the retrieval process.

Does this opacity result in diminished explanatory power, or reduced justification? I'll argue here that it does not, or at least, not to any large degree given the suggested methods of implementation. Let us consider justification first. Would the black boxing of certain steps in the retrieval process result in a weaker justification for the outcomes? That is, would we have greater reason to think that the estimates of trace gas distributions would unknowingly depart from the values that would have been gathered using the traditional retrieval method?

In many sciences, there is (sometimes) legitimate worries that certain AI methods may be trained in ways that the inferential patterns involved are opaque to researchers. This may be extremely problematic in instances where a trained human is being displaced by an AI process that is directly trained on observations. For example, replacing a judge's expertise by an AI trained to predict recidivism for the sake of determining bail or sentencing (Ryberg & Roberts, 2022). One reason to worry is not only that the new process is opaque, but it takes a different form from the previous process. Thus, the end-users' knowledge of how to read results, what the error characteristics are, and the practices of challenging or double-checking a result, may all be different.[4] And this difference creates room for error.

But there is some reason to believe moving from Bayesian optimization to NN techniques does not represent a large departure from existing methods when it comes to retrieval. The difference between the nascent NN techniques and the more traditional ML techniques that use Bayesian optimization is not as large as moving from a human to an algorithm. In fact, the two techniques share some crucial similarities. This is broadly suggested by the way that Zhao et al. describe the function of training for their NN method: they explain the process as the minimization of a loss function, the same basic mechanism at work when Bayesian optimization is operationalized.

In fact, as some have shown, NNs are quite good at estimating the solution to inverse problems. Mohammad-Djafari (2021) points out that many of the steps performed in retrieval, when performed by regularization or Bayesian optimization, can be easily replaced with trained NNs. Certain deep learning NNs can be constructed where each layer is equivalent to one iteration of the search for convergence in more traditional methods. This is not surprising, after all, while I have drawn a distinction here between NNs and other traditional computational techniques like Bayesian optimization, they are both forms of ML.

Still, there are differences between optimization techniques and the proposed MPL-NN technique for retrieval that might be relevant to whether there is a loss of explanation or justification when it comes to remote atmospheric measurement. Bayesian techniques are often most suitable when researchers believe they have a good model but limited or extremely noisy data. A lot of information comes from the model, which is augmented by information about the likely state and device indication. NNs, including the MPL-NNs, are often most suitable when there is a significant amount of data that can be used for training the NN and discovering the relevant relationships. In many applications—especially when trained through unsupervised learning or implemented without constraints—lots if not all the information is coming from the training data. Thus, one might believe the grounds of justification in retrieval has shifted from model to data, and if the relationship identified by the NN is black boxed, that the justification is no longer accessible. After all, it is the first situation—where we have a good model and noisy data—that climate researchers often find themselves.

Climate researchers, however, can deploy the NNs in ways that does not jeopardize their justification or significantly prevent them from understanding what function the NN is estimating. That is, the way that climate scientists train and use the NN can help ensure that they maintain knowledge of what inference the NN performs, and therefore retain their justification. One advantage that climate scientists have is that the retrieval process is extremely well known. For example, all of the steps to obtain the maximum a posteriori solution via Bayesian inference

[4] See Duede (2022) regarding how the belief in the reliability of scientific AI models cannot be reduced to other familiar epistemic categories of justification.

can be written precisely. Since the job of the NN is replacing some of the operators or steps in this process, there isn't a significant shift in how one would justify the outcome. This is in part because the NN does not end up becoming an opaque black box. Because the steps are well known, scientists are able to train the NN to perform the desired operations, and scientists can be reasonably sure that the NN has not learned a different pattern of inference. For example, Zhao et al. train their two-step MPL-NN on synthetic data produced by their theoretically informed retrieval algorithm. One step of the NN performs the equivalent of one operation in the retrieval, while the second step performs a second operation, whose results are known to be linked by a particular equation. Given this setup, researchers can train the NN on synthetic data produced using the legacy retrieval process, which guides the NN towards the same inference.

Therefore, there seems to be very little of what philosophers have called "link uncertainty" (Sullivan, 2022a, b): uncertainty about the connection between a NN inference and the target phenomena. Since scientists can identify the operations that the NN performs and can constrain its learning using data produced by known processes, there is good reason to think that the NN is performing the operation as described. Furthermore, scientists can compare accuracy and error characteristics with the traditional retrieval methods. There is no significant shift in how the process is justified.

It is less clearcut whether claims about explaining data continue to hold with the introduction of NNs into the retrieval process. Lusk's argument about explanation relies on scientists' ability to use the retrieval process to answer counterfactual questions about what indications the instrument would produce given certain atmospheric concentrations. But counterfactual manipulability is an area where AI and NNs sometimes falter. For example, Jebeile et al. (2021) argue that NNs are not robust with respect to manipulability, and this decreases the intelligibility (i.e. our understanding) of the process containing the NN, and in turn, decreases our intelligibility of the process we try to understand with the NN. They support the claim that NNs are not robust with regards to manipulation by citing the existence of adversarial examples: inputs specifically designed to fool AI systems. For example, in image recognition applications of AI, one can add a layer of pixels to an image that can trick an AI into classifying what appears to be a clear picture of a school bus as an ostrich (Demetrio et al., 2022).

However, it is not clear that the existence of such examples is sufficient, or particularly relevant, to undermining claims regarding explanation. There are, as Freiesleben and Grote (2023) point out, and Jebeile et al. recognize, different notions of "robustness" employed in the AI literature. It is unclear why a lack of robustness with regards to adversarial examples should significantly worry the research scientist. Adversarial examples are (by definition) designed by an adversary and often highly artificial exploits. Such examples take advantage of the NN architecture in order to produce an inaccurate output. Climate scientists, however, are unlikely to be (unknowingly) using the contrived inputs or be targeted by malicious actors for manipulation in this way (i.e. in the way a self-driving car might).

Furthermore, whether a NN retrieval scheme is sensitive to input perturbation generally is an empirical question, and it is one that scientists should be able to answer given their knowledge of retrieval and the availability of the traditional algorithm for comparison. That is, the robustness with respect to what Freiesleben and Grote (2023) call "natural distribution shifts" seems assessable. Given that the fundamental operations remain the same between Bayesian and NN retrieval schemes, it is likely that the number or kind of counterfactual questions that can be answered remain the same as well, and this is a testable assumption.

A loss of explanation may come if, for example, the forward model was to play no role in the retrieval or the training of a NN, and thus, there would be no account regarding why or how the instruments indication was related to the atmospheric state, but given the current use of NNs in retrieval, that situation is not likely. The worst-case scenario would be that the NN is using some shortcut to estimate the output of certain steps of the retrieval. This would be problematic if the shortcut result could not be interpreted as an approximation of the correlate step in the retrieval. But again, since the retrieval process is well known, whether the NN result is an adequate approximation can be assessed. It therefore seems to be an empirical question whether explanations of data are maintained when NNs are introduced into retrieval, but there's no obvious reason why they would not be.

Why rely on NNs in retrieval then, if in fact the NNs simply end up performing similar operations? The answer is computational efficiency. The use of NNs changes where the computational load rests in the retrieval process, and what kinds of resources can be used for the computation. The traditional retrieval process requires solutions to radiative transfer equations at each iteration, which makes the process computationally expensive and slow. While training a NN is computationally expensive as well, once trained, retrieval would be much more efficient than the standard process. Also, ML is much easier for scientists to parallelize on multiple CPU's, which provides a significant performance boost.

The impacts of NNs on data assimilation (DA) and reanalysis might be more significant than on retrieval. Since retrieval and DA are similar, it's unsurprising that the lessons from retrieval hold for DA as well. Geer (2021) shows that one can demonstrate equivalences between a specific kind of DA technique (4D-VAR) and a recurrent NN using Bayes nets. Greer also points out several general similarities. Both DA and NNs are attempting to learn about the world via data using an inverse method, the cost function in variational DA is equivalent to the loss function when training NN, both often employ gradient descent, and specifically, the adjoint method often employed in DA is mathematically identical to the NN training technique called backpropagation. They conclude that from a general enough perspective these two approaches can be unified under a Bayesian framework.

While the similarities between the DA methods currently employed and emerging NN methods might be similar, the epistemic status of DA produced products and atmospheric retrievals is very different. Retrievals are often (though not always) thought of as bona fide observations or measurements. DA products, like reanalysis datasets, are however not generally thought of in this way. These products are acknowledged to be empirically informed, and often play the role of observations or measurements, but the climate community is very aware that imperfect models play a significant role in generating these data products. Scientists often include words of caution when a DA produced reanalysis dataset is employed in research.

Wendy Parker (2017) argues that data assimilation systems are a kind of complex measurement system that is in development. She follows Eran Tal (2012, 2013) in thinking about measurement as a model-based activity, where a precondition on the possibility of measurement "is the specification of an abstract and idealized model of the measurement process. To measure a physical quantity is to make coherent and consistent inferences from the final state(s) of a physical process to value(s) of a parameter in the model" (Tal, 2012, p. 17). Parker has a tripartite view of measurement that includes direct, derived, and complex measurement. Direct measurements are those where no symbolic calculation is performed on the outcome of a measurement device, e.g. a rain gauge for the measurement of rainfall depth or a pre-calibrated mercury thermometer. A derived measurement involves one or more layers of inference, usually taking a direct measurement as input and using a well-known relationship (e.g. a law of nature) to arrive at the measured quantity. Parker defines a complex measurement as one that utilizes many direct or derived measurements together to obtain a result that is more informative than any of the measurements independently.

Ordinarily, the results of a computer simulation would not be candidates for measurement, except in special circumstances (see Lusk, 2016), as they lack any kind of physical interaction with the measured quantity. However, it is tempting to see reanalysis data produced using DA as a case of complex measurement because the DA system is ingesting measurements of direct or derived forms that resulted from interactions with physical systems. After all, the whole idea is to use numeric weather prediction models in combination with past data to produce a consistent time series of quantities of interest. In this sense, it looks as if it is a form of complex measurement in the way that Parker specifies.

Parker, however, argues that DA should be seen as something short of a measurement system, and therefore, reanalysis data are also something short of measurements. She claims that data assimilation seems to be a measurement system that is under development. As Parker puts it: "Ultimately, this is because today's data assimilation systems are not subjected to a rigorous process of calibration that provides well-motivated uncertainty estimates" (2017 p294). Calibrating the DA system would require identifying and correcting for sources of error. There are at least two ways of doing this. The first is called "white box calibration" and essentially requires breaking the measuring system down into components and crafting an uncertainty budget, demonstrating how each component contributes to the overall uncertainty. The other is to perform black box calibration, which can be thought of as treating the entire measurement system as a singular component, the outputs of which are compared directly with a set of known quantities in a way that enables statistical determination of an uncertainty range.

Both types of calibration look daunting in the context of DA. For most implementations of DA, scientists rely on the assumption of a perfect model, an assumption known to be false (Farchi et al., 2023). As a consequence, reanalysis products often come with cautions regarding the uncertainties involved, and how the uncertainties may vary even for different data points within the same product. As the ERA5 reanalysis documentation points out "... the accuracy of the analyses is not uniform throughout each assimilation window. If the model and observations are unbiased and their errors follow Gaussian distributions and if the observations are homogeneous in space and time, then the analysis error will be smallest in the middle of the assimilation window." However, it helpfully adds "... none of these assumptions are actually true ..." (Hersbach et al., 2023).

The difficulty comes from the varied sources of error, and the lack of a standard against which to calibrate the system. After all, the observations contain their own errors, which become larger and less well known as one goes further back in the twentieth century. There is also no complete gridded dataset that comes from pure observations; if such data existed, the need for DA would be significantly diminished. This makes the straightforward comparisons between DA products and standards with known uncertainties needed for black box calibration difficult. White box calibration perhaps fares even worse. The way different components of the model, for example, contribute to the error are not well known at this point. The ERA5 documentation indicates that their product accounts for some uncertainties in observations and sea surface temperature, they make plain a significant problem: "Other uncertainties are not accounted for, such as uncertainties in radiative forcing due to greenhouse gases, or systematic errors in the model or the way in which observations are used" (Hersbach et al., 2023).

The hope is that NNs can help with this task of clarifying the uncertainties involved in a DA system. Though speculative, the success of this project could make progress towards a more rigorous quantification of error and thus the establishment of DA products as bona fide measurements. There are different ways that this might be done.

Some, for example, have attempted to train a deep NN in an attempt to learn the entire assimilation process (Arcucci et al., 2021). In this example, the NN is trained through a comparison of the extant observational data (the input into the DA system) and the products of DA (the outputs) to learn the DA process. The thought is that, in theory, a NN can learn the traditional DA process and improve upon it using the ability of NNs to update their models as new data is introduced. That this could result in reduced error compared to traditional DA systems has been proven in theory, and some limited proof of concept experiments have been carried out (Arcucci et al., 2021). Still, there are some potential downsides to such methods, for example, there is no guarantee that the purely NN assimilation system will produce physically consistent data, though the hope is that it will produce data as consistent as the DA products it is trained upon. It's also an open question under what conditions an operational NN assimilation system would outperform a traditional one, or whether such a system could be implemented within existing weather and climate systems for prediction or projection.

Another way NNs might result in better DA techniques is through hybrid assimilation, which attempts to deploy a NN alongside the traditional DA process in order to curtail error. This process starts with a set of observations. The DA system attempts to "update" the observations by finding their optimal states, and filling in areas of sparse observations, using a physically based model that has a fixed set of model parameters. It is essentially asking what the observations should look like given the model, essentially as is done in traditional DA. The products (i.e. updated observations) are then handed to the NN. In this case, the NN has already been trained to learn the DA process, so it is supposed to be a surrogate model of the DA system. The purpose of the surrogate, however, is to update the model based on the data products. That NN then specifies what *the model* should look like given the observations. From this, information can be passed back to the DA system, for example, updated parameters to use. This cycle continues until there is convergence between the two systems (see Brajard et al., 2020). This process can be used to probe and reduce parameter error and other forms of structural model error. Using surrogate models like this has reached a state of maturity in which they perform at least as well as the extant ways of handling model error in operational data assimilation and appears close to operational implementation for research purposes.

The above suggests that NNs might be a virtue for climate scientists when it comes to processing observational data. When it comes to remote sensing, NNs promise enhanced efficiency without much loss of explanatory power due to their opaqueness (at least, if deployed in the way explored above). When it comes to DA, NNs promise to help scientists curtail and quantify error, and in doing so, DA products move one step closer to genuine measurements.

21.3 Climate Models and AI

The kind of models that I'll refer to as climate models—which include general circulation models (GCM) and earth system models (ESM)—attempt to depict the evolution of the Earth's climate system. A modern GCM will contain representations of the ocean, atmosphere, land mass, and sea ice, whereas an ESM may additionally include components that represent atmospheric chemistry, ocean chemistry, and perhaps soil and flora. These components are often divided into different modules or components that are then linked together by a coupler, which specifies how information passes between them (see Parker 2025 for more details). Many of these modules are operationally similar in that they approximate solutions to a set of equations that represent the evolution of Earth's atmospheric systems over successive timesteps. These equations contain much of our theoretical knowledge of the factors that drive climate's behavior. The equations are discretized, and numerical methods are employed to estimate solutions at regular points on a three-dimensional grid. Typically, supercomputers are needed to generate these solutions. Available computer power to perform the estimations has been a significant limitation, but as this power has increased, the resolution of climate models has as well. Our best models currently produce results on a grid of 50–100 km with 50 or more vertical layers.

Climate models are used for a variety of different purposes, including gaining an understanding of the way the Earth system operates (see Held, 2005). Climate models are crucial in providing evidence for anthropogenic climate change. To detect the signal of climate change, we need to have some indication of how the planet would behave in its absence. Yet, there is only one planet, and there is good theoretical reason to think climate change is happening. Thus, climate models provide us with what the counterfactual case would be like, that is, how the planet would naturally behave without anthropogenic interference. Only with this information could we detect a departure from that baseline. Additional support for anthropogenic climate change comes from signatures: telltale signs in models of atmospheric phenomena that is unlikely to occur if anthropogenic interference were not present.

In addition, climate models provide projections of the future climate. To provide these projections, scientists run the models into the future under different emission pathways. These models then provide estimates of how climate behavior may change with ongoing greenhouse gas and land use changes. We can also detect changes in phenomena that are likely to impact human health and well-being, for example, extreme weather events. For these reasons, it's obvious why the results of climate models are so important for policy and addressing the risks that arise from climate change.

Climate models of the sort described above, however cannot do it all. They are relatively coarse grained and there are lots of physical processes and geographic features that are not fully resolved, and the timesteps may remain long (they are often a day) to make runs of the model over the time periods of interest tractable. To gain insight as to how more local climate might behave, scientists run regional variants of climate models. These "regional climate models" (RCMs) employ the same theory-based representations of the climate system, but they cover a smaller area of the globe, often a country or region. Representing a smaller area allows for increased resolution under the same computational constraints when compared with a global climate model. Thus, regional climate models can represent areas with greater detail and are able to resolve smaller geographical features on shorter timesteps than a global model. These models are thus often used to learn how smaller scale features impact local climate and thus are important for informing regional planning.

It is precisely on these tradeoffs between scale of resolution and computational resources that NNs are poised to have the greatest impact. Researchers have focused on two issues that NNs might be helpful to address: parameterization and dynamical downscaling. Let us now examine these in turn.

21.3.1 Parameterization and Neural Networks

Unfortunately, due to computational limitations, one cannot resolve all features of the climate. Those features that cannot be resolved are often called "sub-grid" in that they are too small to be realized on the grid of a climate model. However, scientists are aware that these features are relevant to the climate dynamics that they do want to capture. For example, cloud formation—which has important components that occur on the molecular level—is extremely important for precipitation and storm creation. But such fine-grained processes could not be resolved on the scale of a climate model. But leaving out such processes is likely to result in an inaccurate picture of climate.

Instead of explicitly representing these processes, scientists end up substituting a complex fine-grained picture of their operation with a cruder and coarser one. Scientists divide the way they represent the climate within their model into two broad categories: the dynamics and the physics. For example, in the atmospheric component of a climate model, the dynamics refers to the fluid dynamical and thermodynamical equations that capture large-scale atmospheric behavior. These equations come from physical theory and express the conservation of momentum and mass among other relations. The "physics", somewhat counterintuitively, represents the sub grid processes that impact large scale behavior, but are too small to be represented explicitly. Instead, these processes are parameterized and expressed as functions of the elements that are computable in the dynamics (often temperature, pressure, wind, etc.). In this sense, parameterizations are often simplified representations of complex small-scale processes. They are engineered to be adequate substitutes (Katzav, 2013) for explicit representations of sub-grid processes.

Parameterization is recognized as a particularly difficult problem within climate modeling. After all, parameterizations are seemingly attempts to find a representation that is simple enough to be efficient but rich enough to capture complex and important components of the system. What a parameterization should look like, or which parameterizations are good ones, is often highly uncertain. In addition to uncertainty, and often times because of it, parameterizations are also a central focus for climate model tuning. Model tuning is akin to measurement calibration discussed above: aspects of the model are adjusted so that the model results agree with observations to within some specified degree. Parameterizations are frequently altered to ensure agreement with observations. One chooses these alterations based on how uncertain one is of the accuracy of the parameterization and the kind of observations one is trying to fit. It should be noted that how to tune a parameter is also a contentious process, as the adequacy of particular parameterizations may rest on other parameterizations in the model, to at least some degree.

There are several different methods of parameterization. The simplest is representing the sub-grid process deterministically by finding, often through data-driven investigation, what the average contribution of the process might be, or using a crude physical relationship or heuristic rule. The most complex might be what is called a super-parameterization. Super-parameterization uses a separate, high resolution and short timestep, model of the sub-grid process, which runs alongside a climate model with much coarser resolution and longer timesteps. When the climate model requires information about the sub grid process, the parameterization model provides it. For example, convection, the movement of less dense air through cooler more dense air, is a sub-grid process that is crucial for correctly describing cloud formation, precipitation, and storm formation. Parametrization of cloud related processes is a significant locus of disagreement between climate models and a significant uncertainty related to climate projections (Bony et al., 2015). In a super parameterization scheme, one would employ a fine-grained model of convection and cloud formation. These cloud resolving models are based primarily on physical theory. Each climate model grid box would have its own implementation of the resolving model. But while the climate model might run with a timestep of 1 hour, the resolving model might run with a timestep of 10 or 20 seconds. Each hour, when the climate model requires convection information, the resolving model provides it, and the information from the climate model is then used to constrain the next run of the resolving model.

The virtue of a super parameterization scheme is that it provides a more detailed account of sub-grid processes than other parameterization schemes. Super parameterization is also heavily informed by physical theory, meaning that what is happening in the model is (relatively) transparent or surveyable by scientists interested in those processes. There are downsides, however, that might by now be obvious: running such a small-scale model for a sub-grid process alongside a climate model is computationally costly, and it is infeasible for many of the long term and global investigations that climate models are used for.

Is there a way to harness the benefits that may come from using super parameterization while not incurring the high costs? This is what scientists examining the use of NNs for climate model parameterization hope to achieve. There have been several attempts to harness the power of super parameterization, particularly of cloud resolving models (CRMs), while trying to cut computational costs. In these attempts, scientists attempt to replace the super parameterization with a NN. For example, in Gentine et al. (2018), researchers used the SPCAM-3 climate model (simplified for an aqua-only planet) that employed a super parameterization in the form of a cloud resolving model. They recorded two years of the input and output for each instance of the cloud resolving model, and then divided it according to year. The first year of data was used to train a multilayer NN to see if it could learn the relations that the cloud resolving model was relying on to give information to the climate model. Researchers then coupled the trained NN to a climate model and ran it for a year in an analogous situation to their original two-year SPCAM-3 run. They validated their NN by comparing the results to the second year of the SPCAM-2 data, which the NN had never seen, to determine if it successfully learned from the super parameterization scheme.

In some ways the results are extremely promising. They found good agreement between the climate model using the super parameterization scheme and the one that employed the NN. The most promising result, however, was the efficiency gains of employing the NN: the climate model using the NN parameterization performed ten times faster than the original SPCAM-3 using the super parameterization. But Gentine et al. also point to two obstacles. First, the NN was in no way constrained by physical theory, and thus some physical processes, like energy and moisture conservation, were not respected. Second, given that this NN was trained on an aqua only planet, it is unclear how it would perform in novel situations, for example, areas with significant landmass.

Philosophers express concerns about these precise limitations. For example, Suzanne Kawamleh (2021) examined several attempts to emulate cloud resolving models for the purposes of parameterization. What she finds is that they fail to generalize outside their area of training and may in fact find relations that overfit the trained data. She also argues that by employing NNs, scientists loose access to one of the crucial ways of building confidence in their models: process evaluation (see Kawamleh, 2022). She indicates that "the neural network remembers and reproduces the mapping between the input and output variables of that particular CRM data set instead of learning the general physical or causal relations that underlie the CRM output data" (Kawamleh, 2021, p. 2014).

For Kawamleh, the causal relations and the processes that they give rise to are central to assessing the quality of our climate models and improving their reliability through process evaluation. However, with a NN parameterization these processes are "black boxed" and obscure; we don't know what relations have been learned. Furthermore, if performance in novel situations is bad, we can be assured that whatever relationship was posited by the NN, it is not capturing a set of underlying stable physical relations that hold across contexts. By employing NNs then, scientists limit their ability to examine the climate processes, thus placing the weight of justification for model reliability on predictive success or fit with the data.

The claim that physical processes relevant to assessing the reliability of a climate model are obscured when a NN is employed for parameterization is a fairly general feature of many NNs. Barring the success of explainable AI, which climate scientists are keenly interested in (see Mamalakis et al., 2022), it is likely that NNs will remain black boxes. Thus, it may be hard to point to why they are successful, even when they are, or to discern what relationships they have learned. The question, however, is whether this lack of representational transparency is in fact troubling.

To address the question, it is worth comparing NN parameterizations to the two other kinds of parameterization schemes on offer. When compared to a super parameterization, it seems clear a researcher's ability to survey the way relationships are represented within the NN scheme is much reduced. However, the ability to survey the relationship may not completely prevent scientists from employing process validation. Climate scientists can still analyze the way that a process looks dynamically within the model output that employs a NN parameterization scheme. This is often how climate scientists build confidence in their models via processes: they compare the visualized outputs of the model to other sources (e.g. observations or data assimilation products) to see if characteristic atmospheric patterns are present in both. This kind of pattern matching gives an indication that the model and parameterization scheme are able to capture processes that contribute to atmospheric phenomena of interest. This is one way to build confidence in a model. Even if the parameterization scheme is opaque to researchers, all is not lost. It is also worth noting that super parameterization schemes are too costly to implement for large models and long runs, so even though they might allow for enhanced ability to examine how scheme structure relates to model behavior, the ability to do so is quite limited.

But super parameterizations are not the typical case, and more often climate models employ parameterizations that consist of simplifications or empirical statistical properties to save on computational costs. While these parameterizations are representationally transparent to scientists, they are often inaccurate. Still, such parameterizations may be reliable, not because they are accurate, but because they nicely compensate or balance out other sources of error, including other parameterizations. When compared to this style of parameterization, a NN scheme does have possible advantages. First, it shares many similarities with traditional parameter schemes: they are both (often) data-driven and may not accurately capture physical processes. However, a NN scheme trained via super parameterization could have the ability to capture a more intricate relationship than a rule of thumb or statistic. If so, then it can reasonably be expected to provide better performance if it can be trained for the relevant situation, which in turn could result in more realistic model behavior and better climate projections, without the high computational costs of super parameterization. Thus, relative to the way climate parameterizations presently work, a NN scheme may provide the same level of representational transparency while offering better results.

21.3.2 Downscaling

Downscaling is the process by which information from a GCM is enhanced to produce higher resolution climate information and projections for specific geospatial areas. While the term "downscaling" gets used in many ways throughout the climate literature, there are two main forms of downscaling: dynamical downscaling and statistical downscaling. Dynamical downscaling involves using the coarse information from a GCM as a constraint on a higher resolution RCM, that then "zooms in" on regions of interest. Essentially, the RCM is nested within the GCM to obtain information with higher detail. Much of this kind of work is done through the Coordinated Regional Climate Downscaling Experiment (CORDEX), which uses several RCMs to create localized data and projections for 14 regions spanning most of the globe. Statistical downscaling is much simpler: it involves finding an empirical statistical relationship between the low-resolution information provided by a GCM and higher resolution information obtained by observations, for example. This relationship is then applied to the data from the GCM to obtain the regional resolution information.

Climate scientists are beginning to deploy NNs to emulate the dynamical downscaling process. The training set for the NN is gathered from dynamical downscaling experiments. For example, scientists might obtain 10 years of data for a region via the standard downscaling process, and then use that data to train the NN in conjunction with local land features not present in the GCM. The NN then becomes what these scientists call an emulator: it is supposed to approximate the function of the RCM in downscaling. By now, the advantage of employing a NN in this context should be clear: running the emulator is more computationally efficient than dynamical downscaling using an RCM. It also happens to be a well-understood problem in this domain. NNs have been employed extensively to upscale video: taking low resolution television and film and interpolating the necessary information to bring them up to modern 1080p or 4 k resolution standards. The downscaling problem is essentially the same: a GCM creates a low resolution "film" of the atmosphere, and the job of the NN is to enhance it.

A similar situation to the use of NNs in parameterization is observed when it comes to downscaling: the NN technique seems to have some of the advantages of both previous methods. The NN technique is computationally cheap, like the statistical downscaling method. However, some of the virtues provided by dynamical downscaling can also be retained. In particular, statistical downscaling often assumes that the empirical downscaling relationship is stationary through time, which often leads to poor performance when examining future climate scenarios, as the climate is known to be nonstationary because of anthropogenic change. The dynamical downscaling approach need not assume stationarity, enabling it to perform better when making future projections of the sort requested by policy makers and insurance companies for climate adaptation. A NN downscaling approach trained using a RCM need not assume stationarity, and thus may perform better under changing conditions than statistical downscaling (Hobeichi et al., 2023). NNs therefore may represent a cheap approximation of dynamical downscaling.

Along several epistemic criteria of goodness—for example representational accuracy, empirical adequacy, and physical plausibility—a NN emulator will fall short of full-fledged dynamical downscaling. However, the computational cheapness of the NN approach has one significant advantage: it is possible to perform many more downscaling operations with the same resources. This enables scientists to explore uncertainty with greater confidence. Previously, when making future projections of how regional climate might behave under different scenarios, the computational costs would restrict scientists to making a limited, or only a single, projection. Thus, they could only offer decision makers a "best guess" as to what the future regional climate might tend towards. Using a NN and making downscaling much more efficient, scientists may now run ensembles of downscaling experiments using different RCMs or different GCMs. In doing so, they would be better able to explore the spread of uncertainty in their results and offer decision makers better information about the range of conditions that might occur under a projection scenario. In this way, NNs permit scientists to trade accuracy and plausibility to probe the space of possibility more deeply and gauge uncertainty.

21.3.3 Conclusion: Could AI Replace Current Theory-Based Methods of Climate Projection?

Thus far we have examined the various ways that NNs are being implemented to help improve the capacities of climate observation systems and climate models. The instances we have examined take a hybrid or embedded approach where NNs complement (or are trained by) existing systems that have significant theory-driven components. Could scientists just forego these theory-driven methods and completely start over with only data-driven NNs?

On the spatial and temporal timescales that climate operates, there is skepticism that a purely NN projection system could be successful. First, NNs are most successful when there is ample training data. As was pointed out in the discussion of data assimilation, the available observational data on climate timescales is severely limited. High quality global climate datasets only go back to the late 1950's and the launch of early satellites, and even these may be spotty and incomplete. Given the multi-decadal timescales on which climate can operate, this is not enough data to expect a NN to extrapolate meaningful predictive relationships given current technology.

Additionally, the problem of assessing future climate change is not one that is tailored to the strengths of AI or NNs (see Chantry et al., 2021). As the downscaling discussion above mentions, the atmospheric system is not thought to be stationary. That is, the description of the climate regime our planet once inhabited is unlikely to continue to be accurate into the future because of changes brought on by anthropogenic forcings. Climate change means that the future climate is unlikely to resemble the past. The type of problem that results—inferring from a historical state to a new unseen state—is exactly the kind of problem that AI and NNs struggle with. In short, the future pattern may not be in the historical training data, and if it is not, there is no easy way for the NN to learn it. Furthermore, it is likely that any errors that the NN representation does contain will build up over climate-relevant timescales. As we have seen, climate scientists could train NNs on existing climate models, but it is likely that the existing models would represent the upper limit of accuracy and reliability. Given present technology and knowledge, it looks unlikely that NNs will soon make theory-informed climate modeling obsolete.

There are, however, climate-adjacent modeling tasks where NNs are already demonstrating their promise, and perhaps superiority (Ebert-Uphoff & Hilburn, 2023). These tasks tend to be short to medium term weather forecasts. Nowcasting—making weather and storm forecasts up to 6 hours—seems to be better suited to NN style analysis than global long-term climate projection (Chantry et al., 2021). Nowcasts extrapolate from the current weather patterns to near-future weather, helping to predict the movement and intensity of heat, rainfall, and storms. Such information is often used in disaster response, aviation, and even planning for events like the Olympics. Unlike long-term analyses, the short time scales involved do not permit errors to build up over time, which indicates that the NNs need not identify a physically realistic pattern to successfully nowcast. Furthermore, there is significant weather data from satellite, ground stations, and remote sensing devices available to train a NN, and the availability of this data is only growing as smart devices proliferate. Recent attempts to use NNs for nowcasting look promising: Zhang et al. (2023) have created a system trained on both observations and physical equations that outperforms traditional systems on many forecasting tasks.

AI systems may be useful for medium range forecasts (up to two weeks) as well. Bi et al.'s (2023) Pangu-Weather forecast system has been entirely trained on observational data and seems capable of forecasting quantities like temperature, wind speed, and pressure (though it hasn't been tested on precipitation, which is often most difficult to predict). Like with most of the implementations discussed in this chapter, what is particularly impressive is the computational savings: Bi et al. estimate that their AI powered system produces forecasts 10,000 times faster than numerical weather prediction with a theory driven model.

Here again, we see another instance of what makes climate science special as a case for examining the deployment of NNs. Climate scientists often face tradeoffs between theoretical rigor and computational efficiency. This methodological concern arises when gathering data, dealing with existing climate models, or even considering scrapping existing models and starting over with data-driven technologies. This concern will not soon go away, regardless of the success of AI. As many of the above examples show, NNs promise to capture some of the complexity of our best climate tools, without the computational costs. AI therefore offers a new tool to face this tradeoff, one where hybridization of theory-driven models with data-driven NNs is a promising way forward.

References

Arcucci, R., Zhu, J., Shuang, H., & Guo, Y.-K. (2021). Deep data assimilation: Integrating deep learning with data assimilation. *Applied Sciences, 11*(3), 1114. https://doi.org/10.3390/app11031114

Bi, K., Xie, L., Zhang, H., Chen, X., Xiaotao, G., & Tian, Q. (2023). Accurate medium-range global weather forecasting with 3D neural networks. *Nature, 619*(7970), 533–538. https://doi.org/10.1038/s41586-023-06185-3

Bogen, J., & Woodward, J. (1988). Saving the phenomena. *The Philosophical Review*, 303–352.

Bonavita, M., & Laloyaux, P. (2020). Machine learning for model error inference and correction. *Journal of Advances in Modeling Earth Systems, 12*(12), e2020MS002232. https://doi.org/10.1029/2020MS002232

Bony, S., Stevens, B., Frierson, D. M. W., Jakob, C., Kageyama, M., Pincus, R., Shepherd, T. G., Sherwood, S. C., Pier Siebesma, A., & Sobel, A. H. (2015). Clouds, circulation and climate sensitivity. *Nature Geoscience, 8*(4), 261–268.

Brajard, J., Carrassi, A., Bocquet, M., & Bertino, L. (2020). Combining data assimilation and machine learning to emulate a dynamical model from sparse and noisy observations: A case study with the Lorenz 96 model. *Journal of Computational Science, 44*(July), 101171. https://doi.org/10.1016/j.jocs.2020.101171

Buckner, C. (2019). Deep learning: A philosophical introduction. *Philosophy Compass, 14*(10), e12625. https://doi.org/10.1111/phc3.12625

Chantry, M., Christensen, H., Dueben, P., & Palmer, T. (2021). Opportunities and challenges for machine learning in weather and climate modelling: Hard, medium and soft AI. *Philosophical Transactions of the Royal Society A: Mathematical, Physical and Engineering Sciences, 379*(2194), 20200083. https://doi.org/10.1098/rsta.2020.0083

Creel, K. A. (2020). Transparency in complex computational systems. *Philosophy of Science, 87*(4), 568–589.

Demetrio, L., Biggio, B., & Roli, F. (2022). Practical attacks on machine learning: A case study on adversarial windows malware. *IEEE Security & Privacy, 20*(5), 77–85.

Di Noia, A., & Hasekamp, O. P. (2018). Neural networks and support vector machines and their application to aerosol and cloud remote sensing: A review. In A. Kokhanovsky (Ed.), *Springer series in light scattering* (pp. 279–329. Springer Series in Light Scattering.). Springer. https://doi.org/10.1007/978-3-319-70796-9_4

Di Noia, A., Hasekamp, O. P., Van Harten, G., Rietjens, J. H. H., Smit, J. M., Snik, F., Henzing, J. S., De Boer, J., Keller, C. U., & Volten, H. (2015). Use of neural networks in ground-based aerosol retrievals from multi-angle spectropolarimetric observations. *Atmospheric Measurement Techniques, 8*(1), 281–299.

Duede, E. (2022). Instruments, agents, and artificial intelligence: Novel epistemic categories of reliability. *Synthese, 200*(6), 491. https://doi.org/10.1007/s11229-022-03975-6

Ebert-Uphoff, I., & Hilburn, K. (2023). The outlook for AI weather prediction. *Nature, 619*(7970), 473–474. https://doi.org/10.1038/d41586-023-02084-9

Edwards, P. N. (2010). *A vast machine*. MIT Press.

Farchi, A., Chrust, M., Bocquet, M., Laloyaux, P., & Bonavita, M. (2023). Online model error correction with neural networks in the incremental 4D-Var framework. *Journal of Advances in Modeling Earth Systems, 15*(9), e2022MS003474. https://doi.org/10.1029/2022MS003474

Freiesleben, T., & Grote, T. (2023). Beyond generalization: A theory of robustness in machine learning. *Synthese, 202*(4), 109. https://doi.org/10.1007/s11229-023-04334-9

Geer, A. J. (2021). Learning earth system models from observations: Machine learning or data assimilation? *Philosophical Transactions of the Royal Society A: Mathematical, Physical and Engineering Sciences, 379*(2194), 20200089. https://doi.org/10.1098/rsta.2020.0089

Gentine, P., Pritchard, M., Rasp, S., Reinaudi, G., & Yacalis, G. (2018). Could machine learning break the convection parameterization deadlock? *Geophysical Research Letters, 45*(11), 5742–5751. https://doi.org/10.1029/2018GL078202

Held, I. M. (2005). The gap between simulation and understanding in climate modeling. *Bulletin of the American Meteorological Society, 86*(11), 1609–1614. https://doi.org/10.1175/BAMS-86-11-1609

Hersbach, H., Bell, B., Berrisford, P., Hirahara, S., Horányi, A., Muñoz-Sabater, J., Nicolas, J., et al. (2023). *ERA5: Data Documentation*, November 2023. https://confluence.ecmwf.int/display/CKB/ERA5%3A+data+documentation#ERA5:datadocumentation-Accuracyanduncertainty

Hobeichi, S., Nishant, N., Shao, Y., Abramowitz, G., Pitman, A., Sherwood, S., Bishop, C., & Green, S. (2023). Using machine learning to cut the cost of dynamical downscaling. *Earth's Future, 11*(3), e2022EF003291. https://doi.org/10.1029/2022EF003291

Jebeile, J., Lam, V., & Räz, T. (2021). Understanding climate change with statistical downscaling and machine learning. *Synthese, 199*(1), 1877–1897. https://doi.org/10.1007/s11229-020-02865-z

Katzav, J. (2013). Hybrid models, climate models, and inference to the best explanation. *The British Journal for the Philosophy of Science, 64*(1), 107–129. https://doi.org/10.1093/bjps/axs002

Kawamleh, S. (2021). Can machines learn how clouds work? The epistemic implications of machine learning methods in climate science. *Philosophy of Science, 88*(5), 1008–1020. https://doi.org/10.1086/714877

Kawamleh, S. (2022). Confirming (climate) change: A dynamical account of model evaluation. *Synthese, 200*(2), 122. https://doi.org/10.1007/s11229-022-03659-1

Lloyd, E., Lusk, G., Gluck, S., & McGinnis, S. (2022). Varieties of data-centric science: Regional climate modeling and model organism research. *Philosophy of Science, 89*(4), 802–823.
Lusk, G. (2016). Computer simulation and the features of novel empirical data. *Studies in History and Philosophy of Science Part A, 56*(April), 145–152. https://doi.org/10.1016/j.shpsa.2015.10.005
Lusk, G. (2021). Saving the data. *The British Journal for the Philosophy of Science, 72*(1), 277–298. https://doi.org/10.1093/bjps/axy072
Mamalakis, A., Barnes, E. A., & Ebert-Uphoff, I. (2022). Investigating the fidelity of explainable artificial intelligence methods for applications of convolutional neural networks in geoscience. *Artificial Intelligence for the Earth Systems, 1*(4). https://doi.org/10.1175/AIES-D-22-0012.1
Mohammad-Djafari, A. (2021). Regularization, bayesian inference, and machine learning methods for inverse problems. *Entropy, 23*(12), 1673. https://doi.org/10.3390/e23121673
Parker, W. (2017). Computer simulation, measurement and data assimilation. *The British Journal for the Philosophy of Science, 68*(1), 273–304. https://doi.org/10.1093/bjps/axv037
Parker, W. (2025). Climate science. In *The Stanford encyclopedia of philosophy*, edited by Edward N. Zalta and Uri Nodelman, 2018th ed. Metaphysics Research Lab, Stanford University. https://plato.stanford.edu/archives/fall2023/entries/climate-science/
Reichstein, M., Camps-Valls, G., Stevens, B., Jung, M., Denzler, J., Carvalhais, N., & Prabhat. (2019). Deep learning and process understanding for data-driven earth system science. *Nature, 566*(7743), 195–204. https://doi.org/10.1038/s41586-019-0912-1
Rodgers, C. D. (2000). *Inverse methods for atmospheric sounding: Theory and practice*. World Scientific Publishing Company, Incorporated.
Ryberg, J., & Roberts, J. V. (2022). Sentencing and artificial intelligence: Setting the stage. In J. Ryberg & J. V. Roberts (Eds.), *Sentencing and artificial intelligence*. Oxford University Press.
Schmidt, E. (2023). This is how AI will transform the way science gets done. *MIT Technology Review*, July 5, 2023. https://www.technologyreview.com/2023/07/05/1075865/eric-schmidt-ai-will-transform-science/
Setchell, H. (2020). *ECMWF reanalysis V5*. ECMWF. February 19, 2020. https://www.ecmwf.int/en/forecasts/dataset/ecmwf-reanalysis-v5
Sullivan, E. (2022a). Understanding from machine learning models. *The British Journal for the Philosophy of Science, 73*(1), 109–133. https://doi.org/10.1093/bjps/axz035
Sullivan, E. (2022b). Inductive risk, understanding, and opaque machine learning models. *Philosophy of Science, 89*(5), 1065–1074.
Tal, E. (2012). *The epistemology of measurement: A model-based approach*. PhD Thesis, Ph. D. Dissertation, University of Toronto.
Tal, E. (2013). Old and new problems in philosophy of measurement. *Philosophy Compass, 8*(12), 1159–1173. https://doi.org/10.1111/phc3.12089
Woodward, J. (2003). *Making things happen: A theory of causal explanation*. Oxford University Press. https://books.google.com/books?hl=en&lr=&id=vvj7frYow6IC&oi=fnd&pg=PR7&dq=woodward+making+things+happen&ots=xRkR9fxOVU&sig=KlXJTnwbZi_hqN2iulNgE1gAsL4
Zednik, C. (2021). Solving the black box problem: A normative framework for explainable artificial intelligence. *Philosophy & Technology, 34*(2), 265–288. https://doi.org/10.1007/s13347-019-00382-7
Zhang, Y., Long, M., Chen, K., Xing, L., Jin, R., Jordan, M. I., & Wang, J. (2023). Skilful nowcasting of extreme precipitation with NowcastNet. *Nature, 619*(7970), 526–532. https://doi.org/10.1038/s41586-023-06184-4
Zhao, Z., Xie, F., Ren, T., & Zhao, C. (2022). Atmospheric CO2 retrieval from satellite spectral measurements by a two-step machine learning approach. *Journal of Quantitative Spectroscopy and Radiative Transfer, 278*(February), 108006. https://doi.org/10.1016/j.jqsrt.2021.108006

Dr. Greg Lusk is an Assistant Professor in the Department of Philosophy at Durham University. His research focuses on the philosophy of science and technology, especially the use of models, data, and social values in policy-relevant sciences. His recent work examines the epistemic and ethical dimensions of machine learning in environmental research and decision-making. Greg serves as Co-Director of the Institute of Hazard, Risk and Resilience, as well as Co-Director of the Centre for Humanities Engaging Science and Society.

Chapter 22
Machine Learning in Public Health and the Prediction-Intervention Gap

Thomas Grote and Oliver Buchholz

Abstract This chapter examines the epistemic value of (purely) predictive ML models for public health. By discussing a novel strand of research at the intersection of ML and economics that recasts policy problems as prediction problems, we argue—against skeptics—that predictive models can indeed be a useful guide for policy interventions, provided that certain conditions hold. Using behavioral approaches to policymaking such as Nudge theory as a contrast class, we carve out a distinct feature of the ML approach to public policy problems: the ML model itself may turn into a cognitive intervention. In underscoring the epistemic value of predictive models, we also highlight the importance of taking a broader perspective on what constitutes good evidence for policymaking. Moreover, by focusing on public health, we also contribute to the understanding of the specific methodological challenges of ML-driven science outside of traditional success areas.

Keywords Machine learning · Public health · Prediction · Health economics · Algorithmic decision-making · Evidence-based policymaking

22.1 Introduction

Fueled by successes in image-based diagnostics and risk-prediction models, there is ongoing enthusiasm for using machine learning (ML) models as decision-support tools in clinical medicine (Gulshan et al., 2016; Esteva et al., 2017; Hyland et al., 2020). In comparison, the adequate role of ML applications has yet to be found in the context of public health. This may seem surprising, given that health data from virtually every domain of people's lives are being collected digitally,

T. Grote (✉)
University of Tübingen, Tübingen, Germany
e-mail: thomas.grote@uni-tuebingen.de

O. Buchholz
ETH Zürich, Zürich, Switzerland
e-mail: oliver.buchholz@hest.ethz.ch

J. M. Durán, G. Pozzi (eds.), *Philosophy of Science for Machine Learning*, Synthese Library 527, https://doi.org/10.1007/978-3-032-03083-2_22

potentially improving healthcare policies. A particular obstacle relates to the gap between predictions and interventions: while predicting unknown outcomes from patterns of past observations can be useful, how can policymakers anticipate which interventions are most likely to change the outcome in desirable ways? Besides, ML models perform demonstrably well when they are deployed in environments that mirror their training conditions. Yet they demonstrably do not perform well beyond training conditions (Freiesleben & Grote, 2023). The environment in public health contexts is typically unstable—culminating in prediction errors. In consequence, many researchers argue that purely predictive models are not the right tools to inform public health interventions (Athey, 2017; Hernán et al., 2019).

To overcome the gap between predictions and interventions, the proposed solution is to either supplement predictive ML models with causal inference methods or to embrace a causal attitude towards data science problems in public health altogether (Cui & Athey, 2022; Hernán et al., 2019; Schölkopf et al., 2021). While this view has become increasingly influential, it is not unequivocally shared. For example, Broadbent and Grote (2022) contend that it is misguided to impose causal constraints on ML models in epidemiology, since such constraints could prevent the discovery of novel statistical relationships that have the potential to revise the apparatus of public health concepts. However, their account sidesteps the issue of how the gap between predictions and healthcare interventions can be diminished. We do not try to settle the dispute between predictive and causal modeling cultures in public health in this chapter. Instead, we pursue the relatively modest goal of understanding how (and to what extent) purely predictive ML models can guide public health interventions.

For this purpose, we particularly draw on a study from the field of health economics by Mullainathan and Obermeyer (2022) who are using ML models as a tool to precisely characterize inefficiencies in the healthcare systems by examining how physicians' decisions deviate from predicted risk in the diagnosis of heart attack.[1] The upshot of the study is that physicians systematically overtest low-risk patients while undertesting high-risk patients. By developing a nuanced picture of the microfoundations of physician testing behavior, Mullainathan and Obermeyer conclude that current moral hazard models, such as low-testing regimes, can have perverse effects (see also Hausman, 2021). The approach taken in this study is paradigmatic for a strand of research at the intersection of ML and economics that recasts the design of optimal policies in criminal justice, social welfare, or healthcare as prediction problems (Kleinberg et al., 2015; Kleinberg et al., 2018).

That said, whereas Mullainathan and Obermeyer's (2022) modeling of physicians' decision-making allows ruling out misguided policies, it is unclear how it can translate into the design of more efficient public health interventions—other than providing physicians with algorithmic decision-support tools. This issue is closely

[1] Although analyzing the decisions of individual physicians, the study is adopting a public-health perspective because its ultimate aim is to ground health-related policies on said individual decisions rather than on aggregate views (Mullainathan & Obermeyer, 2022, p. 723).

linked to the granularity of the relevant risk-prediction model. In that respect, there are some striking discontinuities between the ML approach and other strategies of policy-design grounded in the behavioral sciences. A pertinent example here is Nudge theory. Guided by the assumption that decisions can result from two styles of information processing, the objective is to intervene in the decision architecture in a way that counteracts biases and ultimately enables people to act rationally/in desirable ways (Thaler & Sunstein, 2008). By contrast, the ML approach to public health problems is likely to result in interventions that diminish the role of policymakers: the ML model sets the bar for which payoff function to maximize and for what counts as the optimal decision-rule, leaving little room for policymakers. Hence, through a careful analysis of the methodology and its underlying constraints, the hope is to gain a clearer picture regarding the scope of ML approaches to public health problems. In this vein, this chapter also provides important groundwork that allows science and society alike to think more systematically about the role of ML approaches within the context of evidence-based policymaking.

The chapter proceeds as follows: Section 22.2 explains the gap between prediction and intervention in ML approaches to public health by pointing out various factors that might undermine the external validity of purely predictive models. Section 22.3 studies the methodology of the ML approach to policy problems. Of particular importance here is specifying what it means to reframe policy problems as prediction problems, as opposed to matters of causal inference. Furthermore, (dis)continuities between the ML approach inspired by behavioral science will be highlighted. Finally, Sect. 22.4 considers to what extent the ML approach to public health problems manages to overcome the gap between predictions and interventions. For this purpose, we examine the paper by Mullainathan and Obermeyer (2022) as a case-study. To carve out the opportunities and pitfalls of the ML approach, Nudge theory is taken as a contrast class. Overall, we hope that this chapter stimulates further reflection on the proper role of ML approaches in public health, while also broadening the debate on the epistemology and methodology of policymaking.

22.2 The Prediction-Intervention Gap

The objective in (supervised) ML is to facilitate the learning of mathematical functions that achieve high predictive accuracy on new data. Clinical medicine in particular has shown to be a beneficiary of the recent breakthroughs of ML in image-based classification and risk modeling. For instance, using measurements from multiple organ systems as input data, Hyland et al. (2020) developed an ML model that predicts circulatory failure in patients in the intensive care unit (ICU) with an accuracy of up to 90%. ICU patients cannot be monitored continuously by nurses or physicians, and it is exceedingly difficult for them to process the large quantities of data from many patients stored in electronic patient-data management systems. The risk-prediction model might therefore act as an early-warning system for deteriorating patients, potentially enabling timely interventions and decreasing the number of false alerts.[2]

This study is exemplary for dozens of ML applications meant to be used as decision-support tools in clinical medicine. By contrast, studies on public health applications of ML are still scarce—at least when measured in the currency of publications in high-profile (medical) journals. With 'public health', we refer to the societal practice that deals with the health of a country or region, driven by goals that manifest in policy documents and regulations (Munthe, 2008, p. 39; see also Verweij & Dawson, 2007). For example, while many leading research groups have been eager to develop ML-based forecasting models in the course of the SARS-CoV-19 pandemic, national health institutions usually were inclined to stick to stochastic compartmental epidemiological models, such as the Susceptible-Exposed-Infectious-Recovered (SEIR) model to control for the effectiveness of social distancing policies and mask mandates.[3] There is little need to argue that powerful predictive models can be useful for evidence-based policymaking. So why is it that ML applications still play an ancillary role in the methodological repertoire of public health?

Especially when factoring out material constraints—e.g., a lack of available datasets for public health purposes—then the underlying problem can be expressed by a simple slogan: predictions are not interventions. However, behind this slogan lies a multi-layered set of issues. We follow the orthodoxy in ML and define 'prediction' as an estimation of unknown outcomes based on patterns found in past observations (Hardt & Recht, 2022, p. 16). Predictions and interventions can be individuated by their distinct causal roles. Predictions are causally upstream to interventions. If a model predicts an increase in virus spread, then this might guide policies to contain the spread. By contrast, the intervention/policy sets out to change the healthcare system. For a predictive model to guide interventions, it must meet the conditions of robustness and control. The condition of robustness states that the model maintains its predictive performance (to a significant degree) in a deployment scenario regardless of (reasonable) changes in the probability distribution of the data (Freiesleben & Grote, 2023). This is meant to counteract the threat that predictions are based on artifacts. This, in turn, facilitates trust in the model. In contrast, the condition of control states that it must be possible to extract actionable insights from the model about how and when to intervene (see also Cui & Athey, 2022).

Recall the risk-prediction model by Hyland et al. (2020): One reason for its clinical utility is that it is being used in narrowly confined settings. The medical devices measuring patients' organ systems are likely to remain stable

---

[2] In general, applications of this kind are at risk of hiding the ML model's ignorance with respect to one particular patient behind a high average accuracy, thereby running into a setting that resembles the well-known reference class problem. However, see Buchholz (2023b) for an analysis of this problem and of situations in which it might not be that much of a problem after all.

[3] With some exceptions, such as the city of Valencia using ML-based prediction models to forecast virus spread in real-time (Díaz-Lozano et al., 2022). For a comprehensive treatment of ML applications amidst the SARS-CoV-19 pandemic, see Syrowatka et al. (2021) for review.

over time and ICU patient management is highly regulated. Once the predicted risk reaches a predefined threshold, the alarm will go off and a physician will initiate the corresponding treatment. Especially since physicians have background knowledge concerning patients' physiology and appropriate treatment options, there is a straightforward path from predicted risk to intervention. Even still, dataset shifts (e.g., the model gets applied to new demographics or policies in the clinic have changed) might induce malfunctions. However—for the most part—these are controllable and can be mitigated by retraining the model (Finlayson et al., 2021; Freiesleben & Grote, 2023).

22.2.1 Skepticism Against (Purely) Predictive Models

In contrast to the setting studied by Hyland et al. (2020), many environments that are of concern for public health lack clear boundaries and are subject to rapid changes. The predictive performance of an epidemiological model might be impacted by seasonal changes in virus spread or changes in the population's behavior. A case in point is Google's Flu Trends (GFT), trying to predict the prevalence of flu from people's online searches. One reason that led to the cancellation of GFT was a steep increase in online searches on flu, leading the model to drastically overestimate flu prevalence. This was particularly due to flu being prominently discussed in the news in fall 2013 (Lazer et al., 2014). Moreover, some changes to the environment can be even triggered by the deployment of the model itself, e.g., when people strategically adjust their behavior to game its functionality. This issue is commonly discussed as the *performativity* of (ML-based) predictions (Perdomo et al., 2020).[4]

Further complications arise when seeking to extract actionable information from ML models to guide policies. Critics of current ML are keen to point out that ML models often do not achieve high predictive accuracy by registering meaningful variables, but by exploiting spurious associations (Pearl & Mackenzie, 2019, Ch. 10; Geirhos et al., 2020). In this case, any information extracted from the model could mislead policymakers. Another problem results from the challenge to extract meaningful information from highly complex and, thus, opaque ML models in the first place—and from the question to what extent the relevant opacity can be ameliorated by making said models explainable (Creel, 2020; Sullivan, 2022; Boge, 2022).[5] After all, policymakers do not just want to predict an outcome of interest but to understand which adjustments are necessary to steer the outcome in a desirable way. Closely related thereto, for evaluating suitable policies, it is not only important

[4] See also Northcott (2017, p. 633) on the openness of social environments.

[5] Speaking of *relevant* opacity is meant to emphasize the insight put forward, for instance, by Buchholz (2023a) that ML models are not opaque *per se*, but different aspects of them are deemed opaque by different individuals, thus making the issue of explainability a highly granular epistemic endeavor.

to predict a given outcome based on observed data, but also to simulate how the predicted output were to change if a new policy gets released (Athey, 2017).

A widely shared view is that tackling these problems requires a different set of statistical techniques than pure predictive modeling.[6] For instance, Hernán et al. (2019) are skeptical about the premise that predictive models can inform public health policies. In their view, predictive models inform us that a certain decision must be made but they cannot guide the respective decision itself. Ultimately, what matters is that policymakers can tackle "what if"-questions—which, according to Hernán et al., can only be achieved by causal inference techniques (p. 49).

In response to these kinds of arguments, several authors have tried to make progress on a methodological shift that involves either amalgamating or at least complementing ML models with causal inference techniques. What is at stake here is capturing dependencies between the effect variable X and the outcome variable Y that remain invariant across different settings. As argued by Schölkopf et al. (2021), such an approach would lead to models that are more robust and thus better able to generalize to data that have been sampled under different conditions than the original training set. To make steps into this direction, Schölkopf et al. (2021) suggest abandoning the crucial—and, in fact, often unrealistic—*i.i.d.*-assumption of ML according to which all instances in the training and test data must be sampled independently from each other as well as from the identical probability distribution. Instead, they argue that one should impose weaker assumptions, for instance, that "the data on which the model will be applied comes from a possibly different distribution but involving (mostly) the same causal mechanisms" (p. 624). In practice, this could be achieved by using a more diverse training set and by employing techniques of data augmentation both of which would force the model to learn "underlying invariances or symmetries present in the augmented data distribution" (p. 626).[7]

A similar approach is pursued with the stable learning framework that has recently been proposed by Cui and Athey (2022): Instead of assuming that the training and test data come from the same distribution, the framework relies on the assumption that both might be from different so-called environments. The overall aim of stable learning then is to learn the mapping between a potentially larger number of treatment variables and the outcome, thereby determining a model that "can achieve uniformly good performance on any possible environment" (p. 112). Note that, in this context, 'performance' is taken to be *predictive* performance. This reveals how the stable learning framework amalgamates the traditional ML paradigm with a notion of causality according to which a model that captures (parts

[6] There are statistical techniques, for instance linear regression models, that can be used both for making predictions and—provided specific assumptions are met—for inferring causal relationships between variables. Standard ML methods like deep neural networks or support vector machines, however, can exclusively be used for the former.

[7] A common example for data augmentation in the case of image classification is the transformation of training instances by, e.g., rotating images or varying a few pixels.

of) the causal structure of an underlying phenomenon is robust to interventions and, thus, remains stable across different environments.

In a similar vein, Hofman et al. (2021) propose an integrative modeling framework for the purpose of reconciling predictive and explanatory modeling approaches in computational social science. Roughly, their idea is to use explanatory models to identify and estimate causal effects. The validity of the causal assumptions will then be determined by whether they enable predicting outcomes that can be considered, yet again, as being out-of-distribution relative to the training data. To map the explanatory models onto the predictive models, Hofman et al. suggest various methods, ranging from counterfactual sanity checks to sophisticated knowledge extraction techniques (see also Hinton et al., 2015).[8]

22.2.2 *The Epistemic Value of Purely Predictive Models: A Modest Defense*

The picture that emerges against the backdrop of the last section is that predictive models are ill-suited to guide public health interventions, at least unless they are supplemented by causal inference techniques. While it falls outside of this chapter's scope to further engage with the technicalities of the respective causal inference frameworks, we want to argue that it might be premature to abandon the prospect of using purely predictive models for improving public health policies.

Consider a pragmatic concern: Even though the proposals to bridge predictive models with causal inference techniques are highly persuasive, these are still theoretical frameworks. As it stands, we are not aware of any application of the stable learning framework within the context of public health/policy. By contrast, training a powerful predictive ML model can be a matter of weeks, provided that there is enough data, computational resources, and expertise. Hence, until methods that combine ML models with causal inference techniques have been fleshed out, it could be useful to suspend any constraints on using purely predictive models for public health purposes. A substantive argument for skepticisms against causal ML is that the relevant frameworks typically rely on (overly) strong assumptions, such as unconfoundedness in the data, which cannot be presumed for real-world settings (see also Gelman, 2010).

Then, there are methodological considerations: As the environment in public health contexts is oftentimes messy, it can be challenging to identify and estimate the true causal structure. Public health is an undertheorized domain, many of the necessary assumptions to establish causal claims are difficult to test or might only be justified by large-scale randomized controlled trials (RCTs), of which there are not enough (see also Cui & Athey, 2022). The lack of RCTs for public health purposes

[8] For a more detailed discussion of said integrative modelling framework as well as its perks and perils in the social sciences, see Buchholz and Grote (2023).

is closely tied to logistical (conducting RCTs can be very expensive and time-consuming) and ethical considerations (since they entail that one group receives worse treatment in what are often high-stakes settings). Resulting from this is the problem to delineate causal effects from confounders. The fact that explanatory models in the social sciences oftentimes lack predictive power showcases the difficulty to identify the causal structure in social environments. Indeed, it is debatable whether social environments that have a robust causal structure, holding reliably across cases, exist in the first place (Northcott, 2020, 2022). And while simpler epidemiological models like the SEIR model are also not strictly speaking causal, they provide researchers and policymakers with a lot of control, since it is easy to see how manipulations of individual features affect the outcome of interest.

Moreover, although one should be cautious to not naively interpreting the function of a predictive model as indicating the discovered structure, it is reasonable to assume that if an ML model predicts social phenomena at high accuracy, then *some* structure (or at least stable parameters that can guide policy-interventions) will have been found (Mullainathan & Spiess, 2017, p. 98). One particular promise of using ML models is that they enable discovering structures that so far were elusive to researchers and policymakers. Applied to the context of public health, this could include cross-connections between different kinds of data (e.g., electronic health records, mobility data, lifestyle data, and so on) (Broadbent & Grote, 2022). At least in principle, this might be valuable for developing better public health interventions.

However, a prerequisite for assessing how ML models can guide public health interventions is to test them against practical reality: Which phenomena can be predicted at high accuracy—and which cannot? How can researchers ensure the model's external validity? What are inherent methodological constraints? How can policymakers draw actionable insights from the model? What sorts of interventions might the model guide? Answering these questions requires a pronounced understanding of ML approaches to public health.

22.3 The Machine Learning Approach to Policy Problems

In what follows, we turn to a detailed methodological analysis of the ML approach to policy problems. The distinctive feature of this approach—receiving increased attention in economics—is that it tries to solve policy problems by using predictive algorithms, while disentangling policy choices from causal inference.

22.3.1 Prediction Policy Problems

According to a paper by Kleinberg et al. (2015), policy problems differ regarding their methodological requirements: If a policymaker facing a drought must decide whether to invest in a rain dance to increase the chance of rain, then causal inference techniques are necessary. What matters is whether rain dances cause rain. By contrast, if the task is to decide whether to take an umbrella to work to avoid getting wet considering the clouds in the sky, then this is not a causal inference but a prediction problem. Unlike the rain dance example, taking an umbrella has no direct effect on rain. However, to evaluate the utility of taking an umbrella to work, it is important to estimate the chance of rain. Put differently, it is unknown whether it will rain, and a corresponding prediction is needed to decide about taking an umbrella, while the causal relationship between using an umbrella and staying dry in the case of rain is known and, thus, requires no further investigation. The claim is that such prediction policy problems have been neglected so far and that ML models can help in solving them more effectively, when compared to traditional statistical methods such as regression analysis (p. 491).

Kleinberg et al. (2015) do not formally define prediction policy problems but provide a list of illustrative examples for potential empirical applications. Among those are predicting which teacher will have the greatest value added, predicting unemployment spell length to help workers decide on savings rates and job search strategies, targeting health inspections, predicting highest risk youth for targeted interventions, or predicting the creditworthiness of borrowers in lending decisions (p. 494). If we abstract from these examples, the common structure of prediction policy problems is that there is a known loss function, and the task is to minimize error when predicting a target variable Y from input data X. To this list, we can add some conditions that need to be met whenever some problem ought to be solved by ML techniques: There must be enough data to train the ML model, the target variable needs to be narrowly confined as a precondition for keeping measurement error small, and the background conditions need to be stable over time. In consequence, prediction policy problems relate to routine tasks, as opposed to predicting black swan events. For instance, predicting the spread of a novel virus, whose characteristics are not yet well understood, clearly falls outside of their scope.[9]

To showcase the structure of prediction policy problems, consider a lending decision. Here, a bank is confronted with a binary decision whether to grant or deny a loan application based on the predicted creditworthiness of the borrower. Similarly, while the impact of hiring an additional teacher can be well estimated, hiring the *right* teacher requires predicting individual teacher quality from information available at the time of hiring (Mullainathan & Spiess, 2017, p. 102). Still, things might not be as simple. A skeptic might object that assessing the suitability of a teacher for a particular school may largely hinge on a few causal factors—e.g., their professional skills and certain psychological traits—rather than being a matter of some large-scale prediction exercise. After all, how can an ML model predict whether a teacher fits well into a particular school culture? Likewise, without

[9] At a more abstract level, the underlying problem is that the remarkable success of ML models in interpolating from data is not matched in terms of extrapolation. For an intuitive explanation of the difference between interpolation and extrapolation, see also Freiesleben and Grote (2023).

knowing why certain youths are at risk, any targeted intervention might go astray. The point is that many actual examples of prediction policy problems involve more considerate choices than merely finding a decision-rule that facilitates minimizing error in a known payoff function. Somewhat pointedly formulated, it might be argued that many alleged prediction policy problems fall prey to the 'law of the instrument' (Maslow, 1966): suddenly, policymakers have access to vast amounts of data and highly predictive models, with the result that everything turns into a prediction problem.

Moreover, not everything that may seem like a natural candidate for a prediction policy problem is in fact predictable. For example, to assess whether medical spending in the last year of a life is wasteful—amounting to one-quarter of a person's health-costs in total—Einav et al. (2018) used ML techniques to predict mortality among patients. The result being that less than 5% of spending is accounted for by individuals with predicted mortality above 50%. With that in mind, the unpredictability of mortality can also be deemed as evidence that medical spending at the end of life is not wasteful. In a similar manner, a mass collaboration of 160 teams has shown low accuracy in predicting life trajectories from the Fragile Families and Child Wellbeing Study dataset (Salganik et al., 2020). However, what makes said phenomena unpredictable is not yet well-settled; either the available data proves to be insufficient, or the phenomenon of interest happens to be too complex to be operationalized as a prediction target. We will put this issue aside for now, but see Buchholz and Grote (2023) for details.

The programmatic considerations on the nature of prediction policy problems culminated in an influential study, in which Kleinberg et al. (2018) examine how ML models can be used to understand and improve judges' decisions whether to release defendants pre-trial in the US criminal justice system. Importantly, such decisions are generally assumed to solely hinge on the likelihood of the defendant to commit further crimes if they get released. Pre-trial decisions are therefore paradigmatic of a prediction policy problem for ML. The target variable is narrowly confined, and the data are extensive, provided that there are ten million arrests in the US per year. For the study, Kleinberg et al. use a dataset of 758,027 defendants who were arrested in New York City between 2008 and 2013, including detailed information on the defendants—e.g., whether they were released pre-trial and went on committing further crimes in-case. From a methodological perspective, the basic idea of the study is to train an ML model on the dataset for the aims of identifying the optimal decision function for pre-trial decisions. The model then acts as a benchmark, measuring judges' decision-making. One result being that judges are prone to release roughly half of the defendants, predicted to be within the riskiest 1% of defendants. This even applies to strict judges, drawing additional detainees from the whole risk distribution. As an upshot, Kleinberg et al. argue that if strict judges were to decide in accordance with the algorithmically predicted risk, they could achieve the same reduction in crime by only arresting half as many defendants. (p. 240).

With that in mind, a particular merit of the study is that Kleinberg et al. (2018) are very explicit about the challenges in comparing ML-based risk models to judges' decisions: There is no data on whether incarcerated defendants would have committed crimes if they would have been released, deviance from the algorithmically predicted risk can be attributed to different motivational factors on the judge's side, and there are certain kinds of evidence available to judges that the respective ML models cannot register. To counteract these obstacles, Kleinberg et al. use different techniques from the econometrics toolkit. These will be explained in greater detail at a later point in this chapter.

22.3.2 Diagnosing Error in Human Judgement

While paradigmatic for the ML approach to policy problems, the study by Kleinberg et al. (2018) itself can be situated in the tradition of two research strands concerned with understanding inaccuracies in human judgement. The first one is comparative research on two distinct methods of judgment, namely clinical versus actuarial judgement (Dawes et al., 1989). Clinical judgement relates to the combination or processing of information *in the head* of a human decision-maker. By contrast, the actuarial judgement tries to eliminate the human component and conclusions are made based on empirically established relations. Assuming an antagonistic framing between the two methods, various studies in the realm of biomedical or psychiatric diagnoses find that clinical judgement is typically outperformed by simple statistical decision-rules (see also Goldberg, 1970). A necessary condition for a fair comparison between the two methods is that the respective judgements use the same data. Moreover, it must be ensured by way of cross-validation that the accuracy of actuarial methods is not achieved by capitalizing on chance. The influence on the methodology of the ML approach to policy problems is evident—especially with regard to the antagonistic study design. However, one striking difference is that while actuarial methods rely on simple decision-rules, those of ML models tend to be staggeringly complex.

The second research strand are the works of Kahneman and Tversky (1979), Tversky and Kahneman (1974), examining how people's judgement formation about probabilities is systematically led astray by the reliance on availability heuristics, representativeness heuristics, and anchoring heuristics, and how people deviate from the axioms of Expected Utility Theory (EUT) on decisions about monetary payoffs in risky conditions. These findings, which are drawn from controlled experiments, are considered to provide evidence for the irrationality of human decision-making. The impact of Kahneman and Tversky's work can hardly be overstated. It has given rise to the field of behavioral economics, trying to revise the microfoundations of economic theorizing by integrating insights from psychology and the cognitive sciences. Closely related thereto, it animated a behavioral turn in policymaking, using insights from behavioral science as a guide for public policy interventions—most notably in the guise of Nudge theory (Thaler & Sunstein, 2008) and its competitor, Boost theory (Hertwig & Grüne-Yanoff,

2017).[10] Roughly, the basic idea of Nudges is to promote societal changes by way of small manipulations of the decision architecture, thus accounting for people's behavioral biases. Boosts again intervene on people's decision-making competences either by improved information presentation or by providing them with simple yet effective decision-rules.

Provided that the focus of the ML approach to policy problems is on understanding and improving decision-making, it falls squarely into the paradigm of the behavioral turn to policymaking. However, compared to the works of Kahneman and Tversky and the subsequent research in psychology or behavioral economics, one important methodological difference is that there is a shift from controlled experiments towards the analysis of large-scale sets of retrospective data. This methodological shift poses many advantages. Instead of investigating the decision-making of a limited number of research subjects under stylized conditions, it has now become possible to study thousands of actual decisions. However, relying on retrospective data also has some caveats. Most importantly, the observed outcomes might have been affected by many confounding factors, not captured by the data. These confounders must be controlled for by conducting various empirical tests. In a similar vein, it is still unclear how the insights generated by the ML approach to policy problems can be translated into policy interventions. To get a better grip on these issues, we move on to a detailed discussion of the study by Mullainathan and Obermeyer (2022), trying to counteract inefficiencies in medical testing with ML models.[11]

22.4 Predicting Inefficiencies in Healthcare

It is commonly assumed that resources in the healthcare system are spent inefficiently—culminating in overdiagnosis and deteriorating patient outcomes. This problem is particularly pressing in medical testing, where it can be attributed to two causes. On the one hand, medical imaging techniques enable early screenings, leading to diagnoses that do not benefit patients because the diagnosed condition is not a harmful disease (Rogers & Mintzker, 2016). On the other hand, there is the issue of moral hazard, diminishing the incentives to economize on treatment. Moral hazard is intimately linked to informational asymmetries between patients, physicians, and insurance providers, incentivizing physicians to perform expensive medical tests on patients (Hausman, 2021). Beginning with a study by Abaluck et al. (2016), the view is gaining traction that the best way to tackle moral hazard

[10] See Małecka (2021) for a critical analysis of the kind of knowledge provided by the behavioral sciences for public policy.

[11] We chose the study since we deem its methodology to be representative for a lot of research that takes place at the intersection of ML and public health. For a likeminded study, see Hastings et al. (2020).

involves examining physicians' testing intensity conditional on observable risk factors in patients. Understanding why and how physicians allocate resources inefficiently is key to identifying adequate interventions for ameliorating moral hazard.

22.4.1 Case Study: On Testing Patients for Heart Attack

The objective of the study by Mullainathan and Obermeyer (2022) is to investigate how physicians diagnose heart attack. Testing for heart attack involves a blockage in the coronary arteries, which is a costly and invasive procedure. Even though diagnosing heart attack is a standard procedure for physicians in the emergency room, there are many ambiguities in the diagnosis, especially since many benign conditions share symptoms with heart attack. Diagnosing heart attack requires integrating a diverse set of data. For this reason, it can be best understood as a prediction problem.

To study how physicians predict the risk of heart attack, Mullainathan and Obermeyer use a data-corpus of 250,000 emergency visits at a large academic hospital, registering patients' health records, tests given, resulting treatment, and patient outcomes. They then train an ML model to predict the outcome of testing, with the information available to physicians at the time of testing. The data are split into a training and a hold-out set. Importantly, the hold-out set is not used to benchmark physicians' predictions against the ML model but to predict patient subgroups, where physicians might have erred. To validate whether physicians indeed made errors, Mullainathan and Obermeyer consider the patients' health outcomes, assessing whether they show any major adverse cardiac events within 30 days of their visit. The testing efficiency is estimated by way of quality-adjusted life-year (QALY).[12] One reason for looking at patients' outcomes is that there is an informational asymmetry between the physician and the ML model. The model can assess electronic health records but is unable to capture additional information—e.g., self-reports by the patients, how they look, or results from ECGs and X-rays. Overall, the study finds that physicians overtest low-risk patients, while failing to test many apparently high-risk patients (pp. 680–1).

Mullainathan and Obermeyer formulate two hypotheses why physicians under-/overtest: The first one is that—when compared to the ML model—physicians rely on an overly simple risk-prediction model and overregularize by giving too much weight to salient features in turn. The second hypothesis is tied to the fact that health care models have fueled moral hazard by paying for tests, rather than for outcomes. An important finding of the study is that the implementation of existing moral hazard amelioration strategies, such as incentivizing low-testing regimes, can have perverse

[12] See Herlitz (2018) for how QALYs are calculated.

effects: it will lead to a decrease in overtesting among low-risk patients, while at the same time increasing undertesting of high-risk patients.

To rule out that high-risk patients were not tested for other reasons (e.g., they might have been too frail), the average age of the untested population and the fact whether they at least received an ECG were checked for (p. 701). Yet Mullainathan and Obermeyer acknowledge that the evidence provided for undertesting in physicians is only indirect. A potential confounding factor is that patients are first seen by nurses at the triage desk, before being examined by a physician. The involvement of the nurses in turn can influence the downstream decisions made by the physicians. For this purpose, a *natural experiment* is performed, in which the physicians' predictions are plotted at different shifts, assessing whether the error-rate proves to be stable across different triage teams (pp. 704–6).

To better understand why physicians are prone to making testing errors, it was analyzed how models for predicting physicians' testing decisions deviate from models for predicting the actual risk. The starting point here is that physicians' decision-making is boundedly rational: They are unable to process all the information available and compensate for their cognitive limitations by way of biases and heuristics (see also Wheeler, 2020). This is contrasted by the complexity of the optimal ML model that includes more than 16,000 variables. To investigate physicians' decision-making, Mullainathan and Obermeyer fit models at varying levels of complexity.[13] By testing the different models against the hold-out data, they found that the model that best predicts physician choices uses 49 variables, while the optimal risk-prediction model uses 224.

Note that the aim here is solely to shed light on the level of complexity in physicians' decision-making but not to approximate their *actual* decision-rule. In this respect, Mullainathan and Obermeyer make no assumptions regarding the properties of the models during training (p. 711). However, as a next step, they examine whether certain salient variables are overweighted by physicians—e.g., chest pain, age, and sex. This is meant to account for potential biases in physicians' testing-decisions. Here, the strategy involves training a risk-prediction model with a subset of variables, denoting salient symptoms. Despite the restrictions concerning the input data, the model has been trained exactly the same way as the original risk-prediction model, against which it was benchmarked in turn.

Again, the upshot is that the risk from symptoms is particularly predictive of physicians' testing, suggesting that, as a category, these salient symptoms are overweighted. In addition, the model's predictive accuracy has been tested for patients lacking said symptoms (pp. 716–720). The importance of empirical tests and models of varying complexity as a means to understanding ML models is an intriguing methodological insight from the study. Contrary to a suggestion that is commonly put forward within the philosophical debate on ML in science and, thus, somewhat surprisingly, methods that explain the model's behavior post-hoc through summary statistics or by detecting counterfactual dependencies virtually play no

[13] Where model complexity is measured in terms of the variables included in a given model.

role for the study. The iterative process, beginning with theoretical assumptions on the researchers' side, followed by increasingly nuanced empirical tests, might be best captured by Sullivan's (2022) notion of 'link uncertainty' according to which an understanding of and with an ML model is achieved by connecting the model behavior to background theory.

One problem of the study is that it relies on data from a single hospital, which is why there are concerns regarding the transferability of the ML model to novel settings. The testing behavior of the respective physicians may not be representative of physicians across other hospitals, which is why the ML model might overfit to the idiosyncrasies of that particular cohort. As a means to ensure its external validity, the ML model is tested against a nationally representative dataset of Medicare fee-for-service patients. Here, too, it was confirmed that physicians undertest high-risk patients while overtesting low-risk patients. However, as Mullainathan and Obermeyer (2022) point out, the dataset has significant limitations in that it is based on insurance claims, rather than EHR data and includes only few patient information. The heterogeneity and lack of standardization in medical data highlights the obstacles in validating ML models for healthcare purposes, especially since only very few external benchmark datasets exist. This may particularly affect the model's performance for marginalized social groups—traditionally underrepresented in medical data.[14]

22.4.2 From Predictions to Interventions

It is obvious that Mullainathan and Obermeyer (2022) should be commended for many thoughtful methodological choices. Even still, what lessons can be drawn from the perspective of a policymaker interested in diminishing moral hazard in medical testing? Here, the risk-prediction model by Mullainathan and Obermeyer turns out to be a mixed bag. Public health interventions can take many forms, with the most obvious candidate being the release of the policy that changes the incentive structure for physicians. For instance, one might incentivize testing by other means than via changes to the reimbursement schemes or by way of capacity constraints. However, the simultaneous presence of under- and overtesting suggests that such a policy is predestined to miss its mark, since it is insufficiently nuanced. Based on another empirical test, in which the behavior of lower-testing staff is compared to higher-testing staff, Mullainathan and Obermeyer find that a reduction in tests will equally affect low-risk and high-risk-patients. Thus, even though the policy would decrease overtesting, a low-testing regime could create perverse incentives in that it leads to worse outcomes for high-risk patients. The data also indicates that the accuracy does not increase if physicians were to test less (pp. 720–722).

[14] See Seyyed-Kalantari et al. (2021) for a good overview about how biases in medical data culminate in unfair treatment for marginalized social groups.

Similar issues are likely to arise regarding other policy proposals. Hausman (2021) recently argued that the best strategy to diminish moral hazard is to intervene on non-monetary incentives.[15] For example, by imposing non-monetary costs, so-called *ordeals*, on different agents in the healthcare-system, certain choices—e.g., abundant testing—become less attractive.[16] Imposing ordeals increases the likelihood that physicians will conduct due diligence in their medical testing. However, just as in the case of changes to monetary incentives, the result might either be an increase in tests—benefiting undertested high-risk patients, while harming low-risk patients—or the other way around. After all, neither monetary nor non-monetary incentives manage to address the cognitive limitations of physicians that ground inefficiencies in medical testing. As an upshot, one benefit of the risk-prediction model is that it enables understanding why certain policies are not adequate for a given purpose.

That being said, if the aim is to identify appropriate strategies for diminishing moral hazard in medical testing, then the risk-prediction model's level of granularity turns out to be brittle. This is due to instrumental and—more speculatively—metaphysical reasons. Concerning the former, the underlying issue is that although empirical tests facilitate a high-level understanding of the risk-prediction model, the actual decision-rule is still left obscure.[17] It is therefore difficult to derive an epistemically accessible policy that accounts for optimal testing-strategies from the model. Turning to the latter, the challenge is to identify a policy that fits to complicated natural phenomena yet, for practical purposes, must be simple (see also Mullainathan & Obermeyer, 2022, p. 715).

An alternative to policies are cognitive interventions. Unlike policies that aim at solving societal problems by inducing systemic changes, cognitive interventions shift the locus of attention to improving the decision-making capacity of individual physicians.[18] Nudges or Boosts are the natural candidates here (Thaler & Sunstein, 2008; Hertwig & Grüne-Yanoff, 2017). As the study by Mullainathan and Obermeyer (2022) has shown, one source of error in medical testing is that physicians overweight certain salient variables. Although Nudges and Boosts are driven by different research programs, in both cases, the strategy for overcoming inefficiencies

[15] However, note that Hausman's primary aim is overcoming health inequalities that result from moral hazard.

[16] Any measure that does not impose financial costs, but rather non-monetary burdens such as "waiting or filling out forms" (Hausman, 2021, p. 29) can be considered an example for such ordeals.

[17] See Creel (2020) and Boge (2022) for detailed analyses of the opacity problem in ML.

[18] The distinction between systemic and individual interventions is based on work by Chater and Loewenstein (2022). They critically argue that the individualistic framing, typically inspired by the behavioral sciences, distracts from concerted systemic efforts by way of regulation or taxation. Those are deemed more efficient for overcoming issues such as obesity, addiction, or climate change.

in testing might be to counteract the salience of variables by intervening on the information presentation.[19]

Nevertheless, the prospects of using the risk-prediction model by Mullainathan and Obermeyer as a guide for designing Nudges/Boosts are limited. To begin with, their study only highlights *that* a subset of variables/symptoms gets overweighted by physicians, thereby biasing testing decisions. However, this does not explain *why* physicians attach greater importance to particular variables in their decisions. More importantly, if we accept Mullainathan's and Obermeyer's premises, then Nudges and Boosts again prove to be a too simple solution for tackling a complicated phenomenon. However, this does not preclude that Nudges and Boosts can be useful in improving clinical decision-making. Rather, they are unlikely to simultaneously mitigate undertesting and overtesting. Of course, it could also be that physicians' overtesting and undertesting can be tackled by combining systemic interventions and Nudges or Boosts. However, the feasibility of such a mixed strategy is an empirical matter that can hardly be settled in this chapter.

This leaves policymakers with yet another cognitive intervention, namely, using the risk-prediction model as a decision-support tool for physicians. However, there are some crucial differences between the way that the algorithmic decision-support tool works as a cognitive intervention, when compared to Nudges and Boosts. Roughly, the latter seek to fix the cognitive limitations of the physician, while respecting their decisional authority—at least if one subscribes to the view that Nudges do not necessarily bypass the physicians' reasoning processes (Levy, 2019). In contrast, since the decision-support tool is vastly superior to the physician at predicting patients' risk of heart attack, the physician's role is basically to be deferential. The sole reason for not straightforwardly replacing physicians with risk-prediction models is that there are certain kinds of information that are elusive to the risk-prediction model, potentially confounding the risk estimate. Provided that a patient has a red face or reports that they are feeling pain in the chest area, then this information can be used to override the risk-prediction model (Mullainathan & Obermeyer, 2022, p. 723).

And yet, the epistemic rules about when to override algorithmic predictions are not well-understood, especially since the information that risk-prediction models and physicians use is asymmetric in a two-fold way. Both will use a shared pool of information—e.g., a subset of the EHR data. The risk-prediction model, however, will use the comprehensive information of the EHR data, while the physician may consider information from additional diagnostic modalities (see also Ludwig & Mullainathan, 2021, p. 86). In light of this, how to sort out disagreements? Intriguingly, these informational asymmetries pose a challenge for epistemological theories of disagreement—most pertinently the Equal Weight View—that so far are predominantly concerned with cases in which the disagreement between peers is tied to the same body of evidence (Christensen, 2007; see also Grote, 2024).

[19] See Last et al. (2021) for a systematic review of physician-directed Nudges in healthcare. For Boosts, see also Marewski and Gigerenzer (2012).

Empirical investigations of the interplay between physicians and decision-support tools have shown that especially novices are prone to blindly following algorithmic recommendations (Gaube et al., 2021; Tschandl et al., 2020; see also Genin & Grote, 2021).

We will not engage with the question how this interplay can be improved in this chapter; be it via careful epistemological theorizing or, more pragmatically, by way of best-practice models and regulations. For present purposes, what is at stake is that when purely predictive ML models are used for solving policy problems, then this is, in all likelihood, going to result in a novel kind of cognitive intervention. In addition, public health is consequentialist in its basic construction: Provided that the involvement of risk-prediction models promotes the health of populations—understood as an aggregate of individual health levels (Munthe, 2008, p. 40)—there is nothing ethically troublesome *per se* in the automation of medical testing. That being said, if the aim is to identify systemic interventions for fixing moral hazard, causal inference techniques may prove to be indispensable after all. For example, a pronounced account of the conditional probabilities between the effect variables and the target variable potentially facilitates tailoring an intervention that manages to simultaneously address the issues of undertesting and overtesting. With that in mind, we hope that this chapter provides convincing arguments for why even in the absence of causal inference techniques, predictive models should be explored as a tool for guiding public health interventions.

22.5 Conclusion

The chapter's leitmotif was the question why the enthusiasm for applying ML models in clinical medicine has not been echoed in the realm of public health. The answer is rooted in skepticism against purely predictive models, who are claimed to be neither sufficiently robust nor sufficiently informative to guide public health interventions. Against the widely held view that these deficits should be accounted for by blending ML models with causal inference techniques, we have argued that even purely predictive models can be a useful guide for public health interventions. To this end, we considered a novel strand of research that uses ML models for solving prediction policy problems. This research strand, again, can be seen as a part of the behavioral turn in policymaking. However, there are clear discontinuities regarding the methodology and the resulting kinds of interventions, when compared to existing policy-approaches grounded in the behavioral sciences such as Nudges and Boosts. Based on a case-study from health economics, we discussed in detail how predictive models can foster understanding of inefficiencies in medical testing, while also providing tools for improving public health outcomes. However, the utility of ML models is restricted in a two-fold way: First, they can only be meaningfully used for narrowly confined routine tasks. Second, they are most efficient when the adequate intervention is already known in advance.

Alternatively, the ML model itself may turn into a cognitive intervention, providing decision-support for policymakers.

By discussing studies from health economics, we also contribute to the understanding of the specific methodological challenges of ML-driven science outside of *traditional* success areas—e.g., structural biology, climate science, or medical imaging. In particular, we have highlighted the relevance of empirical tests and comparator models for overcoming the opacity of ML models. Our account has also wider-ranging implications for the (philosophical) debate on evidence-based medicine/policymaking. Given their empiricist foundations, the epistemic value of predictive models is largely ignored by existing frameworks. Instead, they are preoccupied over the caveats of RCTs for establishing causal claims, or the complicated liaison between causal and mechanistic evidence (Deaton & Cartwright, 2018; Russo & Williamson, 2007; Marchionni & Reijula, 2019; Shan & Williamson, 2021). In light of the increasing attention that ML models receive in medicine and the social sciences, it is about time to diversify perspectives, reflecting on what constitutes good predictive models for policymaking, and how the relevant evidence can be meaningfully combined with other kinds of evidence (see also Broadbent, 2013; Boumans, 2019).

Acknowledgements TG is supported by the Deutsche Forschungsgemeinschaft (BE5601/4-1; Cluster of Excellence "Machine Learning—New Perspectives for Science", EXC 2064, project number 390727645). TG also acknowledges support by the project "Certification and Foundations of Safe Machine Learning Systems in Healthcare" funded by the Carl Zeiss Foundation.

OB is supported by an ETH Zürich Postdoctoral Fellowship.

References

Abaluck, J., Agha, A., Kabrhel, C., Raja, A., & Venkatesh, A. (2016). The determinants of productivity in medical testing: Intensity and allocation of care. *American Economic Review, 106*(12), 3730–3764.

Athey, S. (2017). Beyond prediction: Using big data for policy problems. *Science, 355*(6324), 483–485.

Boge, F. J. (2022). Two dimensions of opacity and the deep learning predicament. *Minds and Machines, 32*(1), 43–75.

Boumans, M. (2019). Simulation and economic methodology. In W. Dolfsma, D. Wade Hands, & R. McMaster (Eds.), *History, methodology and identity for a 21st century social economics* (pp. 41–50). Routledge.

Broadbent, A. (2013). *Philosophy of epidemiology*. Palgrave Macmillan.

Broadbent, A., & Grote, T. (2022). Can robots do epidemiology? Machine learning, causal inference, and predicting the outcomes of public health interventions. *Philosophy & Technology, 35*(1), 1–22.

Buchholz, O. (2023a). A means-end account of explainable artificial intelligence. *Synthese, 202*, 33.

Buchholz, O. (2023b). The deep neural network approach to the reference class problem. *Synthese, 203*, 111.

Buchholz, O., & Grote, T. (2023). Predicting and explaining with machine learning models: Social science as a touchstone. *Studies in History and Philosophy of Science, 102*, 60–69.

Chater, N., & Loewenstein, G. (2022). The i-frame and the s-frame: How focusing on the individual-level solutions has led behavioral public policy astray. *SSRN Preprint 4046264*. https://papers.ssrn.com/sol3/papers.cfm?abstract_id=4046264
Christensen, D. (2007). Epistemology of disagreement: The good news. *The Philosophical Review, 116*(2), 187–217.
Creel, K. A. (2020). Transparency in complex computational systems. *Philosophy of Science, 87*(4), 568–589.
Cui, P., & Athey, S. (2022). Stable learning establishes some common ground between causal inference and machine learning. *Nature Machine Intelligence, 4*(2), 110–115.
Dawes, R. M., Faust, D., & Meehl, P. E. (1989). Clinical versus actuarial judgment. *Science, 243*(4899), 1668–1674.
Deaton, A., & Cartwright, N. (2018). Understanding and misunderstanding randomized controlled trials. *Social Science and Medicine, 210*, 2–21.
Díaz-Lozano, M., Guijo-Rubio, D., Gutiérrez, P. A., Gómez-Orellana, A. M., Túñez, I., Ortigosa-Moreno, L., et al. (2022). COVID-19 contagion forecasting framework based on curve decomposition and evolutionary artificial neural networks: A case study in Andalusia, Spain. *Expert Systems with Applications, 207*, 117977.
Einav, L., Finkelstein, A., Mullainathan, S., & Obermeyer, Z. (2018). Predictive modeling of US health care spending in late life. *Science, 360*(6396), 1462–1465.
Esteva, A., Kuprel, B., Novoa, R. A., Ko, J., Swetter, S. M., Blau, H. M., & Thrun, S. (2017). Dermatologist-level classification of skin cancer with deep neural networks. *Nature, 542*(7639), 115–118.
Finlayson, S. G., Subbaswamy, A., Singh, K., Bowers, J., Kupke, A., Zittrain, J., et al. (2021). The clinician and dataset shift in artificial intelligence. *The New England Journal of Medicine, 385*(3), 283.
Freiesleben, T., & Grote, T. (2023). Beyond generalization: A theory of robustness in machine learning. *Synthese, 202*(4), 109.
Gaube, S., Suresh, H., Raue, M., Merritt, A., Berkowitz, S. J., Lermer, E., et al. (2021). Do as AI say: Susceptibility in deployment of clinical decision-aids. *NPJ Digital Medicine, 4*(1), 1–8.
Gelman, A. (2010). *Causality and statistical learning*. https://arxiv.org/ftp/arxiv/papers/1003/1003.2619.pdf
Geirhos, R., Jacobsen, J. H., Michaelis, C., Zemel, R., Brendel, W., Bethge, M., & Wichmann, F. A. (2020). Shortcut learning in deep neural networks. *Nature Machine Intelligence, 2*(11), 665–673.
Genin, K., & Grote, T. (2021). Randomized controlled trials in medical AI: A methodological critique. *Philosophy of Medicine, 2*(1). https://doi.org/10.5195/pom.2021.27
Goldberg, A. L. (1970). Man versus model of man: A rationale, plus some evidence, for a method of improving on clinical inferences. *Psychological Bulletin, 73*, 422–432.
Grote, T. (2024). Machine learning in healthcare and the methodological priority of epistemology over ethics. *Inquiry, 68*(4), 1218–1247.
Gulshan, V., Peng, L., Coram, M., Stumpe, M. C., Wu, D., Narayanaswamy, A., et al. (2016). Development and validation of a deep learning algorithm for detection of diabetic retinopathy in retinal fundus photographs. *JAMA, 316*(22), 2402–2410.
Hardt, M., & Recht, B. (2022). *Patterns, predictions, and actions. A story about machine learning*. Princeton University Press, Princeton.
Hastings, J. S., Howison, M., & Inman, S. E. (2020). Predicting high-risk opioid prescriptions before they are given. *Proceedings of the National Academy of Sciences, 117*(4), 1917–1923.
Hausman, D. M. (2021). Ordeals, inequalities, moral hazard and non-monetary incentives in health care. *Economics & Philosophy, 37*(1), 23–36.
Herlitz, A. (2018). Against lifetime QALY prioritarianism. *Journal of Medical Ethics, 44*(2), 109–113.
Hernán, M. A., Hsu, J., & Healy, B. (2019). A second chance to get causal inference right: A classification of data science tasks. *Chance, 32*(1), 42–49.

Hertwig, R., & Grüne-Yanoff, T. (2017). Nudging and boosting: Steering or empowering good decisions. *Perspectives on Psychological Science, 12*(6), 973–986.
Hinton, G., Vinyals, O., & Dean, J. (2015). Distilling the knowledge in a neural network. *arXiv preprint*. https://arxiv.org/abs/1503.02531
Hofman, J. M., Watts, D. J., Athey, S., Garip, F., Griffiths, T. L., Kleinberg, J., et al. (2021). Integrating explanation and prediction in computational social science. *Nature, 595*(7866), 181–188.
Hyland, S. L., Faltys, M., Hüser, M., Lyu, X., Gumbsch, T., Esteban, C., et al. (2020). Early prediction of circulatory failure in the intensive care unit using machine learning. *Nature Medicine, 26*(3), 364–373.
Kahneman, D., & Tversky, A. (1979). Prospect theory: An analysis of decision under risk. *Econometrica, 47*(2), 263–292.
Kleinberg, J., Lakkaraju, H., Leskovec, J., Ludwig, J., & Mullainathan, S. (2018). Human decisions and machine predictions. *The Quarterly Journal of Economics, 133*(1), 237–293.
Kleinberg, J., Ludwig, J., Mullainathan, S., & Obermeyer, Z. (2015). Prediction policy problems. *American Economic Review, 105*(5), 491–495.
Last, B. S., Buttenheim, A. M., Timon, C. E., Mitra, N., & Beidas, R. S. (2021). Systematic review of clinician-directed nudges in healthcare contexts. *BMJ Open, 11*(7), e048801.
Lazer, D., Kennedy, R., King, G., & Vespignani, A. (2014). The parable of google flu: Traps in big data analysis. *Science, 343*(6176), 1203–1205.
Levy, N. (2019). Nudge, nudge, wink, wink: Nudging is giving reasons. *Ergo, 6*, 10.
Ludwig, J., & Mullainathan, S. (2021). Fragile algorithms and fallible decision-makers: Lessons from the justice system. *Journal of Economic Perspectives, 35*(4), 71–96.
Małecka, M. (2021). Knowledge, behaviour, and policy: Questioning the epistemic presuppositions of applying behavioural science in public policymaking. *Synthese, 199*(1), 5311–5338.
Marchionni, C., & Reijula, S. (2019). What is mechanistic evidence, and why do we need it for evidence-based policy? *Studies in History and Philosophy of Science Part A, 73*, 54–63.
Marewski, J. N., & Gigerenzer, G. (2012). Heuristic decision making in medicine. *Dialogues in Clinical Neuroscience, 14*(1), 77–89.
Maslow, A. H. (1966). *The psychology of science: A reconnaissance*. Harper & Row.
Mullainathan, S., & Obermeyer, Z. (2022). Diagnosing physician error: A machine learning approach to low-value health care. *The Quarterly Journal of Economics, 137*(2), 679–727.
Mullainathan, S., & Spiess, J. (2017). Machine learning: An applied econometric approach. *Journal of Economic Perspectives, 31*(2), 87–106.
Munthe, C. (2008). The goals of public health: An integrated, multidimensional model. *Public Health Ethics, 1*(1), 39–52.
Northcott, R. (2017). When are purely predictive models best? *Disputatio, 9*(47), 631–656.
Northcott, R. (2020). Big data and prediction: Four case studies. *Studies in History and Philosophy of Science Part A, 81*, 96–104.
Northcott, R. (2022). Reflexivity and fragility. *European Journal for Philosophy of Science, 12*(3), 1–14.
Pearl, J., & Mackenzie, D. (2019). *The book of why: The new science of cause and effect*. Penguin Books.
Perdomo, J., Zrnic, T., Mendler-Dünner, C., & Hardt, M. (2020). Performative prediction. In *Proceedings of the international conference on machine learning*, pp. 7599–7609.
Rogers, W. A., & Mintzker, Y. (2016). Getting clearer on overdiagnosis. *Journal of Evaluation in Clinical Practice, 22*(4), 580–587.
Russo, F., & Williamson, J. (2007). Interpreting causality in the health sciences. *International Studies in the Philosophy of Science, 21*(2), 157–170.
Salganik, M. J., et al. (2020). Measuring the predictability of life outcomes with a scientific mass collaboration. *Proceedings of the National Academy of Sciences, 117*(15), 8398–8403.
Schölkopf, B., Locatello, F., Bauer, S., Ke, N., Kalchbrenner, N., Goyal, A., & Bengio, Y. (2021). Toward causal representation learning. *Proceedings of the IEEE, 109*(5), 612–634.

Seyyed-Kalantari, L., Zhang, H., McDermott, M., Chen, I. Y., & Ghassemi, M. (2021). Underdiagnosis bias of artificial intelligence algorithms applied to chest radiographs in under-served patient populations. *Nature Medicine, 27*(12), 2176–2182.

Shan, Y., & Williamson, J. (2021). Applying evidential pluralism to the social sciences. *European Journal for Philosophy of Science, 11*(4), 1–27.

Sullivan, E. (2022). Understanding from machine learning models. *The British Journal for the Philosophy of Science, 73*(1), 109–133.

Syrowatka, A., Kuznetsova, M., Alsubai, A., Beckman, A. L., Bain, P. A., Craig, K. J. T., et al. (2021). Leveraging artificial intelligence for pandemic preparedness and response: A scoping review to identify key use cases. *NPJ Digital Medicine, 4*(1), 1–14.

Thaler, R., & Sunstein, C. (2008). *Nudge: Improving decisions about health, wealth, and happiness*. Yale University Press.

Tschandl, P., Rinner, C., Apalla, Z., Argenziano, G., Codella, N., Halpern, A., et al. (2020). Human–computer collaboration for skin cancer recognition. *Nature Medicine, 26*(8), 1229–1234.

Tversky, A., & Kahneman, D. (1974). Judgment under uncertainty: Heuristics and biases: Biases in judgments reveal some heuristics of thinking under uncertainty. *Science, 185*(4157), 1124–1131.

Verweij, M., & Dawson, A. (2007). The meaning of 'public' in 'public health'. In A. Dawson & M. Verweij (Eds.), *Ethics, prevention, and public health* (pp. 13–29). Oxford University Press.

Wheeler, G. (2020). Bounded rationality. In E. N. Zalta (Ed.), *The Stanford encyclopedia of philosophy*. https://plato.stanford.edu/archives/fall2020/entries/bounded-rationality/

Thomas Grote is a research fellow at the Cluster of Excellence: "Machine Learning: New Perspectives for Science" at the University of Tübingen. He is also Co-PI in a project on certification and safety of ML models in healthcare, funded by the Carl-Zeiss Stiftung.

Oliver Buchholz is a postdoctoral research fellow at the Chair of Bioethics at ETH Zurich and an associate member of the Interchange Forum for Reflecting on Intelligent Systems at University of Stuttgart.

If you have any concerns about our products, you can contact us on
ProductSafety@springernature.com

In case Publisher is established outside the EU, the EU authorized representative is:
Springer Nature Customer Service Center GmbH
Europaplatz 3, 69115 Heidelberg, Germany

Printed by Libri Plureos GmbH
in Hamburg, Germany